Key Thinkers on Space and Place

Key Thinkers on
Space and Place

Edited by
Phil Hubbard, Rob Kitchin and Gill Valentine

SAGE Publications
London • Thousand Oaks • New Delhi

Contents

Notes on contributors   ix
Acknowledgements   xi

Editors' introduction   1

1   **Benedict Anderson**   16
Euan Hague

2   **Trevor Barnes**   22
Susanne Reimer

3   **Jean Baudrillard**   27
David Clarke and Marcus Doel

4   **Zygmunt Bauman**   33
David Clarke and Marcus Doel

5   **Ulrich Beck**   40
Lewis Holloway

6   **Brian Berry**   47
Gordon Clark

7   **Homi K. Bhabha**   52
Constantina Papoulias

8   **Pierre Bourdieu**   59
Gary Bridge

9   **Judith Butler**   65
Minelle Mahtani

10   **Manuel Castells**   72
Phil Hubbard

11   **Stuart E. Corbridge**   78
Simon Batterbury

12   **Denis Cosgrove**   84
Keith Lilley

13   Mike Davis   90
     Don McNeill

14   Michael Dear   96
     Don McNeill and Mark Tewdr-Jones

15   Gilles Deleuze   102
     Marcus Doel and David Clarke

16   Peter Dicken   108
     Jon Beaverstock

17   Arturo Escobar   113
     Simon Batterbury and Jude Fernando

18   Michel Foucault   121
     Chris Philo

19   Anthony Giddens   129
     Barney Warf

20   Reginald Golledge   136
     Rob Kitchin

21   Derek Gregory   143
     Barney Warf

22   Torsten Hägerstrand   149
     Robin Flowerdew

23   Peter Haggett   155
     Robin Flowerdew

24   Stuart Hall   160
     Don Mitchell

25   Donna Haraway   167
     Lewis Holloway

26   J. Brian Harley   174
     Keith Lilley

27   David Harvey   181
     Noel Castree

28   bell hooks   189
     Katherine McKittrick

29 **Peter Jackson** ✓   195
    Don Mitchell

30 **Bruno Latour**   202
    Eric Laurier

31 **Henri Lefebvre** ✓   208
    Rob Shields

32 **David Ley** ✓   214
    Paul Rodaway

33 **Doreen Massey**   219
    Felicity Callard

34 **Gearóid Ó Tuathail (Gerard Toal)**   226
    Euan Hague

35 **Gillian Rose**   231
    Minelle Mahtani

36 **Edward W. Said**   237
    Karen Morin

37 **Andrew Sayer**   245
    Andy Pratt

38 **Amartya Sen**   251
    Simon Batterbury and Jude Fernando

39 **David Sibley**   258
    Minelle Mahtani

40 **Neil Smith**   264
    Noel Castree

41 **Edward Soja** ✓   269
    Alan Latham

42 **Gayatri Chakravorty Spivak**   275
    Robina Mohammad and James Sidaway

43 **Michael Storper**   282
    Susanne Reimer

44 **Peter Taylor**   288
    Jim Glassman

45   **Nigel Thrift**   294
     Barney Warf

46   **Waldo R. Tobler**   301
     Robin Flowerdew

47   **Yi-Fu Tuan**   306
     Paul Rodaway

48   **Paul Virilio**   311
     David Clarke and Marcus Doel

49   **Immanuel Wallerstein**   317
     Jim Glassman

50   **Michael J. Watts**   323
     Thomas Perreault

51   **Raymond Williams**   330
     Don Mitchell and Carrie Breitbach

52   **Iris Marion Young**   337
     Felicity Callard

Glossary   344
Index   353

Notes on Contributors

Simon Batterbury is an Assistant Professor of Geography, Department of Geography and Regional Development, University of Arizona, USA

Jonathan V. Beaverstock is Professor in Economic Geography, University of Loughborough, UK

Carrie Breitbach is a PhD candidate at the Department of Geography, University of Syracuse, USA

Gary Bridge is a Senior Lecturer at the Centre for Urban Studies, Unviersity of Bristol, UK

Felicity Callard is a Lecturer in Human Geography, Queen Mary, University of London, UK

Noel Castree is a Reader at the School of Geography, University of Manchester, UK

Gordon L. Clark is Halford Mackinder Professor of Geography, University of Oxford, UK

David B. Clarke is Senior Lecturer in Human Geography, University of Leeds, UK

Marcus A. Doel is Professor of Human Geography, University of Wales, Swansea, UK

Jude L. Fernando is Visiting Assistant Professor of Geography, University of Arizona, USA

Robin Flowerdew is Professor of Human Geography, University of St Andrews, UK

Jim Glassman is an Assistant Professor of Geography, University of British Columbia, Canada

Euan Hague is an Assistant Professor of Geography, DePaul University, Chicago, USA

Lewis Holloway is a Lecturer in Human Geography, Coventry University, UK

Phil Hubbard is a Senior Lecturer, Department of Geography, University of Loughborough, UK

Rob Kitchin is a Senior Lecturer, Department of Geography, National University of Ireland, Maynooth

Alan Latham is a Lecturer in Human Geography, University of Southampton, UK

Eric Laurier is a Research Fellow at the Department of Geography and Geomatics, University of Glasgow, UK

Keith Lilley is a Lecturer at the School of Geography, Queens University, Belfast, UK

Katherine McKittrick is a Postdoctoral Researcher at the Centre for Feminist Research, York University, Canada

Donald McNeill is a Lecturer at the School of Geography, King's College London, UK

Minelle Mahtani is adjunct Professor in the Department of Geography, University of British Columbia, Canada

Don Mitchell is Professor of Geography, University of Syracuse, USA

Robina Mohammad is a Research Fellow, South Asian Studies Programme, National University of Singapore

Karen M. Morin is an Associate Professor of Geography, Bucknell University, USA

Constantina Papoulias is a Senior Lecturer in Social Theory, Nottingham Trent University, UK

Thomas Perreault is an Assistant Professor of Geography, University of Syracuse, USA

Chris Philo is Professor of Geography, University of Glasgow, UK

Andy C. Pratt is a Senior Lecturer in Human Geography, London School of Economics and Political Science, UK

Susanne Reimer is a Lecturer at the Department of Geography, University of Hull, UK

Paul Rodaway is Director for Centre of Teaching and Learning, Paisley University, UK

Rob Shields is Professor of Sociology and Anthropology, Carleton University, Ottawa, Canada

James D. Sidaway is an Associate Professor of Geography, National University of Singapore

Mark Tewdr-Jones is a Reader in Spatial Planning and Governance, Bartlett School of Planning, University College London, UK

Barney Warf is Professor of Geography, Florida State Univesity, USA

Acknowledgements

The editors wish to thank a number of individuals without whom this volume would not have been possible. Firstly, we would wish to acknowledge Sage Publications, particularly Robert Rojek, for their patient support throughout the duration of this project. Secondly, we would wish to thank members of our respective departments for their comments, advice and (always constructive!) criticism, particularly in relation to our choice of key thinkers. In particular, we would wish to thank Brendan Bartley, Jon Beaverstock, Morag Bell, Clare Blake, Proinnsias Breathnach, Paddy Duffy, James Faulconbridge, Sarah Holloway, Caitriona NiLaoire, Dennis Pringle, and John Short. Thirdly, Gill Valentine wishes to acknowledge the Philip Leverhulme Fellowship that enabled her work on this manuscript. Finally, we would like to thank the large number of contributors to this volume, who patiently dealt with our never-ending stream of editorial threats, enticements, cajolements and suggestions with good humour and grace. The intellectual rigour, quality and incisiveness of the contributions is testament to the work that each and every contributor put into this volume, and we hope that each of them is as delighted with the end result as we are. Finally, and most importantly, we would wish to thank those without whom this book would not have been possible: the key thinkers who figure in this text. Without them, human geography – and the social sciences in general – would be a much duller place.

Editors' Introduction

As 1999 slowly but surely gave way to 2000, and we entered a new millennium, a not altogether surprising phenomenon emerged: the media's preoccupation with cataloguing the cultural, economic and social achievements not only of the last century, but the preceding 1000 years. Examples of this encylopedism are legion, including UK critics' lists of the best album of all time, polls of the most significant British figure, polls of the greatest film ever made, and readers' surveys of the most important works of fiction (for anyone interested: The Beatles' *Sgt Pepper's Lonely Hearts Club Band*; Winston Churchill; Orson Welles' *Citizen Kane* and Cervantes' *Don Quixote* – at least according to some polls). It was against this background of post-millennial 'listomania' that we began compiling this volume, which is intended as a comprehensive and critical guide to some of the most important thinkers and intellectuals influencing the contemporary development of spatial theory. From the outset, however, we were determined that this book should amount to more than an exercise in nostalgia, and that rather than looking backward to profile the figures who have done so much to establish geographical thought, this volume would be forward-looking, highlighting those thinkers who are currently doing most to shape the way that we think about space and place – and, by inference, will undoubtedly shape debates about space and place in the immediate future.

Given this remit, this volume is designed to offer a critical discussion of a selection of figures currently dominating debates about space and place. Our selection of 52 key figures will, no doubt, prove contentious (no less so than those lists of the greatest book or album, which

inevitably fuel counter-lists and ripostes from varied quarters). It is certainly not the list everyone would have us pick, and many readers will be surprised by some of those we have included (and equally amazed by some of our omissions: indeed we have already been taken to task by some of our contributing authors and colleagues for our failure to include certain figures!) Yet in compiling this book we have sought to highlight those, who, in our opinion, have contributed significantly to *theoretical* discussions of the importance of space and place in shaping cultural, social, economic and political life in recent years. These include those working in established and fundamentally important intellectual traditions such as positivism, phenomenology, Marxism and feminism as well as those developing new(er) discourses of space and place as they engage with (and develop) poststructural, queer, postcolonial, postmodern and subaltern theory (for the uninitiated, these terms are defined in the Glossary). Indeed, one of our strategies of selection was to include thinkers advocating different conceptions and approaches in order to highlight the diverse ways in which space and place have been theorized.

Given our disciplinary background, it is unsurprising that geographers dominate our list; given the inequalities that characterize academic geography (as well as other forms of intellectual labour – see Sidaway, 2000), it is also unsurprising that white, Anglo-American academics are most numerous. Yet in seeking to recognize the diverse intellectual traditions and ideas that are shaping the way that we conceive of and write about space and place, our list includes many working beyond the Anglo-American academy, and includes several figures who blur the lines between academic thought, scholarly writing and critical praxis. Furthermore, our selection includes sociologists, historians, political theorists, philosophers and psychologists (as well as many who elude easy disciplinary categorization). The fact that nearly half of the thinkers profiled here are not conventionally defined as 'geographers' is acknowledgement of the centrality of space in social theory and the significance of the so-called 'spatial turn' in disciplines such as sociology, cultural studies, and literary studies (see Hubbard *et al.*, 2002). While it is easy for geographers to overstate the extent to which this spatial turn has transformed the social sciences and humanities, the entries that follow demonstrate there are certainly many leading figures across the social sciences and humanities – **Edward Said**, **Stuart Hall**, **Michel Foucault** and **Raymond Williams** to name but four – who have stressed the importance of taking space seriously in the attempt to understanding social and cultural phenomena. Likewise, writing on globalization and the informational society has also placed concepts of space and place at the centre of social, economic and political thought, with influential thinkers as diverse as **Jean Baudrillard**, **Ulrich Beck**, **Anthony Giddens**, **Manuel Castells**, **Amartya Sen** and **Paul Virilio** all offering their own distinctive takes on the importance of (virtual and real) space in contemporary life. Crang and Thrift (2000: 1) consequently suggest that '[s]pace is the everywhere of modern thought'. The consequence of this is that academics outside the discipline have begun to theorize space in ways that have appeal for geographers. This means that their work is being imported into geographical thought in a variety of ways. Conversely, work by geographers is increasingly being used and read by those in other social sciences and humanities. In part, this explains why so many of the theorists profiled in this book would not necessarily consider themselves to be 'geographers', even though their work is inherently geographical or has been adapted and reworked by geographers.

On the other hand, the book profiles a number of thinkers who would certainly identify as geographers. What is evident here, however, is that our choice of key thinkers in the geographical tradition is entirely biased towards human geographers, despite the apparent common ground shared between physical and human geographers as they explore the constitutive role of space–time in the making of the world around us (see Massey, 1999). Yet despite occasional conversations between physical and human geographers (see Raper and Livingston, 1999; Harrison and Dunham, 1998), and sporadic attempts to unite the discipline through the forging of a shared philosophy and method (e.g. Haggett and Chorley, 1969), it remains the case that physical geography has remained relatively untroubled by theoretical debates about the nature of space and place. As **Doreen Massey** (1999) notes, for physical geographers the notion of *absolute space* still predominates, with phenomena seen to pre-exist their location in space. While this version of spatiality still informs certain human geographical writing – see entries on **Brian Berry**, **Reg Golledge**, **Peter Haggett** and **Waldo Tobler** – the more widespread understanding of space among human geographers is that social, economic and political phenomena are the product of spatial-temporal locality, and that the articulation of inter-relations brings space into being. For example, **Nigel Thrift** offers the following definition:

> As with terms like 'society' and 'nature', space is not a commonsense external background to human and social action. Rather, it is the outcome of a series of highly problematic temporary settlements that divide and connect things up into different kinds of collectives which are slowly provided with the means which render them durable and sustainable. (Thrift, 2003: 95)

Hence, while there are physical geographers who are attempting to contribute to unfolding theoretical debates about the nature of space and place (Kent, 2003), most physical geographers have ignored postmodern, postcolonial or poststructural attempts to deconstruct, critique or reconstruct languages of space and place, and have only made marginal contributions to the literatures problematizing concepts such as globalization. As such, our selection of thinkers does not include any who would identify themselves as a physical geographer, but hopefully does not ignore physical geography, as many of the thinkers here offer food for thought for those in the natural as well as the social sciences (for some, notably **Bruno Latour** and **Donna Haraway**, the distinction often made between 'objective' hard science and the 'subjective' social sciences is a problematic one in any case).

Notwithstanding our decision to focus on those who are presently some of the most influential in theoretical debates over space and place, there are still many thinkers – both dead and alive – who act as key reference points in debates over the spatiality of social, economic and political life. As in Elliott and Turner's (2001) excellent *Profiles in Contemporary Social Theory*, our most difficult decision has therefore been selecting who to omit (starting with a long shortlist of several hundred names that had to be pared down to a more manageable 52). In the final analysis, we have attempted to include a *representative* rather than exhaustive selection of names, and while we are keen to stress that each of the thinkers profiled here is relevant to contemporary *theoretical* understandings of space and place, there are of course many others who have made significant interventions in geographic debates through their empirical or practical contributions. Hence, our choice of key thinkers should not be regarded as some barometer of influence for those for whom space and place are central foci of analysis, as it ignores many (and it would perhaps be invidious to mention names here) who have made significant contributions in applied geography, Geographic Information Sys-

tems, policy-oriented studies, action research and cartography, as well as the many whose prime contribution to geography is their empirical research (whether on environmental issues, the economy, social processes, politics, the country or the city). In this sense, our selection of thinkers should not be read as a guide to who's currently hot (and who's not) in human geography (after all, there are plenty of citation analyses around for those who want a guide to which practitioners exercise most influence within, and beyond, the discipline – see Yeung, 2002). Rather, it stands as a user-friendly guide to some of the more important thinkers informing current debates about *space* and *place*. In the following section, therefore, we seek to outline why these concepts are fundamental in theoretical debates in geography and across the social sciences – and begin to show how their definition is variously problematized and clarified by the existence of different traditions of social, economic and political thought – from critical theory to hermeneutics, from feminism to psychoanalysis, and from postmodernism to poststructuralism.

THINKING SPACE AND PLACE

> Geography . . . has meant different things to different people at different times and in different places.
> (Livingstone, 1992: 7)

In popular discourse, space and place are often regarded as synonymous with terms including region, area and landscape. For geographers, however, these twin terms have provided the building blocks of an intellectual (and *disciplinary*) enterprise that stretches back many centuries. Yet, as Livingstone intimates, the theoretical specification of space and place has remained a matter of some dispute, being transformed as new ways of 'thinking geographically' have developed. Rather than reiterate Livingstone's analysis of how the 'geographical tradition' developed and mutated from an era of early modern navigation, through Enlightenment exploration and onto the institutional geographies of the nineteenth and twentieth centuries (see also Heffernan, 2003), we want to focus here on the more recent history of spatial thinking in human geography to illustrate the diverse ways in which space and place are presently conceptualized and analytically employed to make sense of the world.

As noted above, many physical geographers remain fairly uninterested in problematizing the idea that space is straightforwardly empirical, objective and *mappable*. Likewise, until the 1970s, most human geographers considered space to be a neutral container, a blank canvas that is filled in by human activity. Here, space is defined and understood through Euclidean geometry (with x, y and z dimensions) and, for analytical purposes, treated as 'an absolute container of static, though movable, objects and dynamic flows of behaviour' (Gleeson, 1996: 390). This absolute or 'empirico-physical' conception suggested that space can be conceived as outside of human existence; rather than playing an active role in shaping social life, it is regarded as a backdrop against which human behaviour is played out (an idea explicitly addressed in **Torsten Hägerstrand**'s time–space modelling). In the 1950s and 1960s this conception of space was refined by a number of practitioners who sought to restyle geography as a positivist spatial science, seeking to construct theory or 'spatial laws' on the basis of statistical analysis (Robinson, 1998). This was reflected in the publication of texts covering the principles of statistical analysis to geographers (e.g. Gregory, 1963) and, later, those that sketched out the principles of spatial statistics based on regression, clustering and autocorrelation (Abler *et al.*, 1971). For many, the ultimate promise of this progressive process

of statistical testing and theory-building was the construction of predictive spatial models (with **Waldo Tobler**, **Peter Haggett** and **Brian Berry** leading practitioners).

Retrospectively, this period is thus described as representing a pivotal moment in the history of the discipline – geography's 'Quantitative Revolution' (Bird, 1989; Barnes, 2001a) – and while many geographers were not swept up in the enthusiasm for quantification, hypothesis testing and statistical analysis, this new 'scientific' paradigm was nonetheless responsible for ushering in a new conceptualization of space which became widespread among even those geographers resistant to the notion of quantification. In effect, this was to conceive of space as a surface on which the relationships between (measurable) things were played out. Looking toward other disciplines, notably neo-classical economics and physics, this placed emphasis on the importance of three related concepts – direction, distance and connection. In short, it became axiomatic that the relationships between things on the Earth's surface could be explained in terms of these key concepts, and that it was possible to discern regular patterns that could be mapped and modelled (Wilson, 1999). This heralded a new language of spatial physics where human activities and phenomena could be reduced to movements, networks, nodes or hierarchies played out on the Earth's surface.

Reacting against this rabidly objective type of analysis, some geographers took inspiration from psychology, developing a behavioural perspective that explored the role of the conscious mind in shaping human spatial behaviour (see **Reg Golledge**). While this perspective held to the tenets of positivist inquiry, merely replacing concepts of absolute distance with notions of subjective distance, the historical and geographical materialism that emerged in the 1970s ushered in a rather different interpretation of spatiality, whereby space was deemed to be inherently caught up in social relations, both socially produced and consumed. Here, 'new' urban sociologists joined forces with geographers to document the role of urbanization in capitalist society, with **Manuel Castells** and **David Harvey** arguing that the city concretized certain class inequalities. On a different scale, economic geographers (e.g. **Michael Storper**) and those working in the 'localities tradition' (e.g. **Doreen Massey** and **Andrew Sayer**) sought to expose the way that spatial divisions of labour perpetuated capitalist structures, while political theorists (such as **Immanuel Wallerstein**, **Stuart Corbridge** and **Peter Taylor**) wrote of the international division of labour that was secured through particular geopolitical and territorial strategies. Yet it was arguably not until the work of the Marxist theorist **Henri Lefebvre** (1991) that this notion of space as socially produced was convincingly (if sometimes obtusely) articulated.

Lefebvre inferred that absolute space cannot exist because, at the moment it is colonized through social activity, it becomes relativized and historicized space. Insisting that every society and every mode of production produces its own space, he further distinguished between the abstract spaces of capitalism, the sacred spaces of the religious societies that preceded it, and the contradictory and differential spaces yet to come. In outlining this history of space, Lefebvre implied that conceiving and representing space as absolute (as had been common in geography and across the social sciences) was in fact implicated in the production of relativized abstract space (i.e. the space of capitalism). Rejecting this, he proposed a 'trialectics' of spatiality, which explores the differential entwining of cultural practices, representations and imaginations. Moving away from an analysis of things in space, this is an account that sees space as 'made up' through a three-way dialectic between perceived, conceived and lived space (see also **Ed Soja**). Here, place emerges as a particular form of space, one that is created through acts of naming as well as the distinctive activities and imaginings associated with particular social spaces.

For many geographers, place thus represents a distinctive (and more-or-less bounded) type of space that is defined by (and constructed in terms of) the lived experiences of people. As such, places are seen as fundamental in expressing a sense of belonging for those who live in them, and are seen as providing a locus for identity. As with space, within regional and quantitative approaches place was generally conceived in absolute terms, simply as a largely self-contained gathering of people in a bounded locale (territory). This understanding of place was challenged by humanistic geographers who, in the 1970s, sought to supplant the 'people-less' geographies of positivist spatial science with an approach to human geography that fed off alternative philosophies – notably existentialism and phenomenology (Holloway and Hubbard, 2001). Focusing on the experiential properties of space, the writings of David Lowenthal, Anne Buttimer, **David Ley**, Edward Relph and **Yi-Fu Tuan** in particular were of great value in reminding geographers that people do not live in a framework of geometric relationships but a world of meaning. For example, Tuan's (1977) poetic writings stressed that place does not have any particular scale associated with it, but is created and maintained through the 'fields of care' that result from people's emotional attachment. Using the notions of *topophilia* and *topophobia* to refer to the desires and fears that people associate with specific places, his work alerted geographers to the sensual, aesthetic and emotional dimensions of space. The humanistic tradition that these thinkers developed conceptualized place as subjectively defined. As such, what constituted a place was seen to be largely individualistic, although attachments and meanings were often shared (simply put, a place meant different things to different people).

As Thrift (2003) contends, one thing that does seem to be widely agreed is that place is involved with *embodiment*. The humanistic use of methods that evoke the multisensory experience of place (i.e. its visual, aural and tactile elements, as well as its smells and tastes) provides one means by which this bodily geography of place has been evoked, though the relationship between the human body and highly meaningful places is often more complex than even these methods can reveal (Holloway and Hubbard, 2001). Indeed, being 'in place' involves a range of cognitive (mental) and physical (corporeal) performances that are constantly evolving as people encounter place. In **Nigel Thrift**'s work on embodiment, it is suggested that these encounters cannot be adequately registered through language and discourse (hence, his talk of 'non-representational' theory). Stressing the importance of the pre-cognitive nature of being in the world (i.e. the way we intuitively inhabit places that are close and familiar to us without even thinking about it), Thrift alerts us to the practical knowledges and awarenesses that are deployed in everyday life. Other commentators suggest that these skills come easier to some than others, with the geographies of embodiment implicated in the making of class (see **Pierre Bourdieu**), gender (see **Judith Butler**) and racial divides (see **bell hooks**). Either way, place is seen to be made through the rhythms of being that confirm and naturalize the existence of certain spaces (a point made by **Henri Lefebvre** in his *rhythmanalyses* of modern life).

While places have generally been theorized as authentic, close and *lived* spaces, those adopting structuralist and critical approaches have argued that places are complex entities situated within and shaped by forces from well beyond their own notional boundaries. Here, there is a recognition that places should not be romanticized as pre-political entities but that they are shaped by often oppressive institutional forces and social relationships. This is an idea explored extensively by thinkers such as **Doreen Massey** through her notion of a progressive sense of place. For her, a place is the locus of complex intersections and outcomes of power geometries that operate across many spatial scales, from the body to the global. Places are thus constituted of multiple, intersecting social, political and economic relations, giving rise to a myriad of spatialities. Places and the social relations within and between them are the results of particular arrangements of power, whether it is individual and institutional, or imaginative and material. Such a formulation recognizes the open and porous boundaries of place as well as the myriad interlinkages and interdependencies among places. Places are thus relational and contingent, experienced and understood differently by different people; they are multiple, contested, fluid and uncertain (rather than fixed territorial units).

As detailed in the discussion so far, given the different ways space and place have been operationalized, they remain relatively diffuse, ill-defined and inchoate concepts. Yet they also remain fundamental to the geographical imagination, providing the basis of a discipline that is united primarily by its insistence on 'grounding' analyses of social, economic and political phenomena in their appropriate geographical context. In social and cultural geography, this focus on space and place has been further complicated by the adoption of different theoretical and methodological traditions. Crucial here is the continuing influence of two very different strands of geographic enquiry – one the one hand, Marxist accounts that explore the role of culture in the making of spaces of domination and resistance; and, on the other, the landscape studies of Carl Sauer and the Berkeley School (as well as the less celebrated German *landschaft* tradition) with their particular emphasis on 'place-making' (evident in the manner in which ways of life are inscribed on the landscape). Yet far from holding these literatures in abeyance, 'new' cultural geographers have worked with them, creating a productive dialogue between them as they endeavour to examine how the world is invested with cultural meanings: the work of **Denis Cosgrove** on the role of landscape in creating social and cultural orders is a case in point, while **Gillian Rose**'s feminist critique of the landscape motif offered an influential perspective on the gendering of space and place. As Baldwin *et al.* (1999) suggest, cultural geographers accordingly regard both space *and* place as culturally produced, recognizing the importance of both in the making of culture.

The idea that culture not only takes place, but makes place, is now manifest in a bewildering variety of work (including research into how the worlds of money, work, politics and production are enculturated). Reviewing this, Baldwin *et al.* (1999) assert that cultural geography coalesces around two key issues – firstly, the power and resistance played out in the everyday and, secondly, the politics of representation. Such concerns are certainly evident in those texts that were most significant in marking out the contours of a 'new' cultural geography. **Peter Jackson**'s (1989) *Maps of Meaning*, for instance, offered a distinctive take on the cultural politics of place by emphasizing the discursive construction of people and place via language. Here, Antonio Gramsci's notion of hegemony was used to stress that such representations were crucial in the making of social and cultural orders, while **Raymond Williams**' close attention to spatialized language was also an important influence. Drawing on similar theoretical sources of inspiration, as well as more traditional urban sociology, scholars in the so-called LA school (**Michael Dear, Ed Soja, Michael Storper** and **Mike Davis**, among others) showed how such close attention to the material and discursive workings of power could be used to illuminate the 'struggle' for the city. Again, a key assertion was that the meaning of place is fought over in the realms of cultural politics, being fundamental in the making (and remaking) of identity and difference. Writing in the context of Los Angeles, held up as the exemplary postmodern city (and 'capital of the twentieth century'),

such authors developed the idea that the class divides that characterized the modern industrial city were being recast and redrawn in the late capitalist era as capital and culture entwined to produce an entirely new city. Characterized as de-centred, fragmented and carceral, this postmodern city is one where categories of belonging are problematized, and where notions of a politics of difference take on heightened significance (as **Iris Marion Young** shows).

This attention to the making of cultural identities through cultural practices of boundary maintenance also highlights how concepts of place (and space) have been problematized and challenged by postmodern and poststructural theories that emphasize the slipperiness and instability of language. Rejecting universal definitions of 'place', such notions stress that places are real-and-imagined assemblages constituted via language. As such, the boundaries of place are deemed contingent, their seeming solidity, authenticity or permanence a (temporary) achievement of cultural systems of signification that are open to multiple interpretations and readings. Within geography, significant attention has therefore been devoted to the way that some taken-for-granted ways of representing the world (e.g. maps, atlases and aerial photographs) are in fact partial, distorted and selective, offering a particular 'way of seeing': **Brian Harley**'s influential deconstruction of maps, for instance, demonstrating cartography is implicated in the making of the world, not just its representation. Likewise, **Trevor Barnes**' ongoing explorations of the making of economic geographies have done much to demonstrate the way that spatial practices produce different spaces and places. This attention to the contingent nature of space and place has also problematized the taken-for-granted (binary) distinctions that often structure cultural understandings of the world – e.g. the distinction of self and other, near and far, black and white, nature and culture, etc. Most powerfully, perhaps, work on the construction of global North and South, often scripted in terms of an opposition of Oriental and Occidental values, has shown (through the writing of **Edward Said** in particular) that geopolitical processes of power and resistance (including 'global terrorism') rely on spatial metaphors. While geographers may be keen to take potshots at those corporations and individuals most obviously involved in the stigmatization of the South (including those involved in the development 'industry' – see **Amartya Sen** and **Michael Watts**), **Derek Gregory**'s writing on spatial imaginaries of 'Otherness' squarely implicates geographers in this process. In response, there has been a widespread geographical engagement with postmodern ideas about reflexivity, polyvocality and the need to acknowledge the fluid identities of place, not least through the promotion of subaltern studies (as championed by **Gayatri Spivak**).

On occasion, this focus on language and representation has shifted the attention of geographers from the making of social, political and economic worlds to the making of individual subjectivities, though an obvious tension remains between those accounts which focus on the role of spatialized language in the construction of self (via **Michel Foucault**'s ideas about the imbrication of power and knowledge) and those that borrow from psychoanalytical theories (e.g. the work of Melanie Klein, Julia Kristeva, Derek Winnicott and **Judith Butler**) to explore the projection of the self into places that are part real, part fantasy (see **David Sibley**). This psychoanalytical perspective offers yet another take on space and place, whereby the unconscious mind is seen to 'map' itself onto space in ways that have important consequences in the constitution of gender and sexual identities. Here, as **Gillian Rose** (1993) contends, it is argued that the negotiation of the self, and its complex amalgam of desire, anxiety, aggression, guilt and love, take place within and through the material and symbolic geographies of everyday life, with the psyche employing strategies to

sustain its structure and relationship with the world.

Beyond this focus on the contested nature of space and place, elucidating the relationship *between* space and place remains a strong area of interest for geographers, particularly in the literature on *scale* (see **Neil Smith**, 2000). One key strand here is scrutiny of the way places are being transformed through processes of globalization. Though alert to the entwining of local and global, and the creation of cultural hybridity, a key motif in such work has been that of global homogeneity. Claiming that a 'global space of flows' (to use **Manuel Castells**' terminology) is increasingly responsible for disseminating a standardized repertoire of consumer goods, images and lifestyles worldwide, the implication is that 'local' ways of life and place identities are being undermined by the logic of global capital accumulation as space is annihilated by time. Recently, a number of geographers have cited the work of anthropologist Marc Augé (1996), whose discussions of the familiar spaces of the supermarket, shopping mall, airport, highway and multiplex cinema revolve around the idea that these are 'non-places', symptoms of a supermodern and accelerated global society. Drawing obvious parallels with humanistic geographers' work on placelessness, he appears to suggest that there are now many 'non-places' that are solely associated with the accelerated flow of people and goods around the world and do not act as localized sites for the celebration of 'real' cultures. The cultural theorist **Zygmunt Bauman** (2000) similarly writes of these as 'places without place', making an explicit link to the spatial strategies of purification and exclusion that are at the heart of consumer society (simultaneously condemning the shallow and banal sociality evident in so many sites of consumption). As **Peter Taylor** (1999) has spelt out, the implication here is that local place is being obliterated by global space, while on a different scale, several leading commentators have argued for the redundancy of the nation-state in an era where global corporations are key makers of the global economy (as **Peter Dicken**'s work on transnationalism demonstrates). In extreme 'globalist' accounts, as well as in the sometimes apocalyptic writings of **Paul Virilio** and **Jean Baudrillard**, these changes appear to signify not just the 'end of history', but the death of geography.

Exploring the way real and imagined place identities are bound up with the ways in which we experience and represent time and space, **David Harvey**'s (1989) discussion of the condition of postmodernity (rather than supermodernity) offers a more nuanced account of place-making under conditions of globalization. Drawing on the ideas of Lefebvre in particular, Harvey explores how places are constructed and experienced as material artefacts, how they are represented in discourse, and how they are used as representations in themselves, relating these changing cultural identities to processes of time–space compression that encourage homogenization *and* differentiation. In doing so, Harvey points out the contradictory manner in which place is becoming more, rather than less, important in the period of globalization, stressing that the specificity of place (in terms of its history, culture, environment and so on) is crucial in perpetuating processes of capital accumulation. Such arguments have also been addressed by geographers (albeit in a different manner) in the context of locality studies, and the attempt by **Doreen Massey** (1991), as noted above, to interrogate a 'progressive sense of place' has also been influential for those exploring the equation between globalization and place identity. For example, several authors exploring the economic geographies wrought in an era of globalization have sought to explore the tensions between fixity and mobility, noting that place, if anything, is becoming more, rather than less, important in an economy where 'image is everything'. Literatures on economic agglomeration, location and specialism across a wide variety of sectors

(e.g. high-tech industry, advanced producer services, finance and banking) all thus point to the importance of face-to-face contact, quality of life and placed proximity in the creation of new 'global' industries. In the literature on global cities, for example, scholars such as **Peter Taylor**, **Michael Storper** and Sakia Sassen have developed Castells' take on global space of flows by demonstrating that key world cities have become more important in a global era as they are the strategic 'places to be' for those who seek to control the global economy. As **Nigel Thrift**'s work on performance and the 'non-representational' nature of space emphasizes, these are also places where knowledge is embodied and acted upon by those who are, in effect, the 'fast subjects' of global society.

In **Peter Jackson**'s (1999) summation, the emergence of new place identities through hybridization denies any simple equation between globalization and the homogenization of space. Instead, he argues that the meaning (and hence value) of different goods and cultures is created and negotiated by consumers in different places, with the 'traffic in things' across space implicated in the making of social relations. In many ways, this echoes work in anthropology concerning the meaning of material artefacts, but adds a distinctive geographic focus via notions of displacement, movement and speed. Far from asserting the death of place (or, conversely, its resurgence), this points to a geography that is open to notions of difference and the post-structural insistence (expressed forcefully by **Gilles Deleuze**) that the world is constantly being territorialized, de-territorialized and re-territorialized in unexpected ways. For some commentators, the corollary of this is that space and place need to become conceived of as fragile entities, constantly made and remade through the *actor networks* that **Bruno Latour** insists involve people, things, languages and representations. We might speculate that it is through the creation of shared notions of place – and common understandings of space – that networks gain their power. Economic, political and social orders are thus immanent in these networks, being reinforced or remade as 'material' moves through the network and takes different (commodity) forms in different contexts. Hence, there is no 'constitutive outside' which explains an 'inside'; place is not a location whose character can be explained through reference to wider spatial processes. Instead, such perspectives interpret both space and place as entities always becoming, in process and unavoidably caught up in power relations.

INTELLECTUAL AND DISCIPLINARY GENEALOGIES

As should be clear from the above discussion, there are many varying opinions on how to theorize and study the world. In particular, there is much debate between proponents of different theoretical traditions (positivism, Marxism, feminism, poststructuralism and so on) as they seek to develop and use concepts to think spatially. Of course, how such knowledge is produced is itself theorized, with a number of commentators developing disciplinary and conceptual histories that trace out the development and adoption of spatial ideas and approaches (for example, see Bird, 1989; Cloke *et al.*, 1991; Hubbard *et al.*, 2002; Johnston, 1986, 1991, 2000; Livingstone, 1992; Peet, 1998; Unwin, 1992). These most commonly are genealogical projects that seek to explain spatial thinking at the time of writing – mapping the present – by charting the conceptual paths followed by spatial theorists.

The most popular approach to date has been, following Kuhn (1962, 1970), to focus on identifying different geographic traditions that come to dominate spatial thinking through a particular period – becoming the dominant paradigm – and

to document the transition (a paradigm shift) between traditions as new philosophical approaches emerge to challenge previous ways of thinking. Indeed, the pages of academic journals and books are full of debates in which the authors claim that their 'new' way of looking at the world represents the most meaningful, progressive and correct way of doing geography, rejecting existing modes of exploration and explanation out of hand and inviting others to adopt and develop their 'new' approach. These paradigm shifts, Johnston (1996) has argued, are the by-product of generational transitions. He suggests that as new schools of thought emerge, they are embraced at first by younger academics. As the productivity of earlier generations, schooled in different approaches to geography, declines, the emerging generation become co-opted into the geographical establishment, taking over the editing of journals, incorporating their ideas into teaching and writing textbooks. In this way, Johnston (1996) contends that academics of different age cohorts become socialized through different paradigms so that education and training produce generational shifts in ways of thinking about space and place.

It is common for those adopting such a paradigmatic approach to plot the intellectual development of geography (e.g. Johnston, 1996) to argue that positivist spatial science emerged in geography in the late 1950s to challenge – and ultimately supplant – a regional tradition concerned with describing and mapping (see especially the entries on **Brian Berry**, **Torsten Hägerstrand**, **Peter Haggett**, **Waldo Tobler**). This positivist paradigm was itself challenged in the early 1970s by other approaches such as behavioural geography (see entry on **Reg Golledge**), humanist traditions (see entries on **David Ley** and **Yi-Fu Tuan**), and structural approaches, such as Marxism (see entries on **David Harvey**, **Neil Smith** and **Michael Watts**) and feminism (see entries on **Gillian Rose** and **Doreen Massey**). From a paradigm perspective, we might suggest that these dominant ways of thinking about space and place were challenged in the 1990s by postmodern (see entries on **Micheal Dear** and **Ed Soja**) and poststructural perspectives (see entries on **Jean Baudrillard**, **Judith Butler**, **Gilles Deleuze**, and **Michel Foucault**).

However, the notion of paradigm shifts has been subject to critique as it has become more apparent that different approaches to geography are never completely overthrown (Mayhew, 2001; Hubbard et al., 2002). While it is true that institutional arenas of publishing outlets, departments, professional organizations and informal socio-academic networks can reinforce the interests or agendas of particular academic communities, nonetheless there are always dissenting voices. Different ways of thinking about space and place are always concurrent rather than consecutive, even if at particular moments some are more fashionable than others. The danger of a paradigmatic approach to understanding the geographical tradition is that it creates a linear narrative that suggests that spatial thought has developed through unified (and generational) paradigms when in reality consensus has seldom been complete or stable (something that Johnston acknowledges when he employs the paradigm concept). The notion of sequential progress thus creates a false consistency in which contributions that deviate from the dominant narrative are omitted. Noting this tendency, **David Sibley** (1995) has documented the ways in which the geographies and histories of women, people of colour, those in developing countries, and other oppressed groups, have tended to be written from certain dominant positions, thereby silencing their voices and providing selective and partial geographical accounts.

Further, a paradigmatic approach often fails to fully explore the mechanisms by which ideas are constructed and knowledge is generated. As such, they often trace out trajectories of thought while glossing over the nuances in how intellectual ideas are developed within complex social and institutional structures and practices. Indeed, as **Donna Haraway** and **Pierre Bourdieu** explain in their own distinctive manners, spatial thought is not developed in a vacuum, but is rather constructed by individuals (and individuals collaborating) and situated within their own personal and political beliefs, the culture of academia, and institutional and social structures. From this perspective ideas are never 'pure' but rather emerge and become legitimated and contested according to particular material and social contexts.

Accordingly, an understanding of how ideas emerge, how they are adopted and how they evolve, requires an approach that acknowledges the situation and conditions in which they are constructed. The approach adopted in this book – biographical essays on key thinkers – seeks to provide such an analysis. Although such a biographical approach does not reveal a broad historicization of spatial thought, it is very useful for demonstrating the genealogy of intellectual ideas, revealing for example the ways in which personal history affects intellectual development, as the entries for **Edward Said** and **bell hooks** demonstrate. Edward Said's experiences of being born into a Christian-Arab family in Palestine during British administration, and his subsequent fight throughout his adult life for Palestinian self-determination, have undoubtedly shaped his thinking about the relationship between culture and imperialism. Likewise, bell hooks has attributed her attempt to theorize the problems of black patriarchy, sexism and gender subordination to her childhood experiences of growing up as a young black woman in Kentucky during the 1950s and early 1960s (see also Moss, 2001, on autobiographical accounts of the intellectual development of geographers).

Consequently, a biographical approach reveals how individual thinkers draw on a rich legacy of ideas drawn from past generations (as well as the influence of their contemporaries). Indeed, it should be clear from the cross-referencing between entries that no theorist develops their view of the world in an intellectual vacuum. The courses they took as students, discussions with their mentors and colleagues, the texts that they have read and papers they have heard, all expose them to a multitude of ideas that shape their own intellectual development. Such development can be traced across thinkers to reveal a rough genealogy of ideas. For example, **Gillian Rose**'s ideas about the privileging of male ways of conceiving of space and place have been heavily influenced by psychoanalytic and poststructural writings. One major source of inspiration here has been the works of the feminist philosopher **Judith Butler**. Judith Butler, in turn, while again drawing from a diverse set of philosophical texts, has extensively utilized the writings of **Michel Foucault**. Likewise, when developing his critical philosophy Foucault was influenced by (among others) the German philosophers Friedrich Nietzsche and Martin Heidegger. Of course, Gillian Rose is not the end point in this lineage but is rather a node in a complex web of interconnections, with her theorization in turn no doubt providing influence and inspiration for a generation of feminist and cultural geographers. Moreover, Foucault has inspired many other spatial theorists in ways that are quite strikingly different to the performative analyses of Butler and Rose: for example, **Arturo Escobar** has used his writings on power to study international development, while **Brian Harley** cited Foucault extensively in his deconstruction of the map as a spatial language.

Indeed, it is clear from many of the entries that the same source of inspiration can be interpreted and used in different ways. For example, both **Ed Soja** and **David Harvey** draw upon **Henri Lefebvre**'s seminal text *The Production of Space* to develop their own ideas about the workings of capital, but differ in the interpretation and weight they place on Lefebvre's argument. Of course, a particular thinker can also influence different audiences because their own thoughts

have transformed over time as they themselves come into contact with the thoughts of others and develop new lines of argument. For example, David Harvey remains a key influence on spatial science due his book *Explanation in Geography* (1969), which provided a theoretical blueprint for positivist geography. At the same time, he is also a key source of inspiration and ideas for Marxist geographers who draw upon his 1973 book *Social Justice and the City* (and subsequent work), which utilized the writings of Karl Marx to construct structural explanations for socio-spatial inequality. Indeed, his 1982 text *Limits to Capital* remains perhaps the most important statement by a geographer on the uneven production of space under capitalism.

A situated approach to understanding the production of spatial thought also, of course, reveals the extent to which place makes a difference to knowledge creation. For example, groupings of particular scholars in particular universities at particular periods can produce cross-generational schools of thinking. While Paris seems to be so often the locus of social theory (see **Jean Baudrillard**, **Manuel Castells**, **Gilles Deleuze**, **Michel Foucault**, **Henri Lefebvre**; also Gane, 2003), other centres also emerge if we search for key locations in the theorization of space and place. For example, Carl Sauer inspired the Berkeley School of cultural geography that influenced several generations of American geographers; **Stuart Hall** was a key actor in establishing Birmingham's Centre for Contemporary Cultural Studies whose work did much to shape 'new' cultural geography; the 1950s Washington graduate class (including **Waldo Tobler**) are widely acknowledged as fuelling the so-called 'quantitative revolution'; and the writings of **Michael Dear**, **Ed Soja**, **Michael Storper**, **Mike Davis** and colleagues means that southern California is widely acknowledged as the home of postmodern urbanism. On the other hand, the development of an individual's ideas can represent a reaction *against* the place where they are/were located. For example, to return to Gillian Rose, her book *Feminism and Geography* is widely acknowledged to have grown out of her critique of the Cambridge school of geography in which she was educated. A biographical approach thus alerts to the significant role of disciplinary spaces of education, as well as the often neglected sites of the *field*, the *body* and the *act of dissemination* by which knowledge is produced and circulated (Dewsbury and Naylor, 2002; see also Driver, 1995). As such, the biographical approach adopted in this volume focuses on both the *roots* (origins) and *routes* (directions of evolution) of thinking on space of place. While not providing an exhaustive account, the following entries ultimately allow us to discern the many roots and routes – the intellectual genealogies – that explain why space and place have come to mean such different things to different people in different places.

CONCLUSION

At a time when some are rightly suspicious of the concentration of academic power and influence in the higher-education sector (see Short, 2002), and others are seeking to resist the logic of the auditing procedures that relies on measures of individual research output (see Sidaway, 2000), there are some dangers inherent in compiling a list of key thinkers. Yet, as we have shown in our introduction, our intention is not to identify the most important or influential theorists, but to provide a guide to some (but inevitably not all) of those figures who have progressed our *theoretical* understanding – in some important way – of space and place, at the same time as illustrating the diverse traditions of contemporary geographical thinking. While choosing just a few thinkers inevitably

privileges them as key conduits of theorizing and practising geography – and simultaneously marginalizes and silences other thinkers and their theories – it is important to appreciate the ways in which knowledge is produced through intellectual encounters and dialogues (as illustrated in the previous section).

Given our intention to highlight the theoretical contribution these figures have made, the entries here do not offer a thorough or balanced overview of the career of each thinker. Instead, each follows a common format, starting with an overview of each subject's academic scholarship alongside some basic biographic information. While this overview is, of necessity, cursory, it hopefully provides an understanding of how each thinker developed their ideas in particular social, spatial and temporal contexts. This contextual material is followed by a summary of the way that each has conceived of space and place, aiming to identify why each is regarded as an important and influential thinker in debates on space and place. In a final section, each contributor offers a critical reflection on the work of each thinker, outlining some of the key controversies that adhere to that thinker's work (while showing how their work has been adapted by those working in different geographical and theoretical traditions). Each entry concludes with two reference lists, the first being a guide to the thinker's 'key' works. Here, the most important and major works by each thinker are listed, with an emphasis on those works that are most readily and widely available (hence, where there are multiple editions of one book in existence, we have tended to list the most recent English version rather than the first edition). The second reading list contains minor books, paper and chapters (where these are cited in the text), as well as a range of secondary sources. It is our hope that each entry inspires readers to explore these references and develop their own take on the varied geographical imaginations deployed by these key thinkers on space and place.

REFERENCES

Abler, R., Adam, J. and Gould, P. (1971) *Spatial Organisation: The Geographer's View of the World*. New Jersey: Prentice Hall.

Augé, M. (1996) *Non-places: Introduction to an Anthropology of Supermodernity*. London: Verso.

Baldwin, E., Longhurst, B., McCracken, S., Ogborn, M. and Smith, G. (1999) *Introducing Cultural Studies*. Harlow: Prentice Hall.

Barnes, T. (2001a) 'Retheorizing economic geography: from the Quantitative Revolution to the Cultural Turn', *Annals, Association of American Geographers* 91: 546–565.

Barnes, T. (2001b) 'Lives lived and tales told: biographies of geography's quantitative revolution', *Environment and Planning D – Society and Space* 19: 409–429.

Barnett, C. (1998) 'The cultural turn: fashion or progress in human geography', *Antipode* 30: 379–394.

Bauman, Z. (2000) *Liquid Modernities*. Cambridge: Polity.

Benko, G. and Strohmeyer, U. (1997) *Space and Social Theory*. Oxford, Blackwell.

Bingham, N. (1996) 'Object-ions: from technological determinism to towards geographies of relations', *Environment and Planning D – Society and Space* 14: 635–657.

Bird, J. (1989) *The Changing World of Geography*. Oxford: Clarendon Press.

Blunt, A. and Wills, J. (2000) *Dissident Geographies: An Introduction to Radical Ideas and Practice*. London, Prentice Hall.

Cloke, P., Philo, C. and Sadler, D. (1991) *Approching Human Geography*. Liverpool: PCP Press.

Crang, M. and Thrift, N. (2000) 'Introduction', in M. Crang and N. Thrift (eds) *Thinking Space*. London: Routledge, pp. 1–30.

Cresswell, T. (1996) *In Place/Out of Place: Geography, Ideology and Transgression*. Minneapolis: University of Minnesota Press.

Dear, M. (1988) 'The postmodern challenge: reconstructing human geography', *Transactions of the Institute British Geographers* 13: 262–274.

Dear, M. (2000) *The Postmodern Urban Condition*. Oxford: Blackwell.

Dewsbury, J. D. and Naylor, S. (2002) 'Praticising geographical knowledge: fields, bodies and dissemination', *Area* 34: 253–260.

Doel, M. (1999) *Postructuralist Geographies: The Diabolical Art of Spatial Science*. Edinburgh: Edinburgh University Press.

Driver, F. (1995) 'Geographical traditions: thinking the history of geography', *Transactions, Institute of British Geographers* 20: 403–404.

Elliott, A. and Turner, B. S. (eds) (2001) *Profiles in Contemporary Social Theory*. London: Sage.

Gane, M. (2003) *French Social Theory*. London: Sage.

Gibson-Graham, J. K. (2000) 'Poststructural interventions', in T. Barnes and R. Sheppard (eds) *A Companion to Economic Geography*. Oxford: Blackwell, pp. 95–111.

Gleeson, B. (1996) 'A geography for disabled people', *Transactions, Institute of British Geographers* 21: 387–396.

Gregory, D. (1994) 'Social theory and human geography', in D. Gregory, R. Martin, and G. Smith (eds) *Human Geography: Society, Space and Social Science*. London: Macmillan, pp. 78–109.

Gregory, S. (1963) *Statistical Methods and the Geographer*. London: Longman.

Haggett, P. and Chorley, R. J. (1969) *Network Analysis in Geography*. London: Arnold.

Harrison, S. and Dunham, P. (1998) 'Decoherence, quantum theory and their implications for the philosophy of geomorphology', *Transactions of the Institute of British Geographers* 23: 501–514.

Harvey, D. (1989) *The Condition of Post-modernity*. Oxford: Blackwell.

Heffernan, M. (2003) 'Histories of geography', in S. L. Holloway, S. Rice and G. Valentine (eds) *Key Concepts in Geography*. Sage: London, pp. 3–23.

Holloway, L. and Hubbard, P. (2001) *People and Place: The Extraordinary Geographies of Everyday Life*. Harlow: Prentice Hall.

Hubbard, P., Kitchin, R., Bartley, B. and Fuller, D. (2002) *Thinking Geographically: Space, Theory and Contemporary Human Geography*. London: Continuum.

Jackson, P. (1989) *Maps of Meaning*. London: Routledge.

Jackson, P. (1999) 'Commodity cultures: the traffic in things', *Transactions Institute of British Geographers* 24: 95–108.

Johnston, R. J. (1991) *Geography and Geographers: Anglo-American Geography Since 1945*. London: Edward Arnold.

Johnston, R. J. (1996) 'Paradigms and revolution or evolution?', in J. Agnew, D. Livingstone and A. Rogers (eds) *Human Geography*. Oxford: Blackwell, pp. 37–53.

Johnston, R. J. (2000) 'Authors, editors and authority in the postmodern academy', *Antipode* 32: 271–291.

Kent, M. (2003) 'Space: making room for space in physical geography', in S. L. Holloway, S. Rice and G. Valentine (eds) *Key Concepts in Geography*. Sage: London, pp. 109–130.

Kuhn, T. S. (1962) *The Structure of Scientific Revolutions*, 1st edition. Chicago: University of Chicago Press (2nd edition 1970).

Lefebvre, H. (1991) *The Production of Space*. Oxford: Blackwell.

Livingstone, D. (1992) *The Geographical Tradition*. Oxford: Blackwell.

Massey, D. (1991) 'The political place of locality studies', *Environment and Planning A* 23: 267–281.

Massey, D. (1999) 'Space-time, science and the relationship between physical and human geography', *Transactions, Institute of British Geographers* 24: 261–276.

Mayhew, R. J. (2000) 'The effacement of early modern geography (*c.* 1600–1850): a historiographical essay', *Progress in Human Geography* 1 September, 25 (3): 383–401.

Moss, P. (2001) *Placing Autobiography*. Syracuse: Syracuse University Press.

Olwig, K. R. (2002) 'The duplicity of space: German *raum* and Swedish *rum* in English language geography', *Geografiska Annaler* B 84B 1: 1–17.

Peet, R. (1998) *Modern Geographic Thought*. Oxford: Blackwell.

Raper, J. and Livingstone, D. (2001) 'Let's get real: Spatio-temporal identity and geographic entities', *Transactions of the Institute of British Geographers* 26: 237–242.

Relph, E. (1976) *Place and Placelessness*. London: Pion.

Robinson, G. (1998) *Methods and Techniques in Human Geography*. Chichester: John Wiley.

Shields, R. (1991) *Places on the Margin*. London: Routledge.

Short, J. R. (2002) 'The disturbing case of the concentration of power in human geography', *Area* 34: 323–324.

Sibley, D. (1995) *Geographies of Exclusion: Society and Difference in the West*. London: Routledge.

Sidaway, J. (1997) 'The production of British geography', *Transactions, Institute of British Geographers* 22: 488–504.

Sidaway, J. D. (2000) 'Recontextualising positionality: Geographical research and academic fields of power', *Antipode* 32 (3): 260–270.

Smith, N. (2000) 'Scale', in R. Johnston, D. Gregory, G. Pratt and M. Watts (eds) *The Dictionary of Human Geography*, 4th edition. Blackwell: Oxford, pp. 724–727.

Taylor, P. (1999) 'World cities and territorial states under conditions of contemporary globalization', *Third World Planning Review* 21 (3): 3–10.

Thrift, N. (2003) 'Space: the fundamental stuff of geography', in S. L. Holloway, S. Rice and G. Valentine (eds) *Key Conceps in Geography*. Sage: London, pp. 95–108.

Tuan, Yi-Fu (1977) *Space and Place: The Perspective of Experience*. Minneapolis: University of Minnesota Press.

Unwin, T. (1992) *The Place of Geography*. Harlow: Longman.

Urry, J. (1995) *Consuming Place*. London: Routledge.

Wilson, A. (2000) *Complex Spatial Systems. The Modelling Foundations of Urban and Regional Analysis*. Harlow: Pearson.

Yeung, H. (2002) 'Deciphering citations', *Environment and Planning A* 34: 2093–2106.

Phil Hubbard, Rob Kitchin and Gill Valentine

1   Benedict Anderson

BIOGRAPHICAL DETAILS AND THEORETICAL CONTEXT

Benedict Anderson is the author of one of the most important concepts in political geography, that of nations being 'imagined communities'. Guggenheim Fellow and member of the American Academy of Arts and Sciences, Anderson was born in Kunming, China in 1936. Brother of political theorist Perry Anderson and an Irish citizen whose father was an official with Imperial Maritime Customs, he grew up in California and Ireland before attending Cambridge University. Studying briefly under Eric Hobsbawm, Anderson graduated with a First Class degree in Classics in 1957. He moved to Cornell University in 1958 to pursue PhD research on Indonesia. At Cornell he was influenced by George Kahin, John Echols and Claire Holt (Anderson, 1999). In 1965 Indonesia's military leader Suharto foiled an alleged coup attempt by communist soldiers, purged the army, and killed hundreds of thousands of civilians. Working with two other graduate students, Anderson analysed Suharto's version of events, questioning their veracity. Their assessment reached the Indonesian military who in 1967 and 1968 invited Anderson to the country to persuade him of the errors in this monograph, then known as the 'Cornell Paper'. Failing to be convinced, Anderson was denounced by the Indonesian regime. Following formal publication of the original allegations (Anderson *et al.*, 1971), Indonesian authorities refused Anderson's visa applications, barring him from Indonesia for what

became the duration of Suharto's regime. Anderson returned to Indonesia in 1999 following the dictator's death.

Anderson completed his PhD entitled *The Pemuda Revolution: Indonesian Politics, 1945–1946* in 1967 and taught in the Department of Government at Cornell University until retirement in 2002. Editor of the interdisciplinary journal *Indonesia* between 1966 and 1984, Anderson studied topics as diverse as Indonesia's government, politics and international relations (e.g. 1964), human rights (e.g. 1976), and its role in East Timor (e.g. 1980). As an expert on South East Asia, military conflicts between Cambodia, Vietnam and China in the late 1970s stimulated him to analyse the importance of, and political attraction to, nationalist politics. The result was *Imagined Communities – Reflections on the origin and spread of nationalism* (1983, 1991) in which Anderson proposed the theory of 'imagined communities'. Major theoretical approaches, Anderson maintained, had largely ignored nationalism, merely accepting it as the way things are:

> Nation, nationality, nationalism – all have proved notoriously difficult to define, let alone analyse. In contrast to the immense influence that nationalism has exerted on the modern world, plausible theory about it is conspicuously meagre.
> (Anderson, 1991: 3)

Particularly culpable in this respect was Marxism, the relationship between it and nationalism being the subject of debate in *New Left Review* in the 1970s (e.g. Löwy, 1976; Debray, 1977). In this climate, Anderson (1991: 3; original emphasis) argued Marxist thought had not ignored nationalism; rather, 'nationalism has proved an uncomfortable *anomaly* for Marxist theory and, precisely for that reason, has largely been elided, rather than confronted'. *Imagined Communities* was an effort to reconcile theories of Marxism and nationalism, and counter what Anderson envisaged as a skewed context for the assessment of nationalism, namely an almost wholly European focus to the detriment of examining South American 'Creole pioneers' of modern nationalist politics. This distortion, Anderson maintained, continues both within and outside the academy. From case studies of colonialism in Latin America and Indonesia, Anderson (1991: 6) proposed 'the following definition of the nation: it is an imagined political community – and imagined as both inherently limited and sovereign'.

SPATIAL
CONTRIBUTIONS

Anderson's concept of nations being 'imagined communities' has become standard within books reviewing geographical thought (e.g. Massey and Jess, 1995; Crang, 1998; Cloke *et al.*, 2001). The contention that a nation is 'imagined' does not mean that a nation is false, unreal or to be distinguished from 'true' (unimagined) communities. Rather Anderson is proposing that a nation is constructed from popular processes through which residents share nationality in common:

> It is *imagined* because the members of even the smallest nation will never know most of their fellow members, meet them, or even hear of them, yet in the minds of each lives the image of their communion.
> (Anderson, 1991: 6; original emphasis)

This understanding both shapes and is shaped by political and cultural institutions as people 'imagine' they share general beliefs, attitudes and recognize a collective national populace as having similar opinions and sentiments to their own. Secondly,

> The nation is imagined as *limited* because even the largest of them, encompassing perhaps a billion living human beings, has finite, if elastic, boundaries, beyond which lie other nations.
> (Anderson, 1991: 7; original emphasis)

To have one nation means there must be another nation against which self-definition can be constructed. Anderson is thus arguing for the social construction of nations as political entities that have a limited spatial and demographic extent, rather than organic, eternal entities. Further,

> It is imagined as *sovereign* because the concept was born in an age in which Enlightenment and Revolution were destroying the legitimacy of the divinely-ordained, hierarchical dynastic realm ... nations dream of being free ... The gage and emblem of this freedom is the sovereign state.
> (Anderson, 1991: 7; original emphasis)

Anderson argues that the concept of the nation developed in the late eighteenth century as a societal structure to replace previous monarchical or religious orders. In this manner, a nation was a new way of conceptualizing state sovereignty and rule. This rule would be limited to a defined population and territory over which the state, in the name of nationality, could exercise power.

> Finally, it is imagined as a *community*, because, regardless of the actual inequality and exploitation that may prevail in each, the nation is always conceived as a deep, horizontal comradeship. Ultimately it is this fraternity that makes it possible, over the past two centuries, for so many millions of people, not so much as to kill, as willing to die for such limited imaginings.
> (Anderson, 1991: 7; original emphasis)

Nations hold such power over imaginations, claims Anderson, that patriotic calls to arms are understood as the duty of all national residents. Further, in war, national citizens are equal and class boundaries are eroded in the communal struggle for national survival and greatness.

Anderson's second key aspect of the development of nationalism is what he identifies as the role of 'Creole pioneers'. In both North and South America, those who fought for national independence in the eighteenth and nineteenth centuries had the same ancestries, languages and traditions as the colonizing powers they opposed. Anderson (1991: 50) argues these 'Creole communities' developed nationalist politics *before* Europe, because as colonies they were largely self-administrating territorial units. Thus residents conceived of their belonging to a common and potentially sovereign community, a sentiment enhanced by provincial newspapers raising debate about intercontinental political and administrative relationships. Anderson stakes much of his thesis on 'print capitalism.' Drawing on Erich Auerbach and Walter Benjamin, Anderson argues that the standardization of national calendars, clocks and language was embodied in books and the publication of daily newspapers. This generated a sense of simultaneous national experiences for people as they became aware of events occurring in their own nation and nations abroad. Newspapers 'made it possible for rapidly growing numbers of people to think about themselves, and relate themselves to others, in profoundly new ways' (Anderson, 1991: 36). Disparate occurrences were bound together as national experiences as people felt that everyone was reading the same thing and had equal access to information:

> the convergence of capitalism and print technology on the fatal diversity of human language created the possibility of a new form of imagined community, which in its basic morphology set the stage for the modern nation. The potential stretch of these communities was inherently limited, and, at the same time, bore none but the most fortuitous relationship to

> existing political boundaries (which were, on the whole, the highwater marks of dynastic expansionisms).
> (Anderson, 1991: 46)

The worldwide impact of *Imagined Communities* across academic disciplines led to a revised edition in 1991. In this enlarged edition Anderson noted that he had '[become] uneasily aware that what I had believed to be a significantly new contribution to thinking about nationalism – changing apprehensions of time – patently lacked its necessary coordinate: changing apprehensions of space' (1991: xiii–xiv). Utilizing South East Asian examples, Anderson corrected this omission by including chapters addressing the role of national census, museums, constructions of national memories, biographies and maps. Drawing on a 1988 PhD dissertation by Thongchai Winichakul about nineteenth century Siam/Thailand, Anderson (1991: xiv) argued that maps contribute to the 'logoization of political space' and their myriad reproductions familiarize people with the limitations of national sovereignty and community.

Having examined mass communication with his thesis of print capitalism, Anderson subsequently turned to the legacy of migration:

> The two most significant factors generating nationalism and ethnicity are both closely linked to the rise of capitalism. They can be described summarily as mass communication and mass migrations.
> (Anderson, 1992: 7)

Maintaining that nationalist movements were/are often initiated by expatriates, noting again the 'Creole pioneers' of Latin America and financial contributions from overseas to the Irish Republican Army and ethno-nationalist factions in the Balkan Wars of the early 1990s, Anderson assesses:

> It may well be that we are faced here with a new type of nationalist: the 'long-distance nationalist' one might perhaps call him [fn. 'Him' because this type of politics

seems to attract males more than females].
For while technically a citizen of the state
in which he comfortably lives, but to
which he may feel little attachment, he
finds it tempting to play identity politics
by participating (via propaganda, money,
weapons, any way but voting) in the conflicts of his imagined *Heimat* – now only
fax-time away.
(Anderson, 1992: 13)

Translated into dozens of languages and
arguably the most regularly cited scholar
on the topic, Anderson has appeared on
television, addressed committees of the
United Nations and US Congress regarding Indonesia and East Timor, and raised
questions about human rights abuses in
South East Asia (e.g. Anderson, 1976,
1980). He is one of the most influential
scholars of his generation. Although not a
geographer by training or career, issues of
space, territory and place, and his criticisms of nationalist politics, have led to
Anderson's work being widely utilized
within geographical research.

KEY ADVANCES AND
CONTROVERSIES

Imagined Communities received little attention from geographers upon its publication. Largely without review in major
geography journals such as the *Annals of
the Association of American Geographers*,
Anderson's concepts entered geographical
debate through their impact on interdisciplinary studies of nationalism. Yet engagement was typified by comment that
nations are 'imagined communities' – Anderson being cited accordingly. Indeed,
Spencer and Wollman (2002: 37) claim
that such is the regularity with which
articles about nationalism routinely cite
Imagined Communities that Anderson's
conceptualization 'has become one of the
commonest clichés of the literature' the
result being that 'invocation has, in some
cases, been a substitute for analysis'.

Geographers have not been immune to
this (see, *inter alia*, Jackson and Penrose,
1993; Smith and Jackson, 1999).

Prolonged geographical assessments of
Anderson's contentions seem few. For
example, Blaut (1987) does not assess
Anderson's work in his review of Marxist
theories of nationalism, and Short's (1991:
226) *Imagined Country* simply proposes
Anderson's *Imagined Communities* as additional reading. Arguably the most sustained utilization comes from Radcliffe
and Westwood (1996: 2), who examine
how a national imagined community is
'generated, sustained and fractured' in
Ecuador. They maintain that Anderson's
'geographical imagination . . . permits him
to link themes of space, mobility and the
nation', but comment that he fails to fully
acknowledge or develop the implications
of this within his work (Radcliffe and
Westwood, 1996: 118).

Primarily it is postcolonial scholars
that have questioned Anderson's arguments. **Edward Said** (1993) contends that
Anderson is too linear in his explanation
that political structures and institutions
change from dynasties, through the standardizing influence of print capitalism, to
sovereign nations (see also McClintock,
1995). The most vocal critic has been
Partha Chatterjee (1993), who contends
that the imagination of political communities has been limited by European colonialism. Having had specifically
nationalist institutional forms imposed on
them as colonies, upon independence
these areas had no option but to follow
European paths, with Western powers
ready to prevent any seemingly dangerous deviations. 'Even our imaginations',
asserts Chatterjee (1993: 5) 'must remain
forever colonized.' Nationalism and nations, Chatterjee maintains, operate only
within limits formulated in Europe, and
thus they can only be conceptualized
within these European strictures. Anti-
colonial nationalisms thus typically opposed colonialism using the same nationalist arguments as the colonists.
Distinction could not be made through
political or economic conceptualization

due to the European dominance of these venues and thus the limited sovereignty and territory of the colony was already imagined for the colonized by the colonizers. Consequently, anti-colonial nationalism could only be imagined through cultural processes and practices. Here again Chatterjee challenges Anderson, maintaining that although the processes of print capitalism were important, Anderson's formulation of them as standardizing language, time and territorial extent is too simplistic to impose on the diverse, multilingual and asymmetrical power relations of the colonial situation.

A second major critique of *Imagined Communities* comes from a feminist perspective. With a focus on the 'fraternity' experienced by members of a nation (Anderson, 1991: 7), the protagonists in Anderson's conceptions of nationalism are typically assumed to be male. Mayer (2000: 6) argues that Anderson envisions 'a hetero-male project ... imagined as a brotherhood', eliding gender, class and racial structures within and between national communities; and McDowell (1999: 195) demonstrates that although being seemingly neutral, 'the very term horizontal comradeship ... brings with it connotations of masculine solidarity'. Subsequently, McClintock (1995: 353) laments that sustained 'explorations of the gendering of the national imagination have been conspicuously paltry'.

A third challenge comes from Don Mitchell, who argues that as well as *imagining* communities, there must be attention to:

the *practices* and exercises of power through which these bonds are produced and reproduced. The questions this raises are ones about who defines the nation, how it is defined, how that definition is reproduced and contested, and, crucially, how the nation has developed and

changed over time ... The question is not what common imagination *exists*, but what common imagination is *forged*. (Mitchell, 2000: 269; original emphasis)

Anderson's proposal, therefore, is constrained by its narrowness. What does it matter that a nation is an imagined community? The issue must be to show the work needed to produce and maintain that imagination, how this impacts on people's lives and how power to enforce the national community that is imagined shapes behaviours across time and space.

There is much to commend in the concept of imagined communities, but there is a need to explore power relations inherent in the processes Anderson describes and in their material impacts, whether these are founded on gender, racial, ethnic, class, sexual or other aspect of individual identity. Recent work begins to address such challenges. Angela Martin (1997: 90) maintains that although 'intellectuals have been given the power to "imagine" the nation or national community ... the material dimension, or political economy, of nationalism and the nation have been ignored'. Her assessment of late nineteenth-century Irish nationalism argues for a 'corporeal approach to the nation' to interrogate how gender roles were constructed both in the Irish national imagination and how they restricted behaviour in everyday life (Martin, 1997: 91). In turn, Steven Hoelschler's (1999: 538) study of the construction of a Swiss heritage community in New Glarus, Wisconsin, invokes Anderson to explain that specific 'forms of imagining' are utilized by elites to produce place and community identities, and examines how these elite images are contested by non-elite groups. Thus geographers are moving beyond Anderson's thesis to understand both imagined and *material* communities of nations and nationalisms.

ANDERSON'S MAJOR WORKS

Anderson, B., McVey, R. T. and Bunnell, F. P. (1971) *A Preliminary Analysis of the October 1, 1965 Coup in Indonesia.* Ithaca: Cornell Modern Indonesia Project, Publication No. 52.

Anderson, B. (1972) *Java in a Time of Revolution.* Ithaca: Cornell University Press.

Anderson, B. (1983) *Imagined Communities: Reflections on the Origin and Spread of Nationalism.* London: Verso.

Anderson, B. (1990) *Language and Power: Exploring Political Cultures in Indonesia.* Ithaca: Cornell University Press.

Anderson, B. (1991) *Imagined Communities: Reflections on the Origin and Spread of Nationalism* (revised and enlarged edition). London: Verso.

Anderson, B. (1998) *The Spectre of Comparisons: Nationalism, Southeast Asia and The World.* London: Verso.

Secondary Sources and References

Anderson, B. (1964) 'Indonesia and Malaysia', *New Left Review* 28: 4–32.

Anderson, B. (1967) *The Pemuda Revolution: Indonesian Politics, 1945–1946,* Unpublished PhD dissertation. Ithaca, Cornell University.

Anderson, B. (1976) 'Prepared testimony on human rights in Indonesia', in *Human Rights in Indonesia and the Philippines.* Washington, DC: U.S. Government Printing Office, pp. 72–80.

Anderson, B. (1980) 'Prepared testimony on human rights in Indonesia and East Timor', in *Human Rights in Asia: Noncommunist Countries.* Washington, DC: U.S. Government Printing Office, pp. 231–262 and 275–277.

Anderson, B. (1992) 'The new world disorder', *New Left Review* 193: 3–13.

Anderson, B. (1999) 'The spectre of comparisons', *Cornell University College of Arts and Sciences Newsletter* 20 (2); online: http://www.arts.cornell.edu/newsletr/ spring99/spectre.htm (accessed 2 October 2002).

Blaut, J. M. (1987) *The National Question: Decolonising The Theory of Nationalism.* London: Zed Books.

Chatterjee, P. (1993) *The Nation and its Fragments: Colonial and Postcolonial Histories.* Princeton, NJ: Princeton University Press.

Cloke, P., Crang, P. and Goodwin, M. (eds) (2001) *Introducing Human Geographies.* London: Arnold.

Crang, M. (1998) *Cultural Geography.* London: Routledge.

Debray, R. (1977) 'Marxism and the national question', *New Left Review* 105: 25–41.

Hoelscher, S. (1999) 'From sedition to patriotism: performance, place, and the reinterpretation of American ethnic identity', *Journal of Historical Geography* 25 (4): 534–558.

Jackson, P. and Penrose, J. (1993) 'Introduction: placing "race" and nation', in P. Jackson and J. Penrose (eds) *Constructions of Race, Place and Nation.* Minneapolis: University of Minnesota Press, pp. 1–23.

Löwy, M. (1976) 'Marxists and the national question', *New Left Review* 96: 81–100.

Martin, A. K. (1997) 'The practice of identity and an Irish sense of place', *Gender, Place and Culture* 4: 89–113.

Massey, D. and Jess, P. (eds) (1995) *A Place in the World?* New York: Oxford University Press.

Mayer, T. (2000) 'Gender ironies of nationalism: Setting the stage', in T. Mayer (ed.) *Gender Ironies of Nationalism: Sexing The Nation.* London: Routledge, pp. 1–22.

McClintock, A. (1995) *Imperial Leather: Race, Gender and Sexuality in the Colonial Contest.* London: Routledge.

McDowell, L. (1999) *Gender, Identity and Place: Introducing Feminist Geographies.* Minneapolis: University of Minnesota Press.

Mitchell, D. (2000) *Cultural Geography: A Critical Introduction.* Oxford: Blackwell.

Radcliffe, S. and Westwood, S. (1996) *Remaking the Nation: Place, Identity and Politics in Latin America.* London: Routledge.

Said, E. W. (1993) *Culture and Imperialism* New York: Vintage.

Short, J. R. (1991) *Imagined Country: Society, Culture and Environment.* London: Routledge.

Smith, G. and Jackson, P. (1999) 'Narrating the nation – the "imagined community" of Ukranians in Bradford', *Journal of Historical Geography* 25 (3): 367–387.

Spencer, P. and Wollman, H. (2002) *Nationalism: A Critical Introduction.* Sage: London.

Euan Hague

2   Trevor Barnes

BIOGRAPHICAL DETAILS AND THEORETICAL CONTEXT

Trevor Barnes was born in London, England in 1956. Having grown up in Cornwall, he studied economics and geography at University College London between 1975 and 1978. Barnes completed MA and PhD degrees in Geography at the University of Minnesota under the supervision of Eric Sheppard, and from 1983 taught at the University of British Columbia in Vancouver, Canada. Barnes' work extends across theories of economic value; analytical political economy; flexibility and industrial restructuring; and most recently the 'theoretical histories' of Anglo-American economic geography. He also sought to make key statements on the position of economic geography at the end of the millennium through a number of edited volumes (Barnes and Gertler, 1999; Barnes and Sheppard, 2000) as well as reviews of political economy approaches in the journal *Progress in Human Geography* (e.g. Barnes, 1998).

Although perhaps giving the appearance of a relatively divergent set of themes, there are strong threads of continuity running through Barnes' research and writing interests. He has long been captivated by the work of the economist Piero Sraffa, for example. In Barnes' view, Sraffa's terse expositions on value in *The Production of Commodities by Means of Commodities* (via a set of simultaneous production equations) usefully speak both to rigorous abstract theorists as well as to scholars who are more interested in the contextual and the concrete (Barnes,

1989, 1996, chapter 7). Barnes regards anti-essentialist accounts such as Sraffa's as a useful means of critiquing both classical Marxist accounts of the labour theory of value (which essentialize the role and nature of labour power) as well as neo-classical utility theory.

Together with Eric Sheppard (Sheppard and Barnes, 1990), Barnes has sought to ground political economy in space and place through the development of analytical approaches. Engaging with, but also developing a substantial critique of analytical Marxism, such approaches use 'both mathematical reasoning and rigorous, formal statistical testing to determine logically how space and place make a difference both to the definition of social processes and to their relation to the economy' (Barnes and Sheppard, 2000: 5). Although cursory readings (particularly if solely focused upon the use of formal mathematical language) might discern a preference for the abstract over the concrete and contextual, Barnes would refute such a contention. For example, Barnes' engagement with debates surrounding flexible production has drawn upon research on the forestry industry in British Columbia conducted with Roger Hayter (Barnes and Hayter, 1992; Hayter and Barnes 1992). Whereas many accounts of flexibility through the 1980s and 1990s centred on developments in 'new industrial spaces', Barnes and Hayter sought to extend conceputalizations of flexibility through a consideration of 'in situ' restructuring in the context of a marginal resource economy. The theoretical and the political are also closely connected in Barnes' recent use of the work of Canadian economic historian Harold Innis to understand 'the dependency and disruptions' that have emerged in British

Columbia (Barnes, 2001a: 4; see Barnes *et al.*, 2001). Barnes' explicit concern has been to confront the profound devastation of lives and communities wrought by the decline of the forest products sector in British Columbia (Barnes, 2001a).

Over the course of his career, Barnes (1992, 1996, 2002a) has become increasingly interested in tracing the social and political connections that produced the spatial scientific narratives that came to dominate geography – and particularly economic geography – during the 1950s and 1960s. Drawing in part upon studies in the sociology of scientific knowledge (including the work of **Bruno Latour**), Barnes has been keen to read changes in the nature of Anglo-American economic geography as transformations in attitudes towards theory itself. Ironically, the most significant aspect of geography's quantitative revolution was not that it ushered in a set of new methodologies – in fact 'geography had been quantitative from the time of its formal institutionalization as a discipline in the nineteenth century' (Barnes, 2001c: 552) – but rather that it involved a shift in theoretical sensibilities. This is not to say that the practices of geography remained static: computerization and 'even more complex statistical methods' (Barnes, 2001c: 553) became increasingly dominant. New 'scientific' vocabularies were important in the valorization of new technical competencies, but most significantly the quantitative revolution sought to produce foundational understandings of the world in which the truthfulness of representation was guaranteed (Barnes, 2001c: 553).

In researching the connections between and among spatial scientists in North America, Barnes (2001b, 2001c) has been concerned to reflect upon the socially embedded nature of geography's quantitative revolution. Crucially, transformations in geographical thinking emerged as 'local affairs' within particular institutional sites (Barnes, 2001c: 552). Again, his perspective is informed by philosophy of science literatures and particularly by the notion of externalism, or

'the belief that . . . knowledge is intimately related to the local context in which it develops' (Barnes, 2003: 70). This contrasts sharply from an internalist perspective, which presumes that 'there is a deep-seated, autonomous and universal principle that guides theoretical development' (Barnes, 2003: 70).

Barnes' interest in understanding the production of knowledge derives at least in part from a desire to be conscious of the social power and interests that shape such knowledge. In Barnes' (1996: 250) view, 'from the moment we enter the academy, we are socialized into pre-existing networks of knowledge and power that, whether we are conscious of them or not, come with various sets of interests'. Shifting and changing interests are thus inextricably bound up with transformations in knowledge itself. Writing about the use of locational analysis in geography, for example, Barnes reflects upon 'the duration of . . . principles, that is, how long people were willing to continue using and elaborating them, to pass them on and to defend them' (Barnes, 2003: 91). He suggests that the persistence of particular knowledges 'is a social (and geographical) process, and has as much to do with local context as any inherent quality of the principles themselves' (Barnes, 2003: 91).

SPATIAL CONTRIBUTIONS

One of the key contributions provided by Barnes' examination of the histories of economic geography lies in his provision of a more nuanced story of the discipline than narrations of strict succession and progression of knowledge generally would suggest. The notion of a quantitative *revolution* in geography itself obviously implies a move beyond pre-existing theoretical perspectives – and indeed post-spatial science approaches such as

Marxism, feminism, locality studies and accounts of flexible production were an explicit 'attempt to create something different from the past' (Barnes, 1996: 4). However, Barnes goes on to suggest that economic geography through the 1970s and 1980s remained in the grip of a strong Enlightenment ethos that sought certainties and foundations. Despite seeking to distance themselves from both the language and practice of spatial science, most analysts ultimately were unable to escape the legacy of the seventeenth century.

Excavating the subdisciplinary histories of economic geography might at first glance seem a somewhat atavistic project. However, Barnes explicitly argues that:

> Only by understanding ... earlier issues can we both comprehend the shape of contemporary discussions in economic geography and, more important, define a real alternative to the Enlightenment view that hitherto has dominated the discipline.
> (Barnes, 1996: 6)

Barnes characterizes such emergent alternatives as 'post'-prefixed economic geographies that reject the search for a singular order. Exemplary work includes poststructuralist feminist economic geographies (e.g. Gibson-Graham, 1996); feminist work on local labour markets (Hanson and Pratt, 1995; see also Pratt 1999); and development geography informed by postcolonial sensibilities. Most recently, Barnes (2002b: 95) has exhorted researchers to strive for a more 'edgy' engagement with their topic: to '[attempt] to undo formerly fixed conceptual categories of economic geography, and put them together again in different ways, and add new ones as well'. In the same way that he has sought to use the work of Sraffa (among others) to demonstrate the possibility of 'embrac[ing] openness, context and reflexivity', Barnes hopes that other economic geographers similarly will shun 'closure, universals and dogmatism' (Barnes, 1996: 251).

Such a stance correspondingly informs Barnes' own thinking about space and place. Moving beyond singular conceptions of economic space as (for example) a surface or territory, he has sought to argue that 'there is neither a single origin point for enquiry *or* a singular logic, spatial or otherwise' (Barnes, 1996: 250; emphasis added). Elsewhere, contrasting 'first' and 'new-wave' economic theory, he has argued that the former – which leant heavily on the work of von Thünen, Christaller and Weber – demonstrates that 'one should not explain events or phenomena by reducing them to fundamental entities taken as natural, or at least lying outside the social' (Barnes, 2001c: 559). In this sense, Barnes' (2001b, 2001d, 2002a) considerations of the performances of networks of actors (including, for example, economic geographers as well as textbooks) represent attempts to work with and through anti-essentialist conceptualizations of space and place. His work thus has contributed substantially to the reconfiguration of economic-geographical approaches in ways that seek new theoretical understandings of space and place but which at the same time reject a 'single route from here to there' (Barnes, 1998: 101).

KEY ADVANCES AND CONTROVERSIES

One of Barnes' first statements about the importance of knowledge production was in an editorial for the journal *Environment and Planning A* (Barnes, 1993). Taking his cue from emergent debates surrounding the sociology of scientific knowledge (see **Latour**), Barnes argued for a specific examination of the sociological construction of *geographical* knowledge, suggesting that geographers should be more reflexive both about the form and nature of the explanations they use, as well as the strategies they adopt in presenting these explanations to their audiences. Barnes' argument prompted considerable reaction. For example, Bassett (1994) expressed concern about the implications of increased reflexivity in research and writing, arguing (*contra* Barnes) that certain 'rational' or 'foundational' standpoints might be necessary for the achievement of social justice. Interestingly, Barnes utilized a multiply positioned narrative structure to make his case, arguing that 'there are many different ways to make a convincing argument, [but] there is no formal commonality among them' (Barnes, 1994: 1657). Thus Barnes' writing strategy – 'replying' in five different ways – was a deliberate attempt to take seriously a key tenet of the sociology of scientific knowledge: that the meanings of any particular 'reality' are 'constructed within a wider social network of meanings' (Barnes, 1994: 1655).

Certain commentators have been sceptical of Barnes' approach to the history of economic geographies and of his interest in the economic landscapes created through the use of metaphor (Barnes, 1992). Scott (2000: 495), for example, is uncomfortable with Barnes' emphasis upon the subdiscipline's fissures and dislocations, preferring to foreground 'evident continuities' in economic geography. Scott (2000: 495) also is concerned that attention to the textual effects of metaphors is 'rather off target' when compared with a need to address 'the immensely real substantive issues and purposive human practices that have always been and still are fundamentally at stake'. As is visible in his reviews of

geographical work in political economy, however (see particularly Barnes, 1998), Barnes certainly does not eschew a focus on worlds of (for example) production, class divides and labour market change.

Much of Barnes' writing has sought to contest the drawing of lines around the coherent entities of 'economy', 'politics' and 'culture' (see especially Barnes, 2002b). He has, for example, considered the performances of 'classic' economic geography textbooks (via the networks through which they moved) as a means of developing 'a cultural geography of economic geography and economic geographers' (Barnes, 2002a: 15). In narrating multi-layered histories of economic geography, Barnes is critical of attempts to police where different 'types' of geography are allowed to be and where they cannot. He acknowledges:

> There are critics like Harvey (2000) and Storper (2001) who argue that the focus on culture distracts too much from 'the 'hard world' of production and things' (Hall, 1988), and economic geographers would be better off if they devoted their energies to them.
> (Barnes, 2002b: 95)

At the same time, he maintains that economic geographers would do better to overturn and rupture existing categories. For Barnes, recognition that there is no single road to truth is essential in developing critical theories. Further, he argues that we need a range of imaginative approaches particularly because of the potential role they can play in 'reconfigur[ing] the world and our place within it' (Barnes, 2001a: 12).

BARNES' MAJOR WORKS

Barnes, T. J. (1996) *The Logics of Dislocation: Models, Metaphors and Meanings of Economic Space*. New York: Guildford.

Barnes, T. J. and Duncan D. S. (eds) (1992) *Writing Worlds: Texts, Discourses and Metaphors in the Interpretation of Landscape*. London: Routledge.

Barnes, T. J. and Gertler, M. (1999) *The New Industrial Geography: Regions, Institutions and Regulation*. London: Routledge.

Barnes, T. J. and Sheppard, E. (eds) (2000) *A Companion to Economic Geography*. Oxford: Blackwell.

Sheppard, E. and Barnes, T. J. (1990) *The Capitalist Space Economy: Geographical Analysis after Ricardo, Marx and Sraffa*. London: Unwin Hyman.

Secondary Sources and References

Barnes, T. J. (1989) 'Place, space and theories of economic value: contextualism and essentialism in economic geography', *Transactions of the Institute of British Geographers* 14: 299–316.

Barnes, T. J. (1992) 'Reading the texts of theoretical economic geography: the role of physical and biological metaphors', in T. J. Barnes and J. S. Duncan (eds) *Writing Worlds: Texts, Discourses and Metaphors in the Interpretation of Landscape*. London, Routledge, pp. 118–135.

Barnes, T. J. (1993) 'Whatever happened to the philosophy of science?', *Environment and Planning A* 25: 301–304.

Barnes, T. J. (1994) 'Five ways to leave your critic: a sociological scientific experiment in replying', *Environment and Planning A* 26: 1653–1658.

Barnes, T. J. (1998) 'Political economy III: confessions of a political economist', *Progress in Human Geography* 22: 94–104.

Barnes, T. J. (2001a) 'Critical notes on economic geography from an ageing radical. Or radical notes on economic geography from a critical age', *ACME: an International E-Journal for Critical Geographies* (www.acme-journal.org) 1: no pagination.

Barnes, T. J. (2001b) 'Lives lived and lives told: biographies of geography's quantitative revolution', *Environment and Planning D: Society and Space* 19: 409–429.

Barnes, T. J. (2001c) 'Retheorising economic geography from the quantitiative revolution to the cultural turn', *Annals of the Association of American Geographers* 91: 546–565.

Barnes, T. J. (2001d) ' "In the beginning was economic geography" – a science studies approach to disciplinary history', *Progress in Human Geography* 25: 521–544.

Barnes T. J. (2002a) 'Performing economic geography: two men, two books, and a cast of thousands', *Environment and Planning A* 34: 487–512.

Barnes, T. J. (2002b) 'Never mind the economy, here's culture: economic geography goes punk', in K. Anderson, M. Domosh, S. Pile and N. J. Thrift (eds) *Handbook of Cultural Geography*. London: Sage, pp. 89–97.

Barnes, T. J. (2003) 'The place of locational analysis: a selective and interpretive history', *Progress in Human Geography* 27: 69–95.

Barnes, T. J. and Hayter, R. (1992) 'The little town that did: flexible production and community response in Chemainus, BC', *Regional Studies* 26: 647–663.

Barnes, T. J., Hayter, R. and Hay, E. (2001) 'Stormy weather: cyclones, Harold Innis and Port Alberni, BC', *Environment and Planning A* 33: 2127–2147.

Bassett, K. (1994) 'Whatever happened to the philosophy of science? Some comments on Barnes', *Environment and Planning A* 26: 343–360.

Gibson-Graham, J. K. (1996) *The End of Capitalism (As We Knew It): A Feminist Critique of Political Economy*. Oxford: Blackwell.

Hall, S. (1988) 'Brave new world', *Marxism Today* 24–29 October.

Hanson, S. and Pratt, G. (1995) *Gender, Work and Space*. London, Routledge.

Harvey, D. (2000) *Spaces of Hope*. Berkeley: University of California Press.

Hayter, R. and Barnes, T. J. (1992) 'Labour market segmentation, flexibility and recession: a British Columbia study', *Environment and Planning C: Government and Policy* 10: 333–353.

Pratt, G. (1999) 'From Registered Nurse to registered nanny: diverse geographies of Filipina domestic workers in Vancouver, BC', *Economic Geography* 75: 215–36.

Scott, A. J. (2000) 'Economic geography: the great half-century', *Cambridge Journal of Economics* 24: 483–504.

Storper, M. (2001) 'The poverty of radical theory today: from the false promises of Marxism to the mirage of the cultural turn', *International Journal of Urban and Regional Research* 25: 155–179.

Susanne Reimer

3 Jean Baudrillard

BIOGRAPHICAL DETAILS AND THEORETICAL CONTEXT

Jean Baudrillard, born in Reims in 1929, completed a sociology thesis at Université de Paris X, Nanterre, under the direction of **Henri Lefebvre** in 1966. He took up a teaching assistantship at Nanterre in the same year, and was a key intellectual figure during the political ferment of May 1968. Despite his notoriety, Baudrillard has never held a full university post. He failed his *agrégation*, took pleasure in withdrawing from the formalities of university life, and famously provoked French academe with his publication of *Oublier Foucault* [*Forget Foucault*] in 1977, at the height of **Michel Foucault**'s influence. 'Foucault's discourse is a mirror of the powers it describes,' he wrote (Baudrillard, 1987: 10). 'It is there that its strength and its seduction lie, and not at all in its "truth index", which is only its leitmotiv: these procedures of truth are of no importance, for Foucault's discourse is no truer than any other.' In this relatively early text one can already see three crucial aspects of Baudrillard's theoretical disposition: his indifference to truth and power; his dismissal of critique as an act of complicity (mirroring) rather than subversion (undermining); and his preference for a strategy of seduction rather than production.

Though he is often considered a postmodernist, poststructuralist, lapsed Marxist or a French Marshall McLuhan, Baudrillard regards himself as a 'fatal' theorist: 'Pataphysician at twenty – situationist at thirty – utopian at forty – transversal at fifty – and viral and metaleptic at sixty – my complete history' (Baudrillard, 1996b: 83). Mike Gane (1991) rightly notes that Baudrillard's radicalism is evident in his shift from critical theory (in his case, Marxism) to fatal theory. Rather than adopting the critical strategy of comparing theories with reality and turning theories against themselves in order to precipitate their subversion and transgression (e.g. ideological critique, dialectical critique or deconstruction), Baudrillard favours the fatal strategy of pushing theories beyond themselves in order to glimpse what is on the other side: nothing – or what he prefers to call the 'objective indifference' of the world:

> Unlike the discourse of the real, which gambles on the fact of there being something rather than nothing, and aspires to being founded on the guarantee of an objective and decipherable world, radical thought, for its part, wagers on the illusion of the world. It aspires to the status of illusion, restoring the non-veracity of facts, the non-signification of the world, proposing the opposite hypothesis that there is nothing rather than something. (Baudrillard, 1996c: 97–98)

Given that Western thought has been actively producing different versions of the real for millennia, one can appreciate why Baudrillard's work of annulling reality has been so badly misunderstood. To put it bluntly, he does not wish to produce yet another version of reality. Rather, he is interested in what makes the reality that we have conjured up for ourselves continually withdraw and disappear. Like Friedrich Nietzsche, who claimed to have pushed only that which wanted to fall, Baudrillard is fascinated by the complete cycle of appearance *and disappearance*. By focusing only on the 'mode of production' (of truth, power, reality, desire, knowledge, meaning, value, etc.), he argues that critical theory has blinded itself to the other half of the game: the 'mode of disappearance'. Hence Baudrillard's penchant for the *double game* of appearance and disappearance, which he pursues through a strategy of seduction that testifies to the inseparability of perception and deception. (Baudrillard reminds us that to produce is to make appear and move forward [*pro-ducere*], while to seduce is to lead astray and make disappear [*se-ducere*].) And just as modes of production have mutated over space and time (cf. Karl Marx, Michel Foucault and Jacques Lacan), so too have modes of disappearance. This is why Baudrillard (1988c) has characterized his work as a 'double spiral'. 'If it is nihilistic to be obsessed by the mode of disappearance, and no longer by the mode of production, then I am a nihilist' (Baudrillard, 1994a: 162).

Although Baudrillard was influenced by Lefebvre's focus on everyday life in a 'bureaucratic society of controlled consumption', he rejected his humanistic brand of Marxism, written under the signs of 'alienation' and 'dis-alienation'. In seeking to replace the 'tottery Marxist denture with new, sharp and high-tech teeth, better fit to bite critically into the brave new world of consumption' (Bauman, 1992: 153), Baudrillard evidently found Roland Barthes's structuralist-cum-semiotic framework for *Mythologies* and *The Fashion System* inspirational during the writing of *The Consumer Society* and *The System of Objects*. Nonetheless, Baudrillard took Lefebvre's anti-structuralism seriously; rejecting the possibility that structuralism might represent a neutral, scientific and objective approach. What structuralism gave Baudrillard was not a discourse of truth, but a means to explode the 'reality principle' upon which the entire Marxist edifice rested.

In Baudrillard's reading, Marx regarded exchange-value (defined in terms of relations between people misconstrued as relations between things, and exemplified by money) as an artificial supplement to use-value (defined in relation to nature as the self-evident usefulness of things), which under capitalism has all but eclipsed use-value. For Marx, overcoming capitalist alienation and reification therefore required a return to use-value, famously sloganized as 'From each according to his ability; to each according to his need.' For Baudrillard, however, the very *notion* of use-value was an effect of exchange-value. Use-value is not the natural state of things, but a semiological projection onto an indifferent world, as is need. The idea that pre-capitalist or primitive-communist societies were based on use-value and the satisfaction of need is a nostalgic myth: such societies had as little room for use-value as they had for exchange-value. Marcel Mauss's analysis of 'gift exchange', Georges Bataille's 'general economy,' and Thorstein Veblen's account of 'conspicuous consumption' reveal the existence of an entirely different logic, marginalized by capitalism and ignored by Marx: the logic of *symbolic exchange* (Baudrillard, 1993a). In moving beyond Marx, Baudrillard alighted upon 'another, potentially more radical, critique of capitalist consumption as corrosive not of human relations . . . but of more fundamental symbolic exchange relations' (Gane, 1991: 5). Indeed, he offers his fatal theory as a form of symbolic exchange: 'The absolute rule is to give back more than you were given. Never less, always more. The absolute rule of thought is to give back the world as it was given to us – unintelligible. And, if possible, to render it a little more unintelligible' (Baudrillard, 1996c: 105).

Symbolic exchange, then, does not refer to the fact that commodities transmit meanings alongside their material qualities. Certainly, all commodities function as signs, and modern consumers are expected to manipulate these 'commodity-signs' in order to fashion their own identity – thus forming a powerful productive force for the expanded reproduction of capitalism into which countless millions are now routinely socialized. Yet this sign system as a mode of production and seduction, naturalized in the ideology of consumption, disavows the symbolic principle. Baudrillard gives the example of the sun. As a sign, the sun differentiates itself from other weather types: its unequivocal meaning, evident in countless holiday brochures, is 'not rain'. For the Aztecs, however, the sun was not a meaningful sign but an ambivalent force: both beneficent and cruel; saviour and destroyer; giver of life and requirer of death. It demanded a sacrificial offering, a gift that would, in turn, solicit a counter-gift. This logic of reciprocity – which involves challenge, obligation, and the escalation of stakes – is what Baudrillard means by symbolic exchange. For him, fatal theory does not aspire to truth. 'In my opinion, theory is simply a challenge to the real. A challenge to the world to exist' (Baudrillard, 1987: 124).

To refashion Marxism for the 'age of affluence', Baudrillard eagerly employed the insights and lexicon of structuralism to demonstrate how capitalism and consumerism were sign systems, and therefore modes of seduction (Baudrillard, 1975, 1981, 1996a, 1998a). Rather than a structural *return* to Marx à la Louis Althusser, however, Baudrillard found himself *beyond* Marx. Re-reading Marxism in the light of structuralism, Baudrillard began to detect fundamental problems in Marx. His shocking conclusion was that Marxism represented little more than a reflection of capitalism: a *mirror* of production. By remaining caught in the very code of capitalism (the production of meaning and value), the reversal of terms (e.g. use-value over exchange-value, need over desire, work over labour, and production over circulation) changed nothing. Whether it was dressed up in humanistic or structuralist guise, Marxism remained the alibi (mirror) of capital. Or as McLuhan famously put it: 'The medium is the message.' Moreover, Baudrillard argued that insofar as capitalism is a sign system, it can neutralize and

incorporate *any* opposition that continues to invest in meaning and significance – just as the art establishment can incorporate any form of 'anti-art', and politics any form of 'anti-politics' (even 'nonsense' can be taken up by an interpretative gesture). The fundamental lesson Baudrillard drew from this was that it is futile to confront a sign system in terms of its mode of production. The only glimmer of hope is to confront its mode of seduction (Baudrillard, 1990a, 1990b). However, from that point on, Baudrillard's fatal strategy has more often than not been woefully misunderstood (Smith with Doel, 2001).

In parallel to this fatal response to Marxism, Baudrillard has developed a fatal strategy in response to modernity and the Enlightenment more generally. Following Max Weber, Baudrillard holds that modernity involved the 'disenchantment' of the world through the imposition of a 'reality principle'. The pre-modern world was experienced as an enchanted realm in which fate reigned supreme. Modernity took this world to task and attempted to knock it into shape. It aimed to destroy flighty, deceptive appearances in the name of solid, trustworthy reality; to replace ambivalence with equivalence. Modernity's goal was 'an unconditional realization of the world' (Baudrillard, 1996c: 25), which it pursued through 'the radical destruction of appearances, the disenchantment of the world and its abandonment to the violence of interpretation and of history' (Baudrillard, 1994a: 160). But in keeping with Baudrillard's appreciation for the *other side* of the game, he is also attentive to postmodernity – 'the immense process of the destruction of meaning, equal to the earlier destruction of appearances' (Baudrillard, 1994a: 161). It is in this context that Baudrillard (1994a) offers his infamous 'genealogy of the image': from reflection (the servile image); through perversion (the diabolical image); via dissimulation (the duplicitous image); to the simulacrum (the liberated image). In addition, Baudrillard (1990b) offers an account of three 'orders of appearance (or simulacra)', which demonstrates how modernity's commitment to the reality principle has been repeatedly 'notched up a gear': from first-order counterfeiting (a fake reality based on the natural law of value – an inferior copy that is *distinct from* a unique original); via second-order production (a serial reality based on the commercial law of value – a mechanical reproduction that is *equivalent to* its prototype); to third-order simulation (an encoded reality based on the structural law of value – that which is generated by models and codes; that which is always already reproduced – the *hyperreal*). In the third order, the mode of production is enveloped by the mode of seduction. For when everything is real, reality loses its specificity (*just as when all words in a clause are italicized, italics no longer add emphasis*). Ironically, only that which is reproducible is accorded the status of reality. Our reality would be nothing without the mirror that makes it the reality *of* the image. So, it is not that everything becomes unreal, inauthentic and fake (*less* than real). On the contrary, everything becomes hyperreal (*more* real than real) – thus becoming enigmatic (enchanted) once again. 'The irony of the facts, in their wretched reality, is precisely that they are only what they are,' says Baudrillard (1996c: 98). '[N]othing is wholly obvious without becoming enigmatic.' Such is the diabolical 'obscenity' into which realism plunges: '*verifying to the point of giddiness the useless objectivity of things*' (Baudrillard, 1988c: 31–32). Hereinafter, the obscenity of hyperrealism has immunized itself against all second-order critiques, such as Marxism and Freudianism. Once again, Baudrillard maintains that the only way to confront hyperreality is by recourse to the 'superior irrationalism' of the mode of seduction and symbolic exchange.

SPATIAL CONTRIBUTIONS

As one might expect of a former student of Lefebvre, and from his early involvement with the journal *Utopie* (Baudrillard, 2001), Baudrillard has always attempted to think through his ideas in spatialized terms. From his initial examination of the effects of consumerism on domestic space (Baudrillard, 1996a) to his hyperreal tour of America (Baudrillard, 1988b), Baudrillard's interest in space, architecture and the built environment has remained constant. There can be few theorists who have written on the significance of both the construction and destruction of New York's World Trade Center (Baudrillard, 1981, 2002b). Yet Baudrillard's spatial imagination is frequently reduced to a handful of throwaway remarks. Baudrillard's sardonic take on Disneyland features among his best-known comments:

> Disneyland is presented as imaginary in order to make us believe that the rest is real, whereas all of Los Angeles and the America that surrounds it are no longer real, but belong to the hyperreal order and to the order of simulation ... The imaginary of Disneyland is neither true nor false, it is a deterrence machine set up in order to rejuvenate the fiction of the real in the opposite camp.
> (Baudrillard, 1994a: 12–13)

In another discussion of hyperreality, Baudrillard (1994a: 1) draws on a short story by Jorge Luis Borges concerning the fate of a 1:1 scale map of a great Empire: 'the decline of the Empire witnesses the fraying of this map, little by little, and its fall into ruins, though some shreds are still discernible in the deserts.' Baudrillard inverts this second-order simulacrum to offer his most memorable hyperreal allegory:

> The territory no longer precedes the map, nor does it survive it ... [T]oday it is the territory whose shreds slowly rot across

the extent of the map. It is the real, and not the map, whose vestiges persist here and there in the deserts that are no longer those of the Empire, but ours. *The desert of the real itself.*

(Of course, Baudrillard is hardly satisfied with such an analogy, which preserves the critical distance between map and territory – representation and reality – that has actually been abolished: 'Only the allegory of the Empire, perhaps, remains.')

In marshalling the forces of seduction and symbolic exchange against third-order simulacra, Baudrillard has developed a sustained theorization of space and time under conditions of hyperreality. Modernity was committed to *linear* (spatialized) time: time with a sense of direction. Linear time allowed modernity to forget the past, lament the present, and hope for a better future, instating a canonical morality of 'progress'. If the idea of an end (*qua* goal) is a function of linear time, linear time lends another sense to the end. Linear time is the *time of no return*. Every moment that passes is a finality – once it is gone, it is gone for good. In modernity, time is scarce. Likewise with space. In contrast, *cyclical* time – the time of seduction and symbolic exchange – is the *time of eternal return*: the rhythm of the seasons, for instance, gives every reason for supposing that what has come to an end will, in due course, return. In short, cyclical time is manifold. Rather than a scarcity of time and space, there is an *abundance* of spaces and times. And if linear time has always been subject to a secret curvature, the end can no longer be seen as a mere finality: 'in a non-linear, non-Euclidean space of history the end cannot be located' (Baudrillard, 1994b: 110). In other words, finality is illusory, despite modernity's best efforts to instil a sense of direction. What we might call postmodernity amounts to the situation we face *beyond* the attempt to accomplish a final solution. Baudrillard wants us to acknowledge that the journey never ends; that modernity's vain attempt

to guide us to a final destination (*telos*) has plunged us into a strange *hypertelic* universe – characterized, among other things, by hysteresis (the continuation of the effect in the absence of the cause). Space and time no longer provide a stable framework for tracing out an orderly trajectory across the desert of the real. They mutate in strange, unstable and obscene ways, and in so doing become fatal and fractal (a fourth order of simulacra). Such is the 'objective irony' through which radicality has passed into events. Like a cancerous cell or a viral infection, when things are deprived of their finality they proliferate uncontrollably towards the saturation of a limitless space.

KEY ADVANCES AND CONTROVERSIES

Baudrillard's (1987: 126) account of a world in which there is 'no more system of reference to tell us what happened to the geography of things' has been very much misunderstood. His critics fail to appreciate that his is not a celebration of modernity's furtive destiny, but an *exposé* of the murder of reality by its own hand (Baudrillard, 1996c). Baudrillard is not dealing with a case where the dominant position of reality is simply usurped by representations, images and signs (a reversion to a situation where the *appearance* of the world holds sway). On the contrary, the distinction between appearance and reality dissolves in the opposite direction, in the direction of *hyper*reality. The sense of 'simulation' proper to hyperreality does not imply a degree of dissimulation, falsity or deception – which would be to construe obliquely the presence of a solid and durable reality elsewhere. It marks the irruption of an obscene world characterized by 'the total promiscuity of things' (Baudrillard, 1993c: 60). If the common response to this situation is nostalgia, lamentation, or out-and-out denial, Baudrillard (1987: 127) insists that 'The only thing you can do is let it run all the way to the end.'

BAUDRILLARD'S MAJOR WORKS

Baudrillard, J. (1981) *For a Critique of the Political Economy of the Sign.* Trans. C. Levin. St Louis: Telos [1972].
Baudrillard, J. (1990a) *Seduction.* Trans. B. Singer. London: Macmillan [1979].
Baudrillard, J. (1990b) *Fatal Strategies.* Trans. P. Bietchman and W. G. J. Niesluchowski. London: Pluto [1983].
Baudrillard, J. (1993a) *Symbolic Exchange and Death.* Trans. I. H. Grant. London: Sage [1976].
Baudrillard, J. (1993b) *The Transparency of Evil: Essays on Extreme Phenomena.* Trans. J. Benedict. London: Verso [1990].
Baudrillard, J. (1994a) *Simulacra and Simulation.* Trans. S. F. Glaser. Ann Arbor: University of Michigan Press [1981].
Baudrillard, J. (1996a) *The System of Objects.* Trans. J. Benedict. London: Verso [1968].
Baudrillard, J. (1996c) *The Perfect Crime.* Trans. C. Turner. London: Verso [1995].
Baudrillard, J. (1998a) *The Consumer Society: Myths and Structures.* Trans. C. Turner. London: Sage [1970].
Baudrillard, J. (2002a) *Screened Out.* Trans. C. Turner. London: Verso [2000].

Secondary Sources and References

Barthes, R. (1973) *Mythologies.* Trans. A. Lavers. St Albans: Paladin.
Barthes, R. (1983) *The Fashion System.* Trans. M. Ward and R. Howard. New York: Hill and Wang.
Baudrillard, J. (1975) *The Mirror of Production.* Trans. M. Poster. St Louis: Telos [1973].
Baudrillard, J. (1983) *In the Shadow of the Silent Majorities.* Trans. P. Foss, P. Patton and J. Johnston. New York: Semiotext(e) [1978].
Baudrillard, J. (1987) *Forget Foucault: Forget Baudrillard.* Trans. P. Beitchman, N. Dufresne, L. Hildreth and M. Polizzotti. New York: Semiotext(e) [1977].
Baudrillard, J. (1988b) *America.* Trans. C. Turner. London: Verso [1986].
Baudrillard, J. (1988c) *The Ecstasy of Communication.* Trans. B. Schutze and C. Schutze. New York: Semiotext(e) [1987].
Baudrillard, J. (1990c) *Cool Memories.* Trans. C. Turner. London: Verso [1987].
Baudrillard, J. (1993c) *Baudrillard Live: Selected Interviews.* Ed. M. Gane. London: Verso.
Baudrillard, J. (1994b) *The Illusion of the End.* Trans. C. Turner. Cambridge: Polity [1992].
Baudrillard, J. (1995) *The Gulf War Did Not Take Place.* Trans. P. Patton. Sydney: Power Publications [1991].
Baudrillard, J. (1996b) *Cool Memories II, 1987–1990.* Trans. C. Turner. Cambridge: Polity [1990].
Baudrillard, J. (1997) *Fragments, Cool Memories III, 1991–1995.* Trans. C. Turner. London: Verso [1995].
Baudrillard, J. (1998) *Paroxysm: Interviews with Philippe Petit.* Trans. C. Turner. London: Verso [1997].
Baudrillard, J. (2000) *The Vital Illusion.* New York: Columbia University Press.
Baudrillard, J. (2001a) *Impossible Exchange.* Trans. C. Turner. London: Verso [1999].
Baudrillard, J. (2001b) *The Uncollected Baudrillard.* Ed. G. Genosko. London: Sage.
Baudrillard, J. (2002b) *The Spirit of Terrorism and Requiem for the Twin Towers.* Trans. C. Turner. London: Verso [2002]
Baudrillard, J. (2003) *Cool Memories IV, 1995–2000.* Trans. C. Turner. London: Verso [2000].
Bauman, Z. (1992) 'The world according to Jean Baudrillard', in *Intimations of Postmodernity.* London: Routledge, pp. 149–155.
Butler, R. (1999) *Jean Baudrillard: The Defence of the Real.* London: Sage.
Gane, M. (1991) *Baudrillard: Critical and Fatal Theory.* London: Routledge.
Gane, M. (ed.) (2000) *Jean Baudrillard* (4 Volumes). London: Sage.
Kellner, D. (ed.) (1994) *Baudrillard: A Critical Reader.* Oxford: Blackwell.
Levin, C. (1996) *Jean Baudrillard: A Study in Cultural Metaphysics.* Hemel Hempstead: Prentice Hall.
Rojek, C. and Turner, B. S. (eds) (1993) *Forget Baudrillard?* London: Routledge.
Smith, R. G. with Doel, M. A. (2001) 'Baudrillard unwound: the duplicity of post-Marxism and deconstruction', *Environment and Planning D: Society and Space* 19: 137–159.
Zurbrugg, N. (ed.) (1997) *Jean Baudrillard: Art and Artefact.* London: Sage.

David B. Clarke and Marcus A. Doel

4 Zygmunt Bauman

BIOGRAPHICAL DETAILS AND THEORETICAL CONTEXT

Zygmunt Bauman is a sociologist. He was born into a Jewish family in Poland in 1925. The Nazi invasion of 1939 forced his family to flee their homeland. Bauman later fought on the Russian front with the Polish Army. Returning to Poland after the war, he took up a post at the University of Warsaw in the early 1950s – his army career having been abruptly ended by anti-Semitic elements. In 1968, however, he was forced into exile after the communist authorities fabricated political charges against him during a campaign heavily imbued with anti-Semitism. By way of Tel Aviv and Canberra, Bauman finally settled in England (where he had previously held visiting scholarships at the London School of Economics and University of Manchester), taking up a Chair of Sociology at the University of Leeds in 1972. He retired in 1990. At around the same time, the collapse of the Eastern bloc saw him reconciled with his alma mater. Since then, he has been Emeritus Professor of Sociology at both the University of Leeds and the University of Warsaw. Bauman continues to write prolifically, on themes that are often held to reflect his life experience. As Tester (in Bauman and Tester, 2001: 3) cautions, however, 'if we explain Bauman's thought by reference to his biography we are actually making ourselves completely incapable of understanding what he has to say'. Not only does Bauman eschew the cult of personality he holds responsible for undermining public life, he maintains that the human condition is *necessarily* one of exile. We are all outsiders, strangers even to ourselves. Bauman's insights should not, therefore, simply be referred back to their author – the last person one should quiz about the significance of his work.

Even a cursory glance at the sweep of Bauman's *oeuvre* reveals the breadth of his interests. Among other things, he has produced studies of social movements, socialism and class; critical theory, culture and hermeneutics; modernity, ambivalence and the Holocaust; and postmodernity, consumerism and community. Bauman constantly elucidates the dilemmas that human society is forced to deal with in the absence of a guidebook or instruction manual, thus ensuring a multitude of actual and possible solutions, each offering different opportunities in terms of freedom, politics and morality. Bauman's influences are equally varied: Theodor Adorno, Hannah Arendt, **Jean Baudrillard**, **Pierre Bourdieu**, Cornelius Castoriadis, Mary Douglas, Norbert Elias, **Michel Foucault**, Antonio Gramsci, Jürgen Habermas, Emmanuel Lévinas, Claude Lévi-Strauss, Jean-François Lyotard, Karl Marx, Richard Rorty, Richard Sennett, Georg Simmel; and, of course, his wife, Janina, whose *Winter in the Morning* (1986) precipitated the moral turn that began with *Modernity and the Holocaust* (Bauman, 1989). *Modernity and the Holocaust* embodies one of the changes of perspective that characterize Bauman's work, casting previous conceptions in a new light:

What had happened ... to my thinking between *Socialism: the Active Utopia* [1976a] and *Modernity and the Holocaust* [1989] was that the perspective from which I viewed beauty and humiliation

and ugliness widened. Gradually, the victims of economic injustice began to appear to me as a particular case of a much wider, more ubiquitous and stubborn problem of the 'stranger'.
(Bauman, in Bauman and Tester, 2001: 52)

The plight of the worker under industrial capitalism was perceived as but a 'peculiar, and historically limited, form of the social production of outcasts'. Bauman had come to recognize that modernity's obsession with imposing order *necessarily* produced outcasts – requiring that the entire mythology of modernity, not just the ideology of capitalism, be stripped away. He was on the threshold of this realization in addressing the socialist utopia, but it took a fuller analysis of modernity for the pieces finally to fall into place. Recognition that the gas chambers and industrial crematoria of the Holocaust were made in the image of modernity – that order, efficiency, and rationality could just as readily be harnessed to evil as good – triggered this broadening of perspective; although Bauman has examined modernity from all angles (Bauman, 1982, 1987, 1988, 1989, 1991, 1992a). 'Postmodernity' was, for Bauman, a concept offering the opportunity for extending that analysis.

Bauman, like Baudrillard and Lyotard, was never an apologist for postmodernity, as some have erroneously believed. As an *outgrowth* of modernity – the unanticipated result of modernity's impossible bid for mastery – postmodernity was always part of modernity, even if this realization could only take shape as the Owl of Minerva spread her wings and faith in modernity waned. Bauman's dazzling series of conceptions of postmodernity are simultaneously conceptions of modernity, since 'postmodernity' is nothing other than modernity's posthumous form. Although Bauman frequently deploys suggestive contrasts, modernity and postmodernity are not to be regarded as forming an identifiable, historical sequence. Each conception of postmodernity implies contradiction of

modernity without implying its transcendence, typically involving the return of everything that modernity tried to repress. Following Lyotard (1984), Bauman's (1987, 1992b) initial conceptions focused on the modern alignment and postmodern decoupling of power and knowledge – on the shift from a *repressive* form of social control, reliant on power/knowledge, to a *seductive* form, guided by consumerist desires. Later formulations trace the decomposition of a modern world of transient permanence (driven by the deconstruction of mortality) into a postmodern world of permanent transience (characterized by the deconstruction of immortality) (Bauman, 1992a). Or, following Sigmund Freud (1930), modernity is seen to involve relinquishing a degree of freedom in exchange for an increased sense of security, postmodernity seeing that trade-off reversed (Bauman, 1997). Perhaps the most significant aspect of Bauman's engagement with postmodernity, however, has been in terms of *morality* (Bauman, 1993a, 1995).

Modernity, as the Holocaust clearly revealed, 'out-rationalized' (adiaphorized) the moral impulse, unburdening individuals of moral responsibility by delegating it to a higher authority. The bureaucratization of decision-making – the hallmark of modernity, according to Max Weber – promoted the idea that moral responsibility would be taken care of elsewhere: that I am not responsible. So it was that the 'desk killer' became a chief perpetrator of the Holocaust; the 'banality of evil' being redoubled by the 'rationality of evil'. Yet if the dark side of modernity has been exposed, postmodernity should not be prematurely eulogized. It does not necessarily improve matters. Postmodernity 'out-aestheticizes' the moral impulse. It keeps moral sentiment at bay by refusing to 'choose as its points of reference and orientation the traits and qualities possessed by or ascribed to the objects of spacing' (Bauman, 1995: 101). It selects, instead, 'the attributes of the spacing subject (like interest, excitement, satisfaction or pleasure)' as the measure of the world: what matters to us is us, not them. Insofar as obligations interfere with enjoyment, the postmodern world is as inhospitable to morality as the modern world. Following Levinas's maxim that 'ethics comes before ontology', Bauman (in Bauman and Tester, 2001: 54) finds modernity *and* postmodernity guilty of producing moral indifference: 'since in a world with ethics but without ontology there is no "before" or "later", but only "better" and "worse" . . . it is the socially produced reality which needs to justify itself at the tribunal of ethics, instead of usurping the right to decide what is and what is not moral.' Postmodernity is perhaps all the more pernicious, being programmatically geared towards the anaesthetization of politics (Bauman, 1999). The current reduction of politics to little more than managerialism and opinion-polling, and its fragmentation into single-issue politics that are purportedly 'beyond Left and Right', speak of a society that has lost the ability to ensure both 'the autonomy of society (the ability to change things) and the autonomy of its members (the ability to select things that need to be changed)' that 'are each other's indispensable conditions' (Bauman, 2002: 57) and the *sine qua non* of politics. Personal survival under postmodern conditions relies on the *refusal* of social solidarity. In stark contrast to modernity, postmodernity prevents disaffection from building up into collective calls for the system to be overhauled by co-opting dissent: 'Postmodernity enlists its discontents as its most dedicated storm-troopers' (Bauman 1993b: 44).

Bauman has lately dismissed the term 'postmodernity.' It has, he insists, been wrung dry. His preferred term is now, appropriately, 'liquid modernity' – a fluid form resulting from the meltdown of modernity's previously solid, durable nature. For Émile Durkheim, it seemed obvious that society outlasted the individual, offering 'shelter from the horror of one's own transience' (Bauman, 2000: 183). Today, suggests Bauman, bodily mortality no longer seems so fleeting and fickle, given the increasing ephemerality of institutional forms: 'The body . . . has become the last shelter and sanctuary of continuity and duration.' In a fluid situation, one can no longer count on things remaining in place or staying the way they are. Consequently, it makes little sense to behave in a way that regards commitment, loyalty or any other form exhibiting permanence as likely to be forthcoming from the world or as virtues to be nurtured in oneself. 'No strings attached' has become the most sought-after quality in the liquid-modern world; 'until further notice' the only rational expectation. The consequences of this world – our world of globalization (Bauman, 1998b), consumerism (Bauman, 1998a) and individualism (Bauman, 2001a, 2001b) – require constant vigilance, but vigilance itself is under permanent attack. The erosion of the sense that one can do anything about forces operative at a global scale, coupled with the feeling that looking after number one has become a full-time job, is perhaps the greatest threat – and also the greatest challenge – that society has yet faced (Bauman, 2002).

SPATIAL CONTRIBUTIONS

Drawing on Alfred Schütz (Schütz and Luckmann, 1974), Bauman (1993a: 145) maintains that our very sense of objective, physical space derives from the 'phenomenological reduction of daily experience to pure quantity, during which distance is "depopulated" and "extemporalized" '. What we objectify as 'pure space' is but an abstraction, freed of 'content relative to time and circumstance', deriving from 'social space' (from our relations with others). In commonsensical terms, talk of 'social distance' appears reliant on a metaphorical transference from real, 'objective distance'. Yet apprehension of the objective world is itself actually achieved through notions 'coined originally to "map" the qualitatively diversified relations with other humans' (Bauman, 1993a: 145). Social differentiation – stratified by degrees of intimacy and anonymity, familiarity and strangeness – is the basis of all purportedly empty and objective notions of proximity and distance. Modernity, however, tore apart the connection between social interaction and physical proximity. The appearance of the 'stranger' on the modern stage was a direct consequence of the dissociation of social and physical space: the stranger was an alien presence within the lifeworld; a figure proximate in physical space, yet socially distant. While modernity thrived on relations with strangers – particularly in terms of the proliferation of anonymous monetary transactions – their otherness could not but arouse anxiety. Neither friend nor enemy, the stranger defied accepted categories. The proliferation of strangers thus prompted an effort to construct, 'intellectually, by acquisition and distribution of knowledge', an ordered, bounded and mappable 'cognitive space' (Bauman, 1993a: 146) – within which the stranger could be contained, monitored and normalized. Yet if cognitive spacing was driven by such concerns, modernity gave rise to other forms of social spacing: 'aesthetic space is plotted affectively, by the attention guided by curiosity and the search for experimental intensity, while moral space is "constructed" through an uneven distribution of felt/assumed responsibility' (Bauman, 1993a: 146).

Cognitive spacing derives from modernity's desire to master space; to determine a place for everything and ensure that everything is in its place – so that surveillance might readily reveal whatever is 'out of place'. It is a function of knowledge; a tool of the powerful; part and parcel of modernity's desire for order and abhorrence of ambivalence. But since the stranger is necessarily an ambivalent, boundary-straddling character, all efforts of cognitive spacing will ultimately be in

vain. Cognitive spacing employs a repressive, 'anthropoemic' strategy – 'aimed at the exile or annihilation of *the others*' (Bauman, 2000: 101). Yet since the other always lies within, modernity would gradually transfer its allegiance to a seductive, 'anthropophagic' strategy – 'aimed at the suspension or annihilation of *their otherness*' (Bauman, 2000: 101). The roots of this strategy lie in a different response to being in the company of strangers. That archetypal city-stroller, the nineteenth-century Parisian *flâneur* – another footloose character on the modern scene – embodied a diametrically opposed attitude to the proteophobia (fear of strangers) responsible for gestating cognitive space. Among the urban crowd, the *flâneur* managed to keep his distance socially by transforming physical proximity into *aesthetic* proximity – by opening up another dimension, hidden to others. Purely for his own satisfaction, he narrated the world around him without letting reality interfere with the plot, coming to enjoy the strange and unfamiliar. 'One may say that if proteophobia is the driving force of cognitive space – *proteophilia* prompts the efforts of aesthetic spacing' (Bauman, 1993a: 168). Yet the pleasures of aesthetic space would not remain a private affair. The *flâneur*'s ludic world was gradually transformed into a *managed playground*: modernity appropriated the pleasures of *flânerie*, putting them into the service of consumerism. The result was the shift away from a repressive mode of social control (maintaining the order of cognitive space) towards a new seductive mode (guided by the contours of aesthetic space). Neither mode spares much room for morality. Moral space is hardly commensurate with cognitive space. The unconditional demand to be *for* the other is incompatible with the suppression of moral responsibility promoted by the normalizing designs of cognitive space. Aesthetic space is also inhospitable to moral sensibilities. Moral responsibility demands the kind of serious attention that conflicts with the free play of attention on which aesthetic spacing thrives. Yet there is always hope for morality. Being *with* others opens up a possibility for the ethically prior mode of being *for* others. The construction of a space of moral responsibility is never guaranteed. 'But it does happen, daily, and repeatedly – each time that people care, love, and bring succour to those who need it' (Bauman, 1993a: 185). Nevertheless, being *with* – let alone being *for* – others has become subject to a profound transformation. In an era of globalization, 'Proximity no longer requires physical closeness; but physical closeness no longer determines proximity' (Bauman, 2003: 62). Indeed, globalization 'divides as much as it unites; it divides as it unites – the causes of the division being identical with those that promote the uniformity of the globe' (Bauman, 1998b: 2).

The social roots of globalization, Bauman maintains, lie in the liquefaction of modernity's formerly resilient, solid form (e.g. nation-states, industrial society, and cultural hegemony), and in the liquidation of modernity's erstwhile universalizing ambitions (e.g. The Law, The Truth, The Future). Globalization conveys the precise opposite of universalization: effects rather than intentions, consequences rather than undertakings, untameable rather than tameable forces. It implies that 'jungle law' has broken out once more, this time in what **Anthony Giddens** refers to as the 'manufactured jungle' – the wilderness that unintentionally results from modernity's ordering drive. It also implies that the institutions once holding that wilderness at bay have been rendered increasingly impotent, not least the nation-state. Drawing on Michel Crozier's (1964) principle that being in control rests on minimizing uncertainty for oneself while maximizing it for others, Bauman discusses the essential compatibility of the *deterritorialization* of economic activity (its increasing extraterritoriality, aided and abetted by advances in information technology) with the seemingly paradoxical *reterritorialization* of the nation-state (evident in the clamour to form new nation-states from those whose collective identities were formerly subjugated to invented national traditions, now that the traditional tests of economic viability and military capability have been relaxed). Extraterritorial, hypermobile capital has a vested interest in the weak states associated with 'the *morcellement* of the world scene' (Bauman, in Beilharz, 2001: 303). The consequences of this complex spatial restructuring are glossed over by talk of 'time–space compression': '*rather than homogenizing the human condition, the technological annulment of temporal/spatial distances tends to polarize it*' (Bauman, 1998b: 18).

Globalization, Bauman maintains, is best thought of as *glocalization* – which implies more than the simultaneity of deterritorialization and reterritorialization, or the reassertion of place in the midst of space–time compression. It implies a worldwide *restratification* of society based on freedom of movement (or lack thereof). 'Glocalization' polarizes mobility, or polarizes society in terms of differential mobility: 'Some inhabit the globe; others are chained to place' (Bauman, in Beilharz, 2001: 307). 'Glocalization' means *globalization for some; localization for others*. The ability to use time to overcome the limitations of space is the prerogative of the globals. The locals remain tied to place – where, for many, time is increasingly abundant *and* redundant. Localized existence was hardly a problem when this was the norm, and the means of giving meaning to that existence were within reach. Being merely local in a glocalized world, however, is automatically rendered a secondary existence, since the means for giving meaning to existence have been placed out of reach. It is tantamount to confinement without the need for prison walls. The polarization of freedom of movement thus serves to redefine all other freedoms, adding a new dimension to deprivation.

KEY ADVANCES AND CONTROVERSIES

Bauman's reflections on the evolution of modernity, its key characteristics, and subsequent meltdown, offer numerous lessons for those interested in space and place, not least in relation to globalization. Although Bauman (1998b: 1) readily accepts that term's status as 'a fad word fast turning into a shibboleth, a magic incantation, a pass-key meant to unlock the gates to all present and future mysteries', he nonetheless pinpoints its consequences with startling accuracy. For the first time in human history, we face a situation where the consequences of our collective cohabitation on the same planet are inescapable: *'il n'y a pas hors du monde'* (Bauman, 2002: 12). Yet if globalization means that rich and poor, globals and locals, no longer sit at the same distributive table of the nation-state, Bauman's position is valuable in its oblique refusal of what the globalized world likes to regard as a *fait accompli*. 'Bauman', says Beck (2000: 57), 'may be said to overlook himself. For at least in *his* perspective as observer, he *binds together* what he depicts as irrevocably disintegrated in trans-state world society: namely, the framework, the *minima moralia*, which make the poor appear as *our* poor, and the rich as *our* rich.'

BAUMAN'S MAJOR WORKS

Bauman, Z. (1973) *Culture as Praxis*. London: Routledge and Kegan Paul (new edition, 1999).
Bauman, Z. (1987) *Legislators and Interpreters: On Modernity, Post-modernity and Intellectuals*. Cambridge: Polity.
Bauman, Z. (1989) *Modernity and the Holocaust*. Cambridge: Polity.
Bauman, Z. (1991) *Modernity and Ambivalence*. Cambridge: Polity.
Bauman, Z. (1992a) *Morality, Immorality and Other Life Strategies*. Cambridge: Polity.
Bauman, Z. (1993a) *Postmodern Ethics*. Oxford: Blackwell.
Bauman, Z. (1998a) *Work, Consumerism and the New Poor*. Milton Keynes: Open University Press.
Bauman, Z. (1998b) *Globalization: The Human Consequences*. Cambridge: Polity.
Bauman, Z. (2000) *Liquid Modernity*. Cambridge: Polity.
Bauman, Z. (2002) *Society Under Siege*. Cambridge: Polity.

Secondary Sources and References

Bauman, J. (1986) *Winter in the Morning: A Young Girl's Life in the Warsaw Ghetto and Beyond*. London: Virago.
Bauman, Z. (1972) *Between Class and Elite. The Evolution of the British Labour Movement: a Sociological Study*. Manchester: Manchester University Press.
Bauman, Z. (1976a) *Socialism: the Active Utopia*. London: George Allen and Unwin.
Bauman, Z. (1976b) *Towards a Critical Sociology: An Essay on Common-sense and Emancipation*. London: Routledge and Kegan Paul.
Bauman, Z. (1978) *Hermeneutics and Social Science: Approaches to Understanding*. London: Hutchinson.
Bauman, Z. (1982) *Memories of Class: The Pre-history and After-life of Class*. London: Routledge and Kegan Paul.
Bauman, Z. (1988) *Freedom*. Milton Keynes: Open University Press.
Bauman, Z. (1990) *Thinking Sociologically*. Oxford: Blackwell (2nd edition, with T. May, 2001).
Bauman, Z. (1992b) *Intimations of Postmodernity*. London: Routledge.
Bauman, Z. (1993b) 'The sweet scent of decomposition', in C. Rojek and B. S. Turner (eds) *Forget Baudrillard?* London: Routledge, pp. 22–46.

Bauman, Z. (1995) *Life in Fragments: Essays in Postmodern Morality*. Oxford: Blackwell.

Bauman, Z. (1997) *Postmodernity and its Discontents*. Cambridge: Polity.

Bauman, Z. (1999) *In Search of Politics*. Cambridge: Polity.

Bauman, Z. (2001a) *Community: Seeking Safety in an Insecure World*. Cambridge: Polity.

Bauman, Z. (2001b) *The Individualized Society*. Cambridge: Polity.

Bauman, Z. (2003) *Liquid Love: On the Frailty of Human Bonds*. Cambridge: Polity.

Bauman, Z. and Tester, K. (2001) *Conversations with Zygmunt Bauman*. Cambridge: Polity.

Beck, U. (2000) *What is Globalization?* Cambridge: Polity.

Beilharz, P. (2000) *Zygmunt Bauman: Dialectic of Modernity*. London: Sage.

Beilharz, P. (ed.) (2001) *The Bauman Reader*. Oxford: Blackwell.

Beilharz, P. (ed.) (2002) *Zygmunt Bauman* (4 volumes). London: Sage.

Crozier, M. (1964) *The Bureaucratic Phenomenon*. Chicago: University of Chicago Press.

Freud, S. (1930) *Civilization and its Discontents*. Trans. J. Rivière. London: Hogarth Press.

Kilminster, R. and Varcoe, I. (eds) (1996) *Culture, Modernity and Revolution: Essays in Honour of Zygmunt Bauman*. London: Routledge.

Lyotard, J.–F. (1984) *The Postmodern Condition: A Report on Knowledge*. Trans. G. Bennington and B. Massumi. Manchester: Manchester University Press.

Schütz, A. and Luckmann, T. (1974) *The Structures of the Lifeworld*. Trans. R. M. Zaner and H. T. Engelhardt Jr. London: Heinemann.

Smith, D. (1999) *Zygmunt Bauman: Prophet of Postmodernity*. Cambridge: Polity.

David B. Clarke and Marcus A. Doel

5 Ulrich Beck

BIOGRAPHICAL DETAILS AND
THEORETICAL CONTEXT

Ulrich Beck's sociology of 'late modernity' has been influential on geographical thinking on subjects as diverse as environmental hazards, globalization and the changing relationships between individuals and social institutions. Like **Anthony Giddens**, Bec k is broadly concerned with how human social experience is changing as modern industrial societies face periods of uncertainty and restructuring brought about by problems inherent to their constitution. Beck's writing can also be understood as an agenda for new forms of political action, and his influence has thus gone beyond academic circles; like Giddens, Beck is noted for publishing articles for non-academic audiences (e.g. Beck, 2001). Notably, and in contrast with some critical theorists such as **Michel Foucault**, Beck at times presents a rather optimistic account of how people might engage in radical and liberating forms of political action, suggesting that many of the economic, political, social and ecological crises he identifies may be circumvented.

Beck was born in 1944, growing up in Hanover, in the then West Germany. By the mid-1980s he was Professor of Sociology at the small town of Bamberg, later obtaining Chairs in Sociology at the University of Munich and the London School of Economics and Political Science. During this time, his sociology has sought to make sense of a series of seemingly unrelated global phenomena in the latter part of the twentieth and early twenty-first centuries, including the emergence of particular sorts of environmental hazards, global economic instability and the threat of global terrorism. Beck has attempted to theorize these phenomena together, working with terminology that has become iconic in debates over the nature and future of late modernity. Hypothesizing a transition from 'first modernity' to 'second modernity', Beck's central theme has been the emergence of a so-called 'risk society'. It is this notion which underpins his key publications, including *Risk Society: Towards a New Modernity* (1992); *Ecological Politics in the Age of Risk* (1995); *The Reinvention of Politics: Rethinking Modernity in the Global Social Order* (1997) and *World Risk Society* (1999).

Beck's notion of the 'risk society' refers broadly to a sense in which there has been a transition from an industrial society (in which 'natural hazards' could be regarded as fate and 'human-made' hazards could be understood within a frame of calculability that rendered them insurable and thus manageable), to a late-modern society where the hazards produced by the way society operates are incalculable, perhaps unknowable. Risk society is thus still an industrial society, yet the hazards produced by that society take on a heightened importance in human consciousness. Thus:

> Threats from civilisation are bringing about a new kind of 'shadow kingdom' comparable to the realm of the gods and demons in antiquity, which is hidden behind the visible world and threatens human life on this Earth. People no longer correspond today with spirits residing in things, but find themselves exposed to 'radiation', ingest 'toxic levels' and are pursued into their very dreams by the anxieties of a nuclear holocaust.
> (Beck, 1992: 72)

In relation to ecological hazards, Beck argues that scientific and economic 'progress' is overshadowed by forms of risk produced by the very processes involved in such progress (Lash and Wynne, 1992); as he puts it, the 'production of *wealth* is systematically accompanied by the social production of *risks*' (Beck, 1992: 19). These new risks (e.g. pollution, climate change) can be thought about in three ways. First, they result *from* science and technology, rather than being just something to which science can be applied as a solution. Second, the risks produced may have impacts over greater spatial and temporal scales than was the case in earlier industrial society; in particular, these forms of risk affect people and places not directly involved in their causes. Third, they are often not immediately sensible to individuals; indeed, they require the 'sensory organs of science – *theories, experiments, measuring instruments – in order to become visible or interpretable as hazards at all*' (Beck, 1995: 27). However, in the case of risks like climate change, science can be limited in predicting effects and proposing solutions. According to Beck, late-modern, techno-scientific, industrial capitalist society is systematically affected by the fundamental conditions of its establishment.

In addition to ecological threats, Beck's concept of risk society encompasses the economic and political instabilities associated with global change. A key dimension of Beck's sociology is to consider the implications of these multiple risks for social institutions and individual lives. He thus argues that the social and economic stabilities people learned to expect under conditions of 'first modernity' have been challenged by new instabilities associated with various processes of change:

[T]he collective patterns of life, progress and controllability, full employment and exploitation of nature that were typical of this first modernity have now been undermined by five interlinked processes: globalization, individualization, gender revolution, underemployment and global risks (as ecological crisis and the crash of global financial markets).
(Beck, 1999: 2)

Beck (1992) accordingly describes the contemporary social experience as one of 'reflexive modernization', with society continually evolving through a series of institutional changes in response to unanticipated risks. This fundamental change in society is associated with a shift in the role of politics. In first modernity, politics was principally concerned with the social distribution of 'goods' (such as wealth). In late-modern risk society, however, it is increasingly having to be concerned with the distribution of 'bads': biohazards, environmental degradation or the negative effects of economic globalization (e.g. unemployment, labour casualization or disinvestment). For Beck, this requirement to deal with late-modern 'bads' has produced a crisis in many conventional institutions (e.g. national governments) responsible for their management – a crisis related to two characteristics of contemporary risk. First, as already suggested, many of the new risks are unpredictable, perhaps unknowable, and have incalculable long-term and geographically widespread consequences. Although there is an unequal socio-spatial distribution of such risks (the poorest people and places tend to be worst afflicted), these are nevertheless hazards that potentially affect anybody. Second, both the new risks (e.g. radiation leaks, regional financial collapse) and the emergence of other formations, such as multinational corporations or international terrorist networks, have effects that extend beyond the remit of those institutions conventionally bound by national frontiers: they are transnational.

SPATIAL CONTRIBUTIONS

Beck's writing has alerted to the shifting relationships between individuals and society on a variety of scales in relation to a range of issues, and it is unsurprising that his ideas have influenced geographers. Indeed, Beck's notion of reflexive modernization implies a process of constant revision of social activity, at the institutional and individual levels, with both changing how they function in response to the existence of an uncertain and risky world. Both are held to be caught up in a process of consciously reflecting upon more uncertain, risky and hazardous existences, and, in response, changing the ways they function. In part this process is driven by the constant production of highly specialized knowledges by scientific, technical and expert organizations. In turn, this information reflects the ambiguities and complexities surrounding the types of risk that are being dealt with; that is, it is often unclear what exact causal links are responsible for ecological degradation or financial meltdown, and if that is unclear, then it is also unclear who or what is accountable. For example, are there risks involved in introducing genetically modified organisms (GMOs)? If such risks become apparent only over a longer time scale or across national boundaries, who or what should or could take responsibility? Science, industry and politics seem incapable of supplying definitive answers to such questions, yet GMOs are increasingly used in agriculture and enter human food chains. Institutional and individual responses to this confusing and sometimes incomprehensible or contradictory state of affairs and the possible international political conflicts they might engender become part of wider processes of socio-spatial change:

> This is what I mean by talking about 'reflexive modernization'. Radicalised modernization undermines the founda-

tions of the first modernity and changes its frame of reference, often in a way that is neither desired nor anticipated ... A new kind of capitalism, a new kind of economy, a new kind of global order, a new kind of society and a new kind of personal life are coming into being.
> (Beck, 1999: 2)

One particular impact of this process is in the workplace. For Beck (1992), our working lives are one arena in which we experience risk (Allen and Henry, 1997). Late-modern working practices stress flexibility, which is often translated into processes of work casualization, short-term contracts and periods of unemployment. Similarly, globalization has allowed companies to disinvest from some places and reinvest in others, so that the availability of work in particular places is fluid and uncertain. At the same time, institutions allowing collective bargaining (e.g. trades unions), associated with first modernity, have become less significant in a world of transitory and flexible working relationships. Work is thus an important sphere in which institutions and the individual meet under late-modern, risky conditions:

> The new power-play between territorially fixed political actors (government, parliament, unions) and non-territorial economic actors (representatives of capital, finance, trade) is the central element expressed in the political economics of uncertainty and risk ... capital is global, work is local.
> (Beck, 1999: 11)

Such situations lead to another of Beck's key themes. This is the process of 'individualization', a progressive disassociation of individuals from social institutions such as employers, unions, religions or conventional political agencies. Individualization is a structural process, implying a changing relationship between individuals and institutions. Individualization is partly something which is done to people by economic and political institutions that wish to divest themselves of responsibility for those people.

Simultaneously, it is something people want for themselves as part of the process of reflexive modernization, recognizing that conventional social institutions are limited in responding to the risks they face. Individuals are increasingly sceptical of the authority of governing, moralizing or employing institutions, and of ways of aggregating people into social classes or communities, which went relatively unquestioned in first modernity. In late modernity, then, 'People are invited to constitute themselves as individuals: to plan, understand, design themselves as individuals, and should they fail, to blame themselves' (Beck, 1999: 9). Life is thus increasingly lived as an individual project, in which an individual's bonds of love and friendship become more important than the relationships with social institutions which structured their sense of identity in first modernity (Beck and Beck-Gernsheim, 1995). Beck's emphasis is on how individuals' life projects are forged through their personal experiences of fear and risk, and that while this might produce defensive or paralysing reactions (counter-modernization – a reactionary demand for a return to traditional certainties and securities), it might instead reflexively produce new forms of ethical and political engagement with the world. This might include new forms of ethical and political community connecting people in geographically disparate places, and be centred on, for example, ecological issues.

Beck sets these more radical possibilities in the context of what he calls 'subpolitics'. Subpolitics take place in 'sites which were previously considered unpolitical' (Beck, 1999: 93), and implicate individuals and a range of non-governmental institutions in new forms of political practice:

> The concept of 'subpolitics' refers to politics outside and beyond the representative institutions of the political systems of nation-states ... Subpolitics means *'direct'* politics – that is, *ad hoc* individual participation in political decisions, bypassing the institutions of representative opinion-for-

mation (political parties, parliaments) and often even lacking the protection of the law. In other words, subpolitics means the shaping of society from below. Economy, science, career, everyday existence, private life, all become caught up in the storms of political debate ... Crucially, however, subpolitics sets politics free by changing the rules and boundaries of the political so that it becomes more open and susceptible to new linkages – as well as capable of being negotiated and reshaped.
> (Beck, 1999: 39–40)

Beck points to the role of organizations such as Greenpeace in subpolitical debate, arguing that individualization and subpolitical agency are part of a more critical attitude towards scientific, economic and (conventional) political authorities as regards issues like 'development' and 'progress'. They are, too, associated with a re-emerging public sphere; a space for more democratic dialogue about re-shaping society.

In addition to his work on politics, of major significance for geographers is Beck's work on transnationalism and 're-flexive cosmopolitization' (e.g. Beck, 1996, 1999, 2000, 2001). Beck draws a distinction between 'simple globalization', which envisages interconnections between nation-states that otherwise continue to exist as bounded, independent entities; and transnational or cosmopolitan visions of globalization, which, potentially, both disrupt the assumed stability of the state and create new forms of interconnectedness:

> The cosmopolitan gaze opens wide and focuses – stimulated by the post-modern mix of boundaries between cultures and identities, accelerated by the dynamics of capital and consumption, empowered by capitalism undermining national borders, excited by the global audience of transnational social movements, and guided and encouraged by the evidence of world-wide communication ... World-wide public perception and debate of global ecological danger or global risks of a technological and economic nature ('Frankenstein food') have laid open the cosmopolitan significance of fear.
> (Beck, 2000: 79)

This might involve the creation of new, spatially non-contiguous, communities of interest ('risk communities'), and simultaneously affect the nature of political and social life within states. Beck identifies many such globalized communities of interest in the contemporary world: 'Models of post-national risk communities may be found, for example, in the regional ecological treaties . . . in transnational communities, non-governmental organizations, or global movements, such as ecological or feminist networks' (Beck, 1999: 16).

For Beck, the idea of cosmopolitization involves the emergence of new forms of geographical imagination. Individualized human subjects, living and thinking transnationally, are capable of greater awareness of their mutually being affected by global issues, and have the potential to use subpolitical agency in order to influence, and take responsibility for, social change. Thus,

> The ethic of individual self-fulfilment and achievement . . . [might provide] the basis for a new cosmopolitanism, by placing globality at the heart of political imagination, action and organization.
> (Beck, 1999: 9)

Beck has also applied his theoretical position to the risks associated with global terrorism, linking this threat to the other ecological, social and economic issues that exemplify his attitude to late modernity. He thus argues that;

> In the face of the menace of global terrorism (but also of climate change, of migration, of toxins in food, or organized crime) the only path to national security is by way of transnational co-operation. The following paradoxical principle holds true: States must de-nationalise and trans-nationalise themselves for the sake of their own national interest, that is, relinquish sovereignty, in order, in a globalized world, to deal with their national problems.
> (Beck, 2001: unpaginated)

KEY ADVANCES AND CONTROVERSIES

Beck's key achievement has been to formulate a theory of late modernity which is able to encompass a diverse range of phenomena. He has been a social commentator, writing for a larger public-sphere audience, as well as an academic theorist. Simultaneously, Beck's work is partly an agenda for political and social change. However, his attempt to explain everything under the 'risk society' meta-narrative is also problematic. Critique of Beck's sociology has focused on both the underlying principles on which his theories are based, and the potential for radical change and new forms of political activity that he suggests might emerge as responses to, or within, risk society.

First, although Beck's analysis of risk society has become important in responses to contemporary environmental issues, it has been criticized from social constructivist and poststructuralist positions (see Eden, 1998; Whatmore, 2002). Here, critics have claimed that Beck maintains a distinction between the social and the natural that is questioned by other theorists (e.g. **Bruno Latour** or **Donna Haraway**) and unrealistic given the social-natural hybrids (like GMOs) that increasingly populate our world. Wynne (1996), for example, notes Beck's acceptance of a model of the environment (and its hazards) as an externality affecting society. Beck is criticized for neglecting cultural dimensions of social responses to risk, and for accepting 'expert' constructions of environmental problems. Wynne suggests that Beck contributes to reproducing an expert–lay dichotomy in analysis of environmental issues and to exclusion of lay understandings of those issues.

Second, Beck is accused of not paying enough attention to geographical and cultural difference, despite his focus on transnational and cosmopolitan trends. It has been suggested that more consideration needs to be given to how different cultural groups, in different places, make sense of and respond to risks. Risks are not simply matters of fact, but become parts of cultural discourse and embedded in social practice. It is further argued that in this sense, the 'new' risks identified by Beck are not necessarily that different to forms of risk (disease, famine, etc.) not related to industrial society. These hazards too were differentially interpreted and responded to. While Beck does recognize geographical difference in his work, for example in acknowledging that the wealthy are more able to defend themselves against risk, he is accused (see, for example, Bulkeley, 2001) of the global extension of a theoretical framework that is actually based in the specific conditions of Western Europe.

Third, there is criticism of Beck's perceived overoptimism as regards potential new forms of political activism. Here, Beck is accused of uncritically supporting a progressive, Enlightenment view of a move towards a situation of rational consensus, where reflexive modernization produces effective responses to global risks. Redclift (1997), for example, focusing on questions of ecological sustainability, argues that Beck's vision requires the existence of a politically aware and active citizenry, as yet not present. Similarly, Hinchliffe (1997) is critical of notions that alternative political groupings will form in the ways Beck suggests, and that they could necessarily be more politically effective than earlier forms of protest and activism. Hinchliffe imagines a different possible future, in which an increasingly uncertain world produces new forms of authority and control, rather than individual and political liberation. This alternative view of the future of modernity marks a contrast between what is taken as Beck's optimism and other, bleaker, perspectives. For example, Bulkeley's (2001) study of government responses to possible climate change in Australia suggests that far from radical political formations becoming influential, conventional government–industry links are reinforced by the regulatory structures being proposed. Bulkeley is thus critical of Beck for not engaging with how risks and responsibilities remain negotiated within existing arenas of power. Such criticism of Beck's optimism is perhaps somewhat unfair, given that his own understanding of reflexive modernization (e.g. Beck, 1999) makes it clear that the reflexive process only *might* produce the positive effects he suggests. The decisions made by individuals or in institutions might instead be reactionary or selfish, not 'rational' and cosmopolitan. The nature of politics and society in an individualized, yet globalized, future is not certain; indeed for Beck, this uncertainty is characteristic of reflexive modernization and risk society. Thus, we may conclude by citing Beck's assertion that the human condition is now characterized by 'fundamentally ambivalent contingencies, complexities, uncertainties and risks which, conceptually and empirically, still have to be uncovered and understood' (Beck, 2000: 81).

BECK'S MAJOR WORKS

Beck, U. (1992) *Risk Society: Towards a New Modernity.* London: Sage.

Beck, U. (1995) *Ecological Politics in the Age of Risk.* Cambridge: Polity Press.

Beck, U. (1997) *The Reinvention of Politics: Rethinking Modernity in the Global Social Order.* Cambridge: Polity Press.

Beck, U. (1998) *Democracy Without Enemies.* Cambridge: Polity Press.

Beck, U. (1999) *World Risk Society.* Cambridge: Polity Press.

Secondary Sources and References

Allen, J. and Henry, N. (1997) 'Ulrich Beck's *Risk Society* at work: labour and employment in the contract service industries', *Transactions, Institute of British Geographers* 22: 180–196.

Beck, U. (1996) 'World risk society as cosmopolitan society?', *Theory, Culture and Society* 13 (4): 1–32.

Beck, U. (2000) 'The cosmopolitan perspective: sociology of the second age of modernity', *British Journal of Sociology* 51 (1): 79–105.

Beck, U. (2001) 'The cosmopolitan state. Towards a realistic utopia', *http://www.eurozine.com/article/2001-12-05-beck-en.html*

Beck, U. and Beck-Gernsheim, E. (1995) *The Normal Chaos of Love*. Cambridge: Polity Press.

Bulkeley, H. (2001) 'Governing climate change: the politics of risk society?', *Transactions, Institute of British Geographers* 26: 430–447.

Eden, S. (1998) 'Environmental issues: knowledge, uncertainty and the environment', *Progress in Human Geography* 22: 425–432.

Hinchliffe, S. (1997) 'Locating risk: energy use in the "ideal" home and the non-ideal world', *Transactions, Institute of British Geographers* 22: 197–209.

Lash, S. and Wynne, B. (1992) 'Introduction', in U. Beck *Risk Society: Towards a New Modernity*. London: Sage, pp. 1–8.

Redclift, M. (1997) 'Sustainability and theory: an agenda for action', in D. Goodman and M. Watts (eds), *Globalizing Food: Agrarian Questions and Global Restructuring*. London: Routledge, pp. 97–111.

Whatmore, S. (2002) *Hybrid Geographies: Natures, Cultures, Spaces*. London: Sage.

Wynne, B. (1996) 'May the sheep safely graze? A reflexive view of the expert–lay knowledge divide', in S. Lash, B. Szerszynski and B. Wynne (eds), *Risk, Environment and Modernity: Towards a New Ecology*. London: Sage, pp. 44–83.

Lewis Holloway

6 Brian Berry

BIOGRAPHICAL DETAILS AND THEORETICAL CONTEXT

Brian Berry is perhaps the most important of a handful of people who transformed human geography over the second half of the twentieth century. His immediate peers and contemporaries were **Peter Haggett** and **David Harvey**. This group (along with others, including **Reg Golledge**, Peter Gould, and Leslie J. King) effectively initiated and led the 'quantitative revolution' in human geography from the early 1960s through to the 1980s. In this sense, Brian Berry has had an enormous and lasting impact on the discipline, being for nearly 20 years the world's most cited geographer and, no less importantly, the reference point for successive generations of graduate students in North America, the United Kingdom and Australasia. While intimately associated with spatial analysis and quantitative revolution, his work has continued to develop and evolve, encompassing urbanism, public policy, and long waves of innovation and development. The scope of his academic career and research publications is astounding, identifying him as one of the world's leading social scientists.

Brian Berry was born on 16 February 1934 in Stafford, England, to Joe and Gwendoline (née Lobley) Berry. Berry spent his formative early years in the East Midlands and Lincolnshire, areas of England associated with the Berry family for more than 400 years (tracing his roots back to these areas has been one thread of his intellectual and emotional journey). Brian was one of a small number of children from lower middle-class and working-class families who came through the British selective grammar-school system to go on to university in the 1950s. At a time when few children made it through to high-school matriculation, and even fewer went to university, Berry's subsequent progress through what was then a highly stratified and elitist society to outstanding academic success at University College London and the University of Washington (Seattle) was the result of enormous ability, determination and, perhaps paradoxically, opportunity. On completing his PhD at Washington (1958) he joined the University of Chicago before moving to Harvard (1976–1981), becoming Dean of the Heinz School of Public Policy and Management, University Professor at Carnegie Mellon University (1981–1986), and Williams Professor of City and Regional Planning at Harvard University (1976–1981). In addition, he was the youngest social scientist ever elected to the US National Academy of Sciences (1975), and served on the Council of the National Academy of Sciences (1999–2002). In 2002, Brian Berry was the Lloyd Viel Berkner Regental Professor and Professor of Political Economy at the University of Texas at Dallas.

Whereas he is primarily known as a geographer and social scientist, Berry is also an accomplished family historian and genealogist. Since the late 1980s and early 1990s he has published a large number of books, pamphlets and papers on his own family, his wife's family and related families. Characteristically, these projects combine close attention to empirical detail with a sure feel for the historical geography of trans-Atlantic relations. A geographer by intuition and inclination, he has approached research and scholarship with a strong grasp of his own agenda and a clear commitment to the principles he believes to be appropriate in understanding economic and social processes and their place in the broad sweep of history and geography. In turn, unfashionable, fashionable, and now again unfashionable, he has promoted a social scientific and analytical style of research. He stood by his views and opinions in the face of great social turbulence and dissent during the 1970s and 1980s, and restated his agenda in relation to the identity politics and postmodern angst of the 1990s and early years of the twenty-first century. If politically conservative and egalitarian by instinct, he is surely a radical when it comes to poking fun at orthodoxy whatever its origins. By this account, Berry's ambition has given many geographers the space in which to flourish and contribute to the growth of the discipline. It is difficult to imagine how the discipline could have prospered without him providing the hard edges of ambition, commitment and vision.

SPATIAL CONTRIBUTIONS

It is difficult to do justice to Berry's role in the quantitative revolution that swept through human geography in the 1960s and 1970s (though for his own assessment of his role in this intellectual revolution, see Berry, 1993). In this respect, it is worth repeating that Berry's generation took aim at the very core of geography, deliberately setting out to provide new issues and themes, and an analytical approach in contradiction to the cultural and regional geography that dominated the practice of US human geography. His earliest papers in the *Annals of the Association of American Geographers* (Berry, 1958, 1959, 1960) were remarkable for their agenda-setting, their social science

form and format, and their engagement with the processes driving the development of the urban economic landscape. These papers were especially significant given the personal fiefdoms, patronage and prejudice at work in what was a very small and marginal discipline. There is a certain irony, then, in seeing contemporary cultural geographers rediscover many of the leading figures of North American post-war geography who were so active in resisting Berry's call for relevance.

Berry's early work focused upon crucial theoretical issues in urban economic geography, using the tools and methods of the social sciences (especially economics) to develop empirically informed explanations of urban patterns and processes, and led the discipline into contemporary issues including housing, race, planning and US urban policy. Moreover, the remarkable *Geography of Market Centres and Retail Distribution* (1967) had an enormous impact in the social sciences, going beyond geography and the English-speaking world to symbolize the ambition and scope of an urban geography rooted in the 'Chicago School': Wrigley and Matthews (1986) identified it as a citation 'classic' of the twentieth century. At the time, it stood against William Alonso's (1964) own attempt to set the agenda for urban economic research. Whereas Alonso's book sought to provide a theoretical framework consistent with the axioms and methods of neo-classical economics, Berry's book sought to establish empirical principles and regularities that could account for the observed urban hierarchy and patterns of development (see Yeates, 2001, for a sympathetic assessment). In this respect, Berry's approach and his relationship to economics are issues replayed in recent commentaries on the 'new' economic geography associated with Paul Krugman (1991) and others (see also Clark *et al.*, 2000).

By the early 1970s, however, a number of factors conspired to undermine Berry's project, as well as the more general project of spatial analysis and quantitative urban economic geography. One was obviously the arrival in higher education and postgraduate study of the 'baby boom' generation. Ironically, again, whereas Berry, Haggett and Harvey and their colleagues opened opportunities and set broad intellectual horizons at the interface between geography and other social sciences, their work was the stepping-off point for a new generation of would-be academics who hardly understood the intellectual battles fought and won in the late 1950s and 1960s. The 'baby boom' generation were deeply affected by the Vietnam War and the political leaders who were fighting the Cold War with the lives of the baby boom generation. As a result, this generation was also very suspicious of any social science agenda that put analytical rigour ahead of political and economic realities. Indeed, the close association between social science, the war effort and status and progress through US academic disciplines reinforced a ready-made schism between Berry's generation and the baby boom generation.

A related factor that drove human geography away from spatial analysis and the quantitative revolution was the search for social theory that could, firstly, better understand urban and economic geography in relation to capitalist dynamics, and, secondly, appreciate the sources of social and especially racial inequality in US cities – cities increasingly being pulled apart by market forces of differentiation and middle-class distinction. Here, there was fertile ground for dissent and disagreement but equally fertile ground for great misunderstanding and easily-made enmities. Whether by accident or design, the debate between Berry and David Harvey about 'relevance' in leading journals in geography crystallized for a whole generation many of the limits of social science as an analytical priesthood. Just as Brian Berry had been the reference point for a new generation striving to leave behind a conservative and confining geography, David Harvey (1973, 1982) and his form of Marxist social and economic geography (set out in *Social Justice and the*

City and *The Limits to Capital*) were clarion calls for a new generation.

If Berry was disenchanted by the forces of change and dissent, his resolve to contribute to the formation of public policy and urban justice did not waver. Indeed, his papers, reports and books over the period 1975 to 1990 were a testament to his deeply held belief that social science should make a constructive and purposeful difference to the structure and design of cities and the lived experience of those who dwell therein. In 1976, he published two books that set the agenda for urban studies and public policy across the world, namely *Urbanisation and Counter-urbanisation* and *Chicago: Transformations of an Urban System*. Together, these volumes spelt out Berry's approach to understanding inter- and intra-urban hierarchies, issues placed on the national stage while he taught at Harvard University and later while he was Dean of the School of Urban Public Affairs at Carnegie Mellon University. In this way, Brian Berry became truly an advocate and citizen of social science, consistently portraying social science as a means of solving pressing public policy concerns. This was reflected, for example, in his contributions to the National Academy of Sciences (1982) report *Critical Issues for National Urban Policy*. Throughout this period, Berry's brand of social science became widely recognized throughout US social science and in sociology in particular.

Remarkably, from the 1990s onwards Berry embarked on an ambitious and important intellectual project: understanding and empirically specifying so-called long waves of development. Whereas this topic was associated with discredited views attributed to Kondratiev (a Russian economist of the early twentieth century), Brian Berry has been able to remake the issue into his own, bringing to bear his characteristic close attention to empirical detail, respect for the historical record, and sense of the personal and collective histories of English-speaking people (in particular but not exclusively) over the past 400 years. Setting the stage for this intellectual project was his (1991) book *Long-Wave Rhythms in Economic Development and Political Behaviour*. A succession of papers and monographs developing themes related to this project have been published integrating what he has termed 'long-wave macro-history' with urbanization, innovation, technological change and social and political movements. If less appreciated and less well known in geography, these contributions to understanding the broad sweep of history have had a ready audience across the rest of North American social sciences. Here, demography, politics and economics have been brought together in a quite remarkable synthesis of time and space – a genuine historiography of capitalism.

KEY ADVANCES AND CONTROVERSIES

Berry has confronted recent developments in geography and the social sciences, and the challenges posed by postmodernism in its various forms, voices and themes. He has used his position as editor-in-chief of *Urban Geography* to engage and argue with anyone who will listen – especially younger generations of human geographers. Again, characteristically, he has not been afraid to make plain his own views in contradiction to current fads and fashions as well as others' strongly-held and deeply committed views about the proper path of geography as a social science. For example, his attack (Berry, 1998b) upon a paper by **Michael Dear** and Steven Flusty (1998), published in the *Annals of the Association American Geographers,* combines his outspoken commitment to analytical and theoretical clarity with his vision of the proper role and place of social scientists in the world at large. He attacks these authors for not

understanding the implications of their own argument while suggesting that there are analytical methods available for the rigorous study of urban complexity and diversity. Berry's strongly-held views make for entertaining reading. Those who are the targets of his attention may be offended by his scornful remarks, but Berry's project has never been one of currying favour.

In a related commentary, Berry (1998a: 95) reviewed Edward O. Wilson's argument in favour of reconciliation that 'refers to the convergence of facts and theory to form a common network of explanation across scientific disciplines'. Sharing Wilson's concern about the narrowness and reductionist logic that informs much of the social sciences and especially economics, Berry goes on to suggest that Wilson's mandate is one that geography (among other disciplines) should emulate. In doing so, Berry (1998a: 96) sets out his own template for social science – a mode of explanation wherein 'our findings are repeatable and expressed with economy'. Insisting 'mensuration is unambiguous', Berry stresses we are all 'continually engaged in heuristics, fully committed to the delights of scientific discovery' (Berry, 1998a: 96). There is little doubt that the vast majority of human geographers are deeply suspicious of the rational utility-maximizing model that dominates his explanations of human agency and decision-making. But, at present, there is little in the way of consensus about the proper mode of research practice and the ultimate objectives of human geography. His agenda is one of many that coexist in an uneasy coalition of shared disciplinary anxiety.

Brian Berry remains committed to understanding the spatial organization of economy and society. If this means transcending the discipline as it has been remade over past two decades, he argues we should not be squeamish about seeking an integrated science of geographical understanding. This is a challenge to those who have fashioned their careers out of 'playing the game' within the narrow confines of the discipline. Furthermore, it is a challenge to those who would wish to police the borders of the new cultural consensus; as each generation claims intellectual ascendancy, so too do the forces of resistance and transformation. Ironically, it is much the same kind of challenge that Berry made to the discipline more than 40 years ago. And like all challenges that come from people with such integrity and ability, it is a challenge that cannot be easily ignored (*contra* Yeates, 2001). Whether human geography will have any resemblance to his vision of the future remains to be seen.

BERRY'S MAJOR WORKS

Berry, B. J. L. (1958) 'A note concerning methods of classification', *Annals, Association of American Geographers* 48: 300–303.

Berry, B. J. L. (1959) 'Ribbon developments in the urban business pattern', *Annals, Association of American Geographers* 49: 145–155.

Berry, B. J. L. (1960) 'The impact of expanding metropolitan communities upon the central place hierarchy', *Annals, Association of American Geographers* 50: 112–116.

Berry, B. J. L. (1967) *Geography of Market Centres and Retail Distribution.* Englewood Cliffs: Prentice-Hall.

Berry, B. J. L. (1976a) *Urbanisation and Counter-urbanisation.* London: Sage.

Berry, B. J. L. (1976b) *Chicago: Transformations of an Urban System.* Cambridge, MA: Ballinger.

Berry, B. J. L. (1991) *Long-Wave Rhythms in Economic Development and Political Behaviour.* Baltimore: Johns Hopkins University Press.

Berry, B. J. L. (1993) 'Geography's quantitative revolution: initial conditions, 1954–1960 – a personal memoir', *Urban Geography* 14: 434–441.

Secondary Sources and References

Alonso, W. (1964) *Location and Land Use.* Cambridge, MA: Harvard University Press.

Berry, B. J. L. (1998a) 'On consilience', *Urban Geography* 19: 95–97.

Berry, B. J. L. (1998b) 'Wallerstein's middle ground', *Urban Geography* 19: 393–394.

Clark, G. L., Feldman, M. and Gertler, M. S. (eds) (2000) *The Oxford Handbook of Economic Geography.* Oxford: Oxford University Press.

Dear, M. J. and Flusty, S. (1998) 'Post-modern urbanism', *Annals, Association of American Geographers* 88: 50–72

Harvey, D. (1982) *The Limits to Capital.* Chicago: University of Chicago Press.

Krugman, P. (1991) *Geography of Trade.* Cambridge, MA: MIT Press.

National Academy of Sciences (1982) *Critical Issues for National Urban Policy.* Washington DC: NAS (with Royce Hanson and others).

Wrigley, N. and Matthews, S. (1986) 'Citation classics and citation levels in geography', *Area* 18: 185–194.

Yeates, M. (2001) 'Yesterday as tomorrow's song: the contributions of the 1960s "Chicago School" to urban geography', *Urban Geography* 22: 514–529.

Gordon L. Clark

7   Homi K. Bhabha

BIOGRAPHICAL DETAILS AND THEORETICAL CONTEXT

Homi K. Bhabha was born to an English- and Gujarati-speaking Parsi family in the newly independent India of 1949. Parsis, a minority of Persian descent, were instrumental in the emergence of an urban middle class in nineteenth-century Imperial India and in this capacity often functioned as mediators between the Indians and the British. Because of their dispersion – Parsis live in small clusters in a number of different host cultures – they derive their cultural cohesion partly from their Zoroastrian faith and partly from a negotiation of their host cultures' traits. Thus, Parsis offer a striking example of the hybridized, cosmopolitan minorities at the centre of Bhabha's work; indeed Bhabha has often linked his intellectual preoccupations to the specificity of his origins.

Bhabha spent his childhood and early adulthood in Bombay, where he received a BA in English from Bombay University. Shortly after the completion of his undergraduate degree, Bhabha left India to read English at Christ Church Oxford, where he took an MA, MPhil, and DPhil. While at Oxford, he wrote his doctoral thesis on the Trinidadian writer V. S. Naipaul, whose thematization of the 'mimic' identity of colonized populations remained a central reference point in Bhabha's subsequent development. In 1978, Bhabha gained a lectureship at Sussex University, where he remained until 1994. This position was to be decisive for the young academic: in the late 1970s and 1980s the Sussex English department functioned as a privileged site for the importation and appropriation of French and Continental theory into British academic work. Bhabha's work is a palimpsest of this traffic in theory, and his texts resonate with allusions to Derridean deconstruction, Lacanian psychoanalysis and – perhaps most profoundly of all – the work of **Michel Foucault**.

Between the mid-1980s and the early 1990s Bhabha held visiting fellowships at Princeton, Dartmouth College and the University of Pennsylvania. During this time Bhabha produced several articles published in the journals *Critical Inquiry*, *Screen* and *October*, through which he first established himself as a distinct voice within the emergent field of postcolonial studies. In 1986 he wrote the introduction to the English translation of Frantz Fanon's *Black Skin, White Masks*, in which he argued for the salience of the Martiniquan psychiatrist's writings on the psychology of the colonized for contemporary political practice. During the late 1980s Bhabha joined a number of academics in seeking an engaged understanding of the cultural forces shaping Thatcher's Britain. In this context he argued for the symbolic centrality of cultural minorities for national discourse. In the wake of the 1989 *fatwa* on Salman Rushdie for his book *The Satanic Verses*, Bhabha wrote a series of position pieces for the left-affiliated weekly *The New Statesman* in which he castigated the Western liberal representation of Muslim populations prompted by the Rushdie affair. Through these engagements, Bhabha's writings turned more decisively towards discussions of the place of migrant and displaced populations in the geographies of the new world order. His first edited collection, *Nation and Narration* (1990a), brought together a number of scholars who sought to analyse the discourses through which nations find their coherence as collective bodies.

Bhabha's reputation was cemented in 1994 through the publication of *The Location of Culture*, a collection of most of his earlier essays in revised form. In the same year, he became Professor in the Humanities at the University of Chicago. While in Chicago, Bhabha taught art history as well as English and became a regular contributor to the journal *Artforum*. Since his move to the US in 1994, Bhabha has become more substantially engaged with visual culture. His more recent work has moved in two related directions: the potential of a new, vernacular cosmopolitanism (the subject of his forthcoming book, *A Measure of Dwelling*) and the question of minority rights in discussions of multiculturalism. His essays and books have been translated into several languages. Bhabha is currently Professor of English and American Literature and Language, and African American Studies, at Harvard University, a post he has held since 2001.

SPATIAL CONTRIBUTIONS

Bhabha's early essays intervened in the emerging analyses of colonial discourse following **Edward Said**'s installation of this term in *Orientalism*. While Bhabha agreed with Said's mobilization of **Michel Foucault**'s writings on discourse for an analysis of colonial knowledge formations, he countered Said's claim that colonial discourse imposed a firm distinction between European and native identities.

Instead, Bhabha suggested that colonial texts both assumed such a distinction and continuously cast it into doubt. Bhabha proposed that the colonizers' stereotypes of the colonized were characterized by *ambivalence*; in other words, that stereotypes were 'a form of knowledge and identification that vacillates between what is always "in place", already known, and something that must be anxiously repeated' (Bhabha, 1994b: 66). Stereotypes revealed both the authority of colonial discourse and the limits of this authority: their aim was to fix the colonized through a series of characterizations (e.g. the 'lazy African', the 'lascivious Oriental') but their constant reiteration suggested that such a fixity was impossible. In a series of essays published in the mid-1980s, Bhabha extended his psychoanalytic and deconstructive readings to an elaboration of native resistance against colonialism. In so doing, Bhabha returned to the writings of V. S. Naipaul and refashioned them in the light of Frantz Fanon's understanding of the colonized psyche. While Naipaul had produced the seemingly pessimistic diagnosis that the colonized is nothing but a 'mimic man' living on the borrowed culture of the colonizing powers, in Bhabha's hands this diagnosis became tinged with subversive potential. Bhabha claimed that the colonial 'mimic men' represent a paradox at the heart of the imperial mission: to wit, that the imperial project depended both on the colonizer's wish for the native to become Europeanized and on his fear that – in so becoming – the native may become too much like the colonizer. Mimicry, for Bhabha, is thus premised upon the missionary desire to make the colonized 'almost the same *but not quite*' (Bhabha, 1994c: 86). However, the mimicry of the colonized also confronts the colonizer with an uncanny version of himself, a parodic re-enactment of European distinctiveness. Such a re-enactment has the potential to dislodge the very authenticity assumed by the European and to undermine its originality. Bhabha's arguments here anticipate the later discussions of performativity in the work of **Judith Butler**: for Bhabha, the colonized is not a second-rate European but rather a disturbance to the European's dream of authority and authenticity.

Bhabha's early essays thus contained a double challenge to the understanding of colonial identity formations. On the one hand, they involved the claim that Western discourses of Othering inevitably become fractured and split at the point of their application, and that it is precisely these applications that enabled the resistance of the colonized. On the other hand, however, these essays also troubled understandings of the political implications of such resistance. If the colonizer becomes a split, ambivalent figure, then so does the colonized: the latter possesses no authentic self beneath the mask of mimicry bequeathed by the colonizer. Therefore, any political resistance to colonial rule has to be understood not as the oppressed population's straightforward rejection of the colonizer's legacy, but rather as a much more ambiguous process in which that legacy is both refused and desired.

Bhabha later refined his concept of ambivalence, giving it a decidedly spatial cast. From the late 1980s on, and beginning with his articles for *The New Statesman*, Bhabha urged for a more nuanced reading of cultural identities within the postcolonial geographies of migration and diaspora. In these articles – some of his most elegant and lucid writings to date – Bhabha deplored the binary thinking that pitted the cultural spaces of Western liberal states against the minority enclaves of Islamic populations living within them. He claimed that these categories remained blind to the complex dynamics of negotiation through which displaced populations make sense of their lives across contesting cultural values and traditions. For Bhabha, such experiences should not be seen to form a distinct and self-contained cultural *space* but to signal a *process* of 'cultural translation' between traditions (Bhabha, 1989).

In the 1990s, Bhabha continued his elaboration of 'cultural translation' as a process that has inevitably accompanied the installation of modernity through the legacies of colonization. Bhabha called for a postcolonial revision of modernity and a rejection of the social-scientific vocabulary of development and underdevelopment, of First and Third Worlds. In the place of such totalizing explanations, Bhabha called for partial and localized strategies of analysis capable of recognizing the liminality and instability of cultural and political practices.

In accord with the insights of other postcolonial theorists, Bhabha's work insists that we cannot think of the spatialities of modernity outside those produced through imperial projects; that the cultural present inevitably testifies to a sedimentation of colonial histories. But while some postcolonial theorists have claimed that communities and identities are therefore shaped through boundaries between 'us' and 'them', 'self' and 'other', Bhabha has proposed instead that identities are inevitably hybridized, because the spaces of social life are formed through a *rupturing* of boundaries and through-flows of illicit border traffic. Bhabha has employed a number of terms, such as 'hybridity', 'the in-between' 'cultural translation' and 'third space' to describe this rupturing. These terms are neither intended to be synonymous nor do they point to discrete types of spatiality. Rather, they represent Bhabha's attempt to problematize the spatio-temporal coordinates routinely employed by social scientists. The proliferation of such terms in Bhabha's writings enacts the slipperiness of the spatialities he is attempting to evoke, and their irreducibility to conventional types of mapping. In the context of the 'cultural turn' within geography and the social sciences more broadly, the work of Homi Bhabha, along with that of **Gayatri Spivak** and **Edward Said**, has thus provided a methodological compass for geographers engaging with the spatial legacies of colonization (though, as noted in the next section, not always without criticism).

KEY ADVANCES AND CONTROVERSIES

The distinctiveness of Bhabha's work can be gauged by the amount of criticism his essays have generated. One focus of such critcism and debate has been Bhabha's attempt to redefine culture by submitting it to a 'radical' spatialization. What this means is that while for Bhabha culture is primarily spatial, he suggests it cannot be conceptualized through the oppositions between tradition and modernity, or East and West. Instead, Bhabha suggests that a cultural space is the location of shared practices which, while generated in response to particular historical and geographical conditions, cannot simply be said to belong to one discrete culture or another. It is with reference to such practices that Bhabha initially used the term 'third space'. In Bhabha's discussion of the Rushdie affair, 'third space' described the hybrid cultural practices of British Muslims and other displaced populations who negotiate often irreconcilable fragments of different traditions and make their temporary home at their limits. The identities generated within such spatialities cannot be authentically Muslim or authentically British, but are both and neither at the same time. Bhabha further proposed that the presence of migrant populations disturbs normative understandings of social space and of national culture. Hybridized spaces are omnipresent in the shifting demographies of actual nations, produced by the 'wandering peoples who ... are themselves the marks of a shifting boundary that alienates the frontiers of the modern nation' (Bhabha, 1990a: 315).

This definition of 'third space' has proved particularly attractive to geographers who wish to engage with the complex and shifting experiences in contemporary global spaces. Steve Pile, for example, has read Bhabha's 'third space' as a call for an overcoming of dualistic epistemologies. For Pile, Bhabha gestured to the emergence of a new politics through 'a space which avoids the politics of polarity and enables the construction of new radical allegiances to oppose structures of authority' (Pile, 1994: 271). Similarly, **Gillian Rose** has employed Bhabha's work in her discussions of the kinds of collective solidarity that do not rely on exclusionary tactics. Both Pile and Rose have thus praised Bhabha's problematization of space as 'something which must itself be made different' (Rose, 1995: 369).

In tandem with this identification of 'third space' as the cultural location of hybrid communities, Bhabha has also maintained a rather different usage of the term. This second usage, while equally central in Bhabha's texts, has generated a less amiable reception. In the 1988 essay 'The commitment to theory' for example, 'third space' does not refer solely to a spatiality produced by particular populations, but, additionally, to the 'general conditions of language' within which hegemonic discourses are elaborated (Bhabha, 1994a: 36). Briefly, Bhabha uses the term to suggest that every time particular cultural traditions and discourses are performed, the outcome is something in excess of what had been intended, so that the enactment of cultural practices always necessarily displaces them. Here, 'third space' designates both specific hybridized cultural practices and an aspect of all cultural life. 'Third space' is thus both the space between cultures and the non-coincidence of a *single* culture with itself. This generalization of 'third space' has provoked severe criticism from a number of academic locales. Within postcolonial studies, Robert Young (1990) claimed that Bhabha lacked historical and geographical specificity in so far as his formulations made ambivalence and hybridity into conditions inherent in cultural practice rather than specific to particular impositions of colonial power. More recently, Moore-Gilbert (1997) has similarly concluded that Bhabha's uncritical use of psychoanalysis leads him to homogenize discrete geopolitical formations under a Western set of categories. In addition, a number of Marxist critics have castigated Bhabha for his deprioritization of class dynamics in his accounts of spatiality (e.g. Ahmad, 1992).

Bhabha's work has also generated ambivalent responses among geographers engaging with postcolonial theory. His predilection for gnomic pronouncements and his fondness for a rhetoric of exhortation has meant that he quickly became eminently quotable by many academics seeking a theoretical language attentive to the spatial heterogeneities of cultural life. Thus for example, while **Ed Soja** quotes Bhabha extensively in his book *Thirdspace*, he also makes it clear that his work owes more to **Henri Lefebvre** and to the African American cultural critic **bell hooks** than it does to Bhabha (Soja, 1996). While Soja discusses Bhabha's work at length, he suggests that Lefebvre's and hook's perspectives are more immediately engaged with the production of spatiality and the politics of marginalized people respectively.

Indeed, a number of geographers have taken issue with the level of abstraction and generality in Bhabha's claims. **Gillian Rose** has used Bhabha's work while also critiquing Bhabha's inattention to the actual effectivity of political struggles. For Rose, this is a serious problem of Bhabha's methodology. In her otherwise favourable review of *The Location of Culture*, Rose suggested that it is because of his central thesis concerning the ambivalence of hegemonic discourses, that Bhabha fails to adequately address historical specificity. In other words, in so far as Bhabha claims that dominant discourses (such as discourses of 'nation' or 'Empire') contain their own failure within themselves, he cannot throw light on how *some* powerful discourses remain powerful while others fail (see Rose, 1995; Sparke, 1998).

Other geographers have suggested that hybridity is not always related to a failure of dominant discourses – at least not as far as economic discourses are concerned. In other words, while hybridity may upset the cultural ideologies of nationalism, it gives perfect sustenance to the ideologies of capitalism. For example, Katharyne Mitchell's research on Hong Kong entrepreneurs in Vancouver, Canada, has demonstrated how 'hybrid subject positions . . . and liminal and partial sites can be used for the purposes of capital accumulation' (Mitchell, 1997: 534).

These contestations of Bhabha's approach may be symptomatic of wider tendencies characterizing current research on location and social practice. Bhabha's work militates against the critique of 'theory' launched by many social scientists who privilege the local and geographically specific. As geographers turn increasingly to the politics of experience, they castigate Bhabha's use of hybridity as a catch-all category which, in their view, ignores the specificity and irreducibility of different experiences of marginalization. However, Homi Bhabha's approach does not necessarily entail an indifference to politics and to material practices. Rather, by suggesting that such practices are divided and conflictual, Bhabha's writings may facilitate our questioning of the distinctions between theory and politics, or between resistance and complicity, by recommending an attentiveness to the in-between spaces of everyday life. Recent work by Jane M. Jacobs and Lisa Law illustrates this alternative reading of Bhabha (e.g. Jacobs, 1996; Law, 1997).

In her research on Filipina sex workers, Lisa Law's uses 'third space' as a way of thinking the politics of location together with the ambivalence of psychic life. By refusing to dichotomize between Western sex-tourists as oppressors and Asian prostitutes as their victims, Law claims instead that 'comprehending the political economic and colonial realities of sex tourism does not necessarily provide an understanding of the personalized modes of identification which form around issues of race, gender and sexuality' (Law, 1997: 109). Here, 'third space' refers to the dynamics of psychic life, and to the way in which these dynamics engage particular dominant discourses.

Jane M. Jacobs's book *Edge of Empire* (1996) argues that the spatial legacies of British Imperialism cannot simply be seen as the outcome of imperial impositions, or as the effect of native resistance. The analysis of such spatialities, Jacobs claims, requires a double vision. On the one hand, geographers must be attentive to how local dynamics translate imperial projects. On the other hand, they must also perceive the imperial projects themselves as ambivalent formations (split between the imperative to civilize and the desire to violate the other). In other words, Jacobs suggests that it is possible and, indeed, necessary for geographers to hold on to both a general understanding of culture as 'third space' (as proposed by Bhabha) while also remaining attentive to the specificity of experiences of cultural life.

BHABHA'S MAJOR WORKS

Bhabha, H. K. (1989) 'Beyond fundamentalism and liberalism', *The New Statesman and Society* 2 (39): 34–35.

Bhabha, H. K. (1990a) 'DissemiNation: Time, narrative, and the margins of the modern nation', in *Nation and Narration*. London & New York: Routledge, pp. 291–322.

Bhabha, H. K. (1990b) 'The Third Space: Interview with Homi K. Bhabha', in J. Rutherford (ed.) *Identity: Community, Culture, Difference*. London: Lawrence & Wishart, pp. 207–221.

Bhabha, H. K. (1994a) 'The commitment to theory', in *The Location of Culture*. New York & London: Routledge, pp. 19–39 (revision of essay published in 1988).

Bhabha, H. K. (1994b) 'The other question: stereotype, discrimination and the discourse of colonialism', in *The Location of Culture*. New York & London: Routledge, pp. 66–84 (revision of essay published in 1983).

Bhabha, H. K. (1994c) 'Of mimicry and man: the ambivalence of colonial discourse', in *The Location of Culture*. New York & London: Routledge, pp. 85–92 (reprint of essay published in 1984).

Bhabha, H. K. (1994f) 'How newness enters the world: postmodern space, postcolonial times and the trials of cultural translation', in *The Location of Culture*. New York & London: Routledge, pp. 212–235.

Bhabha, H. K. (1995) 'Translator translated: conversation with Homi Bhabha', *Artforum* 33 (7): 80–83, 110, 114, 118–119.

Secondary Sources and References

Ahmad, A. (1992) *In Theory: Classes, Nations, Literatures*. London: Verso.

Jacobs, J. M. (1996) *Edge of Empire: Postcolonialism and the City*. London: Routledge.

Law, L. (1997) 'Dancing on the bar: sex, money and the uneasy politics of third space', in S. Pile and M. Keith (eds) *Geographies of Resistance*. London: Routledge.

Mitchell, K. (1997) 'Different diasporas and the hype of hybridity', *Environment and Planning D: Society and Space* 15: 533–553.

Moore, D. S. (1997) 'Remapping resistance: "Ground for struggle" and the politics of place', in S. Pile and M. Keith (eds) *Geographies of Resistance*. London: Routledge.

Moore-Gilbert, B. (1997) *Postcolonial Theory: Contexts, Practices, Politics*. London: Verso.

Pile, S. (1994) 'Masculinism, the use of dualistic epistemologies and third spaces', *Antipode* 26: 255–277.

Rose, G. (1994) 'The cultural politics of place: local representation and oppositional discourse in two films', *Transactions of the Institute of British Geographers* 19: 46–60.

Rose, G. (1995) 'The interstitial perspective: a review essay of Homi Bhabha's *The Location of Culture*', *Environment and Planning D: Society and Space* 13: 365–373.

Rose, G. (1997) 'Spatialities of community, power and change: the imagined geographies of community arts projects', *Cultural Studies* 11: 1–16.

Routledge, P. (1996) 'The third space as critical engagement', *Antipode* 28: 399–419.

Soja, E. (1996) *ThirdSpace: Journeys to Los Angeles and Other Real-and-imagined Places*. Oxford: Blackwell.
Sparke, M. (1998) 'A map that roared and an original atlas: Canada, cartography, and the narration of nation', *Annals of the Association of American Geographers* 88: 463–495.
Young, R. (1990) *White Mythologies: Writing History and the West*. London: Routledge.
Young, R. (1995) *Colonial Desire: Hybridity in Theory, Culture and Race*. London: Routledge.

Constantina Papoulias

8 Pierre Bourdieu

BIOGRAPHICAL DETAILS AND THEORETICAL CONTEXT

Pierre Bourdieu was born in 1930 in a small village in south-west France. When he finished high school (as a star pupil and star rugby player) he moved to Paris, where he studied philosophy at the elite École Normale Supérieure. After graduating he worked as a teacher for a year before being drafted into the army to fight in the Algerian War between 1956 and 1958. His experiences in Algeria (a conflict over French colonialism that he subsequently described as 'an appalling war') turned him away from philosophy and towards sociology and anthropology, and he remained in Algeria after the conflict to teach at the University of Algiers while conducting fieldwork among the Kabyle of the North East region. In 1960 he returned to France and held posts at the University of Paris-Sorbonne and the University of Lille, and in 1964 he became Director of Studies at the École des Hautes Etudes en Sciences Sociales. In 1968 he launched the Centre for European Sociology and later established the journal *Actes de la Recherche en Sciences Sociales*, devoted to understanding cultural reproduction and class domination. Bourdieu was elected to a Chair at the prestigious College de France in 1981. He died from cancer in January 2002.

Over his distinguished career Bourdieu published more than 40 books and 400 articles. These range from early anthropological texts – such as *The Algerians* (1962) – and related theoretical contributions – most notably, *Outline of a Theory of Practice* (1977a) – through to definitive contributions to our understanding of class reproduction (*Distinction: A Social Critique of the Judgement of Taste*, 1984); the sociology of education (*The Inheritors*, 1979; *Reproduction in Education, Society and Culture*, 1977b) and theorizations of action in everyday life (*Practical Reason*, 1998). Latterly, Bourdieu had been engaged in a critique of the politics of globalization rendered in the voices of those dispossessed by the neo-liberal global regime (*The Weight of the World*, 1999).

Bourdieu's leading theoretical claim is that his work transcends the dualism between explanations that attribute social change and social reproduction to certain overarching structures and theorizations that privilege individual subjective intention or experience. He draws on existentialism to emphasize the importance of practical action and existence (praxis) in understanding the human individual, while rejecting the highly conscious, subjectivist elements of Sartre's existentialism (which sees authentic existence as an act of self-creation). At the same time, Bourdieu's work has affinities with American pragmatism (especially Mead and Dewey) in his social conception of mind and agency and emphasis on the body as a site of socially instilled habit (see Aboulafia, 1999). This response to the dualisms of objectivism and subjectivism is Bourdieu's 'theory of practice'. This is an idea of ongoing, practical everyday activities – articulated through bodies – that contain prior cultural dispositions. This mix of activities and ways of acting tend to encompass distinctive deployments of different forms of capital. Hence, Bourdieu's innovation is to argue that forms of capital include cultural assets: the ability to 'consume' rare objects,

such as certain works of art, might demand skills of taste and appreciation that
are derived from family background and/
or via institutionalized education (hence
his interest in the sociology of education).
Cultural capital involves the embodied
dispositions and resources of 'habitus'
(see below). It includes competences that
are incorporated via education, as well as
symbolic power, which is the power of a
certain class to legitimize their 'natural'
tastes and so dominate the social order.
Cultural and symbolic capital have relations with social capital (based on social
networks) and economic capital.

Different combinations of forms of
capital operate in 'fields' (economic, political or artistic, for example). If social life
is thought of as a 'game' (an analogy
Bourdieu used often) then fields are the
playing board. Embodied dispositions provide the 'feel for the game', which cannot
be explained wholly by the rules of the
game. Crucially, Bourdieu argued that the
feel for the game varies over social space
(a term he used to suggest a 'force field' of
conflicting social power). Moreover, he
suggested that strategies tend to coalesce
in relatively enduring combinations and
embodied dispositions that he termed
habitus. A crucial concept at the heart of
Bourdieu's sociology, habitus is an array
of inherited dispositions that condition
bodily movement, tastes and judgements,
according to class position (Bourdieu,
1984).

SPATIAL CONTRIBUTIONS

As Joe Painter (2000) has pointed out, the
take-up of Bourdieu's ideas in geography
has been patchy. That is not to say that he
is not frequently cited, but that the sustained and direct engagement of geographers with Bourdieu's work is
surprisingly rare. There are some early

skirmishes, most notably Parkes and
Thrift's (1980) use of the idea of 'social
time' from Bourdieu's study of the Kabyle
people in Algeria. In this work Bourdieu
analyses how the cultural and economic
rhythms of Kabyle life constitute time in
that society. Bourdieu's analysis of the
times and spaces of the Kabyle house also
provides **David Harvey** (1989) with an
example of the structuring of space and
time that is non-capitalist. For Harvey,
the Kabyle example shows that time and
space can be orientated socially around
the body in ways that have deep cultural
resonance. This is evident from Bourdieu's now famous description of the
gendering of this domestic space:

> The low, dark part of the house is also
> opposed to the upper part as the female to
> the male. Not only does the division of
> labour between the sexes (based on the
> same principle of division as the organiz-
> ation of space) give the woman responsi-
> bility for most of the objects belonging to
> the dark part of the house, the carrying of
> the water, wood, manure, for instance:
> but the opposition between the upper part
> and the lower part reproduces, within the
> internal space of the house, the opposition
> between the inside and the outside, be-
> tween female space – the house and its
> garden – and male space.
> (Bourdieu, 1990: 273–4)

Harvey ends his consideration of Bourdieu by questioning the degree to which
this socio-cultural time/space can survive
the increasingly global reach of capitalism
– an ordering of time and space that
relies on the abstraction of place in accordance with the logic of capital accumulation.

This vague definition of habitus as
some kind of authentic, culturally rich,
'place', resisting what **Henri Lefebvre**
calls the 'abstract space' of capitalism,
accounts for many of those citations of
Bourdieu that make no direct engagement
with his ideas. Against this, there has also
been a much more explicit exploration of
the links between habitus and place
which, although not located centrally in
geography, does involve geographers and

urbanists and has significant implications for geographers' engagement with Bourdieu's work. One manifestation of this exploration came in the form of a conference in Perth, Australia, and a subsequent edited collection by Jean Hillier and Emma Rooksby, *Habitus: A Sense of Place* (2002). The overall purpose of the collection is to ask if the idea of habitus is still relevant in a rapidly globalizing world. With an impressive list of contributors (including Bourdieu himself), the book tackles this question at a number of levels, from national and sub-national political fields and their relationship to symbolic capital and domination, through discussions of postcolonialism and habitus as resistance, through to the role of the built environment in the reproduction of habitus. Specific signals for future research in geography include Joe Painter's drawing together of the idea of habitus and Foucault's notion of governmentality in the discursive construction of place as a space for politics. Also significant, and deeply intriguing, are the chapters by Steve Pile, Neil Leach and Kim Dovey that start to articulate the connections between habitus and the formation of identity. Leach (2002) and Dovey (2002) explore this relationship with reference to the intellectual field of the architectural profession, showing that this is responsible for reproducing dominant social divisions through forms of 'place-making'. Pile (2002) explores the less predictable role of the *uncanny* in the formation of habitus as sense of place through ideas of ghostliness and being haunted by places in the city.

Recognizing another important departure, Painter (2000) argues that **Nigel Thrift** (1996) in part takes inspiration from Bourdieu's definition of practice as a means of 'going on in the world' but pursues it though non-representational modes of knowing. In Thrift's non-representational geography, practice, practical sense and 'a feel for the game' can never be fully captured and represented. If that is acknowledged then the spatialities of practice can give rise to the possibility of other ways of being in the world. The spatialities of practice have been taken in another direction by Tim Cresswell (1996), who shows that the geographical 'facts of life' (place, territory, landscape) are constitutive of social life because they relate the body to space through habitus. Cresswell demonstrates the power of these geographical ordering devices by exploring reactions to their transgression (for instance, by New York graffiti artists, Greenham Common peace protestors and new-age travellers). Emphasizing geographies of mobility, Cresswell thus offers an elegant geographic elaboration of Bourdieu's ideas (describing Bourdieu's influence on his work in a glowing appreciation of his legacy – see Cresswell, 2002).

Alternative spatialities of practice aside, the concept of habitus has been more routinely (and sometime lazily) deployed by geographers seeking to explain the importance of culture in certain sociospatial contexts. For example, the literature on gentrification has made sporadic reference to a 'gentrification habitus'. The aesthetics of gentrification, which in various forms seek to appropriate the historic core of the city, have been tied to questions about the rise of a new middle class of urbanites, rather than suburbanites (see especially Jager, 1986). Julie Podmore (1998) shows how well the discursive idea of 'loft living' has travelled into different urban contexts, regardless of their stock of loft-type dwellings, suggesting a transnational set of dispositions highly amenable to the loft lifestyle message. Derek Wynne and Justin O'Connor (1998) use research on buyers of new residential property in the centre of Manchester, England, to argue for a set of overlapping cultural practices among this group but ones that reflect middle-brow rather than elite culture (as *Distinction* argues for the comparable class in 1960s France). Sociologists Tim Butler and Gary Robson (Butler, 1997; Butler and Robson, 2001) have made important contributions by arguing that the gentrification habitus is fragmenting to reveal distinct fractions of the new middle class. Their work on London points to the importance of the space of the metropolis in permitting a distinct neighbourhood milieu to develop. They distinguish a corporate type of gentrification (involving highly paid private-sector employees, largely in the financial services industry) in certain neighbourhoods, and community-type gentrification of other areas by the public-sector intelligentsia. Arguments about the fractions of habitus connect to ideas about the gentrification aesthetic as an internally competitive set of aesthetic practices that are constantly on the move (leading, for example, to a 'gentrification premium' for access to certain gentrified districts and dwellings – see Bridge, 2001a). The competitive, expressive and self-conscious nature of the gentrification aesthetic leads to a questioning of Bourdieu's claims for the unconscious nature of habitus and points to a (possibly overly hasty) rejection of rational choice explanations (Bridge, 2001b). Intriguing and important work by Hage (2000) and May (1996) points to strategies of cosmopolitan distinction that espouse multiculturalism but it is an idea of multiculturalism consumed as an aesthetic commodity rather than a reality of everyday life. The inter-relationships between ethnicity and gentrification and strategies of distinction are likely to be a significant area of future research. Such critical engagements with the idea of habitus via understandings of time and space are likely to continue apace in geography.

Bourdieu's ideas have gained more general currency in contemporary human geography as part of the rising concern with the body in the social theory as a whole. This is part of the wider philosophical critique of the idea of the autonomous, cognitively rational actor. Foucault's work considered the body as the site of the operation of disciplinary power. Maurice Merleau-Ponty (1962) saw the body as locus of primary sociality, incarnated activities that exist prior to abstract cognition. From a Marxist perspective, Lefebvre saw the body as the possibility of sensual and expressive 'lived space' in resistance to the instrumentalization of the body in the abstract space of capitalism. Bourdieu's ideas are absorbed as part of this wider set of discourses about the body as a cultural process, and, more specifically in geography, as an orientation in space. The work of Maurice Merleau-Ponty has been especially important in this regard. Bourdieu studied Merleau-Ponty as a student at the École Normal Supérieure. What he develops from Merleau-Ponty is the idea of incarnated activities, in this case embodied dispositions. As well as being the repository of dispositions, the body is also their most evident register. Thus Bourdieu's understanding of class divisions is scripted around the body moving in space – the space it inscribes, its gait, the gestures of the arms and hands, inclination of the head and so on. Robyn Dowling (1998) has called for geographers to pay direct attention to this classing of the body in space. Overall though, Bourdieu's insights have been used less explicitly as part of the theoretical portmanteau that sees the body as a cultural process or as the significant axis of everyday life.

KEY ADVANCES AND CONTROVERSIES

There are long-standing and ongoing debates about the degree to which Bourdieu's canon manages to transcend the dualism of objectivism or subjectivism (for example, Honneth, 1986; Jenkins, 1992). Equally, his idea of habitus has come under fire for being too 1960s, too limited to the French social structure and too static. Another weakness for geographers is the paucity of references to space or place in Bourdieu's work. Nevertheless, although not an explicit theorist of space and place, Bourdieu has had an important influence on the way that geographers think about these things. There is the direct influence of his early anthropology of the Kabyle and later theorizations of practice that relate to 'social' time and space. Then there is a more diffuse, but no less pervasive, set of influences that ripple out from the idea of the body as strategy of capital accumulation: for example, the idea of the body as the locus of class reproduction; the role of the body in the making of social space and the embodied construction of socio-spatial order. Although the classed body has received least attention from geographers, the idea of the body as the site of cultural reproduction has been pursued strongly in certain analyses of the socio-spatial dynamics of gentrification. Future

flowerings of Bourdieu's geographies are in prospect through ideas of practice and everyday spatialities, and the subconscious sensitivities to space and place. There is one other lesson that Bourdieu leaves us with and that is the richness and value of combining theoretical advances with comprehensive and careful empirical work, finding the place for the self-critical voice of the academic, alongside the voices of the subjects of knowledge. This is fitting legacy for such a 'great and original figure of contemporary sociology', as Jacques Derrida put it (*Le Monde*, 24 January 2002), leading the tributes of a French academy and wider society mourning the death of one of its greatest public intellectuals.

BOURDIEU'S MAJOR WORKS

Bourdieu, P. (1962) *The Algerians*. Trans. A. Ross. Boston, MA: Beacon Press.
Bourdieu, P. (1977a) *Outline of a Theory of Practice*. Cambridge: Cambridge University Press.
Bourdieu, P. (1977b) *Reproduction in Education, Society and Culture*. Trans. R. Nice. London: Sage.
Bourdieu, P. (1979) *The Inheritors: French Students and their Relation to Culture*. Trans. R. Nice. Chicago: Chicago University Press.
Bourdieu, P. (1984) *Distinction: A Social Critique of the Judgement of Taste*. London: Routledge and Kegan Paul.
Bourdieu, P. (1988) *Homo Academicus*. Cambridge: Polity.
Bourdieu, P. (1990) *The Logic of Practice*. Cambridge: Polity.
Bourdieu, P. (1991) *Language and Symbolic Power*. Trans. G. Raymond and M. Adamson. Cambridge: Polity.
Bourdieu, P. (1998) *Practical Reason*. Cambridge: Polity
Bourdieu, P. (1999) *The Weight of the World: Social Suffering in Contemporary Society*. Palo Alto, CA: Stanford University Press.

Secondary Sources and References

Aboulafia, M. (1999) 'A (neo) American in Paris: Bourdieu, Mead, and Pragmatism', in R. Shusterman (ed.) *Bourdieu: A Critical Reader*. Oxford: Blackwell, pp. 153–174.
Bridge, G. (2001a) 'Estate agents as interpreters of economic and cultural capital: the 'gentrification premium' in the Sydney housing market', *International Journal of Urban and Regional Research* 25: 81–101.
Bridge, G. (2001b) 'Bourdieu, rational action and the time-space strategy of gentrification', *Transactions of the Institute of British Geographers* NS 26: 205–216.
Butler, T. (1997) *Gentrification and the Middle Classes*. Aldershot: Ashgate.
Butler, T. and Robson, G. (2001) 'Social capital, gentrification and neighbourhood change in London: a comparison of three south London neighbourhoods', *Urban Studies* 38, 12: 2145–2162.
Calhoun, C., LiPuma, E. and Postone, M. (eds) (1993) *Bourdieu: Critical Perspectives*. Cambridge: Polity.
Cresswell, T. (1996) *In Place/Out of Place: Geography, Ideology and Transgression*. Minneapolis: University of Minnesota Press
Cresswell, T. (2002) 'Bourdieu's Geographies: in memorium', guest editorial, *Environment and Planning D: Society and Space* 20: 379–382.

Dowling, R. (1999) 'Classing the body', *Environment and Planning D: Society and Space* 17,5: 511–514.

Dovey, K. (2002) 'The silent complicity of architecture', in J. Hillier and E. Rooksby (eds) *Habitus: A Sense of Place*. Aldershot: Ashgate, pp. 267–280.

Fowler, B. (1997) *Pierre Bourdieu and Cultural Theory*. London: Sage.

Grenfell, M. and Kelly, M. (eds) (2001) *Pierre Bourdieu; Language, Culture and Education – Theory into Practice*. London: Peter Lang.

Guillory, J. (1993) *Cultural Capital*. Chicago: University of Chicago Press.

Hage G. (2000) *White Nation: Fantasies of White Supremacy in a Multicultural Nation*. London: Routledge.

Harker, R., Mahar, C. and Wilkes, C. (eds) (1990) *An Introduction to the Works of Pierre Bourdieu*. London: Macmillan.

Harvey, D. (1989) *The Condition of Post-modernity*. Oxford: Blackwell.

Hillier, J. and Rooksby, E. (eds) (2002) *Habitus: A Sense of Place*. Aldershot: Ashgate.

Honneth, A. (1986) 'The fragmented world of symbolic forms: Reflections on Pierre Bourdieu's sociology of culture', *Theory, Culture and Society* 3: 55–66.

Jager, M. (1986) 'Class distinction and the esthetics of gentrification: Victoriana in Melbourne', in N. Smith and P. Williams (eds) *Gentrification of the City*. London: Allen and Unwin, pp. 78–91.

Jenkins, R. (1992) *Pierre Bourdieu*. London: Routledge.

Leach, N. (2002) 'Belonging: Towards a theory of identification with space', in J. Hillier and E. Rooksby (eds) *Habitus: A Sense of Place*. Aldershot: Ashgate, pp. 281–299.

May, J. (1996) 'Globalisation and the politics of place and identity in an inner London neighbourhood', *Transactions of the Institute of British Geographers* 21: 194–215.

Merleau-Ponty, M. (1962) *Phenomenology of Perception*. Trans. C. Smith. London: Routledge.

Painter, J. (2000) 'Pierre Bourdieu', in M. Crang and N. Thrift (eds) *Thinking Space*. London: Routledge, pp. 239–259.

Painter, J. (2002) 'Governmentality and regional economic strategies', in J. Hillier and E. Rooksby (eds) *Habitus: A Sense of Place*. Aldershot: Ashgate, pp. 115–142.

Parkes, D. and Thrift, N. (1980) *Times, Spaces and Places: A Chronogeographic Perspective*. Chichester: Wiley.

Pile, S. (2002) 'Spectral cities: Where the repressed returns and other short stories', in J. Hillier and E. Rooksby (eds) *Habitus: A Sense of Place*. Aldershot: Ashgate, pp. 219–240.

Podmore, J. (1998) '(Re)reading the "loft living" habitus in Montreal's inner city', *International Journal of Urban and Regional Research* 22: 283–302.

Shusterman, R. (ed.) (1999) *Bourdieu: A Critical Reader*. Oxford: Blackwell.

Swartz, D. (1997) *Culture and Power*. Chicago: University of Chicago Press.

Thrift, N. (1996) *Spatial Formations*. London: Sage.

Wynne, D and O'Connor, J. (1998) 'Consumption and the post-modern city', *Urban Studies* 5–6: 841–864.

Gary Bridge

9 Judith Butler

Judith Butler was born in 1956. She attended Bennington College and then Yale
University, where she received her BA
and PhD in Philosophy. She taught at
Wesleyan and Johns Hopkins universities
before becoming Chancellor Professor of
Rhetoric and Comparative Literature at
the University of California at Berkeley.
Butler has written extensively on questions of identity politics, gender and sexuality, and is largely considered the
originator of modern queer theory, although she herself has resisted this label
for her work (see Gauntlett, 2002). Her
work has critiqued traditional feminist
theory for remaining within the confines
of a male/female binary and has argued
that gendered subjectivity needs to be
understood as a socially constituted and
context-dependent fluid performance.
Her key contributions include *Gender
Trouble* (1990), *Bodies that Matter* (1993)
and *Excitable Speech* (1997), among other
publications.

Although Judith Butler herself has
very little to say about space or place, her
ideas about performativity have been very
influential for a critical geography 'con
cerned to denaturalize taken-for-granted
social practices' (Gregson and Rose, 2000:
434). First, Butler's theorization of gender
has reshaped geographers' understandings of identities/bodies and their spatialities. Second, her notion of performativity
has been recast to theorize the concept of
space. Third, her work has influenced
critical geographers' engagement with
non-representational theory. Fourth, her
conceptualization of performativity has
upset feminist methodological debates
about reflexivity and positionality.

Trying to summarize *Gender Trouble*,
Butler's most influential work, is a challenging task, primarily because Butler
herself has commented on the difficulty
of its style and content (Butler, 1990:
xviii). However, although well acknowledged as a dense text, Butler's (1990)
analysis of gender ontologies forcefully
questions the naturalness of sex/gender
binaries. The essence of her argument is
that gender identities (masculinity and
femininity) are assumed to be grounded
in, and defined in relation to, two 'natural' biological sexes: male and female.
This binary opposition is also the basis of
heterosexual desire (i.e attraction is towards the opposite sex/gender). However,
in *Gender Trouble* Butler challenges the
assumptions that masculine and feminine
gender identities inevitably correspond
with 'male' and 'female' bodies (e.g. 'masculine' may just as well signify a female
body as a male one). Moreover, she argues that the two sexes themselves are
also social constructions so that there is
nothing 'natural' about everybody being
defined in terms of one sex or the other.
Rather than gender identities corresponding to a biological essence or articulating
an authentic or core self, Butler theorizes
them as an effect or 'performance'.

Butler makes this argument by drawing upon the work of linguistic philosopher J. L. Austin (1955) in discussing the
process of subject formation. She starts by
interrogating the category of 'woman' as
the subject of feminism, questioning de
Beauvoir's maxim that one is not born a
woman, but rather, becomes one (de Beauvoir, 1949). Butler (1990) believes that

any identity is the system of a logic of power and language that generates identities as a function of binary oppositions (i.e. self/other), and that seeks to conceal its own workings by making those identities appear to be natural. Notably, she theorizes identity as performative, arguing that:

> ... gender is the repeated stylisation of the body, a set of repeated acts within a highly rigid regulatory framework that congeal over time to produce the appearance of substance, of a natural sort of being ... The effect of gender ... hence, must be understood as the mundane way in which bodily gestures, movements, and styles of various kinds constitute the illusion of an abiding gendered self.
> (Butler, 1990: 33, 140)

In challenging the notion of biology as the bedrock that underlies social categories such as gender, Butler allows for the possibility of transformed gender roles outside the traditional cartographies of discursive patriarchal power relations. Most notably, she uses a reading of drag balls to argue that the parodic repetition and mimicry of heterosexual identities at these events disrupt dominant sex and gender identities because the performers' supposed 'natural' identities (as male) do not correspond with the signs produced within the performance (e.g. feminine body, language and dress). Thus '[B]y disrupting the assumed correspondence between a "real" interior and its surface markers (clothes, walk, hair etc.) drag balls make explicit the way in which *all* gender and sexual identifications are ritually performed in daily life' (Nelson, 1999: 339).

Butler (1990) insists that if we come to grips with the fact that gender is merely an inscription of discursive imperatives, that is to say, an elaborate, socially constructed fabrication, we open up the possibilities of displacing dominant discourses. She explains:

> Because gender is not a fact, the various acts of gender create the idea of gender,

and without those acts, there would be no gender at all. Gender, is thus, a construction that regularly conceals its genesis; the tacit collective agreement to perform, produce and sustain discrete and polar genders as cultural fictions is obscured by the credibility of those productions.
> (Butler, 1990: 140)

In *Bodies That Matter*, Butler (1993) clarifies some of the ideas outlined in *Gender Trouble*, and tackles many of the critiques that emerged from her early analysis of the sex/gender binary. In particular, she further analyses the relationship between performativity and the material body. Performativity is explained in more detail and is linked with the notion of citationality. Butler (1993: 18) herself represents *Bodies that Matter* as 'a poststructuralist rewriting of discursive performativity as it operates in the materialization of sex'. Indeed, Butler (1993: ix) explains to the reader that the book was conceived as a way to respond to the question, 'What about the materiality of the body, Judy?' Butler demonstrates how the unstable sexed body presents a challenge to the boundaries of discursive intelligibility. She also examines the racialization of gender norms by drawing from Nella Larsen's novel *Passing*.

Butler's ideas about sex as an effect rather than a cause is repeated here from *Gender Trouble*, but Butler plays with that argument to articulate how the supposedly 'natural' body can be a 'naturalized effect' of discourse (Salih, 2002). In exploring the limits of performativity, Butler is predominantly concerned with the question of what produces the effect of a stable core of gender. Relying upon a theory of the performative taken from speech act theory and involving ongoing actualization of gender meanings in the present, Butler argues that performativity is not a single or even deliberate act, but rather 'the reiterative and citational practice by which discourse produces the effects that it names' (Butler, 1993: 2).

Her next book, *Excitable Speech: Politics of the Performance* (1997), analysed name-calling or hate speech as both a social injury and the way in which individuals are called into action for political purposes and considered subject categories in the context of language. In this way, it is important to conceptualize Butler's musings as building blocks, where she returns and revisits themes of the sex/gender binary, performativity, citationality, speech theory and subjectivity, among other arenas. Notably, her own understanding of her work has shifted since she wrote *Gender Trouble*. She admits that she has dramatically revised some of her theoretical positionings since its publication, being compelled to revise her arguments because of increasing political engagements, which in turn forced her to reconceptualize the notion of the term 'universality':

> [In *Gender Trouble*] I tend to conceive of the claim of 'universality' in exclusive negative and exclusionary terms. However, I came to see the term has important strategic use precisely as a non-substantial and open ended category.
> (Butler, 1993: xvii)

Butler has also clarified what she means by a genuinely 'radical' act, insisting that 'we are never fully determined by the categories that construct us' and that one can dissent from the norms of society by 'occupying the very categories by which one is constituted and turning them in another direction or giving them a future they weren't supposed to have' (Butler, in Wallace, 1998: 16).

Butler's work has had a profound impact because the idea of gender as performative (an act, a style, a fabrication, involved in a dramatic construction of meaning) has challenged traditional thinking across the social sciences, including within geography.

SPATIAL CONTRIBUTIONS

Butler's notion of performativity has been drawn on by many geographers who have considered its ramifications for understanding the sexed body as it is lived and spatially constituted. This engagement with Butler has reflected the upsurge of interest in the body within the discipline. Callard (1998: 387) observes that: 'Geographers are now taking the problematic of corporeality seriously. "The body" is becoming a preoccupation in the geographical literature, and is a central figure around which to base political demands, social analyses and theoretical investigations.' For example, Butler's work has been used by Cream (1995) to think about how gender is inscribed on the space of the body; by Johnston (1995) to consider how women body-builders destabilize traditional notions of femininity; and by Bell and Valentine (1995) to theorize the transgressive pleasures of body modification. In the context of the workplace, both Crang (1994) and McDowell (1995, 1997) have shown how in restaurants and banking respectively, the performances of particular gender and sexual identities are integral to selling different types of 'interactive' service products. Some feminist geographers have also focused on psychic and emotional processes to examine the ways in which bodies themselves are imagined as spaces, and the spaces which they are imagined as inhabiting (Rose, 1993, 1996, 1999). Other geographers have drawn upon the language of performativity to talk about identities – for example, in terms of the ways in which hegemonic discourses inscribe nationalist identities (Sharp, 1996) and to examine the subject and spatial politics embedded in Cartesian maps (Kirby, 1996).

But perhaps the most sustained engagement by geographers with Butler's work has been in relation to the sexualization of space. In the inaugural issue of the feminist journal *Gender, Place and Culture*, Bell *et al.* (1994) took Butler's understanding that there is a potential for transgressive politics within the parodying of heterosexual constructs, to suggest that a similar argument can be made for the production of 'queer' space. Bell *et al.* (1994) draw on examples of the lipstick lesbian and gay skinhead to question the 'un-naturalness' of not just day-to-day heteronormative space, but also the masculine and feminine identities associated with their practice in those spaces (Bell *et al.*, 1994). In a similar vein, Lewis and Pile (1996), writing about the Rio Carnival, consider what effects bodily performances may have on understandings of sexuality and space. Thus, instead of thinking about space and place as pre-existing sites that occur, these studies have argued that bodily performances themselves constitute or (re)produce space.

However, these attempts to apply Butler's theory of performativity in specific everyday contexts have led to accusations that geographers have been too enthusiastic to engage with Butler and in process have often used the language of performativity uncritically and without regard to its limitations; or have misread and therefore misapplied Butler's ideas (e.g. Walker, 1995; Nelson, 1999; Gregson and Rose, 2000). In particular, Bell *et al.*'s (1994) paper sparked accusations that they had inserted the notion of choice and intentionality into their reading of the performance of lipstick lesbian and gay skinhead identities, thus presupposing human agency when Butler is at pains to deny that there is 'a *doer* behind the deed' (Nelson, 1999: 324) (these criticisms are also repeated of other geographers' attempts to deploy Butler's theorization of performativity). Walker, for example, explains:

> The problem with the rhetoric of intentionality that creeps into Bell *et al.*'s discussion of how we can self-consciously perform genders in subversive ways is not just that it falls back on philosophical

essentialisms about subjectivity. The problem is also that, as Butler argues, an account of subjectivity that relies too heavily on intentionality does not take into account how people are compelled and constrained by the very regulatory norms of gender identity that are the condition of our resistance. This means that many of us do not experience our gender identities as being very fluid or available to choice.
> (Walker, 1995: 76)

Subsequently, other geographers have sought to rework and elaborate on the implications of performativity for geographical understandings of space. Gregson and Rose (2000), for example, draw on two very different research projects, one with community arts workers and projects and another on car-boot sales as alternative spaces of consumption, to develop the complexity and instability of notion of performances and performed spaces.

As the limits of Butler's notion of performativity are interrogated, some geographers are linking it together with nonrepresentational theory (drawing heavily from the work of theorists such as Deleuze and Guttari) to develop a wider set of ideas about performativity and body-practices (Thrift, 2000; Dewsbury, 2000). Most notably, Thrift (1997: 125) uses dance, which he describes as 'a concentrated example of the expressive nature of embodiment', as a way to develop different ideas of the use of performance (see also Thrift, 2000).

Butler's ideas have been applied by geographers in relation to not only theoretical, but also methodological debates. In a review of feminist writing about positionality, Rose (1997) has critiqued feminist geographers' attempts to define their positionality and be reflexive about their relationships with informants. Rather, drawing on Butler, Rose (1997) argues that our positionings in relation to our interviewees are never *a priori*. Rather an interviewer and interviewee fashion a particular performance of self through their interaction and thus it is impossible to identify a transparent, knowable self in the research process (see also Valentine, 2002).

KEY ADVANCES AND CONTROVERSIES

Although hugely influential, Butler's work has also generated many critiques and some scepticism across the social sciences. Most famously, she has been criticized for her obscure and abstract writing style, which she is accused of using to create an 'aura of importance' (Nussbaum, 1999: 6). Indeed, an extract from *Gender Trouble* won an annual Bad Writing Contest sponsored by the journal *Philosophy and Literature*. Geographers in particular have insisted that there is a disturbingly metaphysical quality to Butler's writing, which is often biased in favour of the literary (Brown, 2000; Rose, 1996). Moreover, Butler has also been accused of making allusions throughout her writing to other theorists from very contradictory theoretical traditions (including Foucault, Lacan, Freud, Althusser, etc.) without any attempt to acknowledge, address or resolve these conflicts or incongruities (Nussbaum, 1999).

Butler's own theorization has also been attacked from a number of different perspectives, including its racial blindness. Race is only mentioned in a cursory fashion in discussions of performativity in both *Gender Trouble* and *Bodies That Matter* (Mahtani, 2002), although Butler is quick to note in the anniversary issue of *Gender Trouble* that she never meant to ignore race – rather she is much more interested in asking, 'what happens to [my theory of performativity] when it tries to come to grips with race' (Butler, 1990: xvi).

Another repeated criticism of Butler's theorization is that she deconstructs agency without effectively putting forth a useful alternative to humanist versions of the concept (Nelson, 1999). She is also challenged as to how collective notions of resistance (resistance is only described by Butler in terms of an individualized paraodic performances) and oppression can be conceptualized within her theory of performativity. For example, **Gillian Rose** (1996: 73), has insisted that performances never offer 'an escape from masculinist discourses', while Mahtani (2002) shares similar concerns in her analysis of the performances among 'mixed race' women, insisting that the total sum of the effects of performances are incalculable (Mahtani, 2002). This said, Mitch Rose (2002) has provided an important counterpart to these criticisms by drawing from Butler to effectively demonstrate that practices of both resistance and domination are enactments. Likewise, in developing an alternative reading of the 'place' of performativity in geography, Houston and Pulido (2002) suggest that contemplating performativity as a form of embodied dialectical praxis creates a space for progressive social change.

More generally, Butler has been charged with inspiring a turn within feminism away from the material (poverty, violence, illiteracy, etc.) towards a verbal and symbolic politics that has little connection with the everyday lives of 'real' women. Probyn (1995: 108), for example has cautioned against the celebration of Butler's work, reminding us about the 'nasty realities of the heterosexist and homosocial society where most of us continue to live and be gendered as either women or men'. Within geography, Butler's work has also been evoked as an example of the kind of abstract theorization that has induced a radical shift in the pattern of research and scholarship within 'new cultural geography', away from a concern with the material and the political towards a heightened concern with language, meanings and representation (see Barnett, 1998).

BUTLER'S MAJOR WORKS

Butler, J. (1990) *Gender Trouble*. London: Routledge.
Butler, J. (1992a) 'Contingent foundations: feminism and the question of "postmodernism"', in J. Butler and J. W. Scott (eds) *Feminists Theorize the Political*. New York: Routledge, pp. 3–22.
Butler, J. and Scott, J. W. (eds) (1992b) *Feminists Theorize the Political*. New York: Routledge.
Butler, J. (1992c) 'The body you want', *Artforum* November: 82–89.
Butler, J. (1993) *Bodies That Matter*. London: Routledge.
Benhabib, S., Butler, J., Cornell, D. and Fraser, N. (1995) *Feminist Contentions: A Philosophical Exchange*. London: Routledge.
Butler, J. (1997) *Excitable Speech*. London: Routledge.
Butler, J., Laclau, E. and Zizek, S. (2000) *Contingency, Hegemony, Universality: Contemporary Dialogues on the Left*. London: Verso.
Butler, J., Guillory, J. and Thomas, K. (2000) *What's Left of Theory? New Work on the Politics of Literary Theory*. London: Routledge.

Secondary Sources and References

Austin, J. (1955) *How to Do Things with Words*. Cambridge: Harvard University Press.
Awkward, M. (1995) *Negotiating Difference: Race, Gender and the Politics of Positionality*. Chicago: University of Chicago Press.
Barnett, C. (1998) 'The cultural turn: fashion or progress in human geography?', *Antipode* 30: 379–394.
Bell, D., Binnie, J., Cream, J. and Valentine, G. (1994) 'All hyped up and no place to go', *Gender, Place and Culture* 1: 31–48.
Bell, D. and Valentine, G. (1995) 'The sexed self: Strategies of performance, sites of resistance', in S. Pile and N. Thrift (eds) *Mapping the Subject*. London: Routledge, pp. 143–157.
Brown, M. (2000) *Closet Space: Geographies of Metaphor From the Body to the Globe*. London: Routledge.
Callard, F. (1998) 'The body in theory', *Environment and Planning D: Society and Space* 16: 387–400.
Crang, P. (1994) 'It's showtime: On the workplace geographies of display in a restaurant in southeast England', *Environment and Planning D: Society and Space* 12: 675–704.
Cream, J. (1995) 'Resolving riddles: The sexed body', in D. Bell and G. Valentine (eds) *Mapping Desire*. London: Routledge, pp. 31–40.
de Beauvoir, S. (1949) *The Second Sex*. London: Everyman.
Dewsbury, J. (2000) 'Performativity and the event: Enacting a philosophy of difference', *Environment and Planning D: Society and Space* 18: 473–496.
Gauntlett, D. (2002) *Media, Gender and Identity*. London: Routledge.
Gregson, N. and Rose, G. (2000) 'Taking Butler elsewhere: performativities, spatialities and subjectivities', *Environment and Planning D: Society and Space* 18: 433–452.
Hood-Williams, J. and C.W. Harrison (1998) 'Trouble with gender', *Sociological Review*, 46: 73–94.
Houston, D. and Pulido, L. (2002) 'The work of performativity: Staging social justice at the University of Southern California', *Environment and Planning D: Society and Space* 20: 379–504.
Johnston, L. (1995) 'The politics of the pump: Hard core gyms and women body builders', *New Zealand Geographer* 51: 16–29.
Kirby, K. (1996) 'Cartographic vision and the limits of politics', in N. Duncan (ed.) *Bodyspace: Destabilizing Geographies of Gender and Sexuality*. London: Routledge.
Lewis, C. and S. Pile (1996) 'Women, body, space: Rio Carnival and the politics of performance', *Gender, Place and Culture* 3: 453–472.
Longhurst, R. (2000) 'Corporeographies of pregnancy: Bikini babes', *Environment and Planning D: Society and Space* 19: 453–472.
Mahtani, M. (2002) 'Tricking the border guards: Performing race', *Environment and Planning D: Society and Space* 20: 425–440.
McDowell, L. (1995) 'Body work: heterosexual gender performances in city workplaces', in D. Bell and G. Valentine (eds) *Mapping Desire: Geographies of Sexualities*. London: Routledge.
McDowell, L. (1997) *Capital Culture: Gender At Work in the City*. Oxford: Blackwell.
Nelson, L. (1999) 'Bodies (and spaces) do matter: the limits of performativity', *Gender, Place and Culture* 6: 331–353.
Nussbaum, M. (1999) 'The professor of parody', *http://www.tnr.com/archive/0299/022299/nussbaum022299.html* (accessed 29/12/02).
Probyn, E. (1995) 'Lesbians in space: Gender, sex and the structure of missing', *Gender, Place and Culture* 2: 107–109.

Rose, G. (1993) *Feminism and Geography: The Limits of Geographical Knowledge.* Cambridge: Polity.

Rose, G. (1996) 'As if the mirrors had bled: masculine dwelling, masculinist theory and feminist masquerade', in N. Duncan (ed.) *BodySpace.* London: Routledge, pp. 56–74.

Rose, G. (1997) 'Situating knowledges: positionality, reflexivities and other tactics', *Progress in Human Geography* 21: 305–320.

Rose, G. (1999) 'Performing space', in D. Massey, J. Allen and P. Sarre (eds) *Human Geography Today.* Cambridge: Polity, pp. 247–259.

Rose, G. and Gregson, N. (2000) 'Taking Butler elsewhere: Performativities, spatialities and subjectivities', *Environment and Planning D: Society and Space* 18: 422–452.

Rose, M. (2002) 'The seduction of resistance: Power, politics, and a performative style of systems', *Environment and Planning D: Society and Space* 20: 383–400.

Salih, S. (2002) *Judith Butler.* London: Routledge.

Sharp, J. (1996) 'Gendering motherhood: a feminist engagement with national identity', in N. Duncan (ed.) *BodySpace.* London: Routledge.

Thrift, N. (1997) 'The still point: resistance, expressive embodiment and dance', in S. Pile and M. Keith (eds) *Geographies of Resistance.* London: Routledge, pp. 124–151.

Thrift, N. (2000) 'Afterwards', *Environment and Planning D: Society and Space* 18: 213–255.

Valentine, G. (2002) 'People like us: negotiating sameness and difference in the research process', in P. Moss (ed.) *Feminist Geography in Practice.* Oxford: Blackwell, pp. 116–126.

Wallace, J. (1998) 'What does it mean to be a woman?', *The Times Higher Education Supplement* 8 May 1998.

Walker, L. (1995) 'More than just skin-deep: Fem(me)ininity and the subversion of identity', *Gender, Place and Culture* 1: 71–77.

Minelle Mahtani

10 Manuel Castells

BIOGRAPHICAL DETAILS AND THEORETICAL CONTEXT

Manuel Castells was born in Spain in 1942. He grew up in Barcelona, where he completed his secondary education, going on to study law and economics at the University of Barcelona in 1958–62. As a student activist against General Franco's fascist dictatorship he had to escape to Paris, graduating from the Sorbonne's Faculty of Law and Economics in 1964. Staying on in Paris, he went on to obtain his PhD from the École des Hautes Etudes en Sciences Sociales in 1967, publishing his first article ('Mobilité des enterprises et structure urbaine') in the same year. Based on a statistical analysis of location strategies of high-tech industrial firms in the Paris region, this doctoral work (supervised by the renowned sociologist Alain Touraine) alerted to two issues that would continue to preoccupy Castells over the next three decades – namely, the emergence of new technologies and the changing form of cities. Working in Paris at this time unsurprisingly brought Castells into contact with leading Marxist theorists, including **Henri Lefebvre**, Nicolas Poulantzas and Louis Althusser; it was thus unsurprising that Castells was to be caught up in the revolutionary fervour of May 1968, with one of Castells' own students at Nanterre, Daniel Cohn-Bendit, a key figure in the student uprising. Expelled by the French government for his part in the uprising, Castells spent periods in Chile and Canada before returning to Paris in 1972 after receiving a state pardon (see Susser, 2002).

Castells first major work – *The Urban Question: A Marxist Approach* – was also published in 1972. Undoubtedly a product of the times, it was heralded as a remarkable and pioneering attempt to bring Marxist concepts and perspectives to bear on the 'urban question'. In essence, Castells suggested that many of those writing about the problems and challenges facing cities in the early 1970s (e.g. race riots, poverty, criminality) were locked into an ideologically bankrupt tradition of urban sociology that could not possibly identify the answers to these urban problems. This tradition had its roots in the influential 'urban ecology' school of sociologists at the University of Chicago in the 1920s, and was manifest in the 'factorial ecologies' produced by many sociologists and social geographers in the 1960s. Rejecting this approach, Castells drew on classical Marxist theory and emerging traditions in pluralist political science to outline an urban sociology that emphasized the conflictual production of urban space within capitalist society. In his 'Epistemological introduction' to the volume, Castells noted that it was born out of his astonishment that Marxist theorists had yet to analyse cities in a 'sufficiently specific way' (Castells, 1977: 2). In the event, the book was to have remarkably little influence on those working within the Marxist tradition; its chief legacy was to inspire a generation of geographers to engage with theories of political economy and to serve as a canonical text in the 'new urban sociology'. First published in English in 1977, *The Urban Question* thus stands alongside **David Harvey's** (1973) *Social Justice and the City* as a rallying call for urban researchers to utilize the insights of Marxist theory as a means to explore the urbanization of injustice.

A pivotal figure in the formation of the *International Journal of Urban and Regional Research*, Castells thus found himself at the vanguard of the new urban sociology. By this time, Castells had moved to the Department of City and Regional Planning at the University of California, Berkeley, where he was exposed to a more empirical tradition of sociology than that prevalent in Paris. This was reflected in *The City and the Grassroots* (1983), a comparative study of urban social movements and community organizations based on fieldwork in France, Spain, Latin America and California. This book extended the assertion developed in *The Urban Question* that one of the chief roles of cities in the capitalist-urban era is to provide collective consumption facilities (e.g. leisure, shopping and health facilities) designed to reproduce labour power. Noting the difficulties in providing equitable access to these facilities, *The City and the Grassroots* focused on the formation of protest groups and political movements seeking to improve access to such facilities. Through detailed case studies, Castells suggested that such groups could exploit the tension between labour and capital to transform capitalism. Significantly, Castells expanded his purview in this volume from exploring issues of class-consciousness to highlight the formation of social movements seeking the emancipation of gay and lesbian groups (e.g. one of his case studies explored the formation of gay political identity in San Francisco).

For some, this focus on agency signalled a turn away from Althusser-inspired Marxism towards Weberianism (Merrifield, 2002). Yet Castells' persistent interest in technology was to inspire a return to structural metaphors in a series of influential works on the new *informational* structures of capitalism. In 1989, Castells published *The Informational City*, an analysis of the urban and regional changes brought about by information technology and economic restructuring in the United States (highlighting changes in the San Francisco Bay area). This was a timely volume in so much as it high-

lighted changes in the nature of urban governance that were contributing to the polarized or 'dual' city where poor, immigrant workers 'serviced' a more affluent elite working in hi-tech and knowledge-rich industries. This served as precursor to his hugely ambitious 1400-page, three-volume treatise on *The Information Age: Economy, Society and Culture*, comprising *The Rise of the Network Society* (1996), *The Power of Identity* (1997) and the *End of Millennium* (1998). Integrating his observations on the nature of urban life in both Western and non-Western contexts with an overview of key changes in the organization of contemporary social, economic and political life, this work is in the tradition of works by Bell, Touraine, Polyani and other major social theorists given it seeks to identify a profound shift in the organization of society. Where it perhaps differs is in its attentiveness to the specificities of place. Coupled with *The Internet Galaxy – Reflections on Internet, Business and Society* (2001), Castells' trilogy on informational society has raised his profile significantly in academic, governmental and business circles (where he has been acclaimed as a 'globalisation guru').

SPATIAL CONTRIBUTIONS

Simplifying to the extreme, both Castells' earlier and later works have exercised considerable influence on the trajectory of geographical thought, albeit that it is possible to detect an important change in Castells' own conception of space and place between (for example) *The Urban Question* and *The Information Age*. In the former, Castells was scathing of those commentators who seemingly 'fetishised' the urban, bequeathing a distinct ecology that was somehow independent of capitalist structures. This included criticism of his contemporary, **Henri Lefebvre**, whom Castells alleged granted the city an autonomy and significance that it simply did not possess. Here, he bracketed Lefebvre with members of the Chicago School (especially Louis Wirth), arguing that he naively equated spatial propinquity with social emancipation 'as if there were no institutional organization outside arrangement of space' (Castells, 1977: 90). Developing this point, he outlined the need for a structural reading of the city:

> It is a question of going beyond the description of mechanisms of interactions between activities and locations, in order to discover the structural laws of the production and functioning of the spatial forms studied ... There is no specific theory of space, but quite simply a deployment and specification of the theory of social structure, in order to account for the characteristics of the particular social form, space, and its articulation with other historically given, forms and processes.
> (Castells, 1977: 124)

This structural solution to the 'urban question' thus offered a valuable corrective to the spatial determinism widely evident in urban studies at this time, whereby specific spaces were seen to dictate the lives of those who inhabited them. Yet, in offering this corrective, Castells seemingly went to the other extreme: space simply became a reflection of social process (hence, his claim that 'space, like time, is a physical quantity that tells us nothing about social relations' – Castells, 1977: 442). This is mirrored in Castells' definition of the city as 'a residential unit of labour power, a unit of collective consumption corresponding "more or less" to the daily organization of a section of labour power' (Castells, 1976: 148). In this sense, the city was interpreted as the outcome of the state's provision of collective means of consuming commodities, something that Castells felt could not be assured by capital but was nonetheless essential to the reproduction of capital.

In effect, Castells' radical take on the urban question shook up urban studies through its insistence that the social processes resulting in the production of the city were not distinctly urban, but endemic to capitalist society. Yet it subsequently attracted criticism from others working in the Marxist tradition for its tendency to treat space as a mere container for social relations (see especially Saunders, 1981, as well as **Ed Soja**, 1980, on the 'socio-spatial dialectic'). In fact, Castells was to later backtrack from this social determinism to suggest instead that 'space is not a reflection of society, it is society' (Castells, 1983). In his latter work on the relation between a new 'informational' mode of development and the social structures that constitute it, this was manifest in a focus on the 'transformation of socially and spatially based relationships of production into flows of information and power that articulate the new flexible system of production and management' (Castells, 1989). Stressing that social flows are inevitably also spatial flows, Castells thus offers a different take on the relations between society and space. Central here is his hypothesis that the 'informational society' is underpinned by a new socio-spatial logic – the 'space of flows' – that is truly global in scope:

> The space of flows refers to the technological and organizational possibility of organizing the simultaneity of social practices without geographical contiguity. Most dominant functions in our societies (financial markets, transnational production networks, media systems etc.) are organized around the space of flows ... However, the space of flows does include a territorial dimension, as it requires a technological infrastructure that operates from certain locations, and as it connects functions and people located in specific places.
> (Castells, 2000: 14)

Elucidating the morphology of this space of flows, Castells suggests that it consists of three prominent levels: the infrastructural, the organizational and the managerial. The first is the 'wired world', the hardware equipment linked to software that makes electronic transmissions around the world possible. The second consists of the centres of translation and organization that make the network society operate. In Castells' (2002b) account, it is prominent world cities (e.g. London, Tokyo, New York) that are the most direct illustration of organizational nodes (places where there is also an increasing divide between an informational elite and what Castells terms the 'fourth world' – those residents in world cities who are cut off from global networks of prestige, power and wealth). Castells in fact spends much time discussing the relationships between the infrastructural layer and this hierarchical layer of hubs and nodes, directing little attention to the third layer in a space of flows: the dominant managerial elites who are the key players in the global economy. The neglect of this third layer has led some to suggest that Castells offers a disembodied conceptualization of global process, generating a relatively thin account of identity formation (e.g. van Dijk, 1999; Bendle, 2002). Nonetheless, by hypothesizing the existence of three coterminous networks, Castells (2002a) powerfully demonstrates that space is the 'material support of time-sharing social practices', leading to an era where *timeless time* exists in tension with chronological time – and a space of flows exists in tension with a space of places.

KEY ADVANCES AND CONTROVERSIES

In much the same way that *The Urban Question* revolutionized urban sociology in the 1970s (by suggesting that urban sociologists should not be afraid to engage with Marxism), invocations of Castells' 'space of flows' idea have set new agendas for urban studies. Both empirically and

theoretically, the notion of flow has offered a valuable corrective to sedentary, static and bounded notions of urban process, instead positing that cities are characterized and defined by the flows that pass through them (see Doel et al., 2002). This interpretation is wholly in keeping with the turn in the social sciences towards a 'sociology of fluids' that emphasizes the movement and mobility of social life (see especially Urry, 2000). Likewise, it engages with debates emerging from the globalization literature, not least the idea that globalization is increasingly responsible for disseminating a standardized repertoire of consumer goods, images and lifestyles worldwide. Indeed, one implication of Castells' 'space of flows' idea is that 'local' ways of life are being undermined by the (network) logic of global capital accumulation as place is annihilated by space. In his summation, this means that the world of places – consisting of bounded and meaningful places such as the home, city, region, or nation-state – is being superseded by spaces characterized by circulation, velocity and flow. Elaborating, Castells (1996: 350) points to the proliferation of serialized ahistorical and acultural building projects that undermine the 'meaningful relationship between society and architecture', citing examples ranging from international hotels, airports and supermarkets through to the 'postmodern' office buildings that punctuate the skylines of world cities.

Superficially, there is certainly much evidence to support the idea that a space of flows is supplanting the world of places: take any city pivotal to the articulation of global financial flow, and one can find many sites that exhibit the architectural anonymity. Yet it is also clear that the space of places has not disappeared with the coming of network society. World cities, for instance, might be considered to be nodes in a network but they are also distinctive places. In this sense, they are not just strategic hubs in the global system, disembedded from their region and caught up in the 'space of flows': they are also distinctive 'centres of comprehension' whose pivotal role in the global economy is a function of their social and cultural milieu. For example, the City of London is far from simply being a 'gateway' space or mere hub in networks of global capital, remaining a strongly identifiable place despite its openness to the world (see also Thrift, 1996, on the distinctive social and cultural milieu of London). This type of observation undermines Castells' (1996: 200) assertion that 'a place is a locale whose form, function and meaning are self-contained within the boundaries of physical contiguity' (see also Taylor, 1999, on the distinction of space/place). On a different scale, Crang (2002) insists that many of the spaces Castells cites as non-places (malls, airports, hotels, etc.) are not simply places of homogenized commodity exchange, rationalization and flow translation: they are also emotionally charged places of desire and disgust, inclusion and exclusion, sociality and familiarity.

Notwithstanding these critiques, identifying the network as the organizing principle of the contemporary (informational) society has offered geographers and urban researchers a useful conceptual signpost towards a renewed urban sociology (see Friedman, 2000). In Castells' estimation, this is a sociology that needs to explore new patterns of communication, 'both face-to-face and electronic', and the interaction between 'physical layouts, social organisation and electronic networks' (Castells, 2002a: 399–400). He further insists that this

focus will help tackle 'lingering questions of urban poverty, racial and social discrimination, and social exclusion', contrasting this with the 'futile exercises of deconstruction and reconstruction' characteristic of postmodern urban theory (Castells, 2002a: 404). However, Castells' cavalier dismissal of much contemporary urban theory exposes the structuralist assumptions that continue to underpin (and inevitably constrain) his work. First and foremost, he devotes insufficient attention to how structure is established: it is taken as a given rather than as an ongoing achievement (e.g. he claims that 'societies are organized around human processes structured by historically determined relationships of production, experience and power' – Castells, 1989: 7–8). Moreover, while the concept of the network society may facilitate an increased understanding of some of the consequences of the development of a globalized informational capitalism, some have argued it is not analytically effective for exploring 'the range of risks and threats to which social and political life, and capitalist economic activity itself, have become exposed as a consequence of globalization' (Smart, 2000: 64). Highlighting this, van Dijk (1999) argues that Castells explores the social conflicts between global networks and collective movements, but makes little of the social conflicts that destabilize networks from within. Ultimately, it appears that the network society that Castells evokes so brilliantly is perhaps less structured than he might imagine.

CASTELLS' MAJOR WORKS

Castells, M. (1977) *The Urban Question: A Marxist Approach*. London: Edward Arnold.
Castells, M. (1983) *The City and the Grassroots: A Cross-Cultural Theory of Urban Social Movements*. Berkeley: University of California Press.
Castells, M. (1989) *The Informational City – Information Technology, Economic Restructuring and the Urban-regional Process*. Oxford: Blackwell.
Castells, M. and Hall, P. (1994) *Technopoles of the World: The Making of 21st Century Industrial Complexes*. London: Routledge.
Castells, M. (1996) *The Information Age: Economy, Society and Culture; Volume 1: The Rise of the Network Society*. Oxford: Blackwell.

Castells, M. (1997) *The Information Age: Economy, Society and Culture; Volume 2: The Power of Identity*. Oxford: Blackwell.
Castells, M. (1998) *The Information Age: Economy, Society and Culture; Volume 3: End of Millennium*. Oxford: Blackwell.
Castells, M. (2001) *The Internet Galaxy: Reflections on the Internet, Business, and Society*. Oxford: Oxford University Press.

Secondary Sources and References

Bendle, M. (2002) 'The crisis of identity in high modernity', *British Journal of Sociology* 53 (1): 1–18.
Bromley, S. (1999) 'The Space of Flows and Timeless Time: Manuel Castells's "The Information Age" ', *Radical Philosophy* 97: 6–10.
Castells, M. (2000) 'Materials for an exploratory theory of the Network society', *British Journal of Sociology* 51 (1): 1–24.
Castells, M. (2002a) 'Urban sociology for the twenty-first century', in I. Susser (ed.) *The Castells Reader on Cities and Social Theory*. Oxford: Blackwell, pp. 390–406.
Castells, M. (2002b) 'Local and global: cities in the network society', *Tijdschrift voor Economische en Sociale Geografie* 93 (5): 548–558.
Crang, M. (2002) 'Between places: producing hubs, flows and networks', *Environment and Planning A* 34 (4): 569–574.
Doel, M., Beaverstock, J., Taylor, P. and Hubbard, P. (2002) 'Attending to the world: co-efficiency and collaboration in the world city network', *Global Networks* 2 (2): 96–116.
Friedman, J. (2000) 'Reading Castells: *Zeitdiagnose* and social theory', *Environment and Planning D: Society and Space* 18 (1): 111–120.
Harvey, D. (1973) *Social Justice and the City*. London: Arnold.
Merrifield, A. (2002) *Metromarxism: A Marxist Tale of the City*. New York: Routledge.
Saunders, P. (1981) *Social Theory and the Urban Question*. London: Hutchinson.
Smart, B. (2000) 'A political economy of new times? Critical reflections on the network society and the ethos of informational capitalism', *European Journal of Social Theory* 3 (1) 51–65.
Soja, E. (1980) 'The socio-spatial dialectic', *Annals, Association of American Geographers* 70 (2): 207–225.
Susser, I. (2002) 'Manuel Castells: conceptualising the city in the information age', in I. Susser (ed.) *The Castells Reader on Cities and Social Theory*. Oxford: Blackwell, pp. 1–12.
Taylor, P. (1999) 'Places, spaces and Macy's: place–space tensions in the political geography of modernities', *Progress in Human Geography* 23 (1): 7–26.
Thrift, N. (1994b) 'On the social and cultural determinants of international financial centres', in S. Corbridge, R. Martin and N. Thrift (eds) *Money, Power and Space*. Oxford: Blackwell, pp. 327–325.
Urry, J. (2000) *Sociology beyond Societies: Mobilities for the Twenty-First Century*. London: Routledge.
Van Dijk, (1999) 'The one-dimensional network society of Manuel Castells', *New Media and Society* 1 (1): 127–138.

Phil Hubbard

11 Stuart E. Corbridge

BIOGRAPHICAL DETAILS AND THEORETICAL CONTEXT

Stuart Corbridge was born in 1957 and grew up in the English West Midlands, before gaining a place at University of Cambridge to read Geography at Sidney Sussex College, graduating in 1978. At Cambridge, Corbridge's earliest influence was **Derek Gregory**'s theoretically inspired historical geography, and he also benefited from the reworking of key ideas in development studies by other Cambridge academics, among them Polly Hill, Suzy Paine and Ajit Singh. His PhD, supervised by Bertram Hughes Farmer, combined fieldwork with archival study of a century of tribal politics in Jharkhand, India, paying particular attention to the influence of positive discrimination polices for people labelled as 'tribal' (the *adivasis*). Corbridge began publishing his Indian research in 1982, and his interest in eastern Indian politics and development issues has been sustained over 20 years, moving between questions of agrarian change, development policy and governance. Other research interests, notably his extensive contributions to geopolitics and development theory, have held sway at certain times. These were often linked to collaborations with colleagues in his successive university positions. Work took him from a lectureship in geography at Huddersfield Polytechnic (UK), then to Royal Holloway, London University and Syracuse University in the USA, before returning to Cambridge as a lecturer in South Asian geography (1988–2000). In 2000 he took up professorships in multi-disciplinary departments at the University of Miami, USA, and the London School of Economics.

SPATIAL CONTRIBUTIONS

In Anglophone geography, Corbridge is best known for his sustained analysis of the development process, money and geopolitics. His first book, *Capitalist World Development* (1986), was a critical examination of radical perspectives in development thinking, particularly the then-common view among radical theorists wedded to 'deterministic models of capitalism' (p. 10) that theorized an inevitable conflict of interest between 'metropolitan capitalism and the development of the periphery of the modern world system' (p. 3). Corbridge was at an early stage of his thinking on these issues, but nonetheless challenged the theory of underdevelopment as well as the counterveiling optimism of modernization theorists. His model of development steered a middle course that could best be described as 'cautiously optimistic'. Subsequently he has gone on to write widely on the potential for liberatory development theory and its moral basis (Corbridge, 1993c), rethinking the colonial experience, the dilemmas faced by postcolonial states in the geopolitics of the Cold War, and debt and the financial system (Corbridge, 1993a; Corbridge *et al.*, 1994). Some, but not all, of this work is phrased in the language of post-Marxism and even post-Keynesianism (Corbridge, 1988,

1990, 1994), while his work on India has recently drawn on the work of Chatterjee and Kaviraj, as well as that of **Foucault** and Gramsci. He argues frequently, as in his second book, *Debt and Development* (1993a), that the evident contradictions of global capitalism do not in themselves make the case for transcendental forms of politics that would seek to iron out what John Toye (1993) has called the 'dilemmas of development'. The idea of development is not, therefore, dismissed out of hand, and indeed 'there is ... a strong case for a massive expansion in aid budgets to help rescue people from levels of absolute poverty not of their own making' (Agnew and Corbridge, 1995: 216), even though important political battles have to be waged around development's meanings and practices. Corbridge's preference for talking about capitalist relations of production and their conditions of existence, and his general support for aid flows, sets his work apart from his Marxist critics, and particularly from the post-development theorists like **Arturo Escobar** who challenge overtly the hegemonic power of Western development discourse (Corbridge, 1998b).

The themes raised in *Debt and Development* were followed with further work on international debt and monetary policy. For example, Corbridge's work on inflation follows a Keynesian argument about the need for economic pragmatism and 'rigorous eclecticism' in monetary policy (1994a: 88), and he engages with the geopolitics of monetary transfer and regulation in *Money, Power and Space,* which also traces the imbrications of money with social and cultural networks of power (Corbridge *et al.,* 1994). This collection was soon followed by a book with John Agnew, *Mastering Space* (1995), which is an overview of the global political economy of the past 200 years. In a development of his earlier position, the authors argue that:

> Globalisation is not only a synonym of disempowerment: it creates certain conditions for democratization, de-central-

ization and empowerment as well as for centralization and standardization. Globalization opens as many doors as it shuts. (Agnew and Corbridge, 1995: 219)

As **Gearóid Ó Tuathail**(1995) argues, the book demonstrated this by some deft applications of **Henri Lefebvre**'s arguments, to distinguish between flows of goods and power, the discursive representations that sustain these flows, and the 'imagined geographies' that 'inspire the future organization and articulation of spatial practices and representations of space'.

This nuanced and critical approach to geopolitics suggests the existence of three 'geopolitical orders' over time, each operating with distinct arrays of hegemonic authority. The period 1815–1975, termed the *Concert of Europe,* gave way to 70 years of *Inter-Imperial Rivalry* until 1945, followed by the *Cold War* from 1945 to 1990. The book argues that, as of the mid-1990s, the world order was missing a dominant nation state, and thus 'there is always hegemony, but there are not always hegemons' (Agnew and Corbridge, 1995: 17). The argument here is prophetic: hegemony, used in a Gramscian sense to describe structures that legitimate dominant practices and organize consent, is creating:

> new conditions for 'ordered disorder' by ignoring, tidying away and/or disciplining a group of countries, regions and communities which are not party to a new regime of market-access economics or which threaten it in some way.
> (Agnew and Corbridge, 1995: 193)

Agnew and Corbridge argue that transnational liberalism has emerged as hegemonic, through the breakdown of Keynesian economic policy and the Bretton Woods agreements, as well as through the power of new transnational business and military networks. Consistent with Corbridge's stand on globalization more generally, *Mastering Space* also argues that opposition to hegemonic discourses and practices necessarily accompanies their growth and their increased spatial reach.

The millennial anti-globalization movements and protests were anticipated in
the book, but perhaps not the turns that
some have taken, for example through
Islamic militancy, nor the geopolitical re-
percussions of the post-9/11 events.

A second major contribution is Corbridge's detailed and sustained interrogation of Indian development as an idea and
practice. Particularly since the early
1990s, he has returned to studies of rural
issues in the eastern Indian states of
Bihar, Jharkhand, Orissa and West Bengal. Alongside new interpretations of
ethno-regional politics in Jharkhand, he
has worked on forest citizenship and forest management, the collection and marketing of non-timber forest products, the
impacts of development policy, the politics of compensatory discrimination, and
the question of how empowerment and
poverty are negotiated in the modern
Indian state. His book with John Harriss,
Reinventing India (2000), analyses economic liberalization and Hindu nationalism
as 'elite revolts' that resisted a discourse
of egalitarianism in India's early post-war
years. Divisive religious and economic
practices have replaced the sense of na-
tionhood that guided India at independence and Partition. India's lack of an
effective 'developmental state' is traced to
the failure of national leaders to assert
political control (particularly under Indira
Gandhi), or to slow the rise of heavily
bureaucratic planning. Echoing the argument developed in *Mastering Space*, the
depressing and violent 'centralising in-
stincts' of Hindu nationalism will, it is
suggested, continue to clash with the
aspirations and social movements of
lower castes and subaltern groupings. A
major empirical study of state performance and empowerment issues has recently been completed in five districts of Bihar
and West Bengal, working with geographers Glyn Williams, Manoj Srivastava
and Rene Veron, based on substantial
fieldwork and hundreds of detailed interviews. The project argues that the upward
accountability of the local state (especially
to political parties) is as important as

downward accountability to communities
in this Indian context. Participatory de-
centralization is not a panacea, because
local actors can develop networks of corruption implicating civil society and local
state officials (Corbridge *et al.*, 2003b).
Corbridge has also worked with Sanjay
Kumar on questions of participation and
social capital (Kumar and Corbridge,
2002), and on the links between community, corruption and landscape in Jharkhand (Corbridge and Kumar, 2002).

Corbridge believes that geographical
research and teaching must draw upon,
and contribute to, social sciences and
international studies more broadly, and
he has made significant contributions to
this interface. But while he signals the
importance of spatiality, geography and
the 'power' of space in explanations of
geopolitical change and development, he
does not privilege them. 'Development
geography', and concepts of space and
place that it holds dear, must form part of
ecumenical analysis and broader debate
across a range of disciplines. His conviction that geography forms part of a
broader intellectual canvas has attracted
him to interdisciplinary work in development studies. Aside from the substantive
contributions mentioned above, Corbridge has completed a substantial essay
on the life and work of **Amaryta Sen**
(2002c), edited a *Reader in Development
Studies* (1995), and published a six-volume
reference collection of readings entitled
Development: Critical Concepts (2000a) that
spans the entire range of historical and
contemporary key works in the field.

KEY ADVANCES AND CONTROVERSIES

In the 1980s, Corbridge's work on global
capitalism and the world economy came
to attention at a time when radical geographers like Richard Peet, **Neil Smith** and

David Harvey dominated the field. Corbridge was uneasy with the core tenets and political ramifications of the Marxist geography of the day, and countered with post-Marxist critiques of determinism, some associated with regulation theorists like Alain Lipietz. As a result, *Capitalist World Development* attracted several comradely but also vituperative ripostes, notably from **Michael Watts** (1990), who argued articulately that Marxism offers more to the study of development than Corbridge permits it. In the same vein, Michael Johns (1990: 180) accused him of wielding a 'rather dull polemical axe'.

With these debates now more than a decade old, and with several of the protagonists occasionally now writing or working together (e.g. Corbridge, Thrift and Martin, 1994), Corbridge's arguments have become broadly accepted in development geography, while he himself has moved between post-Marxism and a critical stance on the style and substance of some mainstream development policies. Arguably (and unlike some of his protagonists), his work on the idea and the practice of 'development' is based on many years of grounded field research projects, and this enables him to speak with some authority when questioning its core values and its outcomes in particular places, and when challenging others coming from different viewpoints (notably **David Harvey**; see Corbridge, 1998a).

Corbridge's work has been taken up most directly by several geographers and former students who have worked on India, including Emma Mawdsley and Sarah Jewitt. Since the 1980s, there have been relatively few sustained criticisms of his work in the two domains of world development/geopolitics and Indian development. *Mastering Space* has made an impact among geographers and international relations scholars (it is heavy on criticism of the latter discipline). In his review, **Ó Tuathail** (1995) felt the book steered away from some of the more deconstructive ambitions of critical geopolitics, but a more substantive critique

concerns the validity of the concept of 'hegemony without hegemons', and whether the term ' hegemony' masks as much as it reveals, by glossing over significant differences in the aspirations and development paths of nation-states lumped together in the new neo-liberal world order (Sidaway, 1997). These doubts about the absence of hegemonic nations have emerged with renewed force in the early 2000s, as a result of the USA's aggressive foreign policy, its questionable claims to be acting multilaterally in the 'war' on terrorism, and its renewed efforts to assert a world order using both military strength and a revival of Cold War language and discourse. Could this be heralding a new set of hegemonic practices? In addition, Agnew and Corbridge downplay the possibility of severe economic crisis emerging in the international political economy and the real possibilities of catastrophic violence occurring on a global scale. In their framing of the hegemonic order, Toal feels, the power of 'institutions to regulate the power of dictators, ethnic cleansers or fundamentalism is not evident'. Again, viewed from the early 2000s, these powers have now shown their colours.

Reinventing India (Corbridge and Harriss, 2000) shares some theories and concepts with *Mastering Space*, but advances an innovative thesis when it suggests that a new India is being 'invented' through elite interests that serve the ends of particular classes (Corbridge also made this argument in his account of the politics of India's nuclear bomb – 1999), particularly since economic liberalization after 1991 and the rise of Hindu nationalism. While Singh (2001) uncharitably accuses the authors of undue attachment to concepts of class-based politics, and Chari (2002) feels that the Marxist and Gramscian approach to social change in *Reinventing India* gives scant attention to gender politics and feminist moments in current economic and political struggles, the book explains current political manoeuvres and discourses in a rich language. The concept of failed social revolutions led by elites ('elite revolts') goes a long way to explain India's particular crisis of nationalism and violence (Hall, 2002), and shows how good intentions turn to bad.

In conclusion, Corbridge's work to date rests on several major contributions. His work on post-Marxist development theory, debt and the transitions experienced in rural India has anticipated some of the current thinking on globalization made by eminent international relations theorists and sociologists like **Anthony Giddens**, David Held and Fred Halliday, as well as contributing to rethinking these areas in geography. The links between the Indian state and its citizens has been reconceived, and fleshed out with 20 years' of detailed study. The notion of 'hegemony without hegemons' has contributed to the emerging field of critical geopolitics (Dodds, 2001), although in the conflict-ridden 2000s we are now hearing strident calls from the new hegemon (the USA) to aggressively assert the 'rightness'

of democracy across the boundaries of the nation-state. However, as Corbridge and Harriss (2000) rightly note, the neo-liberal world economy has encountered resistance, for example from the subaltern spaces and resistance to 'elite revolts', eloquently described in *Reinventing India*. Such movements always accompany these hegemonic forces, even if they sometimes lack power. Corbridge shares with **Castells**, **Watts** and **Escobar** an interest in the promotion of alternatives to mainstream development, but some of these alternatives have unfortunately emerged as aggressively fundamentalist or nationalist in their own right (and thus are unpleasant to the sensibilities of Western activists and scholars), and not all have been able to challenge the state or the market with sufficient force. As other biographies in this volume show, development geography will have to remain attentive to the nuances of globalization, and resistant to it.

CORBRIDGE'S MAJOR WORKS

Agnew, J. and Corbridge, S. E. (1995) *Mastering Space: Hegemony, Territory and International Political Economy*. London: Routledge.
Corbridge, S. E. (1986) *Capitalist World Development: A Critique of Radical Development Geography*. London: Macmillan.
Corbridge, S. E. (1990) 'Post-Marxism and development studies: beyond the impasse', *World Development* 18 (5): 623–639.
Corbridge, S. E. (1993a) *Debt and Development*. Oxford: Blackwell.
Corbridge, S. E., Thrift, N. and Martin, R. (eds) (1994) *Money, Power and Space*. Oxford: Blackwell.
Corbridge, S. E. (ed.) (1995) *Development Studies: A Reader*. London: Edward Arnold.
Corbridge, S. E. (ed.) (2000a) *Development: Critical Concepts in the Social Sciences*. London: Routledge [6 volumes].
Corbridge S. E., Jewitt, S. and Kumar, S. (2003a) *Jharkhand: Environment, Development, Politics*. Delhi: Oxford University Press.

Secondary Sources and References

Chari, S. (2002) 'Review of Reinventing India', *Annals of the Association of American Geographers* 92 (2): 349–351.
Corbridge, S. E. (1988) 'The ideology of tribal economy and society: politics in Jharkhand, 1950–1980', *Modern Asian Studies* 22 (1): 1–41.
Corbridge, S. E. (ed.) (1993b) *World Economy*. Oxford: Oxford University Press.
Corbridge, S. E. (1993c) 'Marxisms, modernities and moralities: development praxis and the claims of distant strangers', *Environment and Planning D: Society and Space* 11: 449–472.
Corbridge, S. E. (1994a) 'Plausible worlds: Friedman, Keynes and the geography of inflation', in S. E. Corbridge, N. Thrift and R. Martin (eds) *Money, Power and Space*. Oxford: Blackwell, pp. 63–90.

Corbridge, S. E. (1994b) 'Bretton Woods revisited: hegemony, stability and territory', *Environment and Planning A* 26 (12): 1829–1859.

Corbridge, S. E. (1998a) 'Reading David Harvey: entries, voices, loyalties', *Antipode* 30: 43–55.

Corbridge, S. E. (1998b) ' "Beneath the pavement only soil". The poverty of post-development', *Journal of Development Studies* 34: 138–48.

Corbridge, S. E. (1999) ' "The militarization of all Hindudom"? The Bharatiya Janata Party, the bomb and the political spaces of Hindu nationalism', *Economy and Society* 28: 222–255.

Corbridge, S. E. (2000b) 'Competing inequalities: the Scheduled Tribes and the reservations system in India's Jharkhand', *Journal of Asian Studies* 59 (1): 62–85.

Corbridge S. E. (2002c) Development as freedom: the spaces of Amartya Sen. *Progress in Development Studies* 2: 183–217.

Corbridge, S. E. and Harriss, J. (2000). *Reinventing India: Liberalization, Hindu Nationalism and Popular Democracy*. Cambridge: Polity, and Delhi: Oxford University Press.

Corbridge, S. E. and Kumar, S. (2002) 'Community, corruption, landscape: tales from the tree trade', *Political Geography* 21: 765–788.

Corbridge, S., Veron, R., Srivastava and M., Williams, G. (2003b) 'Participation, poverty and power', *Development and Change* 34: 163–192.

Dodds, K. (2001) 'Political geography III: critical geopolitics after ten years', *Progress in Human Geography* 25 (3): 469–484.

Hall, J. (2002) 'Review of Reinventing India', *Canadian Journal of Sociology Online* July–August. http://www.arts.ualberta.ca/cjscopy/reviews/india.html

Johns, M. (1990) 'Review of Capitalist World Development', *Antipode* 22 (2): 168–174.

Kumar, S. and Corbridge, S. E. (2002) 'Programmed to fail? Development projects and the politics of participation', *Journal of Development Studies* 39 (2): 73–103.

Muralidharan, S. (2001) 'Review of Reinventing India', *Frontline* 18 (19) online http://www.frontlineonnet.com/fl1819/18190730.htm

Sidaway, J. D. (1997) 'Review of Mastering Space', *Transactions of the Institute of British Geographers* NS 22: 130–132.

Singh G. (2001) 'Review of Reinventing India', *Journal of Development Studies* 38 (1).

Toal, G. (1995) 'Political Geography I: Theorizing history, gender and world order amidst crises of global governance', *Progress in Human Geography* 19: 260–272.

Toye, J. (1993) *Dilemmas of Development: Reflections on the Counter-Revolution in Development Theory and Policy*, 2nd edition. Oxford: Blackwell.

Watts, M.J. (1990) 'Deconstructing determinism: Marxism's development theory and a comradely critique of Stuart Corbridge', *Antipode* 20 (2): 142–170.

Simon Batterbury

12 Denis Cosgrove

In the course of an academic career spanning more than three decades, Denis Cosgrove has done perhaps more than any other Anglophone academic geographer to ensure that 'landscape' continues to occupy a central place on the disciplinary 'map'. His work on the idea of landscape, which came to the fore in the 1980s especially, profoundly altered the course of landscape study in human geography through a series of theoretically informed yet empirically grounded books and articles (Cosgrove, 1984/1998a, 1985; Cosgrove and Daniels, 1988). In 1993 he co-founded a new journal – *Ecumene: a geographical journal of environment, culture and meaning* (later *Cultural Geographies*) – providing a forum for research within the humanities' tradition of geographical scholarship. Meanwhile, in the 1990s, Cosgrove's interests moved towards broader issues of 'mapping' and representation, and his work continued to be characterized by its close attention to empirical – and particularly historical – detail while continuing to engage with theoretical ideas from disciplines such as cultural studies, landscape architecture, cartographic history and creative art (Cosgrove, 1996, 1999a, 2001a, 2001b). This blending of empirical and theoretical material is a hallmark of Cosgrove's work, and it is this that has ensured that his writing has had considerable impact on contemporary geographical thought and enquiry despite the focus of much of his work on the historical geographies of Renaissance and Enlightenment Europe.

Cosgrove was born in 1948, and educated in Liverpool at St Francis Xavier College and then St Catherine's College, Oxford. He established his reputation as a leading practitioner of the 'new' cultural geography while in the Geography Division of Oxford Polytechnic (now Oxford Brookes University), where he served as Chairman (*sic*) from 1976 to 1980, and while at the Department of Geography of Loughborough University (from 1980 to 1993). Alongside those humanistic geographers in the UK and US who were concerned about the positivist and quantitative basis of much of human geography, Cosgrove sought to recognize the subjectivity of geographical enquiry, and to situate geography within a broader humanities tradition that acknowledged both charisma and context (Cosgrove, 1989). This concern with 'authorship' and 'authority' in geographical writing has arguably informed Cosgrove's subsequent scholarship (Cosgrove and Domosh, 1993), and while few contemporary geographers might refer to themselves as 'humanistic', the contributions made by Cosgrove to theoretical debates in the early 1980s about the writing and representation of space have endured (see Cosgrove, 2001b). One important reason for this has been Cosgrove's own enthusiasm to engage with postmodern and poststructural thought, marking him out as one of the most important sources of inspiration for those geographers seeking to interrogate the imbrication of knowledge and power. From 1999 resident in the US as Alexander von Humboldt Professor of Geography at UCLA, he continues in his role as a mediator between geography, the humanities and arts, developing further empirical and theoretical insights in cultural and historical geography that will no doubt influence future human geography more widely.

SPATIAL CONTRIBUTIONS

One of Cosgrove's key contributions has been to engage with the geographical concepts of space and place through the motif of *landscape*. Within the context of cultural geography in the UK, Cosgrove's engagement with ideas of writing and reading landscape drew upon the ideas of North American geographers such as Carl Sauer, J. B. Jackson and **Yi-Fu Tuan**, among others (see Cosgrove, 1998b). Taking one of the concepts that was central to the North American geographical imagination, Cosgrove nonetheless departed from the North American 'Berkeley' cultural tradition by exploring not just the cultural processes that shaped landscape but also the constitutive role that landscape plays in shaping those who engage with that landscape, or landscapes. In the early 1980s this was a new and radical way of viewing landscape, for up to that point UK historical geographers were broadly concerned with 'reconstructing' past landscapes from maps, documents and fieldwork in a tradition derived from the work of W. G. Hoskins (as well as M. W. Beresford and M. R. G. Conzen) (see Muir, 1999; Whitehand, 2001). Cosgrove offered a different approach to landscape, one that sought to recognize that 'landscape is not merely the world we see, it is a construction, a composition of the world'; in short, 'an ideological concept' (Cosgrove, 1984: 13, 15). Cosgrove set out this thesis in his first book, *Social Formation and Symbolic Landscape*, introducing the idea that 'landscape is a way of seeing' – a means by which 'Europeans have represented to themselves and to others the world about them and their relationships with it' (Cosgrove, 1984: 1). Looking back on this work in the 'introduction' to the book's second edition, Cosgrove reflects on its

> prime intention ... to press landscape studies, especially in Geography, towards what seemed to be specific new directions: to locate landscape interpretation within a critical historiography, to theorize the idea of landscape within a broadly Marxian understanding of culture and society, and thus to extend the treatment of landscape beyond what seemed ... a prevailing narrow focus on design and taste.
> (Cosgrove, 1998b: xiii)

The book certainly did this, and at its heart was Cosgrove's interest in Renaissance Italy. Here he found the roots of the 'landscape idea' in Western culture, which was (like the book itself) a humanistic enterprise concerned with new ways of imagining and representing landscape, of 'seeing' the land.

Cosgrove's landscape 'as a way of seeing' idea became more influential in geographical discourse following the publication of his paper, 'Prospect, perspective and the evolution of the landscape idea' (Cosgrove, 1985), and the publication of a collection of papers under the title *The Iconography of Landscape*, edited by Cosgrove and Daniels (1988). In both, Cosgrove reinforced the message of *Social Formation*, that landscape was 'a "way of seeing" that was bourgeois, individualist and related to the exercise of power' (Cosgrove, 1985: 45). Simultaneously, he underlined that this was connected with Renaissance humanism and scientific advances in mapping, projection, surveying and cosmology. As a 'way of seeing', landscape is thus 'a composition and structuring of the world' that 'may be appropriated by a detached, individual spectator to whom an illusion of order and control is offered through the composition of space according to the certainties of geometry' (Cosgrove, 1985: 55). 'Landscape' was property, owned by those beholding it; capturing and controlling the land through representations of it as landscape in maps and in paintings – and through fashioning landscapes on the ground using design and architecture. The landscape, then, far from being neutral and inert, has social and cultural meanings, a symbolism – an 'iconography'. It was this 'iconography' of landscape that Cosgrove (1984) sought to elucidate, and this is perhaps the most enduring and celebrated aspect of Cosgrove's work.

In their editorial introduction to the *Iconography of Landscape*, Cosgrove and Daniels began by referring to the landscape as a 'cultural image, a pictorial way of representing, structuring or symbolising surroundings' (Daniels and Cosgrove, 1988: 1). The idea was that in order to understand the material landscape it was necessary not just to make use of 'images and verbal representations of it' for the purposes of illustration, but to recognize that such images were constitutive of 'landscape', that is, they were 'constituent images of its meaning or meanings' (Daniels and Cosgrove, 1988: 1). To explore this idea, the editors capitalized on the ideas of iconography which 'sought to probe meaning in a work of art by setting it in its historical context and . . . analyse the ideas implicated in its imagery' (Daniels and Cosgrove, 1988: 2) – exactly what Cosgrove had been doing with his work on the emergence of landscape as a 'way of seeing' in late-medieval and Renaissance Italy. Conceptually, this approach owed much to art history, and especially the work of Ernst Cassirer and Erwin Panofsky on Christian iconography, while at the same time the anthropologist Clifford Geertz was drawn on to provide some sense of method with his 'thick description' and 'conceptualisation of culture as a "text"' (Daniels and Cosgrove, 1988: 4). There was little advice apart from this on how to interpret landscapes iconographically. What was instead significant about this book was that it opened up new avenues for understanding 'landscape' – including both the 'imagined' landscapes of maps, literature and art, as well as the 'material' landscape (Cosgrove and Daniels, 1988). Indeed, running throughout much of Cosgrove's

work is his concern with 'representation' – from art and architecture in fifteenth- and sixteenth-century Italy, for example, through to cartographic images of the earth and cosmos in studies of Western and Renaissance cosmography (Cosgrove, 1979, 1988, 1992, 2001a, 2001b; Cosgrove *et al.*, 1996; Atkinson and Cosgrove, 1998).

KEY ADVANCES AND CONTROVERSIES

Certainly, Cosgrove has made a key contribution to geographical thought by offering a distinctive take on the interpretation and understanding of landscape. In the light of his work, landscape can no longer be considered an unproblematic geographical concept. Equally, geographers cannot now refer to landscape without referring to Cosgrove – the two becoming synonymous. Yet it would be wrong to cast Cosgrove's contributions to the thinking of space and place simply in these terms, for his approach to landscape has provided a means by which geographers have broached broader conceptual issues – especially in matters of subjectivity, representation, power and authority (see Cosgrove, 1988). These had of course been present in Cosgrove's earlier work on landscape, but in the 1980s and 1990s they were under closer scrutiny under the rubric of a 'new' cultural geography informed by critical social and cultural theory (particularly feminist and postcolonial critiques). Cosgrove addressed these issues of authority and authorship with Mona Domosh, noting the 'crisis of geographical representation' and stressing that 'the very structures of representation are implicated in moral and political discourse' (Cosgrove and Domosh, 1993: 30). This 'crisis' was one in which geographers began to realize that by writing about the world in the ways that they did, they were also reinforcing a particular view of the world: an institutionalized, bourgeois, male-dominated and sometimes imperialist view. Indeed, others at this time were also becoming conscious of geography's history and duplicity in presenting and shaping the world (see Livingstone, 1992; Driver, 2001). Hence, together with the work of **Peter Jackson** (1989) and Susan Smith (1989) on the scripting of race and racism, James and Nancy Duncan on the landscape as text (Duncan and Duncan, 1988; Duncan, 1990), and **Derek Gregory**'s analysis of the production of geographical knowledge (Gregory, 1994), Cosgrove's idea of landscape as 'way of seeing' (and an ideologically infused mode of representation) emerged as a key reference point in the formation of a putative 'new' cultural geography.

Though Cosgrove has been identified as a key practitioner of the 'new' cultural geography, it is important to stress that not all those implicated in the making of this 'new' cultural geography were agreed on its definition (Dear, 1988). Discussions in the early 1990s waved back and forth about its newness and also concerns over its tendency to deal more with abstract theory than with the empirical world. There were those, too, who sought to argue that this 'new' cultural geography, with its concern with text, image and metaphor, was making 'landscape' more unstable and slippery (see Muir, 1999). Indeed, there seemed to be a polarization in the way that historical and cultural geographers were treating landscape – some continuing with their empirically grounded, largely atheoretical work on landscape history; others more inclined to see landscapes as imaginings, in poetry, art and literature (and accused of detaching these representations from their material contexts) (see Whyte, 2002). This polarization was in part a conceptual parting of the ways, between, on the one hand, those geographers with more positivist leanings, and on the other, those concerned with taking on board critical theory. It was an empirical and historical divide too, with the focus of the latter

being especially – indeed almost exclusively – on the 'modern' world, particularly the eighteenth, nineteenth and twentieth centuries. It was here that historical and cultural geographers led the field during the 1990s (Daniels, 1993; Driver, 1992; Ogborn, 1998; Driver and Gilbert, 1999; Graham and Nash, 2000), neglecting earlier periods, particularly the Middle Ages, which during the 1960s and 1970s had been such a focus for landscape work in geography (see Hooke, 2000). Indeed, to read of 'landscape' in recent UK and US geography textbooks makes it all too clear that the landscape as a 'way of seeing' has won the day, for the 'landscape history' tradition that follows in the footsteps of W. G. Hoskins and J. B. Jackson is fleetingly – if ever – mentioned (see Nash, 1999; Seymour, 2000). Where reaction has occurred to this new approach to landscape, it is among those human geographers whose move towards 'non-representational theory' has begun to open up landscapes of 'performance' (see Thrift, 2000; also Nash, 2000).

It could be said then, that Cosgrove's approach to landscape was not only in and of itself highly original and novel, but that it also influenced geographers' approach to space and place more generally, as well as catalyzing disciplinary changes in cultural and historical geography. More recently, while landscape remains an important element of Cosgrove's work as a geographer (Cosgrove, 1993), he has focused increasingly on issues of representation, especially mapping (Cosgrove, 1992; 1999b). Still there is a firm historical basis to his work, and still there is theoretical insight. Thus in his editorial introduction to *Mappings*, Cosgrove (1999b: 9) discusses 'mapping' in its various guises – both literal and metaphorical – probing the 'complex accretion of cultural engagements with the world that surround and underpin the authoring of a map', and questioning how maps 'may be regarded as a hinge around which pivot whole systems of meaning'. This concern with mapping and its meanings connects of course with a growing querying of the 'map' in geographical discourse, largely inspired by **Brian Harley**'s work in the late 1980s on 'deconstructing the map', and the theoretical challenges laid at cartography's door during the 1990s. At the same time, Cosgrove's focus on mapping connects with his long-held interests in Renaissance and Enlightenment geographical knowledges, particularly the representation of the world and globe in cosmology and cosmography (Cosgrove, 2001a, 2001b, forthcoming). While for some Cosgrove's subject matter might appear esoteric, his approach is catholic, and from it he has always spun a larger and forceful argument that provides fuel for debate within and beyond geography – a true Renaissance project.

COSGROVE'S MAJOR WORKS

Cosgrove, D. E. (1984/1998a) *Social Formation and Symbolic Landscape*. London: Croom Helm (second edition with additional introductory chapter, Madison: Wisconsin University Press).

Cosgrove, D. E. (1993) *The Palladian Landscape: Geographical Change and its Cultural Representations in Sixteenth Century Italy*. London: Leicester University Press.

Cosgrove, D. E. (ed.) (1999a) *Mappings*. London: Reaktion Books.

Cosgrove, D. E. (2001a) *Apollo's Eye: A Cartographic Genealogy of the Earth in the Western Imagination*. Baltimore: John Hopkins University Press.

Cosgrove, D. E. and Daniels, S. (eds) (1988) *The Iconography of Landscape: Essays on the Symbolic Representation, Design and Use of Past Environments*. Cambridge: Cambridge University Press.

Secondary Sources and References

Atkinson, D. and Cosgrove, D. E. (1998) 'Urban rhetoric and embodied identities: city, nation and empire at the Vittorio Emanuele II monument in Rome 1870–1945', *Annals, Association of American Geographers* 88: 28–49.

Cosgrove, D. E. (1985) 'Prospect perspective and the evolution of the landscape idea', *Transactions of the Institute of British Geographers* 10: 45–62.

Cosgrove, D. (1979) 'John Ruskin and the geographical imagination', *Geographical Review* 69: 43–62.

Cosgrove, D. E. (1988) 'The geometry of landscape: practical and speculative arts in 16th century Venice', in D. Cosgrove and S. Daniels (eds) *The Iconography of Landscape*. Cambridge: Cambridge University Press, pp. 254–276.

Cosgrove, D. (1989) 'Geography is everywhere: culture and symbolism in human landscapes', in Gregory, D. and Walford, R. (eds) *Horizons in Human Geography*. London: Macmillan.

Cosgrove, D. E. (1992) 'Mapping new worlds: culture and cartography in sixteenth-century Venice', *Imago Mundi* 44: 1–25.

Cosgrove, D. E. (1996) 'The measures of America', in J. Corner and A. MacLean (eds) *Taking Measures Across the American Landscape*. New Haven: Yale University Press, pp. 3–13.

Cosgrove, D. E. (1998b) 'Introductory essay', in D. E. Cosgrove *Social Formation and Symbolic Landscape*, 2nd edition. Madison: Wisconsin University Press, pp. xi–xxxv.

Cosgrove, D. E. (1999b) 'Mapping meanings', in D. E. Cosgrove (ed.) *Mappings*. London: Reaktion, pp. 1–23.

Cosgrove, D. E. (2001b) 'Geography's cosmos: the dream and the whole round earth', in K. Till, S. Hoelscher and P. Adams (eds) *Textures of Place*. Minneapolis: University of Minnesota Press, pp. 326–339.

Cosgrove, D. E. (forthcoming) 'Renaissance cosmography 1450–1650', in D. Woodward (ed.) *The History of Cartography, Vol.3 'Renaissance cartography'*. Chicago: University of Chicago Press.

Cosgrove, D. E. and Domosh, M. (1993) 'Author and authority: writing the new cultural geography', in J. Duncan and D. Ley (eds) *Place/Culture/Representation*. London, Routledge, pp. 25–38.

Cosgrove, D. E., Roscoe, B. and Rycroft, S. (1996) 'Landscape and identity at Ladybower Reservoir and Rutland Water', *Transactions of the Institute of British Geographers* 21: 534–551.

Daniels, S. (1993) *Fields of Vision: Landscape Imagery and National Identity in England and the United States*. Cambridge: Polity.

Daniels, D. and Cosgrove, D. E. (1988) 'Introduction: iconography and landscape', in D. E. Cosgrove and S. Daniels (eds) *Iconography of Landscape*. Cambridge: Cambridge University Press, pp. 1–10.

Dear, M. (1988) 'The post-modern challenge: reconstructing human geography', *Transactions of the Institute of British Geographers* 13: 262–274.

Driver, F. (1992) 'Geography's empire: histories of geographical knowledge', *Environment and Planning D: Society and Space* 10: 23–40.

Driver, F. (2001) *Geography Militan: Cultures of Exploration and Empire*. Oxford: Blackwell.

Driver, F. and Gilbert, D. (eds) (1999) *Imperial Cities: Landscape, Display and Identity*. Manchester: Manchester University Press.

Duncan, J. (1990) *City as Text: The Politics of Landscape Interpretation in the Kandyan Kingdom*. Cambridge: Cambridge University Press.

Duncan, J. and Duncan, N. (1988) '(Re)reading the landscape', *Environment and Planning D: Society and Space* 6: 117–126.

Graham, B. and Nash, C. (eds) (2000) *Modern Historical Geographies*. Harlow: Pearson.

Gregory, D. (1994) *Geographical Imaginations*. Oxford: Blackwell.

Hooke, D. (ed.) (2000) *Landscape – the Richest Historical Record*. Oxford: Society for Landscape Studies.

Hoskins, W. G. (1955) *The Making of the English Landscape*. London: Hodder and Stoughton.

Jackson, P. (1989) *Maps of Meaning. An Introduction to Cultural Geography*. London: Unwin Hyman.

Livingstone, D. N. (1992) *The Geographical Tradition*. Oxford: Blackwell.

Muir, D. (1999) *Approaches to Landscape*. Basingstoke: Macmillan.

Nash, C. (1999) 'Landscapes', in P. Cloke, P. Crang and M. Goodwin (eds) *Introducing Human Geographies*. London: Arnold. pp. 217–225.

Nash, C. (2000) 'Performativity in practice: some recent work in cultural geography', *Progress in Human Geography* 24: 653–664.

Ogborn, M. (1998) *Spaces of Modernity: London's Geographies, 1680–1780*. New York: Guilford Press.

Seymour, S. (2000) 'Historical geographies of landscape', in B. Graham and C. Nash (eds) *Modern Historical Geographies*. Harlow: Pearson, pp. 193–217.

Smith, S. (1989) *The Politics of 'Race' and Residence: Citizenship, Segregation and White Supremacy in Britain*. Cambridge: Polity.

Thrift, N. (2000) 'Afterwords', *Environment and Planning D: Society and Space* 18: 213–255.

Whitehand, J. W. R. (2001) 'British urban morphology: the Conzenian tradition', *Urban Morphology* 5: 103–109.

Whyte, I. D. (2002) *Landscape and History Since 1500*. London: Reaktion.

Keith Lilley

13 Mike Davis

BIOGRAPHICAL DETAILS AND THEORETICAL CONTEXT

One of the few contemporary urbanists who have had a genuine influence outside the academy, Mike Davis is an uncompromising critic of the urban and environmental impacts of American capitalism. He owes this reputation to a prolific set of writings on the US city, particularly his works *City of Quartz* (1990), *Ecology of Fear* (1998), and *Magical Urbanism* (2000), which chart the conflictual evolution of Los Angeles over the 20th century. With his writing conveying a deep sense of political immediacy – much of the material in his books began life as shorter, topical articles – he is nonetheless an urbanist with a deep sense of radical social history.

Born in Fontana, California, in 1946, Davis has worked variously as a meat-packer, lorry driver, and manager of the Communist Party's Los Angeles book-shop. He was a militant in SDS (Students for a Democratic Society), but his formal academic career began in the early 1970s, when an interest in Northern Ireland took him to the UK, where he studied Irish history at the University of Edinburgh, and undertook research on Irish nationalism in Belfast. For the first half of the 1980s, Davis worked full-time for *New Left Review,* in the pages of which would appear many of his most seminal essays. His first book, *Prisoners of the American Dream* – a pessimistic but carefully theorized analysis of the American working class – appeared in 1986, yet this was quickly overshadowed by the appearance

of *City of Quartz* in 1990. By 2001, when he published *Late Victorian Holocausts,* and 2002, which saw the publication of a collection of essays under the title *Dead Cities,* Davis had emerged as one of the most controversial, yet also most original, urbanists in the history of the social sciences.

The reason for this lies partly in the quality and combativeness of his work. Yet the political and academic context was also highly conducive, for two main reasons. Firstly, because of the power of Hollywood in projecting the city around the globe, its automobile culture and its ethnic diversity, Los Angeles emerged over the course of the 1980s and 1990s as something of a warning, or a paradigm, by which the future of *all* cities might be understood. Here, Davis sat alongside the so-called 'LA School', composed of theorists such as **Michael Dear, Ed Soja** and Allen Scott, each of whom offered influential theories of post-Fordism and postmodernism based on readings of the changing geographies of Los Angeles. Secondly, the publication of *City of Quartz* came only a couple of years before the explosion of urban unrest in Los Angeles in 1992, following the acquittal of police officers implicated in the Rodney King beatings. For academics, journalists and politicians seeking to understand the conflagration, Davis's careful – if startlingly rendered – analysis of gangs, land-grabbing power elites, extremist policing and highly politicized conservative homeowners provided the context of an uprising waiting to happen.

Davis's academic career has been far from straightforward, arriving relatively late as a university teacher. Awarded a MacArthur Foundation grant in 1998, he has gone some way to silencing the conservative backlash against his more polemical narratives. In his university teaching role, he has followed through the radical messages of his books and articles with politically engaged fieldwork. Yet it is in his books and articles that he has proved most influential, spawning and provoking a series of interventions that seek to understand the politics of urbanism in our biggest cities.

SPATIAL CONTRIBUTIONS

Geographers are – or perhaps should be – multidisciplinary creatures, and across the range and breadth of his writing Davis has demonstrated an openness to such multiple perspectives. If one were to pigeonhole him, it may be as a social historian, yet it is clear that Davis has had a major impact on urbanism, and particularly the political geography of cities. Here, from urban planning to anthropology to natural history to architectural critique, Davis has opened up a way of seeing the city as a very real terrain of political struggle, both a microcosm and process of contemporary capitalism. And so it is the tenor of his critique, rather than his disciplinary location, that is most characteristic of his work.

In an interview with Adam Schatz, Davis claimed that *City of Quartz* was a fusion of three competing perspectives:

I had this daydream of Walter Benjamin finally coming to LA and sitting in a bar with Fernand Braudel and Friedrich Engels. They decide to write a book about LA and divide it into three projects. Benjamin is going to get at all the complex and lucid fragments about power and memory. Braudel will explore its natural history, the larger world-historical forces that made it possible. And Engels will report on LA's working classes.
(Davis, cited in Schatz, 1997: no pagination)

These three themes – a fierce class-consciousness, a stress on how long-term historical processes impact on everyday life and a strongly humanized, even individualized, account of the impact of capitalism on the lives of Angelenos – are fundamental to an understanding of his world-view. Such a method – which fuses minute archival research, much of it newspaper based, with acerbic observational qualities – has opened up paths to radical geographers hitherto closed off by the overwhelming structuralist aesthetic of much urban theory.

Such a multi-tooled approach has allowed him to explore themes often beyond the imagination of political-economy perspectives, without excluding the power of the latter in shaping daily landscapes and lives. *City of Quartz* is a sustained and multi-layered discussion of elite power in the boosterist-*noir* representations of the city's historians and writers, in the reach of the Catholic Church, in the political clout of the city's suburban homeowners. In *Ecology of Fear* he contrasts the differing 'fire geographies' in luxury Malibu with working-class downtown; he traces the 'literary destruction of Los Angeles' through pulp fiction and disaster movies; and he warns chillingly of the impending revenge of nature on a city always built upon the domination of the natural environment. In *Dead Cities*, he sets out an eye-opening challenge to future urban geographies:

Very large cities – those with a global not just regional environmental footprint – are thus the most dramatic end-product, in more than one sense, of human cultural evolution in the Holocene. Presumably they should be the subject of the most urgent and encompassing scientific enquiry. They are not. We know more about rainforest ecology than urban ecology ... The most urgent need, perhaps, is for large-scale conceptual templates for understanding the city–nature dialectic.
(Davis, 2002: 363)

With this research agenda being tied to a set of dire projections about the future of urban Southern California, Davis has often been dubbed a latterday Jeremiah, an allusion to the Old Testament prophet of the fall of Jerusalem. His apparent pessimism about the fate of LA is further seen as disempowering by some on the Left. In *Metromarxism*, Andy Merrifield argues that 'Davis's Marxism bespeaks an urbanism that lacks public space and denies any sense of collective experience. It's a Marxist urbanism that expresses only contempt for one's own city, perhaps for good reason . . .' (Merrifield, 2002: 171). By contrast to the quirky exuberance of Marshall Berman, apparently nourished by the cultural inflections of being a New Yorker, Merrifield detects a 'brooding, doom-laden undertow' (Merrifield, 2002: 171) to the work of Davis.

However, this charge might be unfair. It is precisely because of Davis's ability to understand the sprawling contours of Los Angeles – so far away from the cohesive morphology of the Paris of many of Merrifield's 'metromarxists' (such as Walter Benjamin, Guy Debord, **Henri Lefebvre** and **Manuel Castells**) – that he has drawn such admiration. Here, what Merrifield calls a 'metropolitan dialectic' is really taken to its logical extent only in the work of Davis, who finds a set of laws and rules in a sprawling megalopolis with little apparent coherence. Furthermore, not all his work has been so downbeat. In 2000, Davis published *Magical Urbanism*, the subtitle of which – 'Latinos reinvent the American city' – captured an air of apparent optimism about the changing demographic of urban America, where LA – along with the likes of Miami and New York – is having its cultural and political identity transformed by migration. Here, Davis outlines the diverse nature of Latino life, from their 'spicing' of the inner city in a dourly suburbanized US to the parlous working conditions that many immigrants labour under. It can be argued that here Davis eulogizes Latino culture to an extent that ignores its diversity and internal tensions. But in many ways, this completes the urban trilogy that started with the white 'power lines'

charted in *City of Quartz* and ended with a vision of a transformed ethnic and class-conscious urban politics.

KEY ADVANCES AND CONTROVERSIES

Given that Davis is best known for his empirical work rather than his theoretical stance on space and place, what have been the aspects of his work that have proved most influential? We can perhaps identify four of these: first, a remarkable ability to weave a political analysis that fuses city and nature, categories often treated in isolation and even antagonism; second, a powerful statement of the centrality of class politics despite the 'cultural' turn away from structural Marxism; third, his work on the 'fortress' or 'carceral' city; fourth, the introduction of what could be called 'activist writing', in the sense that his approach has helped to politicize theory and make accessible and personal some of the bigger forces that shape the contemporary city.

Davis's interest in nature is prevalent, if usually underplayed, throughout *City of Quartz*, but in *Ecology of Fear* and *Dead Cities* it moves centre-stage, echoing a growing personal interest in 'shock' nature. In *Ecology of Fear* he argues that the city's urbanization is out of step with the region's ecosystem, with its hitherto relatively benign earthquake, fire and drought potential. Above all, Davis reveals himself to be increasingly interested in neo-catastrophism, the idea that sudden natural occurrences may have far-reaching consequences for both social *and* natural history. By extension, and parting company with conventional Marxist wisdom that it is wars, not economic crises, that shape modes of production, he has left some critics puzzled, even alienated, by his emphasis on the importance of urban ecology.

Davis is strongly defensive of the importance of class politics and labour exploitation at a time when other perspectives have grown in significance for geographers. His interest in ethnicity is primarily driven by the subordinate position of Latinos or African Americans in the labour market; his exploration of the dramatic contrast between Malibu and downtown LA in terms of fire and state policy is about access to housing and policy influence; his discussion of gang culture is aligned with the Black Panthers and a 'revolutionary lumpenproletariat' which has been fractured by internecine fights over drug-vending territory. Yet in recent work he has moved beyond the urban, and has applied his interest in class exploitation on a grander scale. In *Late Victorian Holocausts* he argues that the droughts and famines that caused mass death in India, Brazil and China in the last quarter of the nineteenth century were exacerbated by the aggressive imperialist capitalism inflicted on a global peasantry (famine occurring at a time of unprecedented grain production in India, for example), which forms a searing indictment of imperialism and the birth of today's 'third world'. Such an understanding fits with Davis's dystopian take on today's 'global cities'.

Davis is perhaps most infamous for the biting renditions of Los Angeles as a 'militarised space'. In 'Fortress L.A.' (chapter four of *City of Quartz*) he sets out eight trends in the ongoing destruction of public space in Los Angeles, his acerbic style in full flow. Utilizing a series of short vignettes, he argues that 'in cities like Los Angeles, on the bad edge of post-modernity, one observes an unprecedented tendency to merge urban design, architecture and the police apparatus into a single, comprehensive security effort' (Davis, 1990: 224). Some of his villains are predictable. The Los Angeles Police Department (LAPD) he presents as 'space police', combining air power (helicopters) with communications systems imported from the Californian military aerospace industries. Others are less obvious. The

inner-city malls of Alexander Haagen, the employment of armies of private security guards that guard the burgeoning gated communities, the designers who contribute the rounded, sleep-proof 'barrel-shaped benches' to ward off the homeless, and the celebrity architect Frank Gehry (a 'Dirty Harry' for his provocative 'Beirutized' Goldwyn library in the city), are all cited as contributors to the creation of spaces of fear in LA.

This – for many, exaggerated – vision is taken to its logical extension in 'Beyond *Blade Runner*', the concluding chapter to *Ecology of Fear*. Taking the famous Chicago School/Burgess 'concentric ring' model of the archetypal structure of the North American city, he provides an inimitable update: 'My remapping takes Burgess back to the future. It preserves such "ecological" determinants as income, land value, class and race but adds a decisive new factor: fear' (Davis, 1998: 363). The Davis diagram represents LA as a fragmented city, with a 'gulag rim' of privatized prisons on the distant reaches of the city, a surrounding ring of gated, affluent suburbia linked orbitally to a series of edge cities, a variegated set of inner rings dominated by suspicious blue-collar communities, 'homeless containment zones', drug- and gang-free parks subject to specific legal regulation, and a downtown *scanscape*, as opposed to *landscape*, dominated by highly developed surveillance technology. In these two chapters, undoubtedly the most frequently quoted parts of his LA work, Davis sets out a plausible account of a city fractured by the 1992 riots.

The Davis trademark is a style of writing that bites through what he sees as the boosterist rhetoric of many orthodox LA historians, such as Kevin Starr (1991). In many ways it fits into a tradition of *noir* writing on the city (Gregory, 1994), where the dark realities of political life are hidden from everyday view, only accessible through the hard-boiled, fearless detective. The opening chapter of *City of Quartz* – the hermeneutic key to understanding the book – is entitled 'Sunshine or noir?', and it is this dialectic – between the promotional myths of property developers and politicians of fertile land and a better life, and the competing reading of a city built upon the exploitation of labour and the environment – that informs his narrative. Such a perspective reveals itself in the Davis written style, as the following memorable example from *City of Quartz* demonstrates:

> Welcome to post-liberal Los Angeles, where the defense of luxury lifestyles is translated into a proliferation of new repressions in space and movement, undergirded by the ubiquitous 'armed response'. This obsession with physical security systems, and, collaterally, with the architectural policing of social boundaries, has become a zeitgeist of urban restructuring, a master narrative in the emerging built environment of the 1990s ... Hollywood's pop apocalypses and pulp science fiction have been more realistic, and politically perceptive [than contemporary urban theory] in representing the programmed hardening of the urban surface in the wake of the Reagan era.
> (Davis, 1990: 223)

This style, which James Duncan (1996) calls the 'Tragic/Marxian' mode of writing, can be accused of being too muscular in its pursuit of narrative coherence and political persuasiveness. While Duncan's own critique may fall guilty of the crimes he sees inherent in the Davis approach, he nonetheless makes valid observations at how the *Quartz* narrative progresses:

> [Davis] clearly speaks for the underclass. We do not hear from them directly, however. In fact, the only people other than Davis that we do hear from are those whom he parades before us to speak their highly edited 'confessions' and then marches out again with a resounding 'guilty'. The implied readership who stand as jury to the author's prosecutor shake their heads sadly at these confessions. However it is clear that were these people not to be found guilty, they would not have been summoned to speak their lines. Such is the methodology of the show trial. While this strong narrative structure may lend illusory coherence to the plot, it may fail to produce the degree of 'reality effect' that an account more open to contradictory evidence might achieve.
> (Duncan, 1996: 261)

The allegation that Davis's writing is powerful but rather tendentious is one that has been voiced by sympathizers and opponents alike. Yet what he does, for students of Marxism grown tired of struggling with abstract theoretical issues, or left doubting the relevance of the Marxian legacy for the contemporary city, is to show that theory can live in the streets of one of the world's most brutally capitalized cities, and make sense of such diverse subjects as gang warfare, urban Catholicism, street furniture and place myth.

So Davis's influence on geography and the urban is perhaps still to be fully realized. It is not impossible that the current direction of his theory – with his emphasis on a radical urban ecology – may fall between several established camps. His empirical work on Los Angeles and elsewhere will continue to motivate and inform a wide range of publics, and his intellectual creativity and originality should also inspire new interest in an urban geography that is too often ignorant of the political. Ultimately, perhaps, the biggest impact of Davis may be through the vibrancy of his writing, suggesting a voice to those academics and activists who see conventional theory too heavy, or too sluggish, to convey the political immediacy of contemporary urban life.

DAVIS' MAJOR WORKS

Davis, M. (1986) *Prisoners of the American Dream: Politics and Economy in the History of the US working class*. London: Verso.

Davis, M. (1990) *City of Quartz: Excavating the Future in Los Angeles*. London: Verso.

Davis, M. (1998) *Ecology of Fear: Los Angeles and the Imagination of Disaster*. New York: Metropolitan Books.

Davis, M. (2000) *Magical Urbanism: Latinos Reinvent the American City*. London: Verso.

Davis, M. (2001) *Late Victorian Holocausts: El Niño Famines and the Making of the Third World*. London: Verso.

Davis, M. (2002) *Dead Cities and Other Tales*. London: Verso.

Secondary Sources and References

Duncan, J. (1996) 'Me(trope)olis: Or Hayden White among the urbanists', in A. D. King (ed.) *Re-Presenting the City: Ethnicity, Capital and Culture in the 21st-century Metropolis*. Basingstoke: MacMillan, pp. 253–268.

Gregory, D. (1994) *Geographical Imaginations*. Cambridge, MA: Blackwell.

MacAdams, L. (1998) 'Jeremiah among the palms: the lives and dark prophecies of Mike Davis', *LA Weekly* 3 December (www.laweekly.com/ink/99/01/news-macadams.php).

Merrifield, A. (2002) *Metromarxism: A Marxist Tale of the City*. New York: Routledge.

Schatz, A. (1997) 'The American earthquake: Mike Davis and the politics of disaster', *Lingua Franca* September (republished on *Radical Urban Theory*, www.rut.com/mdavis/americanearthquake.html).

Starr, K. (1991) *American Dreams: Southern California Through the 1920s*. Oxford: Oxford University Press.

Donald McNeill

14 Michael Dear

BIOGRAPHICAL DETAILS AND THEORETICAL CONTEXT

Michael Dear's principal research interests relate to Los Angeles, and his pursuit of a project of 'postmodern urbanism'. He is founding director of the Southern California Studies Center at the University of Southern California, and this position makes him one of the most influential members of the 'LA School' of urban theory. While well known for his work on homelessness in Los Angeles, it has been his dogged insistence on the need for a new set of theoretical categories to account for capitalist urbanism in the LA region that has given his work some degree of renown, or notoriety.

Dear was born in Treorchy, in the Rhondda Valley of South Wales, UK in 1944. He obtained a BA in Geography from the University of Birmingham, England, in 1966, followed by an MPhil in Town Planning from University College London in 1969. He undertook graduate work at the University of Pennsylvania, obtaining an MA in Regional Science in 1972 and a PhD in 1974. It was while pursuing research at Pennsylvania that he began studying the geography of community opposition to mental health-care facilities. Following his appointment to McMaster University, Hamilton, Ontario, this research work broadened out to include aspects of the organization of health-care delivery systems, and subsequently to an understanding of the political economic dynamics surrounding mental health care. During the 1970s and 1980s, as homelessness became particu-

larly prevalent in the United States and Canada, and in partnership with Jennifer Wolch, Dear authored two books on homelessness and the city (Dear and Wolch, 1987; Wolch and Dear, 1993), and has continued to work on stigmatization and difference with reference to those in institutionalized care (Dear et al., 1997). In the early 1980s, he became founding editor of the journal Society and Space, which has become one of the most cited academic journals in geography and urban studies.

Dear – like many of his associates in the 'LA School' – is profoundly influenced by the nature of Southern Californian capitalism, experimenting with various ways of representing its peculiar landscapes. As with **Ed Soja**, Dear has grappled with the diffuse nature of LA urbanism and urbanization, employing novel representational strategies to capture its heterogeneous geographies. From his earlier work on postmodernism and urbanism in the mid-1980s, through to The Post-modern Urban Condition in 2000, Dear has accordingly concerned himself with issues of urban (dis)order and the ability of contemporary urban theory to adequately explain or represent it. His suggested alternatives – and his manner of expressing them – have aroused fierce controversy within geography, attracting attacks from many sides (see, for example, Dear, 2000; Lake, 1999; and debates in Urban Geography, Political Geography and Annals of the Association of American Geographers). Yet his output includes some highly respected work on very pressing social issues. In Malign Neglect and Landscapes of Despair, both written with Jennifer Wolch, the massive explosion in homelessness in Southern California is analysed. Here, the authors find that these 'service-dependent' people are increasingly concentrated in well-defined geographical areas of cities (including the infamous 'skid rows' of LA itself). Together, these works provide the first comprehensive analysis of these ghettos, their structure and evolution, their benefits and costs and how they might change in the future. Truly interdisciplinary in scope, these works draw upon geography, planning, social work, psychiatry and history, as well as upon the wider debates in social theory. Furthermore, by being situated within a context of neo-liberal urban policy, Dear and Wolch's work demonstrates how welfare cutbacks are profoundly shaping the life experiences of the socially disadvantaged. As such, and far from being an escapee from the pressing realities of social ills, Dear's postmodernism emerges from a politically progressive position.

SPATIAL CONTRIBUTIONS

As a prominent writer on space and social theory, Dear has consistently drawn attention to the need for new representational approaches to grasp the restructuring of contemporary urban space. Grounding such ideas in the context of Los Angeles, yet with an eye on global processes, Dear (writing with Stephen Flusty) insists that this urban disordering is exemplary of wider changes in the nature of space and time:

> Have we arrived at a radical break in the way cities are developing? Is there something called a *post-modern urbanism*, which presumes that we can identify some form of template that defines its critical dimensions? This inquiry is based on a simple premise: that just as the central tenets of modernist thought have been undermined, its core evacuated and replaced by a rush of competing epistemologies, so too have the traditional logics of earlier urbanisms evaporated, and in the

absence of a single new imperative, multiple urban (ir)rationalities are competing to fill the void. It is the concretization and localization of these effects, global in scope but generated and manifested locally, that are creating the geographies of post-modern society – a new time-space fabric.
> (Dear and Flusty, 1998: 50; emphasis in original)

The most comprehensive statement of Dear's takes on postmodern space can therefore be found in *The Post-modern Urban Condition* (2000), a compendium of much of the work that he had published over the previous decade or so. This collection alerts us to three of the most important geographical contributions Dear has made. The first is his significant contribution to an 'LA School' (alongside the likes of Allen Scott, **Michael Storper**, **Ed Soja**, and **Mike Davis**) that posits Los Angeles as the most appropriate successor to Chicago in terms of a model of an ideal type of urban spatial structure. Second, and related to this, is his attempt to provide a new lexicon for describing the new spatial forms of the postmodern metropolis (set out most provocatively in Dear and Flusty, 1998). Third is his attempt to demonstrate the consequences of the 'postmodern turn' in the social sciences for urban theory (and geography more generally).

In relation to the first of these, the influence of the fairly disparate group of scholars known as the LA School on several fields of human geography has been pronounced. As Mike Davis summarizes:

> During the 1980s the 'L.A. School' (based in the UCLA planning and geography faculties, but including contributors from other campuses) developed an ambitious matrix of criss-crossing approaches and case-studies. Monographs focused on the dialectics of de- and re-industrialization, the peripheralization of labor and the internationalization of capital, housing and homelessness, the environmental consequences of untrammeled development, and the discourse of growth. Although its members remain undecided about

whether they should model themselves after the 'Chicago School' (named principally after its *object* of research), or the 'Frankfurt School' (a philosophical current named after its *base*), the 'L.A. School' is, in fact, a little bit of both. While surveying Los Angeles in a systematic way, the UCLA researchers are most interested in exploiting the metropolis, à la Adorno and Horkheimer, as a 'laboratory of the future'. They have made clear that they see themselves excavating the outlines of a paradigmatic postfordism, an emergent twenty-first century urbanism.
(Davis, 1992: 84)

Dear has embodied this Frankfurt–Chicago fusion by serving as the editor of a number of edited collections on the LA region (Dear *et al.*, 1996; Dear, 2001; Leclerc *et al.*, 1999). Despite, the fact that many of these contributors are cautious about the wider applicability of LA-derived models, Dear's own work on LA urbanism frequently exceeds its local focus to make somewhat grander claims about postmodern cities. For example, in his work with Steven Flusty, the original intent to provide a loosely coherent forum for debate on the region seems to have escaped its moorings:

> Yet ultimately, we are comfortable in proclaiming the existence of a Los Angeles School of urbanism . . . at present, the city is now commonly represented as indicative of new forms of urbanism augmenting (and even supplanting) the older, established forms against which Los Angeles was once judged deviant . . . The body of writing about Los Angeles provides alternative models to past orthodoxies on the 'essential' nature of the city and is proving to be more successful than its detractors at explaining the form and function of urbanism in a time of globalization.
> (Dear and Flusty, 2001: 12)

Thus while Dear is aware of the 'danger that a Los Angeles School could become another panoptic fortress from whence a new totalizing urban model is manufactured and marketed' (Dear and Flusty, 2001: 13), the fact is that he has been one of the most ardent promoters of that place's precocity.

The second, related, contribution made by Dear (again, often with Flusty) is his attempt to update the work of the Chicago School in the context of the globalization of urbanism. The basic claim made is as follows:

> The concentric ring structure of the Chicago School was essentially a concept of the city as an organic accretion around a central, organizing core. Instead, we have identified a post-modern urban process in which the urban periphery organizes the centre within the context of a globalizing capitalism . . . Conventional city form, Chicago-style, is sacrificed in favour of a non-contiguous collage of parcelized, consumption-oriented landscapes devoid of conventional centres yet wired into electronic propinquity and nominally unified by the mythologies of the disinformation superhighway.
> (Dear, 2000: 158–160)

While a synoptic approach to explaining the dynamism of contemporary urban form is not unwelcome, Dear creates problems for himself in deriving his models and terms from a very limited number of places (Los Angeles, Las Vegas and Tijuana). Furthermore, he has failed to convince critics that his 'alternative model of urban structure' is anything more than a rather tired, 'top-down' Marxist theory of the city, a spatial manifestation of a tyrannical, polarizing capitalism:

> As the cybergeoisie increasingly withdraw from the Fordist redistributive triad of big government, big business and big labor to establish their own micro-nations, the social support functions of the state disintegrate, along with the survivability of less-affluent citizens.
> (Dear, 2000: 156–157)

Thus, Dear's argument that the contemporary city is an increasingly atomized collection of spaces echoes the concerns of many Marxist urbanists who have chosen a more conventional language of urban alienation and class conflict to describe cities, past and present.

Thirdly, as noted above, Dear's output since the mid-1980s has attempted to engage with the more general debate on postmodernism by applying it to the urban. In so doing he has made significant interventions in human geography's epistemological debates (including one of the first papers to acknowledge the postmodern challenge for the discipline – Dear, 1988). Here, and elsewhere, he has applauded the epistemological challenge of postmodernism:

> For me, the radical opening made possible by post-modernism is both invigorating and sometimes exasperating. On one hand, it has liberated our theoretical discourse and legitimized a wide variety of different voices. We no longer need to rely on implausible doctrines of objectivity to defend our contributions to knowledge, and we can treat truth claims as arguments rather than as unassailable findings ... On the other hand, the Babel of different, newly-enfranchised voices can also be profoundly disorienting.
> (Dear, 2000: 32)

As a means of expanding on this, Dear (2000: 166) argues that a postmodern urban analysis must take account of two things: first, the problem of urban analysis at a time when 'the urban grows increasingly to resemble televisual and cinematic fantasy' and second, the problem of developing a radical politics in postmodern times. He thus continues to be one of the leading flag-bearers for postmodernism in geography (Dear and Flusty, 2002).

KEY ADVANCES AND CONTROVERSIES

Dear's provocative work – not least his frequent use of neologisms – has excited significant disciplinary controversy and repudiation, as evidenced by several high-profile disciplinary debates: in the *Annals of the Association of American Geographers*, where he was labelled as postmodern geography's 'best known Brooks Brothers, Wall Street ad man' (Symanski, 1994: 301); in the journal *Urban Geography*, which included a range of largely critical responses to his and Flusty's paper on 'Postmodern urbanism' (see Beauregard, 1999; Lake, 1999); and a series of short responses in *Political Geography* (e.g. DeFillipis, 2001; Natter, 2001) to Dear's contention that 'vicious personal attacks are a commonplace experience among geographers' (Dear, 2001: 2). In relation to the latter, Dear cited a litany of academic and personal attacks on his work as symptomatic of a broader 'culture of hate' and destructive criticism in university culture (interestingly, some of the commentaries in the *Urban Geography* forum are listed in this latter category). The reasons for such 'gleeful poison' (Dear, 2001: 3) – whether justified or not – undoubtedly stem from the pioneering nature of Dear's work, combined with his at times grandiose claims. As noted above, these include: the suggestion that Los Angeles is a privileged site from which to study contemporary urbanism; his assertion that urban studies needs to be reinvented in a postmodern era; and his insistence that geographers need to adopt new methods and languages if they are to make sense of contemporary relations of society and space. Each of these claims has been met with considerable scepticism in the aforementioned debates, although his attempt to challenge sometimes tired terminology has been applauded:

> Applied to Los Angeles, such well-worn terms as concentric zone and uniform land surface do violence, sacrificing reality to interpretation and the post-modern world to modernist theory. Consequently, we need to break away from the prevailing language of urban analysis. Only by dislocating ourselves discursively can we open up the possibility of new and more valid interpretations.
> (Beauregard, 1999: 396)

Yet even here the problem of developing theories of the urban based on knowledge of one city are apparent: as Beauregard (1999: 398) suggests, the 'recent non-reductionist turn in urban theory' where 'most theorists are less intent on parsimonious descriptions and abstract and general explanations than on evoking rich images of a single city' means that accounts often make general claims on the basis of (often superficial) case studies. Indeed, while he allows portraits of Tijuana and Las Vegas to ornament *The Postmodern Urban Condition,* there is little evidence that Dear is as humble about the limitations of Los Angeles as a model as perhaps he should be.

A second critique of Dear's work is more methodological, suggesting that it is based on an unconvincing and limited engagement with the active, purposeful individuals of the city. For example, while the Chicago School abstractions were elaborated by detailed, area-based ethnographies, Jackson (1999: 401) suggests that Dear's work is 'populated with the cool abstractions of social polarization and fragmentation'. While Dear and Flusty defend this – citing as an example Dear's earlier work on the homeless (written with Jennifer Wolch) – questions remain as to how such complex, contradictory qualitative work feeds into the model that they offer. In some ways, this apparent abstraction or distance from everyday life is magnified by the use of neologisms – invented fusions of existing words and terminology – in order to describe or represent the apparently novel conditions pertaining in LA. Here, what is (one presumes) a well-intentioned collection of terms at times breaches the bounds of seriousness, or worse, creates a jargonistic vocabulary that further distances interpretation from everyday life. The key terms of this postmodern urbanism are thus rendered within a framework of 'Holsteinization', and 'keno capitalism', with new social classes known as 'cyber-geoisie' and the 'protosurps', and with a model of the cities as being composed of 'commudities', 'cyburbia', 'citidel', 'in-beyond', and 'cyberia' (Dear and Flusty, 1998: 60–66). While some of these terms may have explanatory merit, they may also conceal an even worse fate for a postmodern urbanism – a reinvention of outmoded 'modernist' grand theories. For example, Jackson (1999: 401–402) argues that the 'Holsteinization' process, 'the process of monoculturing people as consumers so as to facilitate the harvesting of desires' (Dear and Flusty, 1998: 61) is simply restating the mass-society thesis of the Frankfurt School, where brainwashed consumers passively consume the products forced upon them by corporations.

Thirdly, and related to this, the somewhat arcane terminology and neologisms coined in the Dear and Flusty 'Postmodern urbanism' paper have been fiercely criticized. As Robert Lake argues:

> Far from succeeding in their intended role of provocative hypothesis-generators, Dear and Flusty's neologisms are simplifications at best, stereotypes at worst, and ineluctably modernist in provenance and perspective . . . They proclaim a victory of style over substance, of cleverness over erudition, of word play over content, neither a new vision nor a new city but only a new style. Where 'Post-modern urbanism' proffers the possibility of a radical break in understanding the urban, the danger is that it will only encourage an outbreak of linguistic dexterity as an end in itself. (Lake, 1999: 395)

Without further empirical analysis and theoretical development, therefore, the potential of the neologisms will remain unrealized.

Thus it is by no means certain that Dear's project of postmodern urbanism has sufficient popularity, or critical depth, to be of lasting theoretical significance. As a provocative and widely read commentator on the postmodern condition, his take on the reconfiguration of Los Angeles is an important contribution to contemporary debates on space and place. Ultimately, however, whether Dear will prove more than a parochial critic riding on the wave of his host city's fame remains to be seen.

DEAR'S MAJOR WORKS

Dear, M. J. (1988) 'The postmodern challenge: re-constructing human geography', *Transactions, Institute of British Geographers* NS 13: 262–274.

Dear, M. J. (2000) *The Postmodern Urban Condition.* Oxford: Blackwell.

Dear, M. J. and Flusty, S. (1998) 'Postmodern urbanism', *Annals of the Association of American Geographers* 88: 50–72.

Dear, M. J. and Flusty, S. (2002) *The Spaces of Postmodernity: Readings in Human Geography.* Oxford: Blackwell.

Dear, M. J. and Wolch, J. R. (1987) *Landscapes of Despair: From Deinstitutionalization to Homelessness.* Princeton University Press/Polity Press.

Dear, M. J., Schockman, H. E. and Hise, G. (eds) (1996) *Rethinking Los Angeles.* Thousand Oaks: Sage.

Leclerc, G., Villa, R. and Dear, M. J. (eds) (1999) *Urban Latino Cultures: La Vida Latina en L.A.* Thousand Oaks: Sage.

Wolch, J. R. and Dear, M. J. (1993) *Malign Neglect: Homelessness in an American City.* San Francisco CA: Jossey-Bass.

Secondary Sources and References

Beauregard, R. A. (1999) 'Break dancing on Santa Monica boulevard', *Urban Geography* 20: 396–399.

Davis, M. (1992) *City of Quartz: Excavating the Future in Los Angeles.* London: Vintage.

DeFillipis, J. (2001) 'Hatred and criticism inside and outside of the academy: a response to Michael Dear', *Political Geography* 20: 13–16.

Dear, M. J. (2001) 'The politics of geography: hate mail, rabid referees, and culture wars', *Political Geography* 20: 1–15.

Dear, M. J. and Flusty, S. (2001) 'The resistible rise of the LA School', in M. J. Dear (ed.) *From Chicago to LA: Making Sense of Urban Theory.* Sage: Thousand Oaks, pp. 5–16.

Dear, M. J. and Flusty, S. (1999) 'Engaging post-modern urbanism', *Urban Geography* 20: 412–416.

Dear, M. J. and Taylor, M. (1982) *Not on Our Street: Community Attitudes Toward Mental Health Care.* Leeds: Pion Ltd.

Dear, M. J., Wilton, R., Gaber, S. L. and Takahashi, L. (1997) 'Seeing people differently: the socio-spatial construction of disability', *Environment and Planning D: Society and Space* 15: 455–480.

Jackson, P. (1999) 'Post-modern urbanism and the ethnographic void', *Urban Geography* 20: 400–402.

Lake, R. W. (1999) 'Post-modern urbanism?' *Urban Geography* 20: 393–395.

Natter, W. (2001) 'From hate to antagonism: towards an ethics of emotion, discussion and the political', *Political Geography* 20: 25–34.

Symanski, R. (1994) 'Why we should fear post-modernists', *Annals of the Association of American Geographers* 84: 301–304.

Donald McNeill and Mark Tewdr-Jones

15 Gilles Deleuze

BIOGRAPHICAL DETAILS AND THEORETICAL CONTEXT

Gilles Deleuze was born into a bourgeois and conservative Parisian family in 1925. He was made famous by his catalytic role in the uprising of French students and workers in May 1968 for his promotion of a 'desiring revolution' that would de-stabilize the subjugation of the individual to the social formations that feed off it (e.g. family, school, factory, office, prison, asylum, etc.), and for his part in the emergence of what has become known as poststructuralism. Deleuze died in 1995, having served most of his career as a Professor of Philosophy at Université de Paris VIII, Vincennes. He was a well-regarded teacher, and his classes were reputed to have been notable events. During his life Deleuze rarely left France, not least because he disliked the conditions under which intellectuals travelled: flit-ting across the globe to do what they could do at home – exchange pleasantries and talk, talk, talk. This is a good example of Deleuze's indifference to the academic crowd and his disdain for intellectual fashion.

While Deleuze was not averse to con-versation (which leads who knows where), exchanges (of ideas, opinions, objections, questions and answers, points of view, etc.) were anathema to him. 'We do not lack communication' wrote De-leuze and Guattari in *What is Philosophy?* (1994: 108). 'On the contrary, we have too much of it. We lack creation. *We lack resistance to the present.*' When Deleuze collaborated with others, it was not to communicate opinions or exchange ideas, as if he were at a swap meet. Collabor-ation is transformatory. It is an act of creative de-personalization that opens a 'line of flight' along which one becomes a stranger to oneself. 'We are no longer ourselves' announced Deleuze and Guat-tari (1987: 3) at the outset of *A Thousand Plateaus*. With Félix Guattari, who was a radical psychoanalyst and institutional schizoanalyst, Deleuze undertook a long collaboration of convergence that span-ned three decades. With Alain Badiou, another French philosopher, Deleuze be-latedly entered into a 'collaboration of divergence' over the nature of multiplicity (Badiou, 1999).

Although the scope of Deleuze's inter-ests was truly enormous, he always con-sidered himself to be a philosopher. According to Deleuze, however, there is nothing privileged about philosophy in relation to other activities, each of which is perfectly capable of thinking for itself. There is nothing 'essential' or 'fundamen-tal' about philosophy. It is not the ground of truth, the pinnacle of reason or the font of knowledge. Such claims are at best laughable and are at worst despotic. So, nobody needs a philosopher to think *for* them, but somebody may benefit from a philosopher thinking *alongside* them: a form of collaboration rather than domina-tion (e.g. Deleuze, 1986, 1989). Neverthe-less, philosophy, like every other academic discipline, has not been averse to serving sectarian interests (nations, states, capital, sexism, racism, an-thropocentrism, etc.). Indeed, philosophy has often been conducted as if it were in the service of a totalitarian police state: an endless succession of interrogations, in-quisitions and critical tribunals presided over by investigators, lawyers and legislators. Deleuze was always attentive to the social context of philosophy, and especially how philosophy could be taken up for both repressive and emancipatory ends. This is why he preferred posing problems to asking questions. So, rather than answering to a higher authority (and thereby submitting oneself to truth, reason, logic, etc.), Deleuze's radicalism consists of problematizing the world (and thereby opening up the possibility that something different might happen).

For Deleuze, then, philosophy is neither communication nor contemplation 'because no one needs philosophy to reflect on anything' (Deleuze and Guattari, 1994: 6). Like any other activity, philosophy is creative – and therefore both positive and joyful. 'It is the same in philosophy as in a film or a song: no correct ideas, just ideas' (Deleuze and Parnet, 1987: 9). Needless to say, this constructivist and nonrepresentational attitude 'beyond good and evil' put Deleuze out of step with most other philosophers, intellectuals and radicals, who tended to be enamoured with negativity, resentment and sadness. Deleuze had no time for doubt, reflection and criticism (the paranoiac disposition of modern philosophy), and even less time for lack, scarcity and limitation (the repressive myths of social science). 'What I detested more than anything else was Hegelianism and the Dialectic' (Deleuze, 1977: 112). For Deleuze, philosophy is not critical: it is creative. This is what makes it radical. Moreover, what philosophers create is addressed to both philosophers and non-philosophers, just as music is addressed to both musicians and music-lovers. Such is Deleuze's democratization of philosophy, which is based on friendship rather than ownership, passion rather than possession, and radicalism rather than criticism.

Eschewing the intellectual snobbery that glorifies thinking by letting it lord over existence and practice, philosophy for Deleuze is simply 'a flow among others; it enjoys no special privilege and enters into relationships of current and counter-current, of back-wash with other flows – the flows of shit, sperm, speech, action, eroticism, money, politics, etc.' (Deleuze, 1977: 114). Such is Deleuze's equalitarianism and anti-Platonism, which returns every hierarchy (e.g. the privileging of mind over matter, truth over error, real over possible, ideal over actual, essence over appearance, form over content, subject over object, Being over beings, identity over difference, and original over copy) to the superficial abyss whence it came (cf. Georges Bataille's 'base materialism', Jacques Derrida's 'deconstruction', and Jean-François Lyotard's 'libidinal economy'). In Deleuze's estimation, the surface – or simulacrum – not only ceases to be negative and inferior, it actually comes to take all. Superficiality goes all the way down, so to speak (cf. **Jean Baudrillard**'s 'simulation'). Additionally, while the vertical axis of hierarchy gives the illusion that identity, stability and conservation are the rule rather than the exception, the horizontal axis of the simulacrum is always open to differentiation, instability and transformation. Reading Deleuze is unsettling, then, because nothing is fixed in place. 'On the level of style, the reader cannot fail to be struck by the fact that Deleuze makes systematic use of division', notes Jean-Jacques Lecercle (1985: 97), 'yet the text grows and multiplies in an extremely disquieting manner.'

SPATIAL CONTRIBUTIONS

Even though philosophy is without privilege, Deleuze is keen to insist that it is not without specificity, and this specificity brings philosophy into the orbit of geography. On Deleuze's account, philosophers are responsible for creating concepts: not concepts in general (Ideas, Truths, Universals, etc.), 'globalizing' concepts, 'concepts as big as hollow teeth, THE law, THE master, THE rebel' (Deleuze, in Boundas, 1993: 254); but situated concepts, contextual concepts, contingent concepts; 'localizing' concepts that only have meaning, value and efficacy in relation to some milieu or other. Concepts must be articulated. This is why Deleuze treats concepts as territories and events. 'Deleuze the thinker is, above all, the thinker of the event and always of this event here (*cet evenement-ci*)', wrote Derrida (1998). 'He remained the thinker of the event from beginning to end.' What Deleuze offers us, then, is not a first philosophy (which would be privileged, fundamental, foundational, essential, universal, etc.) but a philosophy of the event – philosophy as an event. His philosophy responds to problems that make him think. For Deleuze, thought cannot be planned out ahead of time. To the contrary, thought is always encountered. This is a remarkably modest disposition that chimes with the spirit of most kinds of materialism.

By attending to events, Deleuze provided a profound reconceptualization of space and time (Deleuze, 1988a, 1994). The Deleuzian 'event' is far removed from the commonsensical notion of something that is simply present to hand and fully given (what Derrida calls the 'metaphysics of presence'). Rather than a self-contained Now that happens once and for all (at time *t*), an event is an untimely Meanwhile or Meantime: a becoming-multiple (*and, but*) rather than a being-one (*is*). So, space-time does not consist of points, but of folds (Deleuze, 1993a). However, although an event is never simply present, it is not at all lacking. It is full (i.e. immanent) – but it is fully *distributed* (cf. Derrida's quasi-concepts of *différance*, dissemination and supplementarity). One is reminded of the 'stuttering' of events in the plays of Samuel Beckett and the 'undecidability' of events in the novels of Franz Kafka. In Deleuze's version of empiricism, then, an event is as much virtual as it is actual. In accordance with everyday usage, an event is not a matter of fact;

it is extraordinary and irruptive. Events are explosive: they unsettle what appears to be given, and breathe life into what appears spent. What Dada and Marcel Duchamp did for art, literature and politics, Deleuze did for philosophy.

Given that modern philosophy has not been especially concerned with events (and still less with spatialization), one can appreciate why Deleuze – looking back to the Stoics – has often been characterized as an outsider, even during his long apprenticeship in the history of philosophy, which gave rise to what he called 'monstrous' caricatures of Henri Bergson, David Hume, Gottfried Leibniz, Immanuel Kant and Baruch Spinoza, among others. These anti-rationalist portraits were not so much bastardizations as 'immaculate conceptions'. Deleuze perfected the duplicitous art of differential repetition (through which one repeats in order to make different and estrange) which he rehearsed in the history of philosophy before unleashing it on the world at large (e.g. a 'clean-shaven Marx' and a 'bearded Hegel'; a 'cracked-I' and a 'stationary trip').

For our purposes, it is noteworthy that during each 'immaculate conception' Deleuze was eager to disclose the geo-philosophy that subtends the history of philosophy: multiplicities in Bergson, cartography in **Foucault**, relations in Hume, folds in Leibniz, territories in Kant, articulations in Spinoza, and differences in Nietzsche. When Deleuze began to write in his own name – in books such as *Difference and Repetition* and *The Logic of Sense* – this spatial awareness was already well developed. When he wrote with Guattari – in books such as *Anti-Oedipus, Kafka, A Thousand Plateaus*, and *What is Philosophy?* – one has the sense that there is geography, nothing but geography: maps, planes, surfaces, strata, spaces, territories, transversals, etc. And although Deleuze's work is often regarded as dry and difficult, he was fond of putting ordinary words to extraordinary use: virtual reality, becoming-woman, black holes, molecular revolution, geology of morals, Body without Organs, smooth space, etc. While this eclectic 'pop philosophy' composed of deterritorialized terms – terms that have been taken from their customary habitats in order to address problems in other domains – has been appreciated by many (not least for its humour), it has obviously given rise to much misunderstanding and even infringement claims. Indeed, Deleuze and Guattari seemed to go out of their way to provoke incredulity when they declared such things as 'Of course there are werewolves and vampires, we say this with all our hearts' (Deleuze and Guattari, 1987: 275).

Since thought is occasioned by events, 'There is no heaven for concepts. They must be invented, fabricated, or rather created and would be nothing without their creator's signature' (Deleuze and Guattari, 1994: 5). To put it bluntly, communication, interpretation, explanation, representation, reflection and criticism are more often than not a waste of time. They divert us from the 'true voyage', which is a pursuance of creative encounters. Little wonder, then, that Deleuze should have been called a 'radically horizontal' thinker: of difference rather than identity; seriality rather than hierarchy; and becoming (and . . . and . . . and . . .) rather than Being and Nothingness (is/is not). As Deleuze and Guattari famously declared at the outset of their delirious by-passing of the Marxian and Freudian impasse in *Anti-Oedipus* (1984: 2), 'A schizophrenic out for a walk is a better model than a neurotic lying on the analyst's couch.' Needless to say, their promotion of 'schizophrenia as process' has elicited considerable criticism and dismay (see Guattari, 1996, 2000, for clarification). Rather than enlist academics in the service of inquisitions and critical tribunals (state philosophy), Deleuze would prefer to take them out into the world for a stroll (nomad thought). Consequently, he was always on the lookout for creative encounters with texts, paintings, music, films, novels, situations, and events. He was always hoping to find something 'overwhelming' – an event – that would force him to think and in so doing estrange himself from himself. Such is his fondness for becoming, transformation and shape-shifting. Life always unfolds on the edge, on the horizon, and from the midst of things. Indeed, Deleuze's devotion to the immanence and singularity of creative encounters can be gleaned from his claim to have had no 'reserve knowledge'. Everything he learnt, he learnt *for* a specific encounter, and once that encounter was exhausted, he would forget everything and have to start all over again from zero. 'Even if it is for the hundredth time, you must encounter each thing as if you have never known it before', insists Paul Auster (1989: 7). 'No matter how many times, it must always be the first time. This is next to impossible, I realize, but it is an absolute rule.'

KEY ADVANCES AND CONTROVERSIES

During the 1990s, Deleuze's ex-centric empiricism, non-dialectical conception of difference, and immanent materialism began to make ripples in human geography (e.g. Doel, 1999). Like salacious gossip or the flu, word is slowly getting around – especially among poststructuralist, post-Marxist, and other so-called 'critical' human geographers – that Deleuze created an extraordinary form of *geo*-philosophy that proffers events for everyone (e.g. *Environment and Planning D*, 1996). Not only does this non-representational geo-philosophy provide an innovative basis for rethinking the nature of space and place, it is also a fully fledged 'thinking space' in its own right: not an abstract space *for* thought (a space of consciousness, representation, reflection, theory, etc., that would claim to be removed from the play of the world), but a concrete space *of* thought (a portion, region or milieu within the play of the world) – an honest-to-goodness *thinking* space. Quite simply, Deleuze, like Samuel Beckett and Georges Perec, had a remarkable ability to enliven even the most barren and derelict of spaces (an exhausted text, a sparsely furnished room, a stretch of pavement . . .). He showed not only that every kind of space teems with life, but that space *itself* is alive. 'He is like a cork floating on a tempestuous ocean: he no longer moves, but is in an element that moves' (Deleuze, 1997: 26).

The enlivenment of our surroundings – which is sometimes called vitalism, nomadism, materialism or immanence – is perhaps Deleuze's greatest gift to geography, a discipline that remains far too enamoured by a grey and dreary conception of the world as so much dead matter: objects, things, and their inter-relationship. We live in the terrible shadow of the eighteenth-century Enlightenment and its brutal dis-enchantment of the world, which was an expression of our collective desire to make the world around us play dead so that it would not resist being subjugated to our will. At the extreme, even people were evacuated of life. More often than not, people are reduced to data, numbers, cases, opinions and experience. Indeed, it is tempting to say that

the real dullards are those who desire a matter-of-fact world contemplated by passionless and disinterested subjects. For what they leave in their wake is a morgue dressed up as a critical tribunal. Like all of the so-called poststructuralist authors (e.g. **Baudrillard**, Cixous, Derrida, **Foucault**, Kristeva and Lyotard), Deleuze can help us to desist from stilling life, to become sensitive to the vivacity of space, and to create new spaces for life and new ways of being. Most of all, he can help us to become worthy of the specificity of geography: enabling the event of space to take place.

We should end by noting that Deleuze had no wish to be a famous intellectual and was truly horrified by the prospect of being a 'master' thinker with followers and disciples. Given his fondness for encounters, estrangement and transformation, Deleuze always favoured what was ex-centric (out of place and untimely), impersonal (collective and transhuman) and imperceptible (the becoming of the event). In turning to Deleuze, then, we should not turn him into a shining star that would reign over us: bright, eternal and heavenly. Rather, we should turn to him as if we were turning towards the event horizon of a black hole. 'I dream not of being invisible, but imperceptible' (Deleuze, 1977: 112).

DELEUZE'S MAJOR WORKS

Deleuze, G. (1983) *Nietzsche and Philosophy*. Trans. H. Tomlinson. New York: Columbia University Press [1962].

Deleuze, G. (1988a) *Bergsonism*. Trans. H. Tomlinson and B. Habberjam. New York: Zone [1966].

Deleuze, G. (1988b) *Foucault*. Trans. S. Hand. Minneapolis: Minnesota University Press [1986].

Deleuze, G. (1990b) *The Logic of Sense*. Trans. M. Lester with C. Stivale. Ed. C. V. Boundas. New York: Columbia University Press [1969].

Deleuze, G. (1993) *The Fold: Leibniz and the Baroque*. Trans. T. Conley. Minneapolis: University of Minnesota Press [1988].

Deleuze, G. (1994) *Difference and Repetition*. Trans. P. Patton. London: Athlone [1968].

Deleuze, G. and Guattari, F. (1984) *Anti-Oedipus: Capitalism and Schizophrenia*. Trans. R. Hurley, M. Seem and H. R. Lane. London: Athlone [1972].

Deleuze, G. and Guattari, F. (1986) *Kafka: Toward a Minor Literature*. Trans. D. Polan. Minneapolis: University of Minnesota Press [1975].

Deleuze, G. and Guattari, F. (1987) *A Thousand Plateaus: Capitalism and Schizophrenia*. Trans. B. Massumi. Minneapolis: University of Minnesota Press [1980].

Deleuze, G. and Guattari, F. (1994) *What is Philosophy?* Trans. G. Burchell and H. Tomlinson. London: Verso [1991].

Secondary Sources and References

Ansell Pearson, K. (ed.) (1997) *Deleuze and Philosophy: The Difference Engineer.* London: Routledge.

Auster, P. (1989) *In the Country of Last Things.* London: Faber & Faber.

Badiou, A. (1999) *Deleuze: The Clamour of Being.* Trans. L. Burchill. Minneapolis: University of Minnesota Press.

Bogue, R. (1989) *Deleuze and Guattari.* London: Routledge.

Boundas, C. V. (ed.) (1993) *The Deleuze Reader.* New York: Columbia University Press.

Boundas, C. V. and Olkowski, D. (eds) (1994) *Gilles Deleuze and the Theatre of Philosophy.* London: Routledge.

Deleuze, G. (1977) 'I have nothing to admit' *Semiotext(e)* 2(3): 111–116 [1973].

Deleuze, G. (1986) *Cinema 1: The Movement-image.* Trans. H. Tomlinson and B. Habberjam. Minneapolis: Minnesota University Press [1983].

Deleuze, G. (1988c) *Spinoza: Practical Philosophy.* Trans. R. Hurley. San Francisco: City Lights [1970, expanded edition 1981].

Deleuze, G. (1989) *Cinema 2: The Time-image.* Trans. H. Tomlinson and R. Galeta. Minneapolis: Minnesota University Press [1985].

Deleuze, G. (1990a) *Expressionism in Philosophy: Spinoza.* Trans. M. Joughin. New York: Zone [1968].

Deleuze, G. (1995) *Negotiations, 1972–1990.* Trans. M. Joughin. New York: Columbia University Press [1990].

Deleuze, G. (1997) *Essays Critical and Clinical.* Trans. D. W. Smith and M. A. Greco. Minneapolis: Minnesota University Press [1993].

Deleuze, G. and Parnet, C. (1987) *Dialogues.* Trans. H. Tomlinson and B. Habberjam. New York: Columbia University Press [1977].

Derrida, J. (1998) 'I'll have to wander all alone', *Tympanum* 1 http://www.usc.edu/dept/comp-lit/tympanum/1/derrida1.html; accessed 15.01.03.

Doel, M. A. (1999) *Poststructuralist Geographies: The Diabolical Art of Spatial Science.* Edinburgh: Edinburgh University Press.

Environment and Planning D: Society and Space (1996) Theme issue on Deleuze and Geography. J. M. Jacobs and M. Morris (eds) volume 14 (4).

Goodchild, P. (1996) *Deleuze and Guattari: An Introduction to the Politics of Desire.* London: Sage.

Guattari, F. (1996) *Chaosophy: Soft Subversions.* Trans. D. L. Sweet and C. Weiner. Ed. S. Lotringer. New York: Semiotext(e).

Guattari, F. (2000) *The Three Ecologies.* Trans. I. Pinder and P. Sutton. London: Athlone.

Hardt, M. (1993) *Gilles Deleuze: An Apprenticeship in Philosophy.* Minneapolis: University of Minnesota Press.

Kaufman, E. and Heller, K. J. (eds) (1998) *Deleuze and Guattari: New Mappings in Politics, Philosophy, and Culture.* Minneapolis: University of Minnesota Press.

Lecercle, J.-J. (1985) *Philosophy through the Looking-Glass: Language, Nonsense, Desire.* La Salle: Open Court.

Massumi, B. (1992) *A User's Guide to Capitalism and Schizophrenia: Deviations from Deleuze and Guattari.* London: MIT.

Patton, P. (ed.) (1996) *Deleuze: A Critical Reader.* Oxford: Blackwell.

Marcus A. Doel and David B. Clarke

16 Peter Dicken

BIOGRAPHICAL DETAILS AND THEORETICAL CONTEXT

Peter Dicken has been one of the most influential economic geographers in the discipline over the last 30 years or more. From the foundations of his key text *Location in Space: A Theoretical Approach to Economic Geography*, with Peter Lloyd (1990), first published in 1972, Dicken has consistently published very high-quality journal articles, book chapters and books that have investigated: global economic geographies of industrial change; transnational corporations in the world economy; economic development in east Asia (with particular focus on business networks and production chains), and, very important-ly, the effects of global economic change on different geographical scales – from the global to the local. Dicken's most significant impact in the academy for both research, and teaching and learning, has been his seminal text *Global Shift*, now in its fourth edition, which has provided a bedrock for scholars studying the uneven geographies of globalization since the publication of its first edition in 1986. Though written to demonstrate the com-plex global articulation of *economic* pro-duction chains, *Global Shift* has proved influential beyond the subdiscipline of economic geography, standing as a key reference in debates concerning the de-clining sovereignty of the nation-state and the formation of a global society.

Born in 1938, Dicken is a 'Manchester man' through and through. He was awarded his Personal Chair in 1988 after joining the department in 1966 following the successful completion of his MA from the same university. He was awarded his PhD from the University of Uppsala, Sweden. Over almost four decades in academia, he has held distinguished re-search and teaching positions at universi-ties in Australia, Canada, Hong Kong, Mexico, Singapore and the United States, and in 1999 became a Fellow of the Swedish Collegium for Advanced Study in the Social Sciences. The quality and policy relevance of his work has resulted in appointments as the Co-Director of European Science Foun-dation Scientific Programme on Regional and Urban Restructuring in Europe (1989–94), and as a consultant advisor to the UNCTAD Commission on Transnational Corporations (1993–94). He has several editorial positions on international journal boards (including *Competition and Change*, *Journal of Economic Geography*, *Global Net-works*, *Review of International Political Econ-omy*) and is the managing editor of *Progress in Human Geography*. As his work on the strategic behaviour of firms and interna-tional patterns of trade and investment criss-crosses with both management stu-dies and international economics, he has forged successful research links outside of geography. For example, a recent and notable research project has been with colleagues from Manchester Business School (Jeffrey Henderson) and the Nation-al University of Singapore (Henry Yeung) investigating global production networks in Britain, east Asia and eastern Europe. Apart from driving global research agendas in economic geography, Dicken has also been committed to supervising graduate stu-dents (e.g. Henry Yeung) and entertaining (large) audiences on the international con-ference circuit.

A year before **David Harvey** pub-lished *Social Justice and the City*, Peter

Dicken and Peter Lloyd (of the University of Liverpool) published *Location in Space: A Theoretical Approach to Economic Geography*. For economic geography, it became a benchmark for the period. No stone was left unturned in the search for explaining the organization of economy, locational analysis, regional economic development in space and the differential (rather than uneven) economic growth rates experienced in North America and Europe. In essence, this text was devoted to explicating the 'economic' in economic geography. Here, Dicken and his co-author were heavily influenced by the classical and neo-classical modelling of locational theorists (e.g. Christaller, 1966; Isard, 1956; Losch, 1954) and, of course, the path-breaking work of **Peter Haggett**'s (1965) *Locational Analysis in Human Geography* (as well as Haggett's co-writing with Richard Chorley, 1967, 1969). But, if we wind the clock on a full 18 years and read the third edition of *Location in Space* (Dicken and Lloyd, 1990), we begin to unravel other influences on Dicken's view of the (economic) world. The neo-classical spirit of *Location in Space* is prevalent, but Dicken and Lloyd rework the interpretation of the 'economic' in space by considering the *political economy* of location and its uneven distribution through time and space. For example, they discuss the key strategic role of transnational corporations in restructuring the world economy (as espoused by Taylor and Thrift, 1983), introduce the notion of 'chains' of business organization and location (as discussed by Porter, 1985), examine geographies of corporate organization and control (following Pred, 1974), investigate geographical 'linkages' in location (see Scott, 1984) and provide a brief thumbnail sketch of Marx's theories on capital and labour (which are drawn from original sources and informed by Harvey (1973, 1982) and others – e.g. Massey, 1984; Scott and Storper, 1986).

In 1986, Dicken published *Global Shift: Industrial Change in a Turbulent World*. In many ways, this text stands as Dicken's definitive statement on geo-

graphies of production: it is certainly his best known and most widely cited work. In this text, Dicken looked afresh at explaining locational change in the world and to assist him in fulfilling this project, he looked beyond the geographical community for inspiration and ideas, especially from those who studied international economics and the strategic behaviour of organizations and transnational corporations. Of significance here is Dicken's appreciation of writers like Hymer (1972) on multinational corporations, Dunning (1980) on why firms engage in international production and Michalet (1980) on international subcontracting. Also significant was Dicken's use of an array of in-depth case studies (e.g. textiles and clothing) and empirics derived from General Agreement on Tariffs and Trade (GATT), the United Nations Centre for Transnational Corporations (UNCTC), the International Labour Organization (ILO) and Organization for Economic Cooperation and Development (OECD) to tease out the political economy of industrial location at global–local scales.

SPATIAL CONTRIBUTIONS

Peter Dicken has always been fascinated by economic restructuring at the global–local scale (Dicken, 2003) and, in particular, investigating the role of the transnational corporation in producing uneven development as transactionally linked chains of production stretch international and global space (e.g. Dicken, 1976, 1986; Dicken and Malmberg, 2001). The contributions that Dicken's work has made to advancing our understanding of the geographies of restructuring within firms, and the ways in which firms impact upon different places around the globe, is considerable. These contributions can be summarized around three main themes.

First, Dicken has provided one of the most incisive and detailed geographical analysis of transnational corporations in the world economy (e.g. Dicken, 1971, 1980, 1994). From the analysis of the foreign direct investment patterns of Japanese firms in Britain (e.g. Dicken, 1990; Dicken and Lloyd, 1980; Dicken and Tickell, 1997), to in-depth studies of different sectors (e.g. automobiles – Dicken, 1987, 1992b) or, latterly, investigations into firms' production chains and networks (e.g. business networks in the Indonesian clothing industry – Dicken and Hassler, 2000), Dicken has been at the forefront of unpacking the organizational role and geographies of the firm in (re)producing uneven development in economy and society. Of significant interest here is Dicken's reading of how firms impinge/impact upon and restructure local economies (and the state) through their organizational strategies and 'footloose' tendencies (i.e. being able to switch capital from one location to another with little friction – see Allen, 1995).

Second, Dicken has brought us a greater understanding of the role, locational behaviour and organization of Japanese capital (and firms) in the West. He has published very detailed analysis of Japanese foreign direct investment in Britain (see above), case study material on Nissan in Washington (Dicken, 1984) and, more importantly, introduced in great depth the organization of Japanese trading companies (soga shosha) to the Anglo-American academy (Dicken and Miyamachi, 1998). Dicken's research interests in Japanese firms and their management and labour practices outside of Japan reflect a much wider interest in economic restructuring in east and Pacific Asia especially (Dicken, 1987; Dicken et al., 1999).

Third, following on from the above, Dicken's work on economic restructuring within east and Pacific Asia has provided rich pickings for those seeking to understand the significance of chains and network structures within not only transnational corporations but, more im-

portantly, within the world economy (Dicken et al., 2001). Dicken's earlier work on business networks in organizations (Dicken and Thrift, 1992) has provided the foundations for us to think about the organization of economy (and places) in terms of functionally integrated linkages, connections and flows from both traditional Western and Asian business systems (Dicken, 2000). Moreover, Dicken has conceptualized chains and networks in economy through territorialization, bringing scale back onto the agenda in studies of both the firm and industrial sectors (Dicken et al., 2002).

KEY ADVANCES AND CONTROVERSIES

Dicken's key advances in the discipline have already been discussed at length: analysing transnational corporations as a barometer for studying global economic change; unpacking the importance of production chains and networks in understanding contemporary patterns in economy; and examining global economic change in relation to a global–local dialectic. The reaction to this work in the discipline has been, on the whole, both welcomed and relatively uncontroversial. Many view Dicken's work as at the vanguard of illustrating the manifestations of economic globalization in contemporary society by providing neatly crafted case studies and explanations for such patterns. Dicken has summarized his key arguments about geography, location and space as follows:

> The basis of my argument is that firms, just like all other forms of social organization, are fundamentally and intrinsically *spatial* and *territorial*. They are spatial in the sense that they are responsive to geographical distance and to spatial variations in the availability of necessary resources and of business opportunities. Such spatiality may have – indeed most

often has – a territorial manifestation. Hence, firms are territorial as well as spatial in the sense that the 'surface' from which firms originate and on which they operate is most commonly made up of a tessellated structure of territorial entities arrayed along a continuum of variable and overlapping scales, including those of political governance . . . For some functions of the firm the territory may be intensely local, for others it may approach the global.

(Dicken, 2002: 12)

Dicken elucidates this global–local perspective in *Global Shift*, from first to fourth editions. *Global Shift* is Dicken's major contribution to geography and encapsulates his advocacy of spatiality. The 2003 edition, *Global Shift: Reshaping the Global Economic Map in the 21st Century*, has excelled itself in this regard. Not only does Dicken interpret the significant ideas offered by the likes of John Allen, Ash Amin, Gordon Clark, Meric Gertler, Ron Martin, Anders Malmberg, Jamie Peck, Erica Schoenberger, **Michael Storper**, **Nigel Thrift** and Henry Yeung, but he

has once again weaved into the text theoretical contributions from a wide spectrum of business and managment sources (e.g. Gereffi, Krugman, Porter, Sklair, Whitley) to provide a richness that could not have otherwise been achieved through the lens of the geographer. For example, Dicken's *geographical* explanation of the organizational capabilities and competences of transnational corporations has been greatly influenced by Bartlett and Goshal's (1998) notions of 'managing across borders', where they distinguish between transnational, multinational, global and international organizational forms of the firm. Herein lies the contribution of Dicken to the discipline and beyond, in that not only have his ideas and approach been a benchmark for other geographers, but also his work and ideas have been accepted in international economics (especially on foreign direct investment) and organizational studies that focus on the strategic behaviour of firms in the world economy.

DICKEN'S MAJOR WORKS

Dicken, P. (1988) 'The changing geography of Japanese foreign direct investment in manufacturing industry: a global perspective', *Environment and Planning A* 20: 633–653.

Dicken, P. (1992a) 'International production in a volatile regulatory environment: The influence of national regulatory policies on the spatial strategies of transnational corporations', *Geoforum* 23: 303–316.

Dicken, P. (1994) 'Global–local tensions: firms and states in the global economy', *Economic Geography* 70: 101–128.

Dicken, P. (1997) 'Transnational corporations and nation-states', *International Social Science Journal* 151: 77–89.

Dicken, P. (2003) *Global Shift: Reshaping the Global Economic Map in the 21st Century* London: Sage. (4th edition) (also see 1st to 3rd editions).

Dicken, P. and Lloyd. P. E. (1981) *Modern Western Society: A Geographical Perspective on Work and Well-Being*. London: Harper and Row.

Dicken, P. and Lloyd, P. E. (1990) *Location in Space: Theoretical Perspectives in Economic Geography*. New York: Harper and Row (3rd edition) (also see 1st and 2nd editions).

Dicken, P. and Miyamachi, Y. (1998) 'From noodles to satellites: The changing geography of Japanese *sogo shosha*', *Transactions of the Institute of British Geographers* 23: 55–78.

Dicken, P. and Thrift, N. (1992) 'The organization of production and the production of organization: why business enterprise matter in the study of geographical internationalisation', *Transactions of the Institute of British Geographers* 17: 101–128.

Dicken, P., Kelly, P. F., Olds, K. and Yeung, H. (2001) 'Chains and networks, territories and scales: towards a relational framework for analysing the global economy', *Global Networks* 1: 99–123.

Secondary Sources and References

Allen, J. (1995) 'Crossing borders: footloose multinationals', in J. Allen and C. Hamnett (eds) *A Shrinking World*. Oxford: Oxford University Press, pp. 55–102.

Bartlett, C. and Ghoshal, S. (1998) *Managing Across Borders: The Transnational Solution*. London: Random House (3rd edition).

Christaller, W. (1966) *Central Places in Southern Germany*. Englewood Cliffs, NJ: Prentice Hall.

Dicken, P. (1971) 'Some aspects of the decision-making behaviour of business organizations', *Economic Geography* 47: 426–437.

Dicken, P. (1976) 'The multiplant enterprise and geographical space: Some issues in the study of external control and regional development', *Regional Studies* 10: 401–412.

Dicken, P. (1980) 'Foreign direct investment in European manufacturing industry: the changing position of the United Kingdom as a host country', *Geoforum* 11: 289–313.

Dicken, P. (1984) 'Washington welcomes Nissan', *Geographical Magazine* 56: 286–287.

Dicken, P. (1986) *Global Shift: Industrial Change in a Turbulent World*. New York: Harper and Row.

Dicken, P. (1987) 'Japanese penetration of the European automobile industry: the arrival of Nissan in the United Kingdom', *Tijdschrift voor Economische en Sociale Geografie* 78: 94–107.

Dicken, P. (1990) 'Japanese industrial investment in the U.K.', *Geography* 75: 351–354.

Dicken, P. (1992b) 'Europe 1992 and strategic change in the international automobile industry', *Environment and Planning A* 24: 11–31.

Dicken, P. (2000) 'Places and flows: Situating international investment', in G. Clark, M. Gertler and M. A. Feldman (eds) *A Handbook of Economic Geography*. Oxford: Oxford University Press, pp. 275–299.

Dicken, P. (2002) 'Placing firms, firming places: grounding the debate on the global corporation', paper presented at the Conference on Responding to Globalization: Societies, Groups, and Individuals, University of Colorado, Boulder.

Dicken, P. and Hassler, M. (2000) 'Organizing the Indonesian clothing industry in the global economy: The role of business networks', *Environment and Planning A* 32: 263–280.

Dicken, P. and Lloyd, P. (1980) 'Patterns and processes of change in the spatial distribution of foreign-controlled manufacturing employment in the United Kingdom 1963–1975', *Environment and Planning A* 12: 1405–1426.

Dicken, P. and Malmberg, A. (2001) 'Firms in territories: A relational perspective', *Economic Geography* 77: 345–363.

Dicken, P. and Tickell, A. (1997) 'Putting Japanese investment in Europe in its place', *Area* 29: 200–212.

Dicken, P., Olds, K., Kelly, P. F., Kong, L. and Yeung, H. (eds) (1999) *Globalization and the Asian Pacific: Contested Territories*. London: Routledge.

Dicken, P., Henderson, J., Hess, M., Coe, N. and Yeung, H. (2002) 'Global production netwroks and the analysis of economic development', *Review of International Political Economy* 9: 436–464.

Dunning, J. (1980) 'Towards an eclectic theory of international production: some empirical tests', *Journal of International Business Studies* 11: 9–31.

Haggett, P. (1965) *Locational Analysis in Human Geography*. London: Edward Arnold.

Haggett, P. and Chorley, R. (eds) (1967) *Models in Geography*. London: Edward Arnold.

Haggett, P. and Chorley, R. (1969) *Network Analysis in Geography*. London: Edward Arnold.

Harvey, D. (1973) *Social Justice and the City*. London: Edward Arnold.

Harvey, D. (1982) *The Limits to Capital*. Oxford: Blackwell.

Hymer, S. (1972) 'The multinational corporation and the law of uneven development', in J. N. Bhagwati (ed.) *Economics and the World Order*. London: Macmillan, pp. 113–140.

Isard, W. (1956) *Location and Space Economy*. Cambridge, MA: M.I.T. Press.

Lloyd, P. and Dicken, P. (1972) *Location in Space: A Theoretical Approach to Economic Geography*. London: Harper and Row.

Losch, A. (1954) *The Economics of Location*. New Haven, CT: Yale University Press.

Massey, D. (1984) *Spatial Divisions of Labour. Social Structures and the Geography of Production*. London: Macmillan.

Michalet, C-A. (1980) 'International subcontracting: a state of the art', in D. Germidis (ed.) *International Subcontracting a New Form of Investment*. Paris: O.E.C.D.

Porter, M. (1985) *Competitive Advantage*. New York: Free Press.

Pred, A. (1974) 'Industry, information and City-System Interdependencies', in F. E. I. Hamilton (ed.) *Spatial Perspectives on Indistrial Organization and Decision Making*. London: Wiley, pp. 105–139.

Scott, A. (1984) 'Industrial organization and the logic of inter-metropolitan location: I. Theoretical Considerations', *Economic Geography* 59: 111–142.

Scott, A. and Storper, M. (eds) (1986) *Production, Work, Territory. A Geographical Anatomy of Industrial Capitalism*. Winchester, MA: Allen and Unwin.

Taylor, M. and Thrift, N. (1983) 'Business organization, segmentation and location', *Regional Studies* 17: 455–465.

Jonathan V. Beaverstock

17 Arturo Escobar

BIOGRAPHICAL DETAILS AND THEORETICAL CONTEXT

Arturo Escobar was born in Manizales, Colombia, in 1952, and first trained as a Chemical Engineer, graduating from the Universidad del Valle in Cali in 1975. After a year of studying Biochemistry, he relocated to the USA, completing a Masters in Food Science and International Nutrition at Cornell in 1978, before enrolling in an interdisciplinary PhD programme at Berkeley. By this time, Escobar's interests had undergone a major shift towards the social sciences and questions of power, international development and planning. Doctoral work, which included a year of fieldwork in Colombia, saw early expression in a brilliant article published in the Indian journal *Alternatives*, where he applied **Foucault**'s notions of power to the study of international development (Escobar, 1984). His theoretical argument – that development should be seen as a discourse of power and control – was new and challenging, and by the early 1990s Escobar was established as a leading thinker among a strong group of 'post-development' theorists including Ashis Nandy, Wolfgang Sachs and James Ferguson (Escobar, 1992; Rahnema and Bawtree, 1997). Having completed three years as a lecturer in Latin American Studies at University of California at Santa Cruz in 1989, Escobar moved firmly into the discipline of anthropology, beginning as assistant professor at Smith College, and then shifting nearby to the University of Massachusetts at Amherst, where he taught for five years. In 2000 he became the Kenan Distinguished Teaching Professor of Anthropology, University of North Carolina. Shorter periods spent teaching in Colombia, the UK and Spain, and frequent speaking engagements worldwide, have exposed different audiences to his work.

Few twentieth-century ideas have sparked such a prolonged controversy as that of Western 'development', and it was Escobar's major monograph, *Encountering Development* (1995) – an elaboration of the work he had been conducting since the 1980s – that has elevated him to the status of post-development icon. For many critics today, development has reached an impasse. Escobar and the post-development theorists have built upon the work of many others to expose how ' ... development was shown to be a pervasive cultural discourse with profound consequences for the production of social reality in the so-called Third World' (Escobar, 2000: 11). In his later work, Escobar has begun to look beyond the failures and limitations of state, market and international aid to a form of social change led by new social movements and progressive nongovernmental organizations.

This second strand to Escobar's work emerged in the mid-1990s, when he conducted a year of fieldwork on the Pacific coast of Colombia, the first of several periods during which he worked with Afro-Colombians (descendants from African slaves brought to mine gold) and their activist organizations and networks in the region (Escobar and Pedrosa, 1996). The coast is a hot-spot of biological diversity, and has been the subject of several resource conservation projects. It is home to Afro-Colombian rights movements with a strong sense of place and territory, and

Escobar's research traverses his belief in the power of these place-based social movements as alternatives to national and Western development efforts, and his interests in nature, which he describes as a 'constructed' category that becomes immersed in discursive and material struggles (over the meaning of biodiversity and sustainability, for example). Some of his work in this region of Colombia is, therefore, framed in the language of political ecology, and geographical concepts of place and territory are critical to his analysis (Escobar 1996a, 1998, 2001).

SPATIAL CONTRIBUTIONS

Escobar believes in the 'task of imagining alternatives' (1995: 14), and his work is stimulating and provocative. Aside from a wide-ranging conversation that has developed around his post-development critique in anthropology and development studies, some geographers have used his analysis as a point of departure for studies on social movements and development alternatives.

Turning first to Escobar's theorizing of development, we find much of interest to geography, even though he himself claims that his engagement with the discipline has been modest (personal communication, 2003). The ideas expressed in *Encountering Development*, for example, highlight some of the spatial outcomes of the hegemonic 'development' discourse since colonial times. The argument is that Western 'development', particularly during the Cold War, lies behind the construction of almost all aspects of social reality in the Third World, in such a pervasive way that 'that even its opponents were obliged to phrase their critiques in development terms: another development, participatory development, socialist development, and so on' (Peet

and Hartwick, 1999: 145). Discourse, he argues, has the power to influence reality, following **Edward Said**'s theory of Orientalism (Said, 1978: 3). The production of knowledge and the planning of development by Western institutions is something that Third World countries and regions find it hard to escape from. The process of dominating, restructuring, and establishing authority progresses in three stages:

1. The progressive identification of Third World problems, to be treated by specific interventions. This creates a 'field of the interventions of power'.
2. The professionalization of development; the recasting of political problems into neutral scientific terms (poverty indicators, for example), leading to a regime of truth and norms, or a 'field of the control of knowledge'.
3. The institutionalization of development to treat these 'problems' and the formation of a network of new sites of power/knowledge that bind people to certain behaviours and rationalities (in rural development discourse, 'produce or perish' became one such norm – Escobar, 1995: 157).

The professionalization of development in the post-World War II period, Escobar argues, incorporated the Third World as research data in 'academic programs, conferences, consultancy services and local extension services and so on' but also poverty, illiteracy and hunger 'became the basis for an industry for planners, experts and civil servants' (1995: 46). Since this industry never stops producing goods in the form of new projects and reports, but actually achieves its targets only rarely, it justifies its own continued existence.

The development discourse has, therefore, created underdevelopment of the Third World in a much more subtle form than colonialism. Escobar's words are purposely harsh, and they suggest an instrumental aim lying behind a universal (Western) paradigm in which many people, including Western geographers, anthropologists and other social scientists are complicit. The very idea of development is framed by the Western geopolitical imagination that seeks to 'subordinate, contain and assimilate the Third World as Other. Such Western imagination is a violation of the rights of the other societies' (Slater, 1993: 421).

As alternatives, Escobar favours the responses of indigenous/autonomous social movements, and localized strategies of development, rather than the radical overturning of Western-dominated geopolitical relations by the state at a macro level, as was proposed by earlier dependency theorists. Like **Stuart Corbridge** and **Michael Watts**, he argues that emancipatory possibilities exist, because criticism of the 'mainstream' can translate into viable alternatives, and the types of new social movements made famous by writers like **Manual Castells** (1983) can create new conversational communities. Place-based social movements need 'territory', in which their livelihoods or life project are largely conducted, but they also operate in a 'region-territory', which is a 'political construction for the defense of the territories and their sustainability' (Escobar, 2001: 162). This can include more extensive networks, including those now offered by internet technology; his examples include the U'wa indigenous group in Colombia who have mobilized against the Colombian government and Occidental Petroleum to oppose oil exploration (Escobar and Harcourt, 2002). Escobar follows many scholars of indigenous rights in stressing the attachment that such groups feel to particular places – 'The struggle for territory is thus a cultural struggle for autonomy and self-determination. This explains why for many people of the Pacific the loss of territory would amount to a return to slavery or, worse perhaps, to becoming "common citizens"' (Escobar, 2001: 162). Social movements are therefore intimately linked to local geographies; culture resides in places, even under conditions of mass globalization.

In his work addressing matters of nature and the environment, Escobar makes further claims that are of interest to those working around the slippery interdisciplinary field of political ecology, to which anthropologists and geographers have contributed in equal measure (Escobar, 1999a, 1998). Escobar's arguments here are complex, but revolve around that view that much of nature is now artificially produced and, in some fashion, deeply imbricated with technology and social relations, and that these relationships are 'hybrid and multiform' (Escobar, 1999a: 1). He defines political ecology as 'the contingent study of the manifold articulations of history and biology and the cultural mediations through which such articulations are necessarily established' (1999a: 3). This broad-ranging definition neatly combines the concerns of 'realists' interested in the material transformations of the natural world by human actions (cultural ecologists, for example, and the majority of geographers calling themselves political ecologists) and those who perceive nature as a historically and socially constructed category.

Political ecology, for Escobar, should be anti-essentialist, in order to situate the complex of meanings of nature–human relations in the larger context of history and power. He identifies three distinct but interlinked nature regimes, and delineates their characteristics. The regime of organic nature is most commonly found in non-industrialized societies, and best analysed through anti-dualist conceptions of nature–culture and local knowledge. Capitalist nature is that which is commodified (as in the case of bioprospecting operations) and governed. The third, techno-nonature, is artificial, newly manipulated through biotechnology and engineering. These three regimes are not a series of linear stages towards modern life – 'They coexist and overlap' (1999a: 5) and raise substantive questions about the power of discourses about nature – even the term 'biodiversity' resulted from a modern scientific world-view, and such activities are protected through networks, actors and strategies (Escobar, 1998). Escobar urges the creation of a balanced position that, according to David Cleveland, 'acknowledges the constructedness of nature – the fact that much of what ecologists refer to as nature is a product of culture – and nature in the real sense, that is existence of an order of nature, including the body' (Cleveland, 1999: 17).

The type of political ecology advocated by Escobar intends to go beyond naturalism, the common philosophical foundation behind nature conservation, because nature and culture are in fact hybridized as 'cultured nature'. This analysis of hybridization, in a non-essentialist and trans-disciplinary way, makes Escobar's study of nature distinctive. It can be glimpsed indirectly in new, critical geographical work that stresses the social construction of nature and the imposition of dominant discourses. Examples here include work on 'fortress conservation' and wilderness protection in Africa (Neumann, 1998) and Central America (Sundberg, 1998), and in critiques of sustainable development (Rocheleau et al., 2001) and bureaucratic forestry (Robbins, 2001). In anthropology, new varieties of political ecology are constantly evolving, with Fairhead and Leach's *Misreading the African Landscape* (1996) offering a counter to dominant Western views of forest loss, and Stone's work on the perils of genetic modification offering empirical data on 'technonatures' in action (2002).

KEY ADVANCES AND CONTROVERSIES

Since Escobar's major argument has been to challenge the language, discourse and project of mainstream development as a failed modernist project, a critical response to these ideas has been all but inevitable, and it has come from several quarters. For instance, a group of critics argue that his attack on development misses the target, since 'the problem is not so much with development, even less so with modernity, than with capitalism' (Escobar, 2000: 12). Richard Peet and **Michael Watts** and most of their contributors to their book, *Liberation Ecologies* (Peet and Watts, 1996), wish to balance the attention given to discourses with analysis of the impact of material transformations, while Peet and Hartwick (1999) and Kiely (1999) are more blunt: capitalist material relations have penetrated all corners of the globe. Thus, development discourse has arisen first and foremost from the spread of capitalism (Fernando and Kamat, 2000). This critique, therefore, sees the post-development project as partial. Progressive change requires the transformation of the social relations that produced and that sustain the discourse, and this needs active intervention against capitalism (i.e. a new and revolutionary form of development itself), and global solidarity.

A related concern is that development – including the actions of the state – has, of course, long contained progressive critical voices and practices, and is far from monolithic in its opposition to local and marginal voices (Gardner and Lewis, 1996; Lehmann, 1997). Lumping together progressive aid agencies or the actions of radical NGOs with the worst technocratic and domineering aid projects practising their craft on their 'clients', and then collectively condemning the totality as an instrument of Western power (as passages in *Encountering Development* come close to doing), denies these differences and denigrates some of those genuinely involved in radical praxis. It also denies the possibility that development anthropology can promote *better* development or state policies, using ethnographic and technical skills (Gardner and Lewis, 1996; Little and Painter, 1995). Discourse analysis may, therefore, tend to throw together diverse and contested positions of development simply because they 'share the same discursive space'. Escobar may, therefore, be practising a form of essentialism after all (Fernando and Kamat, 2000; Kiely, 1999).

A third question concerns the power of place-based social movements. There is much evidence that many local activist struggles are not really about overturning the status quo or challenging the global and national power relations in which they are embedded, but are more concerned with gaining access to development resources, some of them modern and Western – capital, paid labour, education, health, and so on. The argument goes that because capitalism is crisis-ridden, and it is a contradictory and uneven process, its ideology permeates everywhere, including distant social movements and their members. The most convincing critique of post-development in general, and of Escobar's work in particular, is along these lines. It comes from geographer Tony Bebbington in 'Re-encountering Development' (2000). Bebbington tries to resolve the debate between post-development and the neo-liberal interpretations informed by neo-classical economics (pro-market and modernizing), to call for a notion of development that is alternative, critical and practicable. He argues not only that the post-development case has theoretical shortcomings, but that it falls down empirically in the Andes, when put to the test in study of peasant culture and livelihoods. Criticisms of development's failures in the region are too blunt: plenty of cases exist in the Andes where symbols of failed development – high levels of out-migration, increased consumption of Western commodities, and imported knowledge and technologies – have been accompanied by 'increased indigenous control of everything from municipal government, to regional textile markets, to bus companies' and 'assertive and ever more ethnically self-conscious social organizations' (Bebbington, 2000: 496). Thus, alternatives to capitalist landscapes can emerge from all sorts of 'development' activities. Integral to these alternatives have been the work of the state (particularly its enabling of land reform), NGOs and churches. Post-development thinking either denies, or does not examine their influence, and does not seriously address what peasant farmers must do in the short term to make a living and to sustain their communities. Therefore we need to foreground 'problems of livelihood and production as much as problems of politics and power' and 'emphasize negotiation and accommodation as much as resistance' (Bebbington, 2000: 449).

In addressing some of these criticisms, Escobar has voiced some sympathy with the more radical (and poststructuralist) critics of post-development thinking, but he disputes the arguments that assert the 'the primacy of the material over the discursive' – partly on the grounds that insufficient attention is usually given to the role of language and meaning in the creation of reality (Escobar, 2000: 12). Nonetheless, movements form broader networks of power, and thus he sees Marxist geography as essentially correct in arguing for the grounding of place and identity-based social movements in wider political coalitions; he does recognize that social struggles, like those of the U'wa, form networks, 'in their theoretical and practical action – that is, in the production of alternative discourses' (Escobar and Harcourt, 2002: 4), and particularly around gender responsibilities and rights (the focus of some of his recent work). But he also acknowledges the valuable insights provided by a group of 'actor-oriented' sociologists and anthropologists that register power in a different way, tracing it across the interfaces between actors – for example between peasant farmers and NGO workers, exposing their mutual constructions, sense of identity, and actions, uncovered through detailed ethnography (Arce and Long, 2000). Francophone anthropology and work on rural politics has, in fact, taken these debates further, to highlight the subversion of development by peasant groups and the work of a whole class of 'interlocutors' who shuttle between rural communities and the development industry, representing one to the other (Batterbury, 2002; Bierschenk et al., 2000; Olivier de Sardan,

1995). He also has sympathies with Bebbington's (2000) alternative take on livelihoods and development. However, he refuses to respond to a persistent criticism that post-development offers no concrete political and economic programme for change. This, he says, is precisely the normative goal of most development thinking, and it is something he and his colleagues are trying escape from. The future is, then, up to people themselves, and their social movements must elaborate their own paths (Escobar, 2000: 14).

Escobar's thoughts on political ecology and the study of nature have attracted comment, but his interventions in this area are relatively new and untested, and will feature in two forthcoming volumes, one with Enrique Leff. David Cleveland (1999: 18) argues that essentialism is more difficult to escape from than Escobar makes out. For example, he says that 'in the organic regime nature is not manipulated' (Cleveland, 1999: 18), but this ignores the actions of certain pre-modern peoples, who clearly did exploit natural capital to an advanced degree (see Escobar, 1998, which partially addressed this). One of the founders of the field, Piers Blaikie (2000), takes on the implicit populism and support of the underdog in post-development thinking more generally, and perceives the analysis to be lacking in detailed research into development processes and policy. His view is that 'real' natural resource problems exist, and it is unhelpful to de-emphasize this. However, it is also true that the critical geographical studies that do employ discourse analysis have found ways to apply it quite effectively to specific development 'problems' and places, and there is great value in combining this type of analysis with other forms of investigation (Rocheleau et al., 2001; Neumann, 1998; Peet and Watts, 1996), some of them involving a greater degree of ecological analysis than Escobar has chosen, and more attention to the politics of gender (Rocheleau, 1999).

Despite this we should not diminish the power of Escobar's key political message – local culture matters. The grand revisionism in the classical Marxist account of social change gives way to something much more modest and localized in his work – 'grand theorizing is part of the problem' (Escobar and Harcourt, 2002: 4). Understanding local social movements and their actions does provide an important role for anthropology, a discipline that like geography has an ugly past in its dealings with hegemonic discourses and colonialism. Nonetheless, the stronger Marxist and poststructuralist critics of Escobar's work, who argue that the material impact of capitalism must be the main target of any critical interventions, suggest that whatever one's particular ideology of development or wish-list for change, these need to begin somewhere. Even if nature and culture are hybridized, we need to isolate some essential elements within and between these hybrid regimes, and try to rectify them. This should not prove so difficult, we suggest, where it is clearly the case that capital, as the overarching economic system, is directly responsible for social and human injustices. This also means that we look critically at social movements themselves – they are not to be romanticized, and many are far from progressive and can be nationalist, fundamentalist or insensitive to human rights and gender difference (Kiely, 1999). Escobar is, of course, well aware of this (Escobar and Harcourt, 2002).

In conclusion, Escobar's 'journey of the imagination' has taken him a long way forward in 'reconceiving and reconstructing the world from the perspective of, and along with, those subaltern groups that continue to enact a cultural politics of difference as they struggle to defend their places, ecologies, and cultures' (Escobar, 2000: 15). The sentiment is laudable and his work, which remains controversial for the reasons set out above, has helped to foreground local culture and to transform thinking on development hegemonies and practices.

ESCOBAR'S MAJOR WORKS

Alvarez, S. and Escobar, A. (eds) (1992) *The Making of Social Movements in Latin America: Identity, Strategy, and Democracy.* Boulder: Westview Press.

Alvarez, S., Dagnino, E. and Escobar, A. (eds) (1998) *Cultures of Politics/Politics of Cultures: Revisioning Latin American Social Movements.* Boulder: Westview Press. Also published in Portuguese and Spanish.

Escobar, A. (1984) 'Discourse and power in development: Michel Foucault and the relevance of his work to the Third World', *Alternatives* 10 (3): 377–400.

Escobar, A. (1995) *Encountering Development: The Making and Unmaking of the Third World.* Princeton: Princeton University Press. (In Spanish, Bogotá: Editorial Norma, 1998.)

Escobar, A. (1999a) 'After nature: Steps to an anti-essentialist political ecology', *Current Anthropology* 40 (1): 1–30.

Escobar, A. (1999b) *El Final del Salvaje: Naturaleza, Cultura y Política en las Sociedades Contemporaneas.* Bogotá: Instituto Colombiano de Antropología.

Secondary Sources and References

Arce, A. and N. Long (eds) (2000) *Anthropology, Development and Modernities.* London: Routledge.

Batterbury, S. P. J. (2002) 'Discursive review of *"Courtiers en développement: les villages africains en quête de projets"'*, *Environment and Planning D: Society and Space* 20 (1): 20–25.

Bebbington, A. J. (2000) 'Re-encountering development: Livelihood transitions and place transformations in the Andes', *Annals of the Association of American Geographers* 90 (3): 495–520.

Bierschenk, T., Chauveau, J.-P. and Olivier de Sardan, J.-P. (eds) (2000) *Courtiers en développement: Les villages africains en quête de projets.* Paris: Editions Karthala.

Blaikie, P. (2000) 'Development, post-, anti-, and populist: a critical review', *Environment and Planning A* 32 (6): 1033–1050.

Cleveland, D. (1999) 'Comments on Escobar', *Current Anthropology* 40 (1): 17–18.

Escobar, A. (1992) 'Planning', in W. Sachs (ed.) *The Development Dictionary: A Guide to Knowledge as Power.* London: Zed Books, pp. 132–145.

Escobar, A. (1996) 'Constructing nature: Elements for a poststructuralist political ecology', in R. Peet and M. Watts (eds) *Liberation Ecologies: Environment, Development, Social Movements.* London: Routledge, pp. 46–68.

Escobar, A. (1998) 'Whose knowledge, whose nature? Biodiversity, conservation, and the political ecology of social movements', *Journal of Political Ecology* 5.

Escobar, A. (2000) 'Beyond the search for a paradigm? Post-development and beyond', *Development* 43 (4): 11–14.

Escobar, A. (2001) 'Culture sits in places. Reflections on globalism and subaltern strategies of globalization', *Political Geography* 20: 139–174.

Escobar, A and Harcourt, W. (2002) 'Women and the politics of place', *Development* 45 (1): 7–14.

Escobar, A. and Pedrosa, A. (eds) (1996) *Pacífico, Desarrollo o Diversidad? Estado, Capital y Movimientos Sociales en el Pacífico Colombiano.* Bogotá: CEREC/Ecofondo.

Fairhead, J. and Leach, M. (1996) *Misreading the African Landscape: Society and Ecology in a Forest-Savanna Mosaic.* Cambridge: Cambridge University Press.

Fernando, J. F. and Kamat, S. (2000) 'Response to Ray Kylie and Arturo Escobar', *Development* (online forum, http://www.sidint.org/journal/online).

Gardner, K. and Lewis, D. (1996) *Anthropology, Development and the Post-Modern Challenge.* London: Pluto Press.

Kiely, R. (1999) 'The last refuge of the noble savage? A critical assessment of post-development theory', *The European Journal of Development Research* 11 (1): 30–55.

Lehmann, D. (1997) 'An opportunity lost: Escobar's deconstruction of development', *Journal of Development Studies* 33 (4): 568–578.

Little, P. and Painter, M. (1995) 'Discourse, politics, and the development process: Reflections on Escobar's "Anthropology and the Development Encounter" ', *American Ethnologist* 22 (3): 602–616.

Neumann, R. (1998) *Imposing Wilderness: Struggles over Livelihood and Nature Preservation in Africa.* Berkeley University of California Press.

Olivier de Sardan, J.-P. (1995) *Anthropologie et développement: Essai en socio-anthropologie du changement social.* Paris: Editions Karthala.

Peet, R. and Hartwick, E. (1999) *Theories of Development.* New York: Guilford Press.

Peet, R. and Watts, M. J. (eds) (1996) *Liberation Ecologies: Environment, Development, Social Movements*. London: Routledge.

Pieterse, J. N. (1998) 'My paradigm or yours? Alternative development, post-development, and reflexive development', *Development and Change* 29: 343–373.

Rahnema, M. with Bawtree, V. (1997) *The Post-Development Reader*. London: Zed.

Robbins, P. (2001) 'Tracking invasive land covers in India or why our landscapes have never been modern', *Annals of the Association of American Geographers* 91 (4): 637–654.

Rocheleau, D. (1999) 'Comments on Escobar', *Current Anthropology* 40 (1): 22–23.

Rocheleau, D. *et al.* (2001) 'Complex communities and emergent ecologies in the regional agroforest of Zambrana-Chacuey, Dominican Republic', *Ecumene* 8 (4): 465–492.

Said, E. (1978) *Orientalism*. New York: Pantheon Books.

Slater, D. (1993) 'The geopolitical imagination and the enframing of development theory', *Transactions of the Institute of British Geographers* 18: 419–437.

Stone, G.D. (2002) 'Both sides now: Fallacies in the genetic-modification wars, implications for developing countries, and anthropological perspectives', *Current Anthropology*, 43 (5): 611–630.

Sundberg, J. (1998) 'Strategies for authenticity, space, and place in the Maya Biosphere Reserve, Petén, Guatemala', *Conference of Latin Americanist Geographers Yearbook* 24: 85–96.

Simon Batterbury and Jude L. Fernando

18 Michel Foucault

BIOGRAPHICAL DETAILS AND THEORETICAL CONTEXT

Born in 1926 in Poitiers, France, Michel Foucault's education started inauspiciously in local state schools where his achievements apparently left his father less than satisfied. However, spurred on by the promise that learning philosophy would reveal to him 'the secret of secrets' (Sheridan, 1980: 2), he began to do well at the Catholic school to which he was removed, passing his *baccalauréat* with credit. Securing a place at the École Normale Supérieure (ENS), a highly prestigious institute in Paris, he took his *licence de philosophie* in 1948, but quickly became disillusioned that philosophy could not, after all, reveal to him 'the secret of secrets'. Unsure about pursuing either artistic or political lines of development, he turned instead to psychology, taking his *licence de psychologie* in 1952 and commencing research on psychopathology.

The 1950s and 1960s were a heady period in French intellectual and political circles, with heated debates about the challenge to Marxism and structuralism from various strains of existentialism and phenomenology, and certain trajectories – to do with the struggles between 'determinism' and 'freedom' that marked such debates – were to influence Foucault's intellectual development. Seriously reconfiguring his approach to philosophy, psychology and indeed science, while also wandering between jobs on the fringes of academia, Foucault eventually completed a doctoral thesis in 1959 and returned to

the corridors of the academy. Two years later his first major book, *Histoire de la folie*, was published, and so opened the chapters in his life that ended with him being fêted as one of the most celebrated intellectuals of the late twentieth century. In the early 1960s he returned to the ENS as Professor of Systems of Thought, an awkward term that he deliberately selected, a position that he held for many years alongside visiting professorships to institutions elsewhere. In the process he was 'globalized' as his ideas began to reach many different audiences and destinations, bequeathing a 'Foucauldian' (or 'Foucaultian') approach to social inquiry that ultimately earned him the accolade of being the 'Marx' of the twentieth century. He died in 1984, having already contributed enough for Sheridan's (1980: 225–226) conclusion to remain a plausible accounting: 'It is difficult to conceive of any thinker having in the last quarter of our century the influence that Nietzsche exerted over its first quarter. Yet Foucault's achievement so far makes him a more likely candidate than any other.'

Foucault's reputation was made on the basis of a series of powerful theoretical interventions that problematized the production of knowledge. When asked to write a preface to the second unabridged version of his *Histoire de la folie* (Foucault, 1972a), Foucault remarked upon the mass of 'doubles' that were by then 'swarming' around the original text. By this, he meant the many commentaries and criticisms, the collective effect of which was to ease the text from the grip of its author, to provide it with a life of its own in relation to which Foucault – the particular person who had once laboured so hard in its writing more than a decade before – had largely ceased to matter. Consistent with

Foucault's broader intellectual position, wherein writers of all sorts were viewed as occupying predetermined 'speaking positions' rather than being conduits of peculiar inspiration, he accorded himself no special privilege in the production of his own writings. In other words, he did not suppose that he himself – Foucault the person(ality) – was the key with which to unlock the meanings of his work, and certainly he did not reckon his writings to be reducible to him and his intellectual lineage. As such, he might be suspicious of a book introducing 'key thinkers', worrying that it puts too much emphasis on the figure of the author, unwittingly making particular authors too readily the bearers, emblems and origin-points of particular ideas.

Elsewhere, in an introduction to the English edition of *The Archaeology of Knowledge*, he insisted 'Do not ask me who I am and do not ask me to remain the same' (Foucault, 1972b: 17). He then added in a memorable line, 'Leave it to our bureaucrats and our police to see that our papers are in order' (Foucault, 1972b: 17). The suggestion here is that he objected to intellectual 'bureaucrats and police' who wanted to nail down exactly what kind of academic he was. Again, therefore, he might be suspicious of a book such as this one, wondering if it tries to pigeon-hole intellectuals in too straightforward a fashion, risking ossifying them as the necessary 'partners' of particular ideas from which they are not allowed to depart. Yet, if the *Histoire* preface downplays the role of the author, this moment in the *Archaeology* introduction plays it up, since Foucault as the moving locus of creative thought – a maverick thinker wishing to evade the shackles of conventional reasoning – now appears to be lent an agency, indeed a significance, rendering his ideas more than just his past texts and their batteries of critical commentary. At least in this respect, Foucault was apparently inviting us to take him seriously for himself, not for what the 'bureaucrats and police' might say. Putting things in this way might prompt a more favourable response from him to the notion of a book in which a chapter is devoted to him as a 'key thinker'.

Emphasizing these different stances on the author is more than just a nicety, since there are deeper interpretative tensions here regarding discipline and liberty that go to the heart of much, if not all, of Foucault's endeavour. On the one hand, the author who is relatively unimportant in him- or herself, whose words are determined by forces from outside, equates with the broader focus in much of Foucault's work on how human subjects are 'produced': on how their characters, beliefs and conducts are profoundly shaped by the social and institutional settings in which they find themselves, turning them into thoroughly 'disciplined' citizens with little capacity for independent action. In this guise, Foucault appears as a pessimistic theorist, one who can readily explain why the existing orders of society are commonly reproduced, complete with the in-built inequalities that such orders all too often entail. 'In Foucault country', writes Thrift (2000: 269), 'it always seems to be raining'. On the other hand, the author who is the source of creative change, who appears to have the opportunity to move, to shift positions, equates with a second focus on the possibilities opening up to the human subject who is 'self-produced': to individuals who can just occasionally seize a fragment of liberty to imagine and to accomplish things differently, to mobilize the techniques for presenting and achieving in a 'style' differing from that of contemporaries, to pursue 'the art of a life'.

Foucault's first four major texts are usually cast as his *archaeologies*, wherein the ambition is to excavate for critical inspection the 'discourses' (or organized bodies of knowledge) that have emerged within European history as the foundations for both intellectual orthodoxy and practical endeavour. *Histoire de la folie* (1961; translated as *Madness and Civilization*, 1965) and *Naissance de la clinic* (1963; translated as *The Birth of the Clinic*,

1973) probed the discourses present within prevailing understandings of mental ill-health ('madness') and physical illness, revealing how these gave rise to the 'invention' of both the mental hospital ('asylum') and the modern hospital. *Les mots et les choses* (1966; translated as *The Order of Things*, 1970) interrogated the wider discursive formations (or *épistèmes*) present within European conceptions of language, economics and nature, laying bare subtle shifts in the scaffolding of what Europeans have taken as the root 'order' of the world (the supposed links between 'words' and 'things'). *L'archéologie du savoir* (1969; translated as *The Archaeology of Knowledge*, 1972b) reflected still more broadly on the making of knowledge – the text was more a topical investigation of 'what is knowledge?' than a methodological treatise on 'how to produce knowledge?' – and in so doing examined the conjoint temporality and spatiality of statements, discourses and their ordering in the 'archive'. Such texts betrayed the influence of structuralism, notably in the sense that human subjects appear to be 'spoken' by discourses rather than *vice versa*, but by now – and in line with his realization that there were no 'secret of secrets' – Foucault's quest was not for the deeper truths of discourse, nor for the underlying logic of how they mutate, but merely to 'map' their eruption and effects within different phases of European history.

Foucault's next four major texts are usually cast as his *genealogies*, wherein he decided that the real 'object' of his inquiries was less discourse or knowledge and more the mechanics of power, in which case his earlier archaeologies also became available for re-reading as more critical offerings charting how order (conceptual and substantive) arises and is maintained in the human realm. *Surveiller et punir* (1975; translated as *Discipline and Punish*, 1976) ostensibly traced the spread of prisons and reformatories throughout later-eighteenth- and nineteenth-century Europe, but it also interrogated the transition from an older regime of violent power (whereby monarchies terrorized their populaces into obedience through the bloody spectacle of the scaffold) to a modern 'calculus' of power as less an absolute possession and more a subtle – but ultimately more effective – relational play of forces between the state and its subjects. The result here, so Foucault argued, was that the occupants of identifiable spaces (whether closed institutions or national territories) were quietly disciplined as 'docile minds and bodies' compliant with the demands of capital accumulation and civic responsibility. The next three books – *The History of Sexuality Volume One: An Introduction* (1978), *The History of Sexuality Volume Two: The Uses of Pleasure* (1985) and *The History of Sexuality Volume Three: The Care of the Self* (1986a) – furnished 'chapters' within a projected larger survey of how Europe has 'produced' notions of sexuality, of sexual conduct both accepted and shunned, from ancient times through to the present. Foucault wished to demonstrate that these notions have never been fixed, but rather have differed according to the status, class, gender, age and place of the peoples concerned, and have been converted into the objects of discourse (in everything from self-help manuals to confessional whisperings) wherein the possibilities for sexual expression have been curtailed on many occasions but enlarged on others. If *Discipline and Punish* emphasized the shaping of human subjects from without, through anonymous forces inserting individuals into disciplinary apparatuses of one kind or another, *The History of Sexuality* mingled elements of such a perspective with a sense of how individuals could be more knowingly, willingly even, enlisted into their own self-fashioning not just as sexual beings but as agents consciously monitoring their overall conduct (and who appreciated the rules governing 'the (wider) conduct of conduct'). The tensions between discipline and liberty can hence be witnessed in the distinction between *Discipline and Punish* and parts of *The History of Sexuality*, a point to which we must return

because there are also different geographies to be spied in the gaps between
discipline and liberty.

SPATIAL
CONTRIBUTIONS

It is increasingly argued that Foucault's
contribution to social thought amounts to
a thoroughly geographical provocation, in
that he demands sustained alertness to
questions of space, place, environment
and landscape in a manner rarely encountered from someone who is not a professional geographer. Indeed, Elden (2001)
explicitly characterizes Foucault as a practitioner of a 'spatial history', setting him
in a complex intellectual heritage encompassing Heidegger, Nietzsche and
Hölderlin, and concluding as follows:

> In terms of Foucault's own work I have
> argued that Foucault's historical studies
> are spatial through and through, and that
> this is a fundamental legacy of his work to
> those interested in the question of space
> ... Understanding how space is funda-
> mental to the use of power and to histori-
> cal research into the exercise of power
> allows us to recast Foucault's work not
> just as a history of the present but as a
> mapping of the present.
> (Elden, 2001: 152)

It is important to stress that Foucault used
historical research – converting it into
what Dean (1994) calls 'critical and effective histories' – as a means to understand
how 'we' have arrived at where 'we' are
today (in short, how modernity has been
shaped, complete with all of its tangled
inequalities). Through his teasing out of
how space 'works' in history, tracing the
spatial configurations that expose how
power and knowledge operate in countless (mal)treatments of 'the unloved', he
was seemingly able to throw into relief, to
'map', many of the more questionable
contours of the present.

Most abstractly, Foucault advanced a
fierce critique of what he referred to as
the project of *total history*: 'one that seeks
to reconstitute the overall form of a civilization, the principle – material or spiritual
– of a society, the significance common to
all the phenomena of a period, the law
that accounts for their cohesion – what is
called metaphorically the "face" of a
period' (Foucault, 1972b: 9). Such a project was anathema to Foucault, since its
ambitions stand squarely in opposition to
his own belief that 'nothing is fundamental: this is what is interesting in the
analysis of society' (Foucault, 1972b: 16).
As an alternative he advocated a *general
history*, a 'bellicose history' (Lemert and
Gillan, 1982: 39), which militates against
the tidying up of the past to give neat
patterns, steadfastly resisting the rush
from the countless small details of lived
struggle to the grander pronouncements
of historians (particularly those of social
scientists dabbling in the practice of history). Foucault thereby drew this distinction: 'a total description draws all
phenomena around a single centre – a
principle, a meaning, a spirit, a world-
view, an overall shape; a general history,
on the other hand, would deploy the
space of a dispersion' (Foucault, 1972b:
10). With these comments about 'spaces
of dispersion', he appeared to envisage a
spatialized ontology of the social world: a
vision of a space or plane across which all
of the events and phenomena relevant to
the substantive inquiry are 'dispersed'
(see Philo, 1992).

Foucault's own histories were not spatial solely in the philosophical sense of
offering an *a priori* spatialized conceptualization of worldly phenomena, nor because they offered an overview of
changing conceptions of space; although a
history of space in this guise does flicker
through his famous 1967 lecture 'On
other spaces' (see Foucault, 1986b), and
might be said to have a presence elsewhere in his *oeuvre* too (see the discussion
in Elden, 2001: 116–118). Instead, Elden
(2001: 118) argues that Foucault's 'histories are not merely ones in which space

is yet another area analysed, but have space as a central part of the approach itself', meaning that 'rather than merely writing histories of space, Foucault is writing spatial histories'. In one regard, this was simply because of the insistence on bringing details of past phenomena to the fore, as he acknowledged when borrowing from Nietzsche's notion of 'genealogy' as 'grey, meticulous and patiently documentary':

> [It] requires patience and a knowledge of *details*, and it depends upon a vast accumulation of source material. Its 'cyclopean monuments' are constructed from 'discreet and apparently insignificant truths and according to a rigorous method'; they cannot be the product of 'large and well-meaning errors'.
> (Foucault, 1986c: 76–77; emphasis added)

Hutcheon (1988: 120) duly talks about Foucault's 'assault on all the centralizing forces of unity and continuity', and on his requirement – using terms that should immediately intrigue geographers – that 'the particular, the local and the specific' be pursued in place of 'the general, the universal and the eternal' (Hutcheon, 1988: 120). Although he was no straightforward empiricist, it remains the case that taking seriously details in their particularity, specificity and locality is fundamental to Foucault's notion of 'spaces of dispersion'. In another regard, however, his spatial histories furnished more systematic insights into the play of spatial relations in the historical record.

KEY ADVANCES AND CONTROVERSIES

In *The Order of Things*, Foucault suggested an opposition between 'the Same' and 'the Other' (Philo, 1986) that can be taken as framing many of his major historical studies. Firstly, he identified his inquiries into discourse and knowledge as reconstructions of what it is that a given society takes as the Same, as the accepted conventions for thought and action which incorporate both the leading statements of 'experts' (academics, politicians, moralists) and the more taken-for-granted assumptions figuring in the everyday lives of the populace. Secondly, he effectively identified his social histories of 'the mad, the sad and the bad' as reconstructions of what it is that a given society regards as the Other, as the unacceptable mass of activities, people and places slipping beyond the boundaries of what is deemed as 'normal', as the Same, and which thereby necessitate some response of policing, removal or even eradication:

> Ostensibly, [Foucault's] project is to describe the mechanisms of order and exclusion that have operated within European society since the sixteenth century, and above all since the late-eighteenth century. In a motif that recalls Bataille's reflections on Hegel, Foucault sees a conflict in history against which it can define itself, just as every 'master' needs a 'slave'. When such an 'Other' is absent, it must be invented.
> (Megill, 1985: 192)

For Foucault, madness, sickness and criminality, as well as sexual dissidence, were hence to be traced historically in their Otherness to European norms, laying out the shifting bases for their constitution as oft-feared moments of alterity.

In this respect *Madness and Civilization* can be seen as the blueprint, with its sustained attention to how 'Reason' (or the Same) has progressively identified, named, stigmatized and sought to exclude, through either banishment or incarceration, 'Unreason' (or the Other). It is telling to repeat the assessment that Serres offers of this text, since he 'interprets Foucault's categories of inclusion and exclusion in terms of spatial relationships, and . . . views Foucault's concept of Unreason as a "geometry of negativities" ' (Major-Poetzl, 1983: 120). Such a geometry embraces the projections whereby everyday society imaginatively positions itself over and against those phenomena, especially peoples reckoned less than human in their madness, sickness, criminality and so on, that are supposed to transgress the limits of the sanctioned, to push beyond the boundaries or thresholds of the expected, to disorder the ordered. Yet beyond these projections, and paralleling the spatialized vocabulary that he deployed when charting what occurred at 'the level of the imaginary', it is also true that Foucault's spatial sensibility transferred to 'the level of the real' (Elden, 2001: 93). Thus, he was clearly interested in 'the physical divide of segregation and exclusion' that inscribes into bricks and mortar a distancing of the Other from the Same, and for this reason he ended up 'conceiv[ing] of madness and reason, sickness and health in spatial terms, and then examin[ing] the groups that inhabit these liminal areas' (Elden, 2001: 94–95). He thereby paid repeated attention to specific 'liminal areas', notably the bricks-and-mortar solutions of separate institutions such as asylums, hospitals and prisons (see also Philo, 2000, 2002, 2003) designed to confine, to 'reform' and, where appropriate, to 'cure' those displaying signs of such difference.

Cross-cutting these spatial histories are different conceptualizations of power, themselves the basis for a profound theorization of space and power (or power/ knowledge: Gordon, 1980). Most abstractly, Foucault (1976: 215) insisted that power be understood through its 'microphysics' – 'its techniques, procedures, levels of applications, targets' – and hence in a thoroughly relational fashion that subsequent theorists have readily elaborated in terms of 'capillaries', 'transmissions' and 'relays' of power through specific spatial fields (Driver, 1985, 1993; Hannah, 1997). More empirically, Foucault (1976: 141–149) analysed 'the art of distributions' underlying a host of disciplinary mechanisms ushered into being by the nineteenth century, all depending upon the detailed manipulation of space, and he traced the enactment of spatial innovations across all manner of institutions from Bentham's design for an ideal prison, the high-walled 'Panopticon', to the example of an unwalled reformatory at Mettray. There is a danger that many readings of *Discipline and Punish* reduce Foucault's claims to the figure of the Panopticon, and fail to register the significance of his arguments about Mettray (Driver, 1990), but it remains the case that the Panopticon has now become a dramatic spatial provocation for social theorists of power. With its internal spatial arrangements allowing a constant (threat of) inspection, a surveillance that captures inmates in an overall field of visibility while prompting them to convert the external eye of the inspection tower into the internal eye of conscience (Bender, 1987), it is unsurprising that many have found here keys to unlock the broader workings of disciplinary power the length and breadth of modernity.

Leading out of *Discipline and Punish*, Foucault subsequently developed the notion of 'governmentality' (Foucault, 1979), encompassing both the government of populations, wherein states and religions seek to control the processes of life, birth and death, and the government of individuals, especially in their everyday sexual and reproductive conduct. As Dean (1994: 174) explains, governmentality 'defines a novel thought-space across the domains of ethics and politics, of what might be termed "practices of the self" and "practices of government", that weaves them together without a reduction of the one to the other'. As mentioned above when describing his later geneaologies, Foucault became increasingly concerned with questions surrounding 'the conduct of conduct', in the context of which he sought to reconstruct past codes of conduct, notably sexual and political codes, whose effects have inevitably been ones of power (in the sense of laying down the conditions for the successful exercise of power across different domains of human endeavour). The emphasis hence alighted upon individuals who conceive of themselves to be free or at liberty, as opposed to the inmates of institutions who know themselves to be shut away, although the typical Foucauldian twist was to assert that liberty is itself ultimately a discursive effect, a product of a particular power/knowledge nexus, rather than some true social state. This being said, Foucault did appear to grant the human subject a shade more wiggle-room than before, offering the fleeting possibility, as hinted earlier, of the individual being something other than a mere drone of a pre-existing order. Finally, while acknowledging that he said less about space in this later thinking on power, a spatial sensitivity continued to bubble under the surface in what he claimed about countless specific sites – from the confessional to the late-Roman city-state (see Sharp *et al.*, 2000: 16–19 and Endnote One) – that have always been implicated in the persuading of people (or, rather, in people persuading themselves) to take seriously 'the relationship that one ought to have with one's status, one's functions, one's activities, and one's obligations' (Foucault, 1986a: 84).

FOUCAULT'S MAJOR WORKS

Foucault, M. (1965) *Madness and Civilization: A History of Insanity in the Age of Reason*. New York: Random House.

Foucault, M. (1970) *The Order of Things: An Archaeology of the Human Sciences*. London: Tavistock Publications.

Foucault, M. (1972b) *The Archaeology of Knowledge*. London: Tavistock Publications.

Foucault, M. (1973) *The Birth of the Clinic: An Archaeology of Medical Perception*. London: Tavistock Publications.

Foucault, M. (1976) *Discipline and Punish: The Birth of the Prison*. London: Allen Lane.

Foucault, M. (1978) *The History of Sexuality, Volume 1: An Introduction*. New York: Random House.

Foucault, M. (1985) *The History of Sexuality, Volume 2: The Uses of Pleasure*. Harmondsworth: Penguin.

Foucault, M. (1986a) *The History of Sexuality, Volume 3: The Care of the Self*. Harmondsworth: Penguin.

Secondary Sources and References

Bender, J. (1987) *Imagining the Penitentiary: Fiction and the Architecture of Mind in Eighteenth-Century England*. Chicago: University of Chicago Press.

Dean, M. (1994) *Critical and Effective Histories: Foucault's Methods and Historical Sociology*. London: Routledge.

Driver, F. (1985) 'Power, space and the body: a critical assessment of *Discipline and Punish*', *Environment and Planning D: Society and Space* 3: 425–446.

Driver, F. (1990) 'Discipline without frontiers? Representations of the Mettray Reformatory Colony in Britain, 1840–1880', *Journal of Historical Sociology* 3: 272–293.

Driver, F. (1993) 'Bodies in space: Foucault's account of disciplinary power', in C. Jones and R. Porter (eds) *Reassessing Foucault: Power, Medicine and the Body*. London: Routledge, pp. 113–131.

Elden, S. (2001) *Mapping the Present: Heidegger, Foucault and the Project of a Spatial History*. London: Continuum.

Foucault, M. (1972a) *Histoire de la Folie à l'Age Classique*, 2nd edition. Paris: Gallimard.

Foucault, M. (1979) 'Governmentality', *Ideology and Consciousness* 6: 5–21.

Foucault, M. (1986b) 'Of other spaces', *Diacritics* 16: 22–27.

Foucault, M. (1986c) 'Nietzsche, genealogy, history', in P. Rabinow (ed.) *The Foucault Reader*. Harmondsworth: Penguin, pp. 63–77.

Gordon, C. (ed.) (1980) *Power/Knowledge: Selected Interviews and Other Writings, 1972–1977*. Hemel Hempstead: Harvester Press.

Hannah, M. (1997) 'Space and the structuring of disciplinary power: an interpretive review', *Geografiska Annaler* 79B: 171–180.

Hutcheon, L. (1988) *A Poetics of Post-modernism: History, Theory and Fiction*. London: Routledge.

Lemert, C. C. and Gillan, G. (1982) *Michel Foucault: Social Theory as Transgression*. New York: Columbia University Press.

Major-Poetzl, P. (1983) *Michel Foucault's Archaeology of Western Culture: Towards a New Science of History*. Brighton: Harvester Press.

Megill, A. (1985) *Prophets of Extremity: Nietzsche, Heidegger, Foucault, Derrida*. San Francisco: University of California Press.

Philo, C. (1986) ' "The Same and the Other": on geographies, madness and outsiders', University of Loughborough, Department of Geography, Occasional Paper No.11.

Philo, C. (1992) 'Foucault's geography', *Environment and Planning D: Society and Space* 10: 137–162.

Philo, C. (2000) '*The Birth of the Clinic*: an unknown work of medical geography', *Area* 32: 11–19.

Philo, C. (2002) 'Accumulating populations: bodies, spaces, institutions', *International Journal of Population Geography* 7: 473–490.

Philo, C. (2003) *The Space Reserved for Insanity: An Historical Geography of the Mad-Business in England and Wales to the 1860s*. Lampeter: Edwin Mellen Press.

Sharp, J. P., Routledge, P., Philo, C. and Paddison, R. (2000) 'Entanglements of power: geographies of domination/resistance', in J. P. Sharp, P. Routledge, C. Philo and R. Paddison (eds) *Entanglements of Power: Geographies of Domination/Resistance*. London: Routledge, pp. 1–42.

Sheridan, A. (1980) *Michel Foucault: The Will to Truth*. London: Tavistock Press.

Thrift, N. (2000) 'Entanglements of power: shadows?', in J. P. Sharp, P. Routledge, C. Philo and R. Paddison (eds) *Entanglements of Power: Geographies of Domination/Resistance*. London: Routledge, pp. 269–278.

Chris Philo

19 Anthony Giddens

BIOGRAPHICAL DETAILS AND THEORETICAL CONTEXT

Perhaps the leading sociologist of the late twentieth century, Anthony Giddens has enjoyed a long and exceptionally productive career that has made him a scholar of world renown. Born in 1938, he grew up in Edmonton, North London. He was the first person in his family to go to university or college, studying psychology and sociology at the University of Hull. He then pursued an MA in Sociology at the London School of Economics and Political Science, and completed his doctorate in Sociology at the University of Cambridge in 1976. He taught at Cambridge as a Lecturer in Sociology and Fellow of King's College from 1970 to 1984, then Reader (1984 to 1986), and finally as Professor. He co-founded the academic Polity Press in 1985 and has since served as its director, as well as Director of the Centre for Social Research since 1989. From 1997 to 2002 he served as Director of the London School of Economics and Political Science and acted as advisor for Prime Minister Tony Blair, his ideas strongly influencing the development of New Labour. Giddens has held an extensive variety of visiting professorships around the world and has been the recipient of 13 honorary degrees and awards, including several doctorates, and membership of the Russian Academy of Sciences and the Chinese Academy of Social Sciences. His textbook, *Sociology* (1989), now in its fourth edition, has sold over 600,000 copies. His 34 books, which have been translated into 30 languages, and widespread fame have made him a public intellectual and the subject of extensive critical and biographical commentary (Bryant and Jary, 1997; Cassell, 1993; Cohen, 1991; Held and Thompson 1989; Mestrovic, 1998; O'Brien, 1998; Tucker, 1998).

In what may be considered the first stage of his career, Giddens' early works (1971, 1973, 1979) centred on a critical analysis and reconstitution of traditional European social theory. *New Rules of Sociological Method* (1976) offered a critical translation of European social theory for American audiences, making Giddens a rare trans-Atlantic intellectual. He argued that orthodox theory, ranging from Durkheim to Marx to Weber, lacked an adequate theory of the subject as a conscious actor possessed of the capacity to choose and to exert power. Rather, he argued that actors are often portrayed as unwitting dupes, with social change erroneously held to occur 'above their heads' or 'behind their backs.' These works both established his credentials as an insightful analyst of social change and laid the foundations for an original and innovative social theory.

In the 1980s, Giddens undertook what amounted effectively to the second stage of his career, in which he ardently advocated the theory of structuration, a notion that has become widespread and commonplace in the social sciences. This theme begins with the phenomenological recognition that human beings are sentient, knowledgeable agents (i.e. they have consciousness about themselves and their world, however limited). Everyone, in this sense, is a sociologist. Giddens drew here on the rich humanistic and behavioural traditions concerned with perception, cognition and language. Moving beyond the usual definitions of culture as the sum total of learned behaviour or a 'way of life,' structuration theory portrays culture as what people take for granted, i.e. common sense, the matrix of ideologies that allow actors to negotiate their way through their everyday worlds. Culture defines what is normal and what is not, what is important and what is not, what is acceptable and what is not, within each social context. Culture is acquired through a lifelong process of socialization: individuals never live in a social vacuum, but are socially produced from cradle to grave. Structure is seen to consist of the rules and resources that are instantiated in social systems. In their daily lives, actors draw upon these rules and resources, which in turn structure their actions; hence, the structural qualities that generate social action are continually reproduced through these very same actions. The socialization of the individual and the reproduction of society and place are two sides of the same coin; that is, the macrostructures of social relations are interlaced with the microstructures of everyday life. People reproduce the world, largely unintentionally, in their everyday lives and, in turn, the world reproduces them through socialization. In forming their biographies everyday, people re-create and transform their social worlds primarily without intending to do so; individuals are both produced by, and producers of, history and geography. Everyday thought and behaviour hence do not simply mirror the world, they constitute it. History (and geography) is thus produced through the dynamics of everyday life, the routine interactions and transient encounters through which social formations are reproduced. 'Time' is thus not some abstract independent process; it is synonymous with historical change (but not progress) and the capacity of people to make and transform their worlds (Giddens, 1984).

Giddens began an extensive critique of historical materialism in a provocative and well-received trilogy of works (1981, 1984, 1987a). These volumes both articulated his vision of social relations and

extended their scope. In laying out the theory of structuration in detail, he offered a compelling analysis of how capitalism radically changed the fundamental contours of class relations and culture, extending commodity relations into various spheres of life, dramatically accelerating the tempo of production and reproduction, and marking a decisive break from the past. In pre-capitalist (class-divided) societies, he noted, power was exerted primarily through the state as a mechanism to extract surplus value and implement social control; under capitalism and the private appropriation of wealth (class societies), in contrast, these relations occur through the market – with an apolitical patina in which power lay in the private rather than public domain. Simultaneously, control over time and space, which are central to the exertion of power, shifted from the city-states of Italy to the incipient nation-states of northwestern Europe. *The Constitution of Society* (1984), arguably his best and most explicit summary of structuration theory, emphasizes the contingent nature of social life to confront functionalist and evolutionary interpretations of historical change.

Power in all its complex, multiple forms has long played a central role in Giddens' analysis. He is insistent that as transformative capacity, power is intrinsically tied to human agency: 'Power in this relational sense, concerns the capability of actors to secure outcomes where the realisation of these outcomes depends upon the agency of others' (Giddens, 1986: 93). This view differs from traditional Weberian views that imply that power only exists when the resistance of others is overcome. Rather, Giddens maintains that there are two primary structures of domination, including allocative resources (material wealth and technology) and authoritative resources (the social organization of time, space and the body).

In *The Nation-State and Violence* (1987a), Giddens undertook an ambitious and systematic extension of the nation-state as a 'power-container' for the expression of capitalist social relations. Far from being confined to questions of individual behaviour and everyday life, he applied structuration theory to reveal the historical emergence of the nation-state in the transition from older absolutist regimes with loosely defined boundaries, to the commodified relations prevalent under industrial capitalism in which boundaries are rigidly drawn. Thus, the nation-state served as a means for the organization of surveillance and discipline in the transition from authoritarian to bourgeois democracy and the enforcement of market relations. Within this theme, Giddens explores the roles of nationalism, citizenship, capital punishment, money and the commodification of time, as well as the intertwined scales of the nation-state and the world system through military conflict and war.

More recently, in the third stage of his intellectual journey, Giddens' (1990, 1991) work has centred on issues of modernity. We are not in a postmodern era, Giddens says, but rather a period of late modernity characterized by the *post-traditional* nature of society. When tradition dominates, individuals rarely have cause to analyse their own actions because choices are already prescribed by their taken-for-granted world. Under modernity, however, traditions become revealed *as tradition*, and lose much of their power. Society becomes much more *reflexive* and aware of itself. Without clear, unquestioned rules for behaviour, issues of self-identity become problematic. Thus although tradition is dead, modernity is not. Giddens eschews the label 'postmodernity' for what is essentially a contemporary extension of modern capitalism. This transition is manifested at various scales ranging from the private and individual issues of sexuality and intimacy to globalization (Giddens, 2000a). Modernity is fuelled by the growing extensibility of social relations, their stretching across increasingly global scales, particularly through the use of electronic communications.

Most recently, in what might be considered the fourth stage of his project, Giddens has been at the forefront of developing new ideas to inform progressive politics. He helped to popularize the idea of the 'Third Way' (Giddens, 1999, 2000b), which represents the renewal of social democracy in a world where the views of the Marxist-inspired left have become ineffective while those of the neo-liberal right are inhumane. He maintains that a new social democratic agenda is emerging, and calls for a middle ground for the state between the state-directed model of the left and the free-market orthodoxy of the right. Far from being rendered irrelevant, he argues that government still has an important role to play in the process of globalization. The goal for policymakers, however, is not just a traditional leftist policy of expanding national government, but a redefinition in the context of social democracy.

SPATIAL CONTRIBUTIONS

Giddens' works became widely noticed among geographers during the 1980s, a period of considerable intellectual transformation. In the wake of the collapse of positivism, serious debates erupted between two broad schools of thought, including Marxists, many of whom subscribed to structuralist theory, and a broad ensemble of humanist geographers who took as their point of departure (and often destination) human agency detached from any social context (cf. Duncan and Ley, 1982). This 'micro–macro' schism, which was far from unique to geography, was effectively resolved through structuration theory. While Giddens' works are primarily concerned with ontology rather than the epistemology of the social world, structuration theory was nicely complemented by the parallel emergence of critical realism (see **Sayer,** 1992).

Giddens has been well received in geography because he takes space seriously. This concern is reflected at several spatial scales. At the level of the individual and everyday life he is emphatic that the routinized patterns of behaviour through which social reproduction and change occur (mostly unintentionally) are always structured temporally and spatially. In this respect, Giddens drew heavily from an earlier tradition of time geography, which focused upon the constrained capacities of the human body to negotiate the temporal and spatial trajectories of life (see **Torsten Hägerstrand**). Closely related to this element of structuration theory is his sustained interest in regions and places; he argues (Giddens, 1984) that locales are not just passive places but active milieux that influence, and are in turn influenced by, the interactions of actors.

The stretching of such interactions across time and space is not only fundamental to the organization and exertion of power in all its forms, but is itself historically specific and contingent. This process Giddens calls time–space distantiation, a notion very similar to what geographers had called time–space convergence (Janelle, 1969) or time–space compression (Harvey, 1990). The very malleability of time and space revealed them to be not 'natural' or external to social relations, but a product (and producer of) those relations, i.e. as human constructions. Time and space are thus as plastic and mutable as the social structures of which they are a part. Time–space distantiation was understandably of great utility to geographers concerned with flows of people, goods and information as well as the social construction of scale (Wilson and Huff, 1994).

Finally, Giddens is as much a geographer as sociologist in his analysis of globalization (2000a), which he defines as the expansion of the reach, volume and velocity of international exchanges. While globalization is not new, in the contem-porary historical moment the driving force, he argues, is the deployment of information technologies, which have generated a significant round of time–space compression in the late-modern world. From the standpoint of structuration theory, globalization is simultaneously humanly produced and a force largely beyond conscious human control. While this process is undoubtedly dramatically reshaping the role of nation-states, Giddens rejects simplistic assertions that globalization entails the death of this institution.

KEY ADVANCES AND CONTROVERSIES

Giddens' works have played a major role in advancing and popularizing social theory in many social sciences. He drew upon both Marxist and Weberian lines of thought at a time when positivist approaches were falling out of favour, and articulated a coherent, unified vision of social structure and everyday life that included such diverse issues as the phenomenology of everyday life, the emergence of the nation-state and the contingent dynamics of the world system. In this light, he may be seen as part of a broader resurgence of Western (non-structuralist) Marxism, which included renewed attention to the Frankfurt School, Habermas, Gramsci, Sartre, E. P. Thompson, **Lefebvre** and others.

Structuration theory jettisons the long-standing schism between micro-and macro- approaches that long plagued the social sciences; rather, these two dimensions of human life and sets of theoretical perspectives must be seen as complementary. Thus Giddens favours a *duality* between structure and agency, in which they are simultaneously determinant and mutually recursive, rather than a simplistic dualism of opposing forces.

Giddens has been widely criticized for his stance that social structures enable as well as constrain behaviour: structures are what actors take for granted, reproduce and change in the durée of their daily lives. To view structures simply as constraints, Giddens argued, is to reify them as something other than human products. Marxists in particular were vehement in their condemnation of structuration theory as essentially Weberian idealism. Giddens' view of social relations was profoundly important in forcing social science to take seriously the contextual and contingent nature of human consciousness. Social change in this light is neither lawless, in the sense of being anarchical and unpredictable, nor is it so completely subject to laws that the outcomes of action are predictable with confidence; in short, social organization and change are simultaneously structured and contingent.

Giddens' ideas have become enormously widespread, popular and influential, yet he remains best known to geographers for his critical synthesis of social theory rather than more recent works on globalization and 'Third Way' politics. For instance, geographers generally accepted structuration theory, weaving it into the rich stew of ideas that criss-crossed the discipline in the late twentieth century. Leading theoreticians such as **Derek Gregory** and **Nigel Thrift**, for example, employed it extensively in empirical analyses of the British space-economy, including the Yorkshire textile industry (Gregory, 1982) and London bankers (Leyshon and Thrift, 1997) respectively, as well as utilizing it as a tool in the integration of Marxist and humanist traditions within geography (Thrift, 1983). Using structuration theory as a device to sensitize others to the contextual nature of human consciousness – its embeddedness in social relations as well as place – allowed for the construction of a view of regions as contingently produced entities made and remade, largely unintentionally, by human subjects in everyday life. This view drew heavily from the

parallel tradition of time-geography (Pred, 1984), with which structuration theory shared much in common (but offered a markedly superior perspective on social relations and human subjectivity).

Ironically, the geographers who drew upon Giddens the most heavily also criticized his work for an inadequately developed sense of space, including lack of attention to issues of spatial scale; his view of regionalization, for example, is primitive in comparison with the multiscalar approaches of renewed regional geography (cf. Gregory, 1989; Thrift, 1990). Geographers' works moved beyond Giddens in that they came to view regions not simply as passive recipients of broader social changes such as globalization, but as active producers in their own right, constellations of social practices that drew upon, reflected and in turn constituted changing networks of social relations stretching from the intimately local to the global. Thrift (1996: 54–55) criticized Giddens on several grounds, including an unconvincing account of recursivity, an overemphasis on presence that 'never fully considers the ghost of networked others', an 'impoverished sense of the unconscious', and an inadequately developed theory of culture. Gregson (1989) argued that structuration theory is so vague that it is essentially useless in guiding empirical research.

Despite these reservations, Giddens' works played an important role in the formulation of a coherent alternative to both Marxism and humanist geography that drew upon both, emphasizing the social construction of space and its contingent reproduction and transformation. Without his contributions, geography's subsequent experimentation with cultural studies, actor-network theory, discourse analysis and similar lines of thought would have been seriously hampered. While structuration theory is required reading for most graduate students, human geography moved forward into various permutations of poststructuralist analysis, including discourse analysis and Orientalism. Yet these advances were made possible in no small part by the contributions of Giddens and others working within the framework of struc-

turation theory. In this respect, Giddens ultimately and ironically became a victim of his own success.

GIDDENS' MAJOR WORKS

Giddens, A. (1971) *Capitalism and Modern Social Theory; An Analysis of the Writings of Marx, Durkheim and Max Weber.* Cambridge: Cambridge University Press.

Giddens, A. (1973) *The Class Structure of the Advanced Societies.* London: Hutchinson.

Giddens, A. (1976) *New Rules of Sociological Method: A Positive Critique of Interpretative Sociologies.* London: Hutchinson.

Giddens, A. (1979) *Central Problems in Social Theory: Action, Structure and Contradiction in Social Analysis.* London: Macmillan.

Giddens, A. (1981) *A Contemporary Critique of Historical Materialism.* Berkeley: University of California Press.

Giddens, A. (1984) *The Constitution of Society: Outline of the Theory of Structuration.* Berkeley: University of California Press.

Giddens, A. (1987a) *The Nation-State and Violence.* Berkeley: University of California Press.

Giddens, A. (1990) *The Consequences of Modernity.* Stanford, CA: Stanford University Press.

Giddens, A. (1991) *Modernity and Self-identity: Self and Society in the Late Modern Age.* Stanford, CA: Stanford University Press.

Secondary Sources and References

Bryant, C. and Jary, D. (1997) *Anthony Giddens: Critical Assessments.* London: Routledge (4 volumes).

Cassell, P. (1993) *The Giddens Reader.* New York: Macmillan.

Cohen, I. (1991) *Structuration Theory: Anthony Giddens and the Constitution of Social Life.* New York: Macmillan.

Duncan, J. and Ley, D. (1982) 'Structural Marxism and human geography', *Annals of the Association of American Geographers* 72: 30–59.

Giddens, A. (1972) *Politics and Sociology in the Thought of Max Weber.* London: Macmillan.

Giddens, A. (1982) *Profiles and Critiques in Social Theory.* London: Macmillan.

Giddens, A. (ed.) (1986) *Durkheim on Politics and the State.* Cambridge: Polity.

Giddens, A. (1987b) *Social Theory and Modern Sociology.* Cambridge: Blackwell.

Giddens, A. (1989) *Sociology.* Cambridge: Polity Press.

Giddens, A. (1994) *Beyond Left and Right: The Future of Radical Politics.* Stanford, CA: Stanford University Press.

Giddens, A. (1996) *In Defence of Sociology: Essays, Interpretations, and Rejoinders.* Cambridge: Polity Press.

Giddens, A. (1999) *The Third Way: The Renewal of Social Democracy.* Malden, MA: Polity Press.

Giddens, A. (2000a) *Runaway World: How Globalisation is Reshaping our Lives.* New York: Routledge.

Giddens, A. (2000b) *The Third Way and its Critics.* Cambridge: Polity Press.

Giddens, A. and Pierson, C. (1998) *Conversations with Anthony Giddens: Making Sense of Modernity.* Stanford, CA: Stanford University Press.

Gregory, D. (1982) *Regional Transformation and Industrial Revolution: A Geography of the Yorkshire Woollen Industry, 1780–1840.* Minneapolis: University of Minneapolis Press.

Gregory, D. (1989) 'Presences and absences: Time-space relations and structuration theory', in D. Held and J. Thompson (eds) *Social Theory of Modern Societies: Anthony Giddens and his Critics.* Cambridge: Cambridge University Press.

Gregson, N. (1989) 'On the (ir)relevance of structuration theory to empirical research,' in D. Held and J. Thompson (eds) *Social Theory of Modern Societies: Anthony Giddens and his Critics.* Cambridge: Cambridge University Press.

Häagerstrand, T. (1970) 'What about people in regional science?', *Papers of the Regional Science Association* 24: 7–21.

Harvey, D. (1990) 'Between space and time: Reflections on the geographical imagination', *Annals of the Association of American Geographers* 80: 418–434.

Held, D. and Thompson, J. (1989) *Social Theory of Modern Societies: Anthony Giddens and his Critics.* Cambridge: Cambridge University Press.

Hutton, W. and Giddens, A. (eds) (2000) *Global Capitalism.* New York: New Press.

Janelle, D. (1969) 'Spatial organization: a model and a concept', *Annals, Association of American Geographers* 59: 348–364.

Leyshon, A. and Thrift, N. (1997) 'Spatial financial flows and the growth of the modern city', *International Social Science Journal* 151: 41–53.

Mestrovic, S. (1998) *Anthony Giddens: The Last Modernist.* London: Routledge.

O'Brien. M. (1998) 'The sociology of Anthony Giddens', in Pierson, C. (ed.) *Making sense of Modernity* Cambridge: Polity Press.

Pred, A. (1984) 'Place as historically contingent process: Structuration and the time-geography of becoming places,' *Annals of the Association of American Geographers* 74: 279–297.

Sayer, A. (1992) *Method in Social Science: A Realist Approach*, 2nd edition. London: Routledge.

Thrift, N. (1983) 'On the determination of social action in space and time', *Environment and Planning D: Society and Space* 1: 23–57.

Thrift, N. (1990) 'For a new regional geography 1', *Progress in Human Geography* 14: 272–279.

Thrift, N. (1996) *Spatial Formations.* Thousand Oaks, CA: Sage.

Tucker, K. (1998) *Anthony Giddens and Modern Social Theory.* Boulder: Sage.

Wilson, D. and Huff, J. (eds) (1994) *Marginalized Places and Populations: A Structurationist Agenda.* Westport, CT: Greenwood.

Barney Warf

20 Reginald Golledge

Reg Golledge was born in Australia in
1937. He completed his BA and MA in
Geography at the University of New England, Australia, before taking up a lectureship in Geography at the University of
Canterbury, Christchurch, in New Zealand. In 1964 he moved to North America,
to take a position as a Research Assistant
at the University of Iowa. Drawing influence from colleagues at Iowa (notably
Harold McCarty), from geographers such
as Julian Wolpert and Peter Gould and
from psychologist Jean Piaget, Golledge's
PhD (1966) combined learning theory and
probabilistic modelling to analyse the
marketing of pigs. After a year as an
Assistant Professor at the University of
British Columbia, Vancouver, in 1966
Golledge took up a post at Ohio State
University, where he stayed until 1977. It
was during his time at Ohio that he rose
to prominence as a key proponent of
behavioural geography, a perspective that
holds to the idea that human activity can
only be understood in relation to people's
imperfect and partial knowledge of the
world.

Always keen to collaborate with academics both within Geography and other
disciplines, after arriving at Ohio, Golledge started to work with geographers
such as Les King, Kevin Cox, Larry Brown
and John Rayner, psychologists such as
Paul Isaacs and Jim Wise, and mathematicians Joseph Kruskal and Doug Carroll
(both at Bell Labs) on issues relating to the
modelling of spatial knowledge, and specifically spatial choice and decision-making, a topic that has remained a consistent
focus for his entire career. His first landmark paper, published with Briggs and
Demko (1969), used multidimensional
scaling to 'map' paired-comparison distance estimates, arguing that the resulting
configuration provided a 'mental map' of
how the city appears to people.

Over the next several years, Golledge
developed a consistent and coherent theoretical framework to support his view
that the best way to understand the geographical world was to understand how
people cognized the world around them
and made choices and decisions on the
basis of such knowledge. This was accompanied by a sustained engagement with
cognitive and experimental psychology
and the adaptation of quantitative techniques (e.g. non-metric multidimensional
scaling and hierarchical clustering). This
emphasis on quantification led to Golledge's work being described as 'analytical' behavioural geography, as
distinguished from a more phenomenological approach being developed by
others (see Saarinen et al., 1984).
Nonetheless, Golledge was a key figure in
the active promotion of a broad range of
behavioural approaches through his writing, as well as through organizing conference sessions, taking part in debates and
supporting behavioural work through his
editorship of Geographical Analysis (1973–
78) and Urban Geography (1978–84). This
work resulted in the highly cited and
influential edited collections, Behavioural
Problems in Geography: A Symposium
(1969, edited with Kevin Cox), Environmental Knowing (1976, edited with Gary
Moore), Cities, Space and Behaviour (1978,
written with Les King) and Behavioural
Problems in Geography Revisited (1981,

edited with Kevin Cox). While the first of these arguably established behavioural geography as a mainstream approach in human geography, the latter were written at a time when behavioural geography was coming under attack from both humanists and structuralists. Through these works, Golledge thus became one of behavioural geography's staunchest defenders, providing strong rebuttals of critiques of the behavioural perspective (see Golledge, 1981; Couclelis and Golledge, 1983).

In 1977, Golledge moved to the University of California, Santa Barbara, where he has remained since. Again, quickly building new interdisciplinary links with psychologists, mathematicians and computer scientists, he started to build what was to become the Research Unit on Spatial Cognition and Choice, continuing his development of analytical behaviouralism. In 1984 he lost his sight. This impairment, which initially seemed to threaten his academic career (Golledge, 1997), instead started a remarkable collaboration with psychologists Jack Loomis and Roberta Klatzky which has continued up until the time of writing. Over a series of related projects, they applied what had been Golledge's work to date to visual impairment, seeking on the one hand to understand how people with visual impairments come to understand spatial relationships and use this knowledge to navigate, and on the other to apply their findings to the development of orientation and navigation systems, culminating in a Personal Guidance System, designed by Loomis, that combines the use of Global Positioning System (GPS) and a Geographic Information System (GIS), and uses a virtual auditory/sound interface as output. Continuing his defence and promotion of behavioural approaches, in 1987 Golledge published *Analytical Behavioural Geography*, updated in 1997 as *Spatial Behaviour – A Geographic Perspective*. He has been active in the National Centre for Geographic Information Analysis (NCGIA), organizing and participating in several themes that apply behavioural

approaches to GIS. The recipient of many awards and honours, in 1999 Golledge became the President of the Association of American Geographers, using his presidential address to call for a policy-relevant geography underpinned by a behavioural approach (Golledge, 2002).

SPATIAL CONTRIBUTIONS

Golledge's key role in the study of place and space has been his contributions to the development of analytical behavioural geography. Behavioural geography developed throughout the late 1960s and early 1970s out of a dissatisfaction with the stereotyped, mechanistic and deterministic nature of many of the quantitative models being developed at that time, and a realization that not everyone behaved in a spatially rational manner. As such, it was a direct challenge to the seemingly 'peopleless' geographies of spatial science.

Behavioural geographers argued that space is not experienced and understood in a similar manner by all individuals. Instead, it was posited that each individual potentially possesses a unique understanding of their surroundings, and that this understanding is shaped by mental processes of information gathering and organization (Gold, 1980). Consequently, it was argued that it is misleading to analyse human spatial behaviour in relation to the objective, 'real' environment because people do not conceive of (and experience) space in this way. It was suggested that a more productive approach would be to focus on the way that people act in relation to how they cognize the world around them. Such a focus would explain why human behaviour did not fit the patterns sometimes anticipated in models of spatial science (see entries on **Haggett, Berry**). At its core then, behavioural geography is based upon the belief that the explanatory powers and understanding of social scientists can be increased by incorporating behavioural variables, along with others, within a framework that seeks to comprehend and find reasons for overt spatial behaviour, rather than describing the spatial manifestations of behaviour itself (Golledge, 1981).

By the early 1970s, divisions within behavioural geography started to emerge as to how best to theorize and measure spatial behaviour, with on the one hand the development of a phenomenological-humanist approach (exemplified in research by Lowenthal, Seamon and **David Ley**) and on the other an analytical, scientific-positivist approach (of which Golledge was the chief proponent). While both approaches were united in believing that 'we must understand the ways in which human beings come to understand the geographical world in which they live' and that 'such understanding is best approached from the level of the individual human being' (Downs, 1981), increasingly their alliance fractured, so that by the end of the 1970s they had developed into largely separate ventures (see Saarinen *et al.*, 1984). In the humanist branch of behavioural geography, the search for scientific laws was replaced by an interpretative and reflective search for meaning and how humans come to understand and act in the world. Golledge rejected such conceptualizations, and in particular the subjective and unscientific nature of data collection and analysis. Instead, he advocated an analytical and scientific examination of the thoughts, knowledge and decisions that underpin human action (Golledge and Rushton, 1984), using questionnaires and adapting measures from cognitive psychology such as perceptual tests and rating scales as a means to measure people's ability to remember, process and evaluate spatial information. The findings from these studies were used to test models of spatial choice and decision-making in relation to issues such as way-finding, residential location, industrial agglomeration, tourist behaviour, migration, and so on. Here, geographic space is conceptualized as absolute and given (thus knowable and mappable), but analytically it is how this space is cognized that is considered most important.

Golledge's contribution to analytical behavioural geography cannot be underestimated. Over the course of his career he has developed a systematic programme of research that has consistently sought to deepen and strengthen the theoretical and methodological underpinnings and empirical scope of behavioural geography. So, for example, he has engaged in wider ontological and epistemological debates within the discipline of geography, seeking to tighten and advance behaviouralism's theoretical tenets and to promote it to a wider audience. He has developed a number of specific theories concerning the development and structuring of spatial knowledge, processes of spatial choice and decision-making (in different contexts – transportation, residential choice), and environmental learning with regards to different populations (adults, children, developmental disabilities, visual impairment, men/women). Some of these theories, such as the anchor-point model of spatial knowledge, have been widely engaged with by cognitive and environmental psychologists (see Couclelis *et al.*, 1987). He has pioneered, developed and tested a whole series of behavioural measures and analytical techniques including multidimensional scaling, psychometric testing, sketch maps, distance and direction estimates (see Golledge and Stimson, 1997, for review), and championed a move away from the (psychology) laboratory to real world environments, challenging psychologists in particular to model spatial behaviour in naturalistic settings. Finally, he has sought to apply his research findings to real world issues such as planning, transportation modeling, and, perhaps most successfully, the development of orientation and communication devices for people with visual impairments (notably tactile maps, a personal guidance system, and haptic soundscapes).

KEY ADVANCES AND CONTROVERSIES

While some researchers have used Golledge's ideas to build up a large body of behavioural research (see Golledge and Stimson, 1997), and others have sought to extend his theoretical insights by making explicit links to cognitive science and environmental psychology (see Kitchin and Freundschuh, 2000; Kitchin and Blades, 2001), his work – and behavioural geography more generally – has come under a sustained critique from the late 1970s onwards. As a consequence, Cloke et al. (1991) described behavioural geography as a largely forgotten element of human geography. John Gold (1992) identified three reasons why behavioural geography has not been fully embraced by the geographic fraternity (especially in the UK). First, due to structural changes in the education sector in the late 1960s early 1970s, young behavioural geographers failed to secure posts and thus a critical mass failed to develop. Second, as social issues came to the fore during the 1970s, behavioural geography was perceived to be inappropriate for examining them. Third, the philosophical bases of behavioural geography, particularly of the analytical variety, were heavily criticized by other researchers from different traditions.

Both humanists and structuralists criticized analytical behavioural geography – and thus the approach being advocated by Golledge – for its positivistic allegiances. They argued that instead of offering a viable alternative to the positivistic, spatial science, behavioural geography just shifted emphasis so that many of the criticisms levelled at positivism still applied. As such, Cox (1981) argued that the emergence of behavioural geography was evolutionary rather than revolutionary. Further, both groups criticized analytical behavioural geography for over-emphasizing empiricism and methodology at the expense of worthwhile issues and philosophical content (Gold, 1992). For example, Cullen (1976) argued that analytical behavioural geographers blindly borrowed from the scientific paradigm, which then determined the nature of the problems to be investigated, so that the independent–dependent variable format was overused. Ley (1981: 211) argued that the allegiance to the scientific paradigm led to a preoccupation with measurement, operational definitions and highly formalized methodology, so that 'subjectivity has been confined to the straitjacket of logical positivism'. As such, Golledge's work offered an inadequate and mechanistic understanding of human behaviour.

While structuralists critiqued the reduction of human spatial behaviour to cognition, thus failing to take into account the influence of wider social, economic and political factors on peoples' everyday geography (Cox, 1981), humanistic geographers disputed the dichotomy between subject/object and fact/value and argued that research which accepted these dichotomies would only provide clues to everyday life, failing to 'conceive of life in its wholeness or for that matter of individuals in their wholeness' (Eyles, 1989: 111). They argued that the subject and object could not be separated because of the intervening consciousness which imposes its own interpretations upon the objective world and thus affects behaviour (Cox, 1981). Subject/object, fact/value become infused and inseparable and need to be investigated as such, so that the methods used by analytical behavioural geographers are invalid as they assume that the investigator and investigated have the same meanings. Consequently, it was argued that Golledge's theorizing ignored the contours of experience and reduced individuals to crude automatons (Thrift, 1981), systematically detached from the social contexts of their actions, and thus meanings. Ley (1981) further argued that behavioural geography adopts a naturalist stance that sees no essential discontinuity between people and nature and gives human consciousness little theoretical status.

In addition, Walmsley and Lewis (1993) cautioned that behavioural geographers needed to be aware of the dangers of psychologism; that is, the fallacy of explaining social phenomena purely in terms of the mental characteristics of individuals. By concentrating upon the individual, they noted that behavioural geography is susceptible to the trap of building models inductively, beginning at the level of the individual, so that outcomes can only be treated as the sum of parts (Greenburg, 1984). This is a particularly salient point because one of the main criticisms of behavioural geography has been its one-dimensional look at environmental behaviour at the expense of economic, political and social considerations. Indeed, Gold (1992: 240) has argued that the attempt 'to straitjacket *all* areas within a strictly psychological paradigm' is one of the fundamental reasons for the disillusionment with behavioural approaches.

This latter point is well illustrated in critiques of Golledge's (1993) work on disability. While acknowledged as pioneering, the use of behavioural theory to articulate a geography of disability drew fierce criticism from other geographers, notably Brendan Gleeson (1996) and Rob Imrie (1996). They attacked Golledge's vision in relation to his conception of disability, the ontological and epistemological bases of his research, and his lack of ideological intent. In relation to the first, they note that Golledge adopts a medical understanding of disability in which the problems facing disabled people are seen as a function of their impairment (rather than how society treats them). This in turn positions disabled people as subjects within the research, perpetuating the dichotomy between expert researcher and passive research subject. Moreover, it fails to acknowledge the exclusionary practices of society and the role of social, political and economic processes in the reproduction of disabling environments. Thus for Gleeson and Imrie, Golledge's geography of disability falls into trap of ablesm – the reduction of disability to functional limitations and an acceptance that if we can make disabled people more like able-bodied people, their problems will be significantly reduced. Golledge is accused of reducing the problems faced by disabled people to technical issues that can be solved with technical solutions, thus depoliticizing the problems that disabled people face. This decontextualizes disability, placing it outside of the historical and spatial transformations within which modern relations are embedded. Instead, Gleeson and Imrie suggest a more fruitful approach is to engage with disabled people in their quest for emancipation by exposing the oppressive structures of society.

Despite widespread criticism, Golledge has been fervent in his rebuttals of the perceived shortcomings of analytical behavioural geography (see Golledge, 1981, 1986; Couclelis and Golledge, 1983; Golledge and Stimson, 1997) and it is fair to state that behavioural geography continues to be widely practised within human geography, particularly in North America, where links with cognitive and environmental psychology have been forged (see Gärling and Golledge, 1993; Golledge, 1999; Kitchin and Freundschuh, 2000; Kitchin and Blades, 2001). That said, it is clearly no longer considered at the cutting edge of geographical theory and praxis, despite the efforts of Golledge to re-inspire a return to its ideas (Golledge, 2002).

GOLLEDGE'S MAJOR WORKS

Cox, K. R. and Golledge, R. G. (1969) *Behavioural Problems in Geography: A Symposium.* Evanston, IL: Northwestern University Press.

Golledge, R. G. (2002) 'The nature of geographic knowledge', *Annals of the Association of American Geographers* 92 (1): 1–14.

Golledge, R. G. (1993) 'Geography and the disabled: a survey with special reference to vision impaired and blind populations', *Transactions of the Institute of British Geographers* 18: 63–85.

Golledge, R. G. and Spector, A. N. (1978) 'Comprehending the urban environment: theory and practice', *Geographical Analysis* 9: 403–426.

Golledge, R. G. and Stimson, R. J. (1987) *Analytical Behavioural Geography.* London: Croom Helm.

Golledge, R. G. and Stimson, R. J. (1997) *Spatial Behaviour: A Geographic Perspective.* New York: Guildford Press.

Golledge, R. G., Briggs, R. and Demko, D. (1969) 'The configuration of distances in intraurban space', *Proceedings of Association of American Geographers* 1: 60–65.

Moore, G. T. and Golledge, R. G. (eds) (1976) *Environmental Knowing.* Stroudsberg: Dowden, Hutchinson and Ross.

Secondary Sources and References

Cloke, P. Philo, C. and Sadler, D. (1991) *Approaching Human Geography.* Liverpool: PCP Press.

Couclelis, H. and Golledge, R. G. (1983) 'Analytical research, positivism and behavioural geography', *Annals of the Association of American Geographers* 73: 95–113.

Couclelis, H., Golledge, R. G., Gale, N. and Tobler, W. (1987) 'Exploring the anchor-point hypothesis of spatial cognition', *Journal of Environmental Psychology* 7: 99–122.

Cox, K. R. (1981) 'Bourgeois thought and the behavioural geography debate', in K. R. Cox and R. G. Golledge (eds) *Behavioural Problems in Geography Revisited.* Evanston, IL: Nortwestern University Press, pp. 256–279.

Cox, K. R. and Golledge, R. G. (eds) (1981) *Behavioural Problems in Geography Revisited.* Evanston, IL: Northwestern University Press.

Cullen, I. (1976) 'Human geography, regional science, and the study of individual behaviour', *Environment and Planning A* 8: 397–409.

Downs, R. M. (1981) 'Maps and mappings as metaphors for spatial representation', in L. S. Liben, A. Patterson and N. Newcombe (eds) *Spatial Representation and Behaviour Across the Life Span.* New York: Academic Press, pp. 143–166.

Eyles, J. (1989) 'The geography of everyday life', in D. Gregory and R. Walford (eds) *Horizons in Human Geography.* London: Macmillian, pp. 102–117.

Gärling, T. and Golledge, R. G. (eds) (1993) *Behaviour and Environment: Psychological and Geographical approaches.* London: North Holland.

Gleeson, B. J. (1996) 'A geography for disabled people?', *Transactions of the Institute of British Geographers* 21: 387–396.

Gold, J. R. (1980). *An Introduction to Behavioural Geography.* Oxford: Blackwell.

Gold, J. R. (1992) 'Image and environment: the decline of cognitive-behaviouralism in human geography and grounds for regeneration', *Geoforum* 23: 239–247.

Golledge, R. G. (1981) 'Misconceptions, misinterpretations, and misrepresentations of behavioural approaches in human geography', *Environment and Planning A* 13: 1315–1344.

Golledge, R. G. (1996) 'A response to Gleeson and Imrie', *Transactions of the Institute of British Geographers* 21: 404–410.

Golledge, R. G. (1997) 'On reassembling ones life: overcoming disability in the academic environment', *Environment and Planning D: Society and Space* 15: 391–409.

Golledge, R. G. (ed.) (1999) *Wayfinding Behaviour: Cognitive Mapping and Other Spatial Processes.* Baltimore: Johns Hopkins University Press.

Golledge, R. and Rushton, G. (1984) 'A review of analytical behavioural research in geography', in D. Herbert and R. Johnston (eds) *Geography and the Urban Environment.* London: Croom Helm.

Greenburg, D. (1984) 'Whodunit? Structure and subjectivity in behavioural geography', in T. F. Saarinen, D. Seamon and J. L. Sell (eds) *Environmental Perception and Behaviour: An Inventory and Prospect.* Research paper 209, Department of Geography, University of Chicago.

Imrie, R. F. (1996) 'Ableist geographies, disablist spaces: Towards a reconstruction of Golledge's geography and the disabled', *Transactions of the Institute of British Geographers* 21: 397–403.

King, L. and Golledge, R. G. (1978) *Cities, Space and Behaviour.* Englewood Cliffs, NJ: Prentice Hall.

Kitchin, R. M. and Blades, M. (2001) *The Cognition of Geographic Space*. London: IB Taurus.

Kitchin, R. M. and Freundschuh, S. (eds) (2000) *Cognitive Mapping: Past, Present and Future*. London: Routledge.

Ley, D. (1981) 'Behavioural geography and the philosophies of meaning', in K. R. Cox and R. G. Golledge (eds) *Behavioural Problems in Geography Revisited*. Evanston, IL: Northwestern University Press, pp. 209–230.

Saarinen, T. F., Seamon, D. and Sell, J. L. (eds) (1984) *Environmental Perception and Behavior: An Inventory and Prospect*. Research Paper 209, Department of Geography, University of Chicago.

Thrift, N. (1981) 'Behavioural geography', in N. Wrigley and R. Bennett (eds) *Quantitative Geography in Britain*. London: Routledge and Kegan Paul.

Walmsley, D. J. and Lewis, G. (1993) *People and Environment*. Harlow: Longman.

Rob Kitchin

21 Derek Gregory

BIOGRAPHICAL DETAILS AND THEORETICAL CONTEXT

Born in England in 1951, Derek Gregory was raised in Bromley, Kent. He attended a local grammar school and won a scholarship to become a Fellow at Sidney Sussex College, Cambridge, in 1969. He completed his BA, MA, and PhD (1981) at Cambridge and became University Lecturer there. In 1989, he moved to the University of British Columbia in Vancouver, Canada, as Professor of Geography. He is the recipient of the Lund University Killam Research Prize and the Killam Teaching Prize.

Since 1976, Gregory has published a series of highly visible and well-received volumes that have been widely influential. His works began to appear during a period of substantial flux and transition within geography. In the 1970s and 1980s, the decline of positivism opened a space for competing perspectives, particularly various species of Marxism and humanism, raising a multiplicity of complex questions pertaining to the simultaneously determinant roles of society and space, individual agency and social structures, and human–nature interactions. Gregory's MA thesis, *Ideology, Science, and Human Geography* (1978b), has been extensively cited for its elegant synopses and critiques of structuralism, positivism, and reflexivity within the discipline. The volume *Recollections of a Revolution* (Billinge *et al.*, 1984) revealed the positivist coup as peopled, not simply the inevitable triumph of rational logic over empiricism but the product of active human actors embedded in a changing, politicized matrix of academic and social circumstances. Gregory offered parallel, scathing critiques of traditional location theory (1982b), systems theory (Gregory, 1980), diffusion theory (Gregory, 1985c), and humanistic geography (Gregory, 1981a), the latter of which he accused of being preoccupied with the 'casual interrogation of the obvious.'

Simultaneously, Gregory (1984b, 1989c) played, along with other theorists such as **Nigel Thrift**, an influential role in introducing **Anthony Giddens'** theory of structuration to geographers, a view that resolved the long-standing schism between overly deterministic perspectives that emphasized social structures over individual behaviour and idealist views that began, and typically ended, with the consciousness of the mind. Thus, Gregory argued, geographical research, in both historical and contemporary contexts, should seek to recover the thoughts, desires and meanings of actors, but avoid collapsing its understanding into those same motivations, for social (and spatial) relations forever escape the intentions of their creators. The construction of geographies is thus a 'sensuous swirl of contingency and determination' (Gregory, 1982a: 246). Coupled with erudite evaluations of the alternatives, structuration theory became *de rigueur* for growing legions of graduate students in human geography in that decade and the next.

Gregory's early works introduced social theory into historical geography (1976, 1978a), a field hitherto characterized by its sterility and stubborn empiricism. Following the footsteps of E. P. Thompson, he has long been concerned with class struggle in England during the Industrial Revolution (1984a). His interest in the dynamics of urbanization was evident in his doctoral dissertation (Gregory, 1982b), which applied structuration theory to the Yorkshire woollen industry to shed light on the time–space compression wrought by modernity. In addition to generating a corpus of work that demonstrated to geographers as a whole the importance of taking history seriously, Gregory's unearthing of the origins of modernity allowed him to disclose its historically specific nature, despite the claims of Enlightenment science, both positivist and Marxist, to universal applicability. Indeed, the historicization of modernity led him to examine its crisis and decline in the contemporary historical moment (Gregory, 1991).

Social Relations and Spatial Structures (Gregory and Urry, 1985) marked an important step forward in the ascendancy of social theory within geography. With contributions by major theoreticians within geography and sociology (including, among others, **Massey, Sayer, Soja, Harvey**, Walker, **Giddens**, Pred and **Thrift**), the volume pointed to the growing intersections between the two fields and the seriousness with which non-geographers took space, setting the stage for the 'spatial turn' of the 1990s.

SPATIAL CONTRIBUTIONS

By the late 1980s, Gregory's contributions had surpassed critiques of existing perspectives to offer important alternatives. Making explicit the limitations of modernist interpretations led to flirtations with postmodernism and its sympathy to a renewed regionalism, although he has been reluctant to let himself be identified too closely with appellations drawn from the poststructuralist school(s).

In 'Solid geometry' (Gregory, 1988a), he took to task structuralist views that held to the possibility of considering space, time and society independently. Spatial structures are much more than the passive manifestations of structural transformations that occur aspatially, a widespread but artificial and analytically misleading dichotomy reproduced, for example, in the famous 'socio-spatial dialectic' offered by **Ed Soja** (1980). In contrast, Gregory argued, geographies are active participants in the process of social change, and to appreciate their significance the discipline needed a new repertoire of tools ranging from time-geography to the constitutive phenomenology of a reformulated humanism. Structuration theory offered a particularly ripe opportunity to make these advances, recovering for geography a central role in the reproduction of social relations without advocating a return to the spatial formalism of chorology or positivism.

Geographical Imaginations (1994a) was perhaps Gregory's *magnum opus* and the beginning of an important new stage in his contributions to geographic thought. In remarkably subtle and sophisticated ways, Gregory explored how place and space are implicated in the humanities and social sciences. His argument focused on how seemingly different paradigms reflect particular 'scopic regimes', ways of seeing and knowing the world that appear consistent and complete within their own frames of reference and 'natural' to their subscribers. Buttressed by the philosophical ideas of Descartes and Kant, Western social science, including geography, has long subscribed to the notion of a detached, all-knowing, objective observer. This assumption, increasingly questionable in an age of mounting relativism, forms the foundation for the particular epistemological standpoint that Gregory calls the 'world-as-exhibition'. This perspective, arising during the Renaissance and the ensuing march of modernity, is common to various Marxist and positivist conceptions of space, and confers a particular status, and thus power, to the knower as a rational, presumably male, all-knowing ego. The world-as-exhibition is thus a quintessentially modern and modernist way of enframing the world, a position he labels 'ocularcentrism', one also, and not coincidentally, well suited to the historical process of commodifying social life and space. This view was endlessly replicated in various Weberian (both Max and Alfred), empiricist, neoclassical and instrumentalist conceptions of geographic relations, and denied the inescapable situatedness of knowledge. Within geography, it was reflected in the triumph of abstract space over lived experience, a position made possible only if representations of the world are held to be detached from the world they reflect.

However, Gregory maintains, rather than comprise some inherent 'truth' independent of historical experience, the world-as-exhibition can be revealed, through the deployment of various postmodernist, poststructuralist and postcolonial perspectives, as but one scopic regime among many. Towards this end, he invokes an astounding galaxy of diverse authors: Georg Simmel and Walter Benjamin, Captain Cook and Allen Pred, Vidal de la Blache and **Ed Soja**, Immanuel Kant and Edmund Husserl, Frederic Jameson and **Donna Haraway**. In contrast to the arrogance of modernism, he upholds the poststructuralist position that any standpoint is incomplete and situated, linked to a power interest and refracted through various prisms of social position. 'Knowing one's place' is thus as much a geographical as an epistemological standpoint. Jettisoning the world-as-exhibition opens the door to novel and creative epistemic encounters, including feminism and subalternity, that take seriously largely marginalized voices and their positionality within the topographies of power. Because the legitimacy and purchase of representational systems is derived from their connections to the institutionalized systems of commodity production and consumption, the periodic restructuring of capitalism ineluctably initiates concomitant changes in symbolic systems. Thus, the crisis of representation so widespread in the late twentieth century – and the search for new forms of

meaning – were spawned by the massive rounds of time/space compression unleashed by contemporary globalization. He offered case studies that draw upon the perspectives of local inhabitants rather than arrogant academic flyovers. Gregory's work thus paved the way for poststructuralist theorizing within geography, noting its linkages to and departures from earlier traditions of thought within the discipline. Unsurprisingly, it generated considerable heat and debate (cf. Deutsche, 1995; Harvey, 1995; Katz, 1995).

Gregory's career also marked a distinct change in the style of academic writing, for he has earned a reputation for his clever word choice, literary allusions and elegant phraseology without resorting to hyperbole. Harvey (1995: 161), for example, notes his 'legendary skills as "phraseur"'. Like the human subjects about whom he writes, and whose moist and pungent worlds he seeks to recover, Gregory's textual strategies are deliberately, self-consciously poetic and, at times, challenging, and have provided a model for numerous other geographers to emulate as they discarded the desiccated, boring and sterile language of positivism.

Gregory has also been a co-editor in all four editions of *The Dictionary of Human Geography* (1981, 1986, 1994, 2000). A series of increasingly voluminous tomes, these works have provided concise but informative summaries of the discipline's terminology and concepts, making clear for numerous graduate students (and no doubt faculty) the implications of words and phrases whose meanings have often been lost in the obscurantist gobbledygook of academics seeking to impress other academics.

KEY ADVANCES AND CONTROVERSIES

Gregory's career has been characterized by a long-standing interest in the mutually constitutive relations between knowledge, power and space. In the tradition of Western (non-structuralist) Marxism, he portrays social relations and geographies as the contingent (if often unintended) outcomes of human actors whose networks are structured, but not predetermined, by rotating regimes of power and knowledge. Following Habermas, language for Gregory occupies a pivotal position in the acquisition, interpretation, transmission and representation of information about people, context and places.

Recently, in light of the influential perspective launched by **Edward Said** (1978), Gregory's trajectory has focused on Orientalism as a discourse intimately intertwined with the European penetration of non-Western spaces (Gregory, 1999). He contends that geography as a way of knowing space – the active 'geographing' of various parts of the globe – was part and parcel of the Western administrative and political control of such regions, which including the inventory of use values as well as the ideological legitimation of these relations. The colonial world-as-exhibition thus sustained the naturalization of Western dominance and parallel implication of non-Western inferiority. Drawing upon Foucauldian notions of how knowledge and power intersect in temporary, mutually interpenetrating constellations, Gregory maintains that Western notions of space were vital parts of the colonial imaginary: the ways in which space was demarcated and brought into Western frames of understanding drew critical boundaries between identities, self and other, and underpinned particular regimes of power and knowledge. Colonialism was thus as much a cultural and ideological project as an economic and political one, and geography has been deeply Eurocentric in ways that continue to shape the contemporary constitution of the discipline.

The specific instance that comprises the empirical focus of this postcolonial project concerns European and American representations of Egypt in the late nineteenth and early twentieth centuries

(Gregory, 1995a, 2002a). In the nineteenth century, with the Rosetta Stone deciphered and Egyptology all the rage in Europe, Egypt occupied an important geographical and ideological position in the evolving self-conception of the West: the ancient, stagnant, senile culture, simultaneously proximate and distant, which could be rendered sensible through the application of Western rationality, the 'empire of the gaze'. Reworking the observations of travellers and administrators centred in Cairo and traversing the River Nile, Gregory unpacks the 'imaginative geographies' that their writings revealed, tracing them to the patriarchical, sexualized and often racist imagery that pervaded Western views of the Arabic 'Other'. The opening of Verdi's opera *Aida* to a British audience in Cairo in the mid-nineteenth century, for example, revealed the 'micro-physics of imperialism': newspaper accounts recalled how the audience was made to feel 'as if they were in Egypt', i.e. the idealized land of their fantasies rather than the sordid, dirty and impoverished reality that their own policies had helped to construct. Space, power and identity were thus fused in an inseparable skein as Egypt was 'geographed' by a panopticonic foreign authority. Grounding Orientalism spatially reveals, then, how textual and discursive practices can have profound material consequences, allowing as they do the appropriation of space by rendering it meaningful to those capable of exerting control: the world-as-exhibition is always an exhibition for *someone*.

Perhaps in reaction to Harvey's (1995) criticism that Gregory fails to take nature sufficiently seriously, his postcolonial turn has led him to facilitate a creative engagement between the literatures concerned respectively with Orientalism and the social construction of nature (Gregory, 2001). During the Western conquest of the Other, for example, he noted how the developing world was enframed within Eurocentric ways of knowing that depicted 'the tropics' as either fecund Gardens of Eden inhabited by child-like innocents or disease-ridden swamps populated by savages. In Egypt, the desert sand dunes – utterly foreign to conquerors from Britain – were first portrayed as irrational ('like a pack of wolves') before being subjected to the process of rational scientific dissection, i.e. aeolian geomorphology.

Gregory's contributions to geography are both conceptual and methodological. For close to 30 years he has repeatedly and emphatically stressed the importance of historical context in academic work, a move that thwarted the aspirations of Enlightenment modernists to construct universal interpretations independent of time and space. In line with structuration theory, he has remained steadfast in his insistence that times and spaces are made by ordinary folk in everyday life, and that the outcomes are always contingent and never predetermined, even if, *contra* overly structuralist interpretations, they are likewise never random or chaotic. Along with **Nigel Thrift**, his works have firmly established in the discipline the need to take the intentions and context of actors seriously in the explanation of spatial relations.

Gregory's works decisively demonstrate the need to understand colonialism and capitalism in discursive terms, and he has been an astute observer of the roles played by language in the process of representation. In so doing, he has helped geographers to problematize the question of intellectual inquiry, both that of academics and those whom they study, wrenching it from its perch as a passive exercise in interpretation and recasting it as an active, often muddy, process of constructing meaning. Gregory has remained sensitive throughout his career to the politics of representation, the ways in which all forms of understanding are inescapably entwined with temporally and spatially varying regimes of power-knowledge: discourses simultaneously reflect and constitute the relations that they depict and that give rise to them. He thus was a significant contributor to the contemporary fusing of power, knowledge and space.

Finally, Gregory has played a pivotal role introducing geographers and theories of postcolonialism to one another, helping to awaken the discipline to the profound but often invisible ways in which geography as discourse was complicit in the systematic othering of non-European peoples. While not everyone accepts his views unproblematically, few geographers have contested his contributions to the discipline, which have helped it to reach new levels of theoretical sophistication.

GREGORY'S MAJOR WORKS

Duncan, J. and Gregory, D. (eds) (1999) *Writes of Passage: Reading Travel Writing.* London: Routledge.

Gregory, D. (1978b) *Ideology, Science and Human Geography.* London: Hutchinson.

Gregory, D. (1988a) 'Solid geometry: Notes on the recovery of spatial structure', in R. Golledge, H. Couclelis, and P. Gould (eds) *A Ground for Common Search.* Santa Barbara: Santa Barbara Geographical Press.

Gregory, D. (1989a) 'Areal differentiation and post-modern human geography', in D. Gregory and R. Walford (eds) *Horizons in Human Geography.* Totowa, NJ: Barnes and Noble.

Gregory, D. (1989b) 'The crisis of modernity? Human geography and critical social theory', in R. Peet and N. Thrift (eds) *New Models in Geography.* London: Unwin Hyman, pp. 348–385.

Gregory, D. (1994a) *Geographical Imaginations.* Cambridge, MA: Blackwell.

Gregory, D. (1995d) 'Imaginative geographies', *Progress in Human Geography* 19: 447–485.

Gregory, D. and Urry, J. (eds) (1985) *Social Relations and Spatial Structures.* New York: St. Martin's.

Secondary Sources and References

Barnes, T. and Gregory, D. (eds) (1997) *Reading Human Geography: The Poetics and Politics of Inquiry.* London: Arnold.

Billinge, M., Gregory, D. and Martin, R. (eds) (1984) *Recollections of a Revolution: Geography as Spatial Science.* London: Macmillan.

Deutsche, R. (1995) 'Surprising geography', *Annals of the Association of American Geographers* 85: 168–175.

Gregory, D. (1976) 'Rethinking historical geography', *Area* 8: 295–299.

Gregory, D. (1978a) 'The discourse of the past: Phenomenology, structuralism and historical geography', *Journal of Historical Geography* 4: 161–173.

Gregory, D. (1980) 'The ideology of control: Systems theory and geography', *Tijdschrift voor Economische en Sociale Geographie* 71: 327–342.

Gregory, D. (1981a) 'Human agency and human geography', *Transactions of the Institute of British Geographers* 6: 1–18.

Gregory, D. (1982a) 'A realist construction of the social', *Transactions of the Institute of British Geographers* 6: 1–18.

Gregory, D. (1982b) *Regional Transformation and Industrial Revolution: A Geography of the Yorkshire Woollen Industry, 1780–1840.* Minneapolis: University of Minneapolis Press.

Gregory, D. (1982c) 'Solid geometry', in P. Gould and G. Olsson (eds) *A search for common ground.* London: Pion.

Gregory, D. (1984a) 'Contours of crisis? Sketches for a geography of class struggle in the early Industrial Revolution in England,' in A. Baker and D. Gregory (eds) *Explorations in Historical Geography.* Cambridge: Cambridge University Press.

Gregory, D. (1984b) 'Space, time and politics in social theory', *Environment & Planning D: Society & Space* 2: 123–132.

Gregory, D. (1985a) 'Suspended animation: The stasis of diffusion theory,' in D. Gregory and J. Urry (eds) *Social Relations and Spatial Structures.* New York: St. Martin's.

Gregory, D. (1989c) 'Presences and absences: Time–space relations and structuration theory', in D. Held and J. Thompson (eds) *Critical Theory of Modern Societies.* Cambridge: Cambridge University Press.

Gregory, D. (1991) 'Interventions in the historical geography of modernity: Social theory, spatiality and the politics of representation', *Geografiska Annaler* 73B: 17–44.

Gregory, D. (1994b) 'Lefebvre, Lacan and the production of space,' in G. Benko and U. Strohmayer (eds) *Spatial History.* Dordrecht: Kluwer.

Gregory, D. (1995a) 'Between the book and the lamp: Imaginative geographies of Egypt, 1849–50', *Transactions of the Institute of British Geographers* 20: 29–57.

Gregory, D. (1995b) 'Commitments: The work of theory in human geography', *Economic Geography* 72: 73–80.

Gregory, D. (1995c) 'A geographical unconscious: Spaces for dialogue and difference', *Annals of the Association of American Geographers* 85: 175–186.

Gregory, D. (1999) 'Scripting Egypt: Orientalism and the cultures of travel', in J. Duncan and D. Gregory (eds) *Writes of Passage: Reading Travel Writing.* London: Routledge.

Gregory, D. (2001) '(Post)colonialism and the production of nature', in N. Castree and B. Braun (eds) *Social Nature: Theory, Practice, and Politics.* Oxford: Blackwell.

Gregory, D. (2002a) 'Emperors of the gaze: photographic practices and productions of space in Egypt (1839–1914)', in J. Schwartz and J. Ryan (eds) *Picturing Place.* London: IB Tauris.

Gregory, D. (ed.) (2002b) *Lefebvre, Spatiality and Capitalism.* London and New York: Routledge.

Gregory, D., Johnston, R., Haggett, P., Smith, D. and Stoddart, D. (eds) (1981b) *The Dictionary of Human Geography.* Oxford: Blackwell.

Gregory, D., Martin, R. and Smith, G. (eds) (1994) *Human Geography: Society, Space and Social Science.* London: Macmillan; Minneapolis: University of Minnesota Press.

Gregory, D. and Walford, R. (eds) (1989) *New Horizons in Human Geography.* London: Macmillan.

Harvey, D. (1995) 'Geographical knowledge in the eye of power: Reflections on Derek Gregory's *Geographical Imaginations*', *Annals of the Association of American Geographers* 85: 160–164.

Johnston, R. J., Gregory, D. and Smith, D. (eds) (2000) *The Dictionary of Human Geography*, 4th edition. Cambridge: Blackwell.

Katz, C. (1995) 'Major/minor: Theory, nature, and politics', *Annals of the Association of American Geographers* 85: 164–168.

Said, E. (1978) *Orientalism.* New York: Vintage.

Soja, E. (1980) 'The socio-spatial dialectic', *Annals of the Association of American Geographers* 70: 207–225.

Barney Warf

22 TORSTEN HÄGERSTRAND

BIOGRAPHICAL DETAILS AND THEORETICAL CONTEXT

Stig Torsten Erik Hägerstrand was born in 1916. He grew up in rural Sweden and studied at Lund University, Sweden, obtaining his BA in 1944 and his PhD in 1953. His initial impressions of geography were negative: 'lectures in regional geography were abominably boring . . . Geography appeared not as a realm of ideas or a perspective on the world but as an endless array of encyclopedic data' (Hägerstrand, 1983: 244). Matters improved when he undertook a regional study around Asby in southern Sweden, based on the biographies of every inhabitant from 1840 to 1940. He later realized that this large set of unique information showed stability over time and could be described in quantitative terms. At Lund he was also introduced to the location theories of von Thünen and Christaller by Edgar Kant (see Hägerstrand, 1983). This introduction was to spark off a realization that probability theory could be used to study settlement patterns. It was this interest in the 'science of locations' that was to culminate in his renowned models of spatial diffusion.

Most of Hägerstrand's career was spent in Lund's Department of Social and Economic Geography, where he was appointed as Associate Professor in 1953, taking over as Chair in 1957 (see Pred, 1967, for an account of Hägerstrand's academic progression). In addition to academic research and teaching, Hägerstrand also played a major role in planning for the Swedish national government. His work included the redivision of Sweden into new local government units, the reorganization of the provincial government, the national land- and water-use plan and the formulation of a national settlement strategy. He remains Professor Emeritus at Lund.

Hägerstrand's publications on innovation diffusion led to his taking on a leading role in the 'quantitative' revolution in the 1950s and 1960s, and his later development of time geography made him a leader in the behavioural geography that came to prominence in the 1970s. At a time when many Anglophone geographers paid little attention to scholars outside Britain and North America, Hägerstrand's impact did much to ensure that geography in Sweden became and remained visible at a time when French and German geography was (no doubt unjustly) widely ignored. This contribution was celebrated through the publication of a *Festschrift* for Hägerstrand in 1981 (edited by Pred), which includes a bibliography of his published writings.

SPATIAL CONTRIBUTIONS

Hägerstrand was responsible for introducing several important ideas into human geography, many of these contributions coming from his thesis and early publications. His early study of migration in Sweden (1957) included some interesting generalizations about the inverse relationship of migration to distance and how it changed over time, based on detailed and painstaking individual biographies of residents of the small town of Asby. His work on the decline of migration with distance proved a starting point of the important tradition of quantitative research on distance and human interaction (e.g. Olsson, 1965). He also identified the historical decline in the importance of distance. The impact of the analysis was strengthened considerably by his innovative use of map projections, notably his depiction of the space relevant to migration decisions in terms of an azimuthal logarithmic distance map centred on Asby.

Empirically, he introduced the simulation of diffusion to geography, particularly the concept of wave or contagious diffusion, which he originally applied to agricultural innovation but which was extended by others to encompass many other geographical phenomena, including the spread of the ghetto (Morrill, 1965) and the settlement of Polynesia (Levison et al., 1973). The concept of 'mean information field' was an important idea at the heart of Hägerstrand's conceptualization of diffusion; this was a matrix of probabilities arranged around a central location. A person located at the centre was regarded as receiving messages from one of the surrounding places with probability given by the mean information field. In the context of the diffusion of agricultural innovations, this could be regarded as generating the probability at each time period of a farmer being aware of a particular innovation (such as a new agricultural technology or form of cultivation). Another important introduction into geography associated with this work is the use of simulation to shed light on what would have happened if a simple contagious process had determined the uptake of an innovation. The use of random numbers to model how the process of innovation diffusion unfolded was new to geography but proved popular subsequently in a wide range of fields, especially as it became clear that simulation remained an option when mathematical characterization of a process was too difficult. Simulations of a spatial process could be mapped and compared with reality, giving at least an intuitive idea of whether the observed pattern could have resulted from the process simulated.

Hägerstrand was also one of a number of quantitative geographers who moved on to emphasize the importance of individual behaviour in geography. This developed into the idea of 'time geography' (Hägerstrand, 1973), which attracted a large number of geographers in Sweden and elsewhere (Carlstein et al., 1978; Parkes and Thrift, 1980). Gregory (2000b) explains Hägerstrand's basic framework in terms of four basic propositions. First, space and time are finite resources on which individuals have to draw in order to realize projects. Second, realization of any project is subject to constraints of three types: capability constraints, which limit what is physically possible for people to do given their location and time resources; coupling constraints, which determine where, when and for how long individuals can meet; and authority constraints, which impose conditions of access to particular time–space domains. Third, these constraints delineate what is possible for individuals to fulfil as particular projects. Fourth, the 'central problem for analysis' is usually competition between projects for 'open space–times'.

Practising time geography involves the detailed dissection of individuals' movements over very short time periods, looking at the paths in space that people traverse over time, including daily or weekly space–time paths. There may be trade-offs between expenditures of time and other resources. Interesting perspectives about individuals' activities can be appreciated, for example, by looking at people who meet at a particular time but who may have made very different types of journey to get there. It may also be interesting to look at household members who may all be together at the beginning and end of the day but may have followed very different space–time paths in the meantime. Developments of these ideas, often expressed in diagrammatic form, incorporate concepts like space–time prisms, which describe the set of times and locations that can be reached by somebody who has to be at particular places at times before and after.

KEY ADVANCES AND CONTROVERSIES

Like **Peter Haggett** and others involved in the spread of the new ideas known as the 'quantitative revolution', Hägerstrand was more concerned with theory and generalized models than empirical description. He was also interested in processes operating in space and over time, perhaps in contrast to the more purely geometric (and static) approaches characteristic of much location theory. One such process was diffusion. In contrast to earlier work on diffusion conducted in the tradition of Berkeley School and Carl Sauer, which tended to view diffusion as a phenomenon at the level of a society or culture, his perspective focused on individuals located at specific points in space making decisions to adopt (or not to adopt) an innovation. When he moved into behavioural geography (Hägerstrand, 1970), these individuals could be regarded not just as 'economic man' but as decision-makers operating with incomplete knowledge and imperfect ability to calculate the benefits of alternative actions. Hägerstrand also acknowledged that decisions were usually made in conditions of uncertainty.

Geographical diffusion theory developed from Hägerstrand's work, partly through the application of his diffusion models to other phenomena, such as fashions or rumours. There was also work separating out different forms of diffusion. Hudson's work (1969) on diffusion within a central place hierarchy was certainly important here, but the field developed in terms of increasingly realistic models of diffusion in a range of different concepts. At the same time, there was also a convergence of interest between geographical and sociological work on diffusion.

The other difference between Hägerstrand's studies and earlier work on diffusion was the concern with general models, expressible in quantitative terms, rather than the specific times and places where particular innovations could be assumed to have occurred. Indeed, Hägerstrand's models of diffusion were explicitly probabilistic. The simulation methodology meant that an array of different outcomes was presented, rather than one 'true' deterministic outcome. This resulted in a view of the world, shared by other quantitative geographers like Curry (1964) and Dacey (1966), where what happened was one of a (large) number of possible eventualities. Geographers were encouraged to adopt a deductive approach to inference, where the issue was to test whether a given state of affairs could have been generated by the assumptions of a simple model. This approach could be (and was) criticized when the assumptions of the model were totally unrealistic (for example, in classical central place theory). Hägerstrand's work, however, was grounded in detailed collection of data, preceding the development of general theory.

Gregory (2000a) gives a succinct account of Hägerstrand's work on diffusion and some of the reactions to it. An important development has been the derivation of more complex models of spatial diffusion, especially through the work by Cliff, Haggett and others on the modelling of epidemics (see Haggett et al., 1977, chapter 7). Blaikie (1978) criticized diffusion research for its preoccupation with spatial form and space–time sequence, while Gregory (1985) criticized the unwillingness of diffusion theory to engage with social theory which clarifies the conditions under which diffusion occurs and the consequences of the diffusion.

Hägerstrand's ideas about time geography were followed up by several other authors, in Sweden and elsewhere. The study of time constraints emphasized differences in resources that confirmed other aspects of social class and difference. In particular, time geography has drawn attention to gender differences in the availability of time resources, especially to the time-related problems of mothers with young children. Other studies have picked up on other aspects of the role of time in geography, such as Thrift's study (1977) of the spread of standardized time in nineteenth-century England through the influence of the railways, and the development of factorial ecology to incorporate time as well as space (Parkes and Thrift, 1980, chapter 8). In addition to geographers, the concept of time geography attracted the attention of sociologists in the 1980s (Giddens, 1989) and, as such, was influential in increasing the status of geography among sociologists. Many of the terms and concepts associated with time geography have accordingly remained an important part of human geography's theoretical landscape up to the present (see May and Thrift, 2001).

Nonetheless, a common criticism of time geography has been its 'physicalism' – the emphasis on people's locations in space/time and the constraints affecting which projects are possible for them to undertake. Some critics, like Rose (1977), suggest that this downplays the role of attitudes, motives and choices (some of which may of course be irrational or unconscious in derivation). Rose's criticisms are addressed by Parkes and Thrift (1980: 275–277), who argue that they arise either from misconceptions of what time geography is, or from the fact that the approach has not been able to explain all aspects of the social use and allocation of time and space. From a feminist perspective, Rose (1993) likewise criticizes time geography for its conception of the person as a basic 'elementary particle' for time-geographic studies from which all social and cultural identities, such as race, gender and sexuality, have been erased. Hoppe and Langton (1988) have made the point that time geography has tended to confine itself to the micro-scale and short term, with little regard for the changing structures and contexts in which they operate. Accordingly, they have tried to develop time geography to cast light on macro-scale processes operating over longer time spans.

Hägerstrand himself (1983: 254) reports and acknowledges criticisms of his time-geography work on the grounds that it 'neglected and ran over the more important part of human existence: the internal realms of experience and meaning'. For example, he admits the truth of criticisms from Anne Buttimer (in Hägerstrand, 1983: 254) to the effect that his time-geography diagrams seemed to omit crucial dimensions of human temporality – images and perceptions of time, and bio-ecological rhythms. As this last point implies, Hägerstrand's perspective does not exclude humanistic insight into the human condition. Indeed, Van Paassen (1981) presents a humanistic analysis of Hägerstrand's thinking, relating it to the ideas of Vidal de la Blache. This often-overlooked humanistic element to Hägerstrand's work is perhaps best encapsulated in the title of his (1970) paper, 'What about people in regional science?' Unlike other quantitative geographers, Hägerstrand's ideas have therefore been of interest to geographers of different persuasions (including humanistic geographers such as Buttimer and Pred) and remain an important component of social and cultural geography's conceptual toolkit.

HÄGERSTRAND'S MAJOR WORKS

Hägerstrand, T. (1952) 'The propagation of innovation waves', *Lund Studies in Geography, Series B* 4: 3–19.

Hägerstrand, T. (1953) *Innovationsforloppet ur Korologisk Synpunkt.* PhD thesis, Lund.

Hägerstrand, T. (1957) 'Migration and area', in D. Hannerberg, T. Hägerstrand and B. Odeving (eds) *Migration in Sweden: A Symposium*, Lund Studies in Geography, Series B, No. 13.

Hägerstrand, T. (1967a) *Innovation Diffusion as a Spatial Process.* Chicago: University of Chicago Press.

Hägerstrand, T. (1967b) 'On Monte Carlo simulation of diffusion', in W. L. Garrison and D. F. Marble (eds) *Quantitative Geography: Part I. Economic and Cultural Topics.* Northwestern University Studies in Geography 13: 1–32.

Hägerstrand, T. (1970) 'What about people in regional science?', *Papers of the Regional Science Association* 24: 7–21.

Hägerstrand, T. (1973) 'The domain of human geography', in R. J. Chorley (ed.) *Directions in Geography.* London: Methuen, pp. 67–87.

Hägerstrand, T. (1982) 'Diorama, path and project', *Tijdschrift voor Economische en Sociale Geographie* 73: 323–339.

Secondary Sources and References

Blaikie, P. M. (1978) 'The theory of the spatial diffusion of innovations: a spacious cul-de-sac', *Progress in Human Geography* 2: 268–295.

Carlstein, T., Parkes, D. N. and Thrift, N. J. (eds) (1978) *Making Sense of Time* (3 volumes). Chichester: Wiley.

Curry, L. (1964) 'The random spatial economy: an exploration in settlement theory', *Annals of the Association of American Geographers* 54: 138–146.

Dacey, M. F. (1966) 'A compound probability law for a pattern more dispersed than random and with areal inhomogeneity', *Economic Geography* 42: 172–179.

Giddens, A. (1989) *Sociology.* Cambridge: Polity Press.

Gregory, D. (1985) 'Suspended animation: the stasis of diffusion theory,' in D. Gregory and J. Urry (eds) *Social Relations and Spatial Structures.* London: Macmillan, pp. 296–336.

Gregory, D. (2000a) 'Diffusion', in R. J. Johnston, D. Gregory, G. Pratt and M. Watts (eds) *The Dictionary of Human Geography*, 4th edition. Oxford: Blackwell, pp. 175–177.

Gregory, D. (2000b) 'Time-geography', in R. J. Johnston, D. Gregory, G. Pratt and M. Watts (eds) *The Dictionary of Human Geography*, 4th edition. Oxford: Blackwell, pp. 830–833.

Hägerstrand, T. (1975) 'Space, time and human conditions', in A. Karlqvist, L. Lundqvist and F. Snickars (eds) *Dynamic Allocation of Urban Space.* Windsor: Saxon House, pp. 3–14.

Hägerstrand, T. (1976) 'Geography and the study of interaction between society and nature', *Geoforum* 7: 329–334.

Hägerstrand, T. (1978a) 'A note on the quality of life–times', in T. Carlstein, D. N. Parkes and N. J. Thrift (eds) *Making Sense of Time, Volume 2: Human Activity and Time Geography.* Chichester: Wiley, pp. 214–224.

Hägerstrand, T. (1978b) 'Survival and arena: on the life-history of individuals in relation to their geographical environment', in T. Carlstein, D. N. Parkes and N. J. Thrift (eds) *Making Sense of Time, Volume 2: Human Activity and Time Geography*, Chichester: Wiley, pp. 122–145.

Hägerstrand, T. (1983) 'In search for the sources of concepts', in A. Buttimer (ed.) *The Practice of Geography.* Harlow: Longman, pp. 238–256.

Hägerstrand, T. (1984) 'Presence and absence: a look at conceptual choices and bodily necessities', *Regional Studies* 18: 373–380.

Haggett, P., Cliff, A.D. and Frey, A. (1977) *Locational Analysis in Human Geography*, 2nd edition. London: Arnold.

Hoppe, G. and Langton, J. (1988) 'Time-geography and economic development: the changing structure of livelihood positions on farms in nineteenth-century Sweden', *Geografiska Annaler B* 68: 115–137.

Hudson, J. C. (1969) 'Diffusion in a central place system', *Geographical Analysis* 1: 45–58.

Levison, M., Ward, G. R. and Webb, J. W. (1973) *The Settlement of Polynesia: A Computer Simulation.* Minneapolis: University of Minnesota Press.

May, J. and Thrift, N. (eds) (2001) *Timespace: Geographies of Temporality*, London: Routledge.

Morrill, R. L. (1965) 'The negro ghetto: problems and alternatives', *Geographical Review* 55: 339–361.

Olsson, G. (1965) 'Distance and human interaction: a migration study', *Geografiska Annaler B* 47: 3–43.

van Paassen, C. (1981) 'The philosophy of geography: from Vidal to Hägerstrand', in A. Pred (ed.) (1981) *Space and Time in Geography: Essays Dedicated to Torsten Hägerstrand.* Lund Studies in Geography, Series B, No. 48: 17–29.

Parkes, D. N. and Thrift, N. J. (1980) *Times, Spaces and Places: A Chronogeographic Perspective*. Chichester: Wiley.

Pred, A. (1967) 'Postscript', in T. Hägerstrand (1967) *Innovation Diffusion as a Spatial Process*. Chicago: University of Chicago Press, pp. 299–324.

Pred, A. (ed.) (1981) *Space and Time in Geography: Essays Dedicated to Torsten Hägerstrand*. Lund Studies in Geography, Series B, No. 48.

Rose, C. (1977) 'Reflections on the notion of time incorporated in Hägerstrand's time-geographic model of society', *Tijdschrift voor Economische en Sociale Geografie* 68: 43–50.

Rose, G. (1993) *Feminism and Geography: The Limits of Geographical Knowledge*. Cambridge: Polity.

Thrift, N. J. (1977) 'The diffusion of Greenwich Mean Time: an essay in social and economic history', *Working Paper* 192, School of Geography, University of Leeds.

Robin Flowerdew

23 Peter Haggett

BIOGRAPHICAL DETAILS AND THEORETICAL CONTEXT

Peter Haggett was born in 1933 and he grew up in rural Somerset, England. He later attributed his interest in geography to his experience of touring his home area on foot and by bicycle – for the geographer, he argued, 'locomotion should be slow; the slower the better' (Haggett, 1965). He studied at St Catherine's College, Cambridge, where his contemporaries included Peter Hall, Michael Chisholm, Ken Warren and Gerald Manners. He received his BA from Cambridge in 1954, and his PhD in 1970. In his doctoral research, concerned with forestry in Brazil, he began to think about locational analysis and to develop an interest in quantitative approaches and modelling. After a short period at University College London (1955–57), he moved to Cambridge, first as University Demonstrator in Geography and from 1962 as University Lecturer in Geography. He moved to Bristol as Professor of Urban and Regional Geography in 1966, where he spent the rest of his career. At Bristol, he played an important role in university administration, where he was Pro-Vice-Chancellor from 1979 to 1982 and Acting Vice-Chancellor from 1984 to 1985. He was awarded a CBE in 1993. Since retirement, he has maintained his activity in geographical research and publication.

Haggett's career can be seen as mirroring a wider series of transformations in how geography is studied. At the start of his career, human geography was clearly defined as a descriptive 'art'. Hartshorne's discussions on the nature of geography (1939, 1959) were the major statements attempting to define the discipline. Human geography was mainly taught on a region-by-region basis, with each region described in a standard form, implicitly or explicitly showing how the human geography arose from the natural environment. It was an atheoretical discipline concentrating on description with little serious attempt to measure the patterns observed and even less attempt to explain them. Haggett was one of the geographers most responsible for changing the nature of geography, its methods, its theories and (to a degree) its subject matter. Nevertheless, he maintained interests in many traditional branches of the subject; indeed, Chorley (1995) describes him as 'a quantitative, regional, historical, and economic geographer with biogeographical interests'.

A short essay like this can do little more than scratch the surface of Haggett's contributions to human geography; fortunately, two books have appeared which give far more information about his biography and geographical concerns. Haggett himself produced a memoir, *The Geographer's Art* (1990), intended to convey his approach to, and passion for, the subject. A few years later, Cliff *et al.* (1995) produced a *Festschrift*, which includes papers by leading geographers on most of Haggett's major interests, and more personal memoirs by Chorley (1995) and **Nigel Thrift** (1995).

SPATIAL CONTRIBUTIONS

It is hard to overestimate Peter Haggett's contribution to human geography from the 1960s onwards. With the physical geographer Richard Chorley, he organized a series of symposia at Madingley Hall, outside Cambridge, resulting in the very influential *Frontiers in Geographical Teaching* (Chorley and Haggett, 1965) and the even more influential *Models in Geography* (Chorley and Haggett, 1967). Both books consisted of a series of chapters discussing new approaches in the different branches of geography, many of them written by the editors' Cambridge colleagues. Not all the chapters in these books represented major departures from the established ways of studying the subject, but most, especially Haggett's own (Haggett and Chorley, 1967; Haggett, 1967), were influenced by a new and powerful set of ideas that became pivotal in human geography's putative 'quantitative revolution'.

The term 'quantitative revolution', however, does not do justice to the changes that Haggett and his colleagues tried to bring about (and largely succeeded) in the way geography was studied. Certainly geographers made increasing use of numbers, measurements and computers (Haggett, 1969) rather than relying on verbal descriptions, but the changes went way beyond this. The use of statistical analysis, largely unknown in geography beforehand, became increasingly important. Geographers began to think in terms of hypotheses and how to test them, many adopting the hypothetico-deductive approach regarded as typifying the scientific method (Harvey, 1969; Amedeo and Golledge, 1976). Geographers also developed the idea of constructing theory, sometimes based on working out the consequences of a set of assumptions.

One of the most important aspects of Haggett's work was to align human geography with the social sciences rather than the arts, and to increase its respectability as an intellectual discipline. Most fundamentally of all, he introduced ideas, concepts and readings from a wide range of other disciplines, starting the outward-looking tradition that is arguably one of geography's greatest strengths today. For instance, Haggett did much to popularize the locational theories of von Thünen (1826) and Christaller (1933), little known previously in geography but highly relevant to Haggett's concepts of spatial analysis as applied to agricultural land use and urban location respectively. The hugely influential *Locational Analysis in Human Geography* (Haggett, 1965) thus laid the foundations for a completely new approach to human geography, organized not by region or by systematic subfield, but by geometric form. As manifest in the chapter headings – 'Movement', 'Nodes', 'Networks', 'Hierarchies', 'Surfaces' – this book introduced to geography a whole new langauage of spatial science. Simultaneously, it showcased new concerns and exemplars, including a few hitherto little-known geographers, like **Torsten Hägerstrand**, and many economists, like Weber, Losch and Isard, together with engineers, systems theorists and philosophers of science. An expanded second edition was produced in 1977 (Haggett *et al.*, 1977), while another joint publication with Chorley appeared in 1969, in which methods for studying networks in both human and physical geography were discussed (ranging from transport networks to fluvial systems). With Cliff *et al.* (1975), Haggett also studied space–time diffusion in economic systems, such as spatio-temporal changes in unemployment. His work also stressed the practical uses of such models, in particular through the development and application of spatial forecasting (Chisholm *et al.*, 1970; Haggett, 1973). In all these books, Haggett adopted an abstract but elegant approach, using mathematical and statistical methods to tackle a wide range of examples from all over the world.

A contribution of a different kind was made by his influential textbook *Geography: A Modern Synthesis* (first edition 1972; subsequent editions 1975, 1979, 1983; rewritten as *Geography: A Global Synthesis*, 2001). Widely adopted as a student text in Britain and elsewhere, it was important in upholding the unity of geography at a time when many human geographers were turning their backs on physical geography (and *vice versa*). Many of Haggett's ideas filtered through to school geography, especially in Britain. His use of the beach as a means of explaining many of what he regarded as basic geographical concepts gave the book an appealing and innovative start, although the attempt to introduce the 'G-scale' as a way of expressing a variety of spatial scales proved unsuccessful in the longer term.

Starting from a largely accidental linkage with a group of epidemiologists in Geneva (see Haggett, 2000, pp. 9–10), Haggett's later career became increasingly concerned with medical geography, in particular the diffusion of infectious diseases (see Cliff and Haggett, 1989; Cliff *et al.*, 1981, 1986). By careful choice of disease and study area (measles in Iceland was a particularly successful choice), he was able to find out a great deal about the spread of such diseases, including the ability to predict the return time and proportion of population likely to be infected. Haggett (2000) provides an excellent and (as with all his work) well-written discussion of the major themes of his work in the field, much of which takes diffusion studies (Hägerstrand, 1969) as a starting point. Considerable conceptual and mathematical development was needed, however, to apply the ideas to disease diffusion, adapting the structure of the model to fit in with the characteristics of the specific disease and the way in which it spread. Although much of his work has established the spatio-temporal characteristics of epidemics in small and relatively closed geographical areas, he

has developed an interest in processes operating at a global scale over a longer time period. He has also studied ways in which the spread of epidemics can be controlled. With collaborators, he produced an *Atlas of Disease Distributions* (Cliff and Haggett, 1988) – in fact far more than an atlas – and *The International Atlas of AIDS* (Smallman-Raynor, Cliff and Haggett, 1992). He has also acted as a consultant to many scientific and medical bodies, including the World Health Organization.

KEY ADVANCES AND CONTROVERSIES

As stated above, Haggett was a great innovator in British human geography. His work was one of the main foundations of the definition of geography as spatial science. His adoption of the scientific method, later pilloried as positivism, was also seminal for a generation of British geographers. Both notions were poorly received by establishment geography at the time (for example, see David Stoddart's account of the reception of a 1964 Royal Geographical Society lecture presenting a statistical model for the distribution of forest cover in Brazil, cited in Thrift, 1995: 381–382). However, they achieved greater acceptance among the younger generation of scholars (an unusually numerous group because the British university system was expanding rapidly at the time). **Peter Taylor** (1976) provides an interesting perspective on these debates. He argued that a shift of paradigm was aided by the use of language that was difficult for an older generation to come to terms with (certainly true of statistical arguments) but not too difficult for the younger ones to master.

Positivism and spatial science held sway for much of the 1970s, but criticisms of positivism and quantification began to emerge before too long (Mercer, 1984; Gregory, 1978). Such criticisms came from two main quarters. First, an increasingly vocal group of Marxist geographers, augmented by some of Haggett's earlier followers, like **David Harvey** and **Peter Taylor**, gained impetus from the wider critique of capitalism, the military-industrial complex and the Vietnam War. Their commitments to unobservable structures as the basis of social formations, and to dialectic rather than scientific method as a guide to reasoning, were opposed to the positivist tradition that Haggett had brought into geography. A second group of geographers reacted against the abstract models that Haggett had made fashionable in geography, stressing the importance of artistic and humanistic insights and the limitations of idealized models of the individual like 'economic man' in adequately representing thinking, feeling people (see **David Ley** and **Yi-Fu Tuan**).

The idea of geography as spatial science also came under attack at the same period. One element of this attack was from geographers whose interests did not lie in this area. These included people interested in the traditional geographical concern of the relationships between people and environment, and others interested in development studies grounded in specific places and their experiences. More philosophically grounded objections appeared, such as Sack's (1974) rejection of the idea that space could be a subject of study in itself. Both Marxists and humanists, not to mention feminists and postmodernists, tended to reject or at least play down the spatial perspective, either because it was unimportant in their view to the major social trends or because abstract models of spatial structure did not apply to the real people and events they were interested in.

Haggett's ideas are, however, far from dead. Despite the unfashionability of modelling, quantification and positivism for the last two decades, work of this kind continues to flourish, arguably being returned to prominence by the growth of geographical information science (GIS) in the 1990s. Similarly, spatial analysis remains as an important if minority interest in contemporary human geography, if only because there are many applications to commercial and public location problems. The importance of space as a key concept in geography has also revived in recent years, with increased attention being devoted to the social construction of space and the role of space in mediating social relations and processes. Haggett himself, while unfailingly polite and respectful of work done under new paradigms, remains consistent in his interests and approaches, and continues to be an enthusiastic and effective advocate of his own ideas.

HAGGETT'S MAJOR WORKS

Chorley, R. J. and Haggett, P. (eds) (1965) *Frontiers in Geographical Teaching.* London: Methuen.

Chorley, R. J. and Haggett, P. (eds) (1967) *Models in Geography.* London: Methuen.

Cliff, A. D., Haggett, P., Ord, J. K., Bassett, K. A. and Davies, R. B. (1975) *Elements of Spatial Structure: A Quantitative Approach.* Cambridge: Cambridge University Press.

Cliff, A. D., Haggett, P. and Ord, J. K. (1986) *Spatial Aspects of Influenza Epidemics.* Pion: London.

Haggett, P. (1965) *Locational Analysis in Human Geography.* London: Arnold.

Haggett, P. (1990) *The Geographer's Art.* Oxford: Blackwell.

Haggett, P. (2001) *Geography: A Global Synthesis.* Harlow: Prentice Hall.

Haggett, P. and Chorley, R. J. (1969) *Network Analysis in Geography.* London: Arnold.

Secondary Sources and References

Amedeo, D. and Golledge, R. G. (1975) *An Introduction to Scientific Reasoning in Geography*. Chichester: Wiley.

Chisholm, M. D. I., Frey, A. E. and Haggett, P. (eds) (1970) *Regional Forecasting*. London: Butterworth.

Chorley, R. J. (1995) 'Haggett's Cambridge: 1957–1966', in A. D. Cliff, P. R. Gould, A. G. Hoare and N. J. Thrift (eds) *Diffusing Geography: Essays for Peter Haggett*. Oxford: Blackwell, pp. 355–374.

Christaller, W. (1933) *Die zentralen Orte in Suddeutschland*. Jena.

Cliff, A. D., Gould, P. R., Hoare, A. G. and Thrift, N. J. (eds) (1995) *Diffusing Geography: Essays for Peter Haggett*. Oxford: Blackwell.

Cliff, A. D. and Haggett, P. (1988) *Atlas of Disease Distributions*. Oxford: Blackwell.

Cliff, A. D. and Haggett, P. (1989) 'Spatial aspects of epidemic control', *Progress in Human Geography* 13: 315–347.

Cliff, A. D., Haggett, P., Ord, J. K. and Versey, G. R. (1981) *Spatial Diffusion: An Historical Geography of Epidemics in an Island Community*. Cambridge: Cambridge University Press.

Gregory, D. (1978) *Ideology, Science and Human Geography*. London: Hutchinson.

Hägerstrand, T. (1969) *Innovation Diffusion as a Spatial Process*. Chicago: University of Chicago Press.

Haggett, P. (1967) 'Network models in geography', in R. J. Chorley and P. Haggett (eds) (1967) *Models in Geography*. London: Methuen, pp. 609–668.

Haggett, P. (1969) 'Geographical research in a computer environment', *Geographical Journal* 135: 500–509.

Haggett, P. (1972) *Geography: A Modern Synthesis*. London: Harper and Row.

Haggett, P. (1973) 'Forecasting alternative spatial, ecological and regional futures: problems and possibilities', in R. J. Chorley (ed.) *Directions in Geography*. London: Methuen, pp. 219–235.

Haggett, P. (2000) *The Geographical Structure of Epidemics*. Oxford: Clarendon.

Haggett, P. and Chorley, R. J. (1967) 'Models, paradigms and the new geography', in R. J. Chorley and P. Haggett (eds) (1967) *Models in Geography*. London: Methuen, pp. 19–41.

Haggett, P., Cliff, A. D. and Frey, A. E. (1977) *Locational Analysis in Human Geography*, 2nd edition. London: Arnold.

Hartshorne, R. (1939) *The Nature of Geography*. Washington: Association of American Geographers.

Hartshorne, R. (1959) *Perspective on the Nature of Geography*. Chicago: Rand McNally.

Harvey, D. (1969) *Explanation in Geography*. London: Arnold.

Mercer, D. C. (1984) 'Unmasking technocratic geography,' in M. Billinge, D. Gregory and R. Martin (eds) *Recollections of a Revolution: Geography as Spatial Science*. London: Macmillan, pp. 153–199.

Sack, R. D. (1974) 'The spatial separatist theme in geography', *Economic Geography* 50: 1–19.

Smallman-Raynor, M., Cliff, A. D. and Haggett, P. (1992) *International Atlas of AIDS*. Oxford: Blackwell.

Taylor, P. J. (1976) 'An interpretation of the quantification debate in British geography', *Transactions, Institute of British Geographers NS* 1: 129–142.

Thrift, N. J. (1995) 'Peter Haggett's life in geography,' in A. D. Cliff, P. R. Gould, A. G. Hoare and N. J. Thrift (eds) *Diffusing Geography: Essays for Peter Haggett*. Oxford: Blackwell, pp. 375–395.

von Thünen, J. H. (1826) *Der isolierte Staat in Beziehung auf Landwirtschaft und Nationalokonomie-I*. Hamburg.

Robin Flowerdew

24 Stuart Hall

BIOGRAPHICAL DETAILS AND THEORETICAL CONTEXT

A prolific essayist and editor, prominent member of the British New Left, guiding influence in the development of the Birmingham Centre for Contemporary Cultural Studies (CCCS) during the 1970s, prominent teacher at the Open University in the 1980s and 1990s, and globally sought-after lecturer on questions of culture and identity, Stuart Hall has shaped the multinational project of cultural studies in decisive ways, while also influencing public debate in Britain about the changing politics of race and class. In geography his influence has been felt through the adoption of CCCS interests in subcultures, hegemony and resistance by the 'new cultural geography' and in the recent disciplinary turn to questions of identity, postcoloniality and diaspora.

Hall was born in Kingston, Jamaica, in 1932 to middle-class parents. In class formation and (mixed) colour, Hall's family 'played out . . . culturally . . . the conflict between the local and the imperial in the colonized context' (Chen, 1996: 484–485). Hall says he was 'the blackest member of the family' and so 'I always had the identity in my family of being the one from outside, the one who didn't fit, and the one who was blacker than the others . . .' (Chen, 1996: 484–85). This sense of marginality has been crucial to the development of Hall's theoretical and political interventions.

Hall went to Oxford University in 1951 as a Rhodes Scholar. Despite his sense of outsider-ness in England, Hall was a chief architect of the British New Left. In response to the Soviet invasion of Hungary and the British invasion of Suez in summer, 1956, and with the support of the Socialist Society (which 'had always opposed stalinism [sic] and imperialism' (Chen, 1996: 493)), Hall and others started the *Universities and Left Review* (*ULR*) as a forum for the political and intellectual debates unleashed by the events of 1956. The 1956 invasions had other decisive impacts on Hall. They led him, counter-intuitively, to the study of Marx (Hall, 1996); and to reconsider his decision to write a PhD dissertation on Henry James. Instead, he moved to London, to supply-teach during the day and edit the *ULR* at night. Simultaneously, he began a serious study of film (Whannel and Hall, 1964). In 1960, the *ULR* merged with the *New Reasoner* to form the *New Left Review* (*NLR*). Hall served as its first editor. The *NLR*, like its parent journals, was closely affiliated with the Campaign for Nuclear Disarmament (CND). When the decline of CND as a social movement (in 1961) suggested that a new editorial direction was necessary, Hall left to take a post teaching media studies and popular culture at Chelsea College, University of London. His position at Chelsea was likely the first ever in mass-media studies in Britain (Chen, 1996: 497).

In 1964, Richard Hoggart invited Hall to the University of Birmingham to manage Hoggart's new Centre for Contemporary Cultural Studies (CCCS). When Hoggart left CCCS in 1968, Hall became the acting director. In 1972 he became permanent director. He left CCCS in 1979. The years under Hall's direction were the CCCS's most productive and most innovative. Staff and students engaged in wide-ranging theoretical exploration while also developing innovative empirical studies of the socialization of working-class youth, the importance of 'style' to subcultural formation and practices of resistance, the ideological role of the mass media, and the policing of race and unrest. Hall served a central role as organizer of many of these efforts, and as editor of the studies that resulted, but he was also highly prolific in his own right, producing some 60 academic and popular essays over the course of his career at CCCS.

The highly collaborative, and deeply political, nature of the work at CCCS led to innumerable ideological and practical battles. In particular, 'the question of feminism' (Chen, 1996; Brunsdon, 1996; Women's Study Group, 1978) led to what Hall (1996: 268) irritably describes as an 'interruption' in the work of the Centre. Women members of the Centre struggled through the 1970s to promote their issues and theoretical and political concerns (Brunsdon, 1996: 278–279). Of the feminist 'interruptions', Hall says:

> I was checkmated by feminists; I couldn't come to terms with it, in the Centre's work. It wasn't a personal thing . . . It was a structural thing. I couldn't any longer do any useful work, from that position. It was time to go.
> (Chen, 1996: 500)

Hall left CCCS in 1979 to become Professor in the Sociology Department of the Open University (OU), where he remained until his retirement in 1999. Concurrent with this move, Hall aligned himself with the magazine *Marxism Today*. In regular contributions to the magazine, Hall explored the rise of Thatcherism as a hegemonic project, seeking to learn political and strategic lessons from its success. Hall's theorization of Thatcherism and the 'New Times' it augured owed much to the study of Gramsci, but also to the development of Althusserian theories of ideology, overdetermination and the constitution of identity (Hall, 1988a, 1988b).

Much of this Althusser-inspired work is highly technical, highly abstract and difficult in language. And it is indicative of a more general move in Hall's theoretical and political work. While the OU provided new teaching opportunities, it also meant that Hall was no longer engaged in the type of collaborative work typical of CCCS. In particular, there was less opportunity for the sort of strongly empirical work that had grounded the Centre's theoretical endeavours. Not coincidentally, then, Hall's work subtly moved from focusing on questions of culture as systems of meaning and politics related to the social reproduction of social formations, to one of understanding complex cultural processes central to the construction of *identity* (Hall and du Gay, 1996). This has been particularly apparent in Hall's work on new identities understood as a *condition* of neo-liberal globalization.

SPATIAL CONTRIBUTIONS

In order to understand Hall's impact in geography, it is essential to first understand his impact in British cultural studies and politics. Hall's work at CCCS sought to construct cultural theory that moved beyond traditional literary studies, yet which did not succumb to the 'structuralist-functionalist methodology' of classical sociology (Hall, 1980a). In particular he showed how 'meaning' was constructed (through the media and otherwise) and how it shaped people's lives. In an influential essay called 'Encoding/Decoding', Hall (1980b) argued that any media 'text' circulated through several linked 'moments'. Besides the moment of production when dominant meanings were encoded in the text, there were also the moments of circulation, distribution, consumption and reproduction, when meanings were decoded and when *new* meanings were encoded. In the process of circulation and distribution, critics not only decode the intended meanings, but add layers of meaning to the text itself; likewise 'readers' (or listeners or viewers) both decode the text and transform its meaning based on their own frames of reference. Sometimes readings will be in accordance with the intended reading; other times, new meanings will be 'negotiated' between readers and texts; and still other times, resistant, counter-hegemonic readings will arise. The process of decoding and re-encoding is strongly influenced by class or other social positions, since these create both differential access to (discursive or other) resources for decoding, and differential desires and needs when reading. This implies the need for both a sociological and a geographical analysis of media production and reception in the construction and transformation of meaning. Such analyses, Hall argued, could uncover the 'maps of meaning' that constitute culture as lived experience.

'Maps of meaning' were the means through which people made sense of the world, and that these maps were themselves a site of social contestation. One of Hall's (and CCCS's) primary arguments was that the *modes* of contestation through which maps of meaning are produced and reproduced are highly varied. Resistance can appear, for example, in the development of a new subcultural *style* that takes the products and practices of the dominant culture and reshapes them into something new that gives a subcultural group an *identity*. In a landmark study (Hall and Jefferson, 1976), Hall and his collaborators at CCCS argued that subcultural style and resistance was always formed in relation to hegemony, that is, in a dialectical relation to dominant society's need to gain the consent, or coerce the acquiescence, of subordinate classes. The negotiation of these relations of domination is always contentious (even if conflict is sometimes sublimated). But perhaps most importantly, the negotiation of subcultural identity within a hegemonic order requires that subcultural

groups *'win space . . . to mark out and appropriate "territory" '* (Hall and Jefferson, 1976: 45). The negotiation of hegemony requires the production of space.

In the 1980s, Hall elaborated the theory of hegemony within the context of Thatcherism (right-wing conservative politics espoused by the British Prime Minister of the time, Margaret Thatcher). Hall argued that the 'authoritarian populism' (Hall, 1989) of Thatcherism worked in dialectical relationship to a capitalism 'increasingly characterized by diversity, differentiation and fragmentation' (Hall and Jacques, 1989: 11). 'Populism' in this context referred to a new populism of the flexible marketplace and the new opportunities it seemed to open up; 'authoritarianism' referred to the new social conservativism that Thatcher harnessed to the rise of a punitive state. 'Authoritarian populism' thus named the conjuncture of forces that shaped social and cultural life in Britain in the 1980s. It also indicated the way that Thatcher had successfully won the consent of large portions of the working class to her administration's project of political and social restructuring (Hall, 1988a).

To make this argument, Hall drew from Althusser (1969) to develop the idea of 'interpellation'. 'Interpellation' refers to a subject being 'hailed' into a particular subject position through the working of ideology. Subjects were 'interpellated' through the 'over-determined' political, social and economic restructuring that gave rise to new ideologies, new maps of meaning that structured the moments of encoding and decoding possible at any particular time and place.

Hall's decreasing interest in the actual materialist basis of social change is clear in the above. Instead, Hall became more animated by questions of subjectivity and identity than with understanding social and political transformations *per se*. In particular, he deepened his commitment to understanding the *experience* of race, ethnicity and subjectivity. Hall argued that *diaspora* was the defining process of our times and understanding our experience of it was politically vital. The questions he asked thus became ones of how to account for identity formation and why identity formation matters (Hall and du Gay, 1996). In these terms, 'diaspora' provided a window on what Hall called the 'articulation' of the subject in postmodern and postcolonial global circumstances. During the 1980s and into the 1990s, the concept of 'articulation' rather than 'interpellation' came to be Hall's predominant metaphor for understanding subjectivity:

> [A] theory of articulation is both a way of understanding how ideological elements come, under certain conditions, to cohere within a discourse, and a way of asking how they do or do not become articulated, at specific conjunctures . . . [T]he theory of articulation asks how an ideology discovers its subject rather than how a subject thinks the necessary and inevitable thoughts which belongs to it . . .
> (Hall, in Grossberg, 1996: 141–142)

The legacy of Althusser's idealist, hyper-antihumanism is still obvious, but Hall, unlike Althusser, does at least suggest some room for a 'subject' to do the 'articulating'. Indeed, Hall argued that the theory of 'articulation' allowed for an understanding how 'ideology empowers people, enabling them to make some sense . . . of their historical situation, without reducing that [sense] to their socio-economic or class location or social position' (Hall, 1996: 142).

'Making sense' was a function of cultural politics. In 'New ethnicities', Hall (1996: 442–443) argued that 'black cultural politics' (and by extension other identity-based politics) underwent a shift in the 1980s 'from a struggle over the relations of representation to a politics of representation itself'. This is important because:

> how things are represented and the 'machineries' and regimes of representation in culture do play a *constitutive*, and not merely a reflexive, after-the-event role. This gives questions of culture and the scenarios of representation – subjectivity,

identity, politics – a formative, not merely an expressive, place in the constitution of social and political life.
(Hall, 1996: 443)

This politics of representation, Hall argued, was – or had to be – inflected through a further politics of heterogeneity (Hall, 1996: 444). To make this argument, Hall returned to one of his primary tropes, arguing that black cultural politics as a politics of representation required an 'awareness of the black experience as a *diasporic* experience' and thus an understanding of 'the consequences this carries for the process of unsettling, recombination, hybridization, and "cut-and-mix" – in short, the process of cultural *diaspora-ization* (to coin an ugly term) which it implies' (Hall, 1996: 447). To put that another way, throughout Hall remained interested in 'maps of meaning', but now at a much higher level of abstraction.

Perhaps the earliest sustained engagement with Stuart Hall's (and CCCS's) work was by **Peter Jackson**, both in his work on the geographies of racism and, especially, in his book *Maps of Meaning* (1989). The title is taken, of course, from one of Hall's signature phrases and the book itself is a sustained engagement with, and geographical development of, the issues and theoretical approaches that animated CCCS. The work of CCCS was also central to such key works of the 1990s as Cresswell's (1996) *In Place/Out of Place*, which established the importance of geographical research on social and spatial transgression and resistance. But interestingly, work by CCCS and Hall barely figure at all in the 1997 volume *Geographies of Resistance* (Pile and Keith, 1997). Such a volume, however, is simply unthinkable without the decisive impact that Cultural Studies has made in geography: the study of resistance in its myriad forms (popular culture, style, out-and-out opposition, etc.) now no longer requires an elaborate justification, either through recourse to the foundational work of the Centre or through the explicit development of theories that establish 'culture' as

– cultural landscape
Can explore intangible values + dynamic tangible values

a prime ground for 'resistance'. This can be taken as some evidence of the very success of CCCS's projects: to study subcultural resistance is now just part of the regular intellectual agenda in geography.

This same can be said of studying geographies of subjectivity and identity. In *Mapping the Subject*, Pile and Thrift (1995) draw extensively on Hall's theories of diaspora as decisive in the structuring of subjectivity. For Pile and Thrift (1995: 10), quoting Hall (1995, 207), diasporic subjects 'represent new kinds of identities – new ways of "being someone" in the late modern world' that entail people learning 'to think of themselves, of their identities and their relationship to culture and to place in ... more open ways.' Understanding subjectivity in this way, Pile and Thrift cleverly argue, challenges us 'to consider the subject anew' and to make a commitment:

> To map the subject; a subject which is in some ways detachable, reversible and changeable; in other ways fixed, solid and dependable; located in, with and by power, knowledge and social relations. This map, this subject, this book are not the same: they seek new paths, new performances, new politics.
> (Pile and Thrift, 1995: 11–12)

The source for Pile and Thrift's argument is an essay that Hall (1995) wrote for an Open University textbook call *A Place in the World* (Massey and Jess, 1995). This essay is now widely cited in geography being perhaps Hall's most frequently cited work in the field. In this essay, Hall (1995: 208) brings together many of his arguments about the ways that culture is constructed and contested and the way that a focus on diaspora helps us to see how cultures must always be understood as 'complex combination[s] of *continuities and breaks, similarities and differences*: what Gilroy ... calls a conception of tradition as the "changing same" '. Hall is speaking here specifically of the black diaspora but he intends his point to be more encompassing than that. Or more accurately, it is the black diaspora (and

others like it) that is the model: *it* is a truer representation of culture and identity (and their formation) than static concepts of culture and identity. Geography is critical. Hall argues that we need to understand that it is *routes* rather than *roots* that are determinant. He writes:

> From the *diaspora* perspective, identity has many imagined 'homes' (and therefore no one single homeland); it has many different ways of 'being at home' – since it conceives of individuals as capable of drawing on different maps of meaning and locating them in different geographies at one and the same time – but it is not tied to one, particular place.
> (Hall, 1995: 207)

Doreen Massey (1994) concurs, and argues that one of the key political tasks ahead of us (as citizens of a diasporic world and as geographers) is to forge a 'global sense of place', a map of meaning that takes interconnectedness rather than separatism, routes rather than roots, as its foundation. For Hall, as for Massey, this means that we need to understand the way that 'culture' entails a politics of emplacement and displacement. Hall draws directly on Massey and other geographers to argue that culture, identity and maps of meaning must always be understood in relation to geography, even, or perhaps especially, when trying to understand the current 'revival of ethnicity' and other 'more "closed" definitions of culture, in the face of what they see as the threats to cultural identity which globalization in its late-twentieth-century forms represents' (Hall, 1995: 200–201).

KEY ADVANCES AND CONTROVERSIES

What is evident here is a confluence of concerns between the work of many geographers and Hall's arguments about culture and identity, as each opens up new avenues of inquiry for the others. This is clear in Hall's (1991b) and Massey's (1994) concurrent interest in geographies of the home. But it is also clear in the growing interest in poststructuralism and postcoloniality in geography. Hall's poststructuralist theory of articulation *requires* theories of space and spatiality – the very metaphor of 'joining up' makes that clear – even as his encounter with the metaphor of language requires attention to the different spatialities of different kinds of texts (see Barnes and Duncan, 1992). And geography's growing concern with geographies of the body owes a considerable debt to Hall's (e.g. 1991a, 1991b) theories of identity (Teather, 1999: 11). Current theories of identity and subjectivity, and of the globalization process that give these their shape, are indebted, as is Hall's, to a poststructuralism borne of Althusser's condescending antihumanism that understands the sort of subjectivity that allows us to act and to think and to be political as only 'a fiction, an inevitable failure' (Chambers, 1994: 26). But these theories are radically inadequate to the political and intellectual demands that face the discipline, and, more importantly, the geopolitical world.

As Hall would be the first to argue, the development of his own work, and hence its relationship to our various projects in geography, is overdetermined. His work and his concerns have arisen from particular 'conjunctural' moments, and have become 'articulated' in specific, but not at all necessary ways. The identities of both Hall and of the geographies influenced by him have as much to do with their routes through a set of historical and political debates and projects (over the nature of globalization, the formation of subjectivity, the nature of determination, or the rise of popular culture) as they do with their roots in specific disciplinary formations. But what remains unclear is whether these routes will take either Hall or geographers working with and through his ideas away from the Althusserian idealism and hyper-antihumanism that has come to animate Hall's own project of uncovering the determinations of identity and subjectivity, and return it, in new and more productive ways, to its roots in a much more thoroughly materialist politic of culture that Hall advocates but which his own theories of diasporic 'New Times' tend constantly to ignore (see Hall, 1996).

HALL'S MAJOR WORKS

CCCS (1982) *The Empire Strikes Back: Race and Racism in 70s Britain.* London: Hutchinson.

Hall, S. (1975) *Television as a Medium in Relation to its Culture.* Birmingham: CCCS.

Hall, S. (1988a) *The Hard Road to Renewal: Thatcherism and the Crisis of the Left.* London: Verso.

Hall, S. (1997) *Representation: Cultural Representation and Signifying Practices.* Thousand Oaks: Sage.

Hall, S., Critcher, C., Jefferson, T., Clarke, C. and Roberts, B. (1978) *Policing the Crisis: Mugging, the State ad Law and Order.* London: Macmillan.

Hall, S., Hobson, D., Lowe, A. and Willis, P. (eds) (1980) *Culture, Media, Language: Working Papers in Cultural Studies (1972–1979).* London: Hutchinson.

Hall, S. and Jacques, M. (eds) (1983) *The Politics of Thatcherism.* London: Lawrence and Wishart.

Hall, S. and Jacques, M. (eds) (1989) *New Times: The Changing Face of Politics in the 1990s.* London: Lawrence and Wishart.

Hall, S. and Jefferson, T. (eds) (1976) *Resistance Through Rituals: Youth Subcultures in Post-War Britain.* London: Hutchinson.

Secondary Sources and References

Althusser, L. (1969) *For Marx*. London: Allen Lane.

Barnes, T. and Duncan, J. (eds) (1992) *Writing Worlds: Discourse, Text, and Metaphor in the Representation of Landscapes*. London: Routledge.

Brundson, C. (1996) 'A thief in the night: Stories of feminism in the 1970s at the CCCS', in D. Morley and K-H. Chen (eds) *Stuart Hall: Critical Dialogues in Cultural Studies*. London: Routledge, pp. 276–286.

Chambers, I. (1994) *Migrancy, Culture, Identity*. London: Routledge.

Chen, K-H. (1996) 'The formation of a diasporic intellectual: An interview with Stuart Hall', in D. Morley and K-H. Chen (eds) *Stuart Hall: Critical Dialogues in Cultural Studies*. London: Routledge, pp. 484–503.

Cresswell, T. (1996) *In Place/Out of Place: Geography, Ideology, and Transgression*. Minneapolis: University of Minnesota Press.

Gilroy, P. Grossberg, L. and McRobbie, A. (eds) (2000) *Without Guarantees: In Honour of Stuart Hall*. London: Verso.

Grossberg, L. (1996) 'On postmodernism and articulation: An interview with Stuart Hall', in D. Morley and K-H. Chen (eds) *Stuart Hall: Critical Dialogues in Cultural Studies*. London: Routledge, pp. 131–150; first published in *Journal of Communication Inquiry* (1986) 10: 45–60.

Hall, S. (1980a) 'Cultural Studies and the Centre: Some problematics and problems', in S. Hall, D. Hobson, A. Lowe, and P. Willis (eds) *Culture, Media, Language: Working Papers in Cultural Studies (1972–1979)*. London: Hutchinson, pp. 15–47.

Hall, S. (1980b) 'Encoding/Decoding', in S. Hall, D. Hobson, A. Lowe, and P. Willis (eds.), *Culture, Media, Language: Working Papers in Cultural Studies (1972–1979)*. London: Hutchinson, pp. 128–138; first published as 'Encoding and decoding in the media discourse', Stencilled Paper 7 (1973). Birmingham: CCCS.

Hall, S. (1988b) 'The toad in the garden: thatcherism among the theorists', in C. Nelson and L. Grossberg (eds) *Marxism and the Interpretation of Culture*. Champaign: University of Illinois Press, pp. 35–57.

Hall, S. (1989) 'Authoritarian populism', in B. Jessop *et al.* (eds) *Thatcherism*. Cambridge: Polity Press, pp. 99–107.

Hall, S. (1991a) 'Old and new identities, old and new ethnicities', in A. King (ed.) *Culture, Globalization and the World System*. London: Macmillan, pp. 41–68.

Hall, S. (1991b) 'The local and the global: Globalization and ethnicity', in A. King (ed.) *Culture, Globalization and the World System*. London: Macmillan, pp. 19–39.

Hall, S. (1995) 'New cultures for old', in D. Massey and P. Jess (eds) *A Place in the World? Places, Cultures and Globalization*. Oxford: Oxford University Press, pp. 175–213.

Hall, S. (1996) 'New ethnicities', in D. Morley and K-H. Chen (eds) *Stuart Hall: Critical Dialogues in Cultural Studies*. London: Routledge, pp. 441–464; first published in *ICA Documents 7: Black Film, British Cinema* (1989).

Hall, S. (1996) 'Cultural Studies and its theoretical legacies', in D. Morley and K-H. Chen (eds) *Stuart Hall: Critical Dialogues in Cultural Studies*. London: Routledge, pp. 262–275; first published in L. Grossberg *et al.* (eds) (1992) *Cultural Studies*. London: Routledge, pp. 277–286.

Hall, S. and du Gay, P. (eds) (1996) *Questions of Cultural Identity*. Thousand Oaks: Sage.

Hall, S., Maharaj, S., Campbell, S., and Tawadros, G. (2001) *Modernity and Difference*. London: Institute of Visual Arts.

Jackson, P. (1989) *Maps of Meaning*. London: Unwin Hyman.

Massey, D. (1994) *Space, Place, and Gender*. Minneapolis: University of Minnesota Press.

Massey, D. and Jess, P. (eds) (1995) *A Place in the World? Places, Cultures and Globalization*. Oxford: Oxford University Press.

Morley, D. and Chen, K-H. (eds) (1996) *Stuart Hall: Critical Dialogues in Cultural Studies*. London: Routledge.

Pile, S. and Keith, M. (eds) (1997) *Geographies of Resistance*. London: Routledge.

Pile, S. and Thrift, N. (1995) 'Introduction', in S. Pile and N. Thrift (eds) *Mapping the Subject: Geographies of Cultural Transformation*. London: Routledge, pp. 1–12.

Teather, E. (ed.) (1999) *Embodied Geographies: Spaces, Bodies and Rites of Passage*. London: Routledge.

Whannel, P. and Hall, S. (eds) (1964) *Studies in the Teaching of Film Within Formal Education*. London: BFI.

Women's Study Group (1978) *Women Take Issue*. London: Hutchinson.

Don Mitchell

25 Donna Haraway

BIOGRAPHICAL DETAILS AND THEORETICAL CONTEXT

As a thinker and writer whose extensive and wide-ranging work combines theoretical innovation with political commitment, Donna Haraway has become a key figure in debates surrounding knowledge, science, technology, nature and culture throughout the social sciences (including human geography). Haraway was born in 1944 in Denver, Colorado, attending Colorado College, where her undergraduate studies combined zoology, English and philosophy. Following her graduation in 1966, she spent a year in France, where she developed an existing interest in political activism, and on her return to the US, she participated in anti-Vietnam war protests and engaged with the civil rights movement. Simultaneously, she began a PhD in Biology at Yale University. A switch in direction from experimental biology to the study of the history and philosophy of biology was pivotal in the emergence of Haraway's philosophy, allowing her to expose the social production of scientific knowledge. Contrasting conventional ideas of science as objective knowledge of an external reality with the idea that science is simultaneously material and discursive, Haraway began to develop a powerful critique of science as a masculinist endeavour, concerned with the mastery of nature.

After working at the University of Hawaii, where she completed her thesis, and at Johns Hopkins University, Haraway moved to the University of California, Santa Cruz, in 1979. Here, she joined the new Program in the History of Consciousness, teaching and writing in the areas of feminist studies and science studies, and becoming Professor of the History of Consciousness and Women's Studies. Throughout, her work has continued to transcend conventional academic disciplinary boundaries, bringing biology into dialogue with primatology, poststructural and postmodern theory, feminist and science and technology studies, cultural theory, and more, in a series of influential and innovative papers and books. The titles of two particularly influential essays (revised and reprinted in Haraway's 1991 book *Simians, Cyborgs and Women: the Reinvention of Nature*) give an indication of her interests – 'Manifesto for cyborgs: Science, technology and socialist feminism in the 1980s' (1985) and 'Situated knowledges: The science question in feminism as a site of discourse on the privilege of partial perspective' (1988). Other major works include *Primate Visions: Gender, Race and Nature in the World of Modern Science* (1989), a deconstruction of dualistic categories of, for example, human/animal and nature/culture; *Modest_Witness@Second_Millennium: FemaleMan_Meets_Oncomouse*(TM). *Feminism and Technoscience* (1997), an analysis of the hybrid and social nature of a series of cybertechnological and biotechnological entities and of the rhetoric and mythology surrounding their existence; and *How Like a Leaf* (2000), an interview in which Haraway discusses the influences on her intellectual development.

Haraway's academic career, her eclectic interests and experience of working in scientific as well as social scientific disciplinary areas, have made her an important contributor to interdisciplinary or post-disciplinary debates on space and place. Key facets of her philosophy – especially her concern with how boundaries between different categories of knowledge and object are constructed and maintained – are informing geographical debates about the production of space and spatial knowledges. Of particular importance is her questioning of the conventional distinctions drawn between those things that are considered 'natural' and those seen as 'cultural', arguing instead for recognition of the 'hybridity' of things and developing metaphorical terminologies for referring to entities that are simultaneously natural and cultural, natural and technological, human and nonhuman, material and semiotic.

SPATIAL CONTRIBUTIONS

Haraway's critique of scientific objectivity has deeply affected debate and practice within human geography. She argues that modern science attempts to perform what she calls the 'god-trick' (Haraway, 1991) of displacing the (necessarily subjective) human observer of the world in favour of supposedly objective observational technologies which claim to produce a truthful account of the world. Here, the metaphorical use of scientific 'vision' is significant (see Gregory, 1994; Barnes and Gregory, 1997), with Haraway arguing that science attempts to create a 'view from nowhere', a mythical objectivity external to social conditions. Visual technologies designed to observe the world at the largest scale (e.g. satellite-mounted earth observation systems) and smallest scale (e.g. electron microscopes) serve to illustrate attempts to 'see' everything, and to present nature as a set of observable, factual, manageable phenomena:

> Vision in this technological feast becomes unregulated gluttony; all perspective gives way to infinitely mobile vision, which no

longer seems just mythically about the god-trick of seeing everything from nowhere, but to have put the myth into ordinary practice.
(Haraway, 1991: 189)

For Haraway, however, such a god-trick perpetuates a distinctively masculinist and exploitative understanding of the world. For example, 'feminine' nature becomes something to be observed, rationally explained and mastered. Thus, the 'view from nowhere' dissimulates the way in which science is actually a cultural, power-laden process: it is not value-free:

> [A]ccording to the 'Modern Constitution' of scientific practice, a standard trick has been to generate representational systems of meaning that hide or obscure the subjectivity of the physical body. Objectivity has been socio-materially constructed by placing the situated and embodied character of all knowledge-and-practice, including the Observer, in the background.
(Gren, 2001: 212)

Two related themes in Haraway's writing emerge from this understanding of scientific enquiry: first, an emphasis on overcoming the scientific separation of the natural and the cultural (leading to a conceptualization of things as hybrid, 'material-semiotic' or 'cyborg'); and second, a concept of 'situated knowledge' which is positioned against the supposedly objective knowledge of modern science.

In relation to the first of these, Haraway has explored how science reproduces a dualism of nature and culture: science, and the objective nature it purports to study, are seen as other to a socio-cultural realm of subjectivity and meaning. Along with others working on the sociology of science, Haraway has instead described the ways in which science is a social and cultural process, so that the nature known through science is a cultural artefact, constituted through both the practices and technologies of scientific research and the language used to describe and explain scientific findings – but only in part. Opposing forms of social constructivism that suggest the world is simply a cultural or linguistic construct, Haraway has also been concerned to maintain a sense of the materiality or corporeality of nature (including humans, who are embodied subjects). Thus, things are never simply either natural or cultural. Neither are they simply either material or linguistic (see Demeritt, 1994). Entities such as genes, living organisms, disease microbes, cancer cells or electrons are ineluctably all of these things; they have a 'real' or material presence, which, significantly, produces agency (i.e. the ability to do things) within the complex and heterogeneous systems in which they exist, and are simultaneously conceptualized within linguistic or semiotic systems which constitute them as known entities within cultural systems. The term 'material-semiotic entity' is employed by Haraway to encompass this sense of simultaneity in describing such objects; material-semiotic entities are hybrids of what are conventionally considered as nature and culture.

While science is theorized as a socio-cultural process, for Haraway society and culture have, particularly in the period since World War II, become increasingly scientific and technological. In this context, she deploys a further metaphor – that of the cyborg – to overcome dualistic thinking. The idea of the cyborg, drawing upon the imaginaries of science fiction, is that of the biological body as supplemented or conjoined with technological apparatus in ways that extend its capabilities. For instance, recent biotechnology, genetic engineering, cloning and so on point to the ways that technoscience increasingly works with biological materials in ways that erode conventionally held distinctions between the human and the nonhuman, between the biological and the technical and again, between nature and culture. The figure of the cyborg is key in Haraway's writing, both as an attempt to describe the complex ways in which humans and other organisms are increasing-

ly caught up in machine-like assemblages of material, biological and communications technologies, and as a metaphor used to describe alternative forms of political and social identity that do not conform to conventional models of identification (by questioning, for example, prevailing categorizations of gender, race or sexuality).

For Haraway, there are negative and positive aspects to the emergence of cyborgs in the contemporary world. While she associates them with capitalist exploitation of the planet and oppression of some human and nonhuman beings, she also sees in them a liberatory potential because of how they undermine conventional dualisms and categories. Here, Haraway wishes to deconstruct the categories associated with race, gender, age, sexuality, etc., which become seen as 'natural' and given, and are associated with specific social roles and cultural representations. Such a deconstruction again has possibilities for liberation from assigned roles through the emergence of new, hybrid identities and understandings of embodiment and subjectivity as the dynamic, fluid products of social discourse and practice, rather than pre-given and fixed. Hence, a 'cyborg politics' which brings into question dualisms associated with gender, race and so on might be of value to those seeking to work towards the emancipation of subjugated groups, and has, for example, been adopted by geographers such as **Gillian Rose** as part of projects of writing feminist geographies concerned with how women are socially and spatially marginalized and oppressed. The figure of the cyborg is thus used by Haraway to signify a seditious political identity, an identity continually being reforged in processes of subverting totalizing and dualistic theory and practice.

The concept of 'situated knowledge' again makes use of the metaphor of vision, and works with the notion of cyborg identity to suggest an alternative to the myth of scientific objective knowledge. For Haraway, situated knowledges are those of embodied, located subjects (using Haraway's geographical terminology, they are views from somewhere, rather than views from nowhere). They are necessarily partial perspectives, are geographically and historically specific, and are in a continual process of structuring and being structured by social conditions. While situated knowledges cannot simply be added together to produce totalizing world-views, 'shared conversations' (Haraway, 1991) are possible within the complex networks of connection binding people together, producing meeting points of shared solidarity around which layers of difference nevertheless persist. Such shared conversations are political and ethical, potentially affecting our ways of being in the world through the meeting, but not the resolution, of differently situated knowledges. As Haraway writes (1991: 195), 'I am arguing for politics and epistemologies of location, positioning and situating, where partiality and not universality is the condition of being heard to make rational knowledge claims'.

In contrast to the association of the god-trick of objective scientific vision with male, white, Western domination of nature and other people, situated knowledges can be the knowledge perspectives of subjugated, marginalized groups – women, people of colour, and so on. Again, there is an association here with wider feminist theory, and also with (for example) postcolonial theory. For different people, then, understandings of nature, technology, society, etc. can vary, and this multiplicity of situated knowledges must be recognized and valued. Here, for example, Haraway has been instrumental in the emergence of ecofeminist perspectives in the sociology of science and technology and in relation to environmental issues (e.g. Instone, 1998). However, she cautions against simply romanticizing the perspectives and knowledges of 'other' groups; they incorporate other biases and power relations, and hence are never 'innocent'. In addition, they may be difficult to access:

'Subjugated' standpoints are preferred because they seem to promise more adequate, sustained, objective, transforming accounts of the world. But *how* to see from below is a problem requiring at least as much skill with bodies and language, with the mediations of vision, as the 'highest' techno-scientific visualizations. (Haraway, 1991: 191)

Yet, Haraway does not adopt an anti-science position. Instead, she argues that we must create *better* accounts of the world: since science is a social activity, alternative scientific praxis can be undertaken that does not reproduce the regimes of domination, and illusion of objectivity, of conventional science. Situated knowledges, including hypothetical situated *scientific* knowledges, differ from conventional scientific knowledge in reflexively recognizing how the conditions of their production affect their conceptualization of particular objects of study. They might thus also be described as 'objective' knowledges, dealing with the 'facts' of a situation, while recognizing that from other somewheres the facts might be rather different. For Haraway, situated knowledges are bound into the notion of taking responsibility for what we 'see', for how we have learned to see, for the associated ethical positions we take towards the world and its entities, and for our action in the world. While she ascribes agency to nonhuman entities, Haraway nevertheless emphasizes that 'it is people who are ethical, not these nonhuman entities' (Haraway, 2000: 134).

KEY ADVANCES AND CONTROVERSIES

The 'implosion' or collapse of dualistic and ontological categories in Haraway's writing has, as part of geographers' more general engagement with what **Nigel Thrift** (1996) calls 'non-representational theory', stimulated reconceptualizations of nature, culture, agency and ethical community. In her later work, Haraway (1997) goes beyond her writing on cyborgs (hybrids of human and technology) to insist on the agency of a much wider range of nonhuman actors. Here, there are similarities to the actor-network theory associated with writers like Michel Callon, **Bruno Latour** and John Law, which disrupts realist accounts of the world in favour of understandings of the complex material-semiotic relationships existing between things in heterogeneous networks or assemblages. For Haraway, nonhuman (and human) technoscientific material-semiotic entities must be understood in ways that transcend conventional categories of analysis, and that juxtapose categories, spaces and entities such that the 'cleanliness' and orderliness associated with the technical is dissolved in a proliferation of interpenetrated 'sticky threads'. As she argues:

> Any interesting being in technoscience, such as a textbook, molecule, equation, mouse, pipette, bomb, fungus, technician, agitator, or scientist, can – and often should – be teased open to show the sticky economic, technical, political, organic, historical, mythic, and textual threads that make up its tissues. 'Implosion' does not imply that technoscience is 'socially constructed', as if the 'social' were ontologically real and separate, 'implosion' *is* a claim for heterogeneous and continual construction through historically located practice, where the actors are not all human.
>
> (Haraway, 1997: 68)

Extrapolating away from the realm of technoscience, such 'teasing open' might also be applied to other complex objects and representations, such as the global financial systems that draw simultaneously on discourses of transnationalism and technoscience (see Chernaik, 1999). Such discourses, importantly, 'are not just "words", they are material-semiotic practices through which objects of attention and knowing subjects are constituted' (Haraway, 1997: 218). Similarly, in relation to environmental systems, ecofemin-

ist politics based on partial, positioned knowledge, might reformulate approaches to issues like environmental degradation, with environments imagined in terms of specific conjunctions of 'nature' and 'culture' (Instone, 1998). At smaller geographical scales, the situated agency of nonhuman actors is being drawn upon by geographers interested in rethinking the active and semiotic contributions of, for example, trees (Cloke and Jones, 2002) and nonhuman animals (Whatmore, 2002; Wolch, 1998) to networks that also include human actors.

Haraway's work has also had implications for understandings of ethics in geography. Contrasting with ethical models that ascribe rights to autonomous individuals, Haraway's dissolution of centred human subjectivity in favour of a focus on the material-semiotic connectivities between human and nonhuman actors encourages an understanding of heterogeneous ethical communities (Whatmore, 1997). Here, the relationships between entities are implicated in situated moral understandings (Lynn, 1998) and a conceptualization emerges of ethics as situated praxis, rather than as an aspatial moral framework (Whatmore, 2002). As with Haraway's notion of shared conversations, this idea of ethical community implies a continuous process of negotiating partial understanding and solidarity. It also suggests ways that ethical community might extend over space in different modes to conventional understandings that have often relied on the close proximity of autonomous ethical subjects. Instead, ethical community can be understood as connecting and enfolding different spaces within networks of ethical connection (e.g. through the material-semiotic exchange of money, goods, people and information on a global scale).

Although it is clear that geographers have worked sympathetically with Haraway's theorization in relation to a wide range of topics, criticisms have nevertheless arisen. Key among these has been a sense that, where her analytical strategy has involved the juxtaposition of different categories of thought in order to decon-struct and make problematic the status of technoscientific and other human and nonhuman entities, it is unclear exactly how such entities move or exist simulta-neously across such categories. For example, in relation to possibilities for new understandings of ethical commu-nity:

> ... although Haraway's account of hybrid-ity successfully disrupts the purification of nature and society and the relegation of 'nonhumans' to a world of objects, it is less helpful in trying to 'flesh out' the 'material' dimensions of the practices and technologies of connectivity that make the communicability of experience across dif-ference, and hence the constitution of ethical community, possible. These di-mensions require a closer scrutiny of over-lapping life-practices and corporeal processes, for example those mediated by food, energy, disease, birth and death, than Haraway has so far admitted.
> (Whatmore, 1997: 47)

As such, 'Haraway does not consider as fully as she might do the transformation of actors, from the social to the non-social, the human to the nonhuman, the micro to the macro, and back again' (Michael, 1996: 42). Similarly, Haraway's accounts of hybridity and the cyborg are criticized as apparently reliant on the combination of pre-existing entities into a new entity. Such accounts raise the prob-lem of who or what conducts this combi-nation (see Whatmore, 2002), an issue recognized by Haraway in writing of the

ambiguities inherent in, and essential to, her understanding of such entities. Cy-borgs and hybrid actors make 'thoroughly ambiguous the difference between natu-ral and artificial, mind and body, *self-developing and externally-designed ...*' (Haraway, 1991: 152; emphasis added). Even so, for Doel (1999), notions of hy-bridity and situatedness are 'unbecoming' and 'sedentary', relying on the addition of units and embedded (even if only tempor-ary) locatedness, rather than fully submit-ting to a world of ephemera, dislocation, promiscuity and becoming.

Despite such criticism, Donna Har-away's influence on current human geo-graphical theory has been significant, providing a powerful means of reconcep-tualizing human subjectivity, and human relationships with the nonhuman and technological worlds, in terms that allow challenges to conventional modes of cat-egorization and that provide often opti-mistic accounts of the emancipatory potential of cyborg, hybrid and multiple political identities and the recognition of nonhuman relational agency, while main-taining the analytical and political value of concepts of class, race, gender and so on. At the same time, her work opens up ways of approaching space that empha-size material-semiotic connectivity, and present possibilities for making sense of micro- and macro-scale entities in ways that do not simply reduce them to the usual categories of natural or cultural, human or nonhuman, subject or object.

HARAWAY'S MAJOR WORKS

Haraway, D. J. (1976) *Crystals, Fabrics and Fields: Metaphors of Organicism in 20th Century Developmental Biology.* New Haven, CT: Yale University Press.

Haraway, D. J. (1989) *Primate Visions: Gender, Race and Nature in the World of Modern Science.* London: Routledge.

Haraway, D. J. (1991) *Simians, Cyborgs and Women: The Reinvention of Nature.* London: Free Association Books.

Haraway, D. J. (1997) *Modest_Witness@Second_Millennium. FemaleMan_Meets_Oncomouse™. Feminism and Technosci-ence.* London: Routledge.

Haraway, D. J. (2000) *How Like a Leaf: An Interview with Thyrza Nichols Goodeve.* London: Routledge.

Secondary Sources and References

Barnes, T. and Gregory, D. (1997) *Reading Human Geography: The Poetics and Politics of Inquiry*. London: Arnold.

Chernaik, L. (1999) 'Transnationalism, technoscience and difference: the analysis of material-semiotic practices', in M. Crang, P. Crang and J. May (eds) *Virtual Geographies: Bodies, Space and Relations*. London: Routledge, pp. 79–91.

Cloke, P. and Jones, O. (2002) *Tree Cultures: The Place of Trees and Trees in their Place*. Oxford: Berg.

Demeritt, D. (1994) 'The nature of metaphors in cultural geography and environmental history', *Progress in Human Geography* 18 (2): 163–185.

Doel, M. (1999) *Poststructuralist Geographies: The Diabolical Art of Spatial Science*. Edinburgh: University of Edinburgh Press.

Gregory, D. (1994) *Geographical Imaginations*. Oxford: Blackwell.

Gren, M. (2001) 'Time-geography matters', in J. May and N. Thrift (eds) *Timespace: Geographies of Temporality*. London: Routledge, pp. 208–225.

Instone, L. (1998) 'The Coyote's at the door: revisioning human-environment relations in the Australian context', *Ecumene* 5 (4): 452–467.

Lynn, W. (1998) 'Contested moralities: animals and moral value in the Dear/Symanski debate', *Ethics, Place and Environment* 1: 223–242.

Michael, M. (1996) *Constructing Identities: The Social, the Nonhuman and Change*. London: Sage.

Thrift, N. (1996) *Spatial Formations*. London: Sage.

Whatmore, S. (1997) 'Dissecting the autonomous self: hybrid cartographies for a relational ethics', *Environment and Planning D: Society and Space* 15: 37–53.

Whatmore, S. (1999) 'Hybrid geographies: rethinking the "human" in human geography', in D. Massey, J. Allen and P. Sarre (eds) *Human Geography Today*. Cambridge: Polity Press, pp. 22–39.

Whatmore, S. (2002) *Hybrid Geographies: Natures Cultures Spaces*. London: Sage.

Wolch, J. (1998) 'Zoöpolis', in J. Wolch and J. Emel (eds) *Animal Geographies: Place, Politics and Identity in the Nature-Culture Borderlands*. London: Verso, pp. 119–138.

Lewis Holloway

26 J. Brian Harley

BIOGRAPHICAL DETAILS AND THEORETICAL CONTEXT

It may be that some might think it odd to see J. B. Harley appearing here as a 'key thinker' on space and place. For some Harley will be thought of mainly as an eminent and highly prolific cartographic historian with a strong empirical approach to the history of mapping and map-making. For others, however, this aspect of Harley's work will be less well known, particularly perhaps to undergraduate students of geography whose encounter with Harley may be just through his later, more theoretically informed approach to maps and mapping, his attempts to 'deconstruct the map' of the late 1980s and early 1990s (Harley, 1988a, 1988b, 1989a, 1989b). So Harley's contribution to geographical thinking and knowledge is twofold – partly empirical and partly theoretical, depending on whether Harley's academic work is being examined in its early or later phases. As an historical geographer whose academic career spanned more than 30 years (born in 1932, he died in 1991), Harley witnessed many changes in how geographers go about their work, particularly the change in approach referred to as 'the cultural turn', a move within human geography from a broadly positivist approach to a more critical philosophy and epistemology (see Philo, 2000). While some historical geographers of his age and tradition did not themselves make this 'turn', Harley not only embraced it, and the new currents in geographical thinking that were so characteristic of the late

1980s, but also played a key role in steering human geography through this 'cultural turn'. In this sense, Harley quite rightly appears in a volume of key thinkers on space and place, for he changed the way geographers and cartographers view maps and map-making.

Harley's later papers have been brought together in a posthumous edited volume with introductory pieces by Laxton and Andrews which offer very useful overviews of his life, his work and his scholarly contributions (Harley, 2001; see Daston, 2001; Edney, 2001). In his preface, Laxton remarks on the enduring impact of Harley's latter work, what he calls Harley's 'philosophy of cartographic history', a way of approaching maps and mapping that 'has been influential beyond the confines of map history' and which continues 'to excite students from a variety of disciplines', not just students of geography (Laxton, 2001: ix). Yet it is worth looking back, too, to Harley's earlier career and the very important part he played in shaping geographical discourse as an historical geographer working in a tradition that owed much to H. C. Darby's work on 'Domesday geographies'. Indeed, Harley's PhD thesis, undertaken at the University of Birmingham under Harry Thorpe's supervision, was a study of the demographic and agricultural development of medieval Warwickshire (Harley, 1960), a study that might at first appear to be a long way from the work he later did on the philosophy of cartographic history, showcasing Harley as a cartographer, mapping out from taxation accounts the spatial patterns of population and economy of medieval Warwickshire. Far from being simply a commentator on mapping, therefore, Harley had practical experience of making maps. While his interest in medieval settlement and landscape continued during the 1960s (Harley, 1961, 1964a), it was to be his developing work on early-modern and modern British maps and map-making that positioned J. B. Harley as a respected academic cartographic historian, with contributions to numerous articles and papers on cartographers and surveyors, including those of the British Ordnance Survey.

Harley's engagement with Ordnance Survey maps and mapping took up most of his academic career, from the later 1960s through into the 1980s. This is the material by which some would define the significance of Harley's contribution to geography, with his detailed historical analyses of the Ordnance Survey's surveyors and cartographers of the early nineteenth century (e.g. Harley, 1982), his numerous bibliographical notes on reprints of first edition one-inch Ordnance Survey maps (e.g. Harley, 1986), as well as his advice 'manuals' on how to use early Ordnance Survey maps in local history and topographical study (Harley, 1964b, 1972, 1975). For those concerned with a broadly empirical historical geography and cartographic history, these contributions were invaluable. It was during the 1980s, though, at the same time as Harley was co-editing the magnificent *History of Cartography* volumes for the University of Chicago Press, that he began to deal more with the meaning of maps and mapping as a language of power. This was of course a time when other historical geographers were looking more critically at other kinds of geographical knowledge, and the complicity of past geographers in, say, colonialism and imperialism, where geographical knowledges – including maps – were being used to meet political ends. As UK and US historical geographers generally began to engage more with critical theory, particularly under the influence of French poststructuralists such as **Michel Foucault** and Jacques Derrida, uncertainty rose over the nature of geography – its philosophical and epistemological assumptions – and questions began to be

raised about those aspects of geography that had, under positivism, always seemed somehow so secure, like landscapes and maps. This growing unease was to fuel human geography's 'cultural turn', as social, cultural and historical geographers (in particular) started to take more critical, theorized views of the world around them. Harley was one of them.

SPATIAL CONTRIBUTIONS

Harley was not alone in the later 1980s in beginning to look more critically at maps and map-making (Wood, 1992). But what makes Harley's 'philosophy of cartographic history' particularly important in geographical thinking is the way that he adopted the new currents of critical theory and used them to make sense of the maps he had long been studying empirically. By so doing, Harley put cartography and cartographic history firmly back on geography's curriculum. Through his critical engagement with the 'science' of cartography, and his querying of the map's authority, Harley made maps and mapping a fashionable subject again in academic geography. His approach was to draw upon those same theoretical arguments that were being used by human geographers during the 1980s in order to 'argue that we need to regard . . . maps as text rather than as "objective" technical representations which can be considered apart from the social implications of our discourse' (Harley, 1989a: 80). For Harley (1989b), this meant 'deconstructing the map' – the title of one of his most cited papers – dealing not only with the ways in which maps are constructed but also how maps and mapping have agency in the world – the purposes they serve. The phrase 'politics in maps, maps in politics', coined by **Peter Taylor** (1992) in his posthumous tribute to Harley, sums up nicely the approach that Harley was developing in his 'philosophy of cartographic history'.

It was really only in four key papers that Harley set out his new and critical philosophy of cartographic history, though just before his death he had drafted plans for 'a book of his own published or forthcoming essays to be called *The New Nature of Maps*' (Harley, 1988a, 1988b, 1989a, 1989b; Laxton, 2001: ix). These papers more or less dealt with the same issue, that 'cartography is seldom what cartographers say it is' (Harley, 1989b: 1). To demonstrate this Harley (1992: 232) sought 'to show how cartography also belongs to the terrain of the social world in which it is produced', and in this sense the arguments he was developing about cartography were cognate to those that Cosgrove and Daniels (1988) were developing at around the same time about landscape (see **Denis Cosgrove**). Indeed, one of Harley's key papers was a contribution to their *Iconography of Landscape* (Harley, 1988a). In these papers Harley tackled cartography on a number of different fronts. These are perhaps most clearly elucidated in 'Deconstructing the map', a paper that has been reprinted three times since it was first published, one of which Harley himself modified slightly (Harley, 1992; see also Harley, 2001: 149–68).

In trying to 'deconstruct' the map, one tactic Harley used was to expose the 'rules of cartography' (Harley, 1992: 233). Here he was primarily drawing upon Foucault's notion of 'discourse', as other historical geographers were likewise doing at the time (Driver, 1993; Philo, 1992). One of the 'rules of cartography' that Harley drew geographers' attention to was the 'positivistic epistemology' of map-making – a belief upheld by cartographers that their approach was 'scientific' and 'untainted by social factors' (Harley, 1992: 234). This Harley argued against, saying 'we have to read between the lines of technical procedures or of the map's topographical content' and that when we do, what becomes clear is that the 'rules of cartography' are seen to be influenced by the rules 'governing the cultural production of the map'; that is, they are 'related to values, such as those of ethnicity, politics, religion or social class, and they are also embedded in the map-producing society at large' (Harley, 1992: 236). The rules of cartography and 'the rules of society' are thus 'mutually reinforcing the same image' of the world (Harley, 1992: 237). The 'ideology' of the map was not just in its image, therefore, but in the very modes of its production, in its 'mask of a seemingly neutral science' (Harley, 1992: 238). Having established this, another of Harley's tactics was to then bring in the arguments of Derrida to deconstruct the map, to look for the slippages and contradictions within maps and map-making that actually undermine their authority and reason, and expose their ideologies further. To this end, Harley (1992: 238) suggests that 'maps are a cultural text: not one code but a collection of codes'; a text that is rhetoric, that silences as well as informs (see also Harley, 1988b). As he put it, 'all maps state an argument about the world, and they are propositional in nature', and that includes even those topographical maps – such as the British Ordnance Survey – which on first sight might not appear to be 'ideological', and not just the maps traditionally thought of as 'distorted' or 'propaganda'. As Harley states, it is 'all maps', not just some, that 'are rhetorical' (Harley, 1992: 242). Harley thus saw maps as documents that deceive and dupe, and that cartographers in their work tried to cover their tracks by projecting their maps as ever more scientific and true. He also tried other tactics to expose maps and map-making.

Following Foucault, Harley dealt with the 'agency' of maps, how 'the map works in society as a form of power-knowledge' (Harley, 1992: 243). With this tactic, Harley was making the point that maps convey particular geographical knowledges to meet the needs and requirements of those who are doing the mapping or having the maps made. Maps are a key aspect of geography's history as a discipline, like fieldwork, and during the late 1980s and early 1990s, historical and cultural geographers increasingly drew attention to the relationship between geography and capitalism, and geography and imperialism (Livingstone, 1992; Gregory, 1994; Driver, 2001). Maps were a part of this complicity of geography, for in their explorations of the eighteenth- and nineteenth- and twentieth-century world(s) geographers drawing maps were not only charting territories but, by mapping them, were bringing distant lands under the control, scrutiny and 'gaze' of colonial powers such as Britain and France, aiding their governments' colonization of the world and its peoples. This 'agency' of the map is what Harley saw as an 'exercise of power', where 'cartographers manufacture power: they create a spatial panopticon' through which the land and its people can be watched over, controlled, subjugated (Harley, 1992: 244). Not only that, these maps circulated around the globe, and in so doing conveyed geographical knowledge that was itself crucial in projecting a particular view of the world. Thus, as Harley (1992: 245) put it, 'the power of the map maker was generally exercised not only over individuals but over the knowledge of the world made available to people in general'. Hence late Victorian maps of the British Empire not only showed where territorial dominions lay, but also, in schools and government departments, these maps communicated 'to people in general' Britain's imperial might, a knowledge of British imperialism. Harley dealt with this topic of 'maps, knowledge and power' in his paper in Cosgrove and Daniel's (1988) *Iconography of Landscape*, where he considers some examples of 'the specific functions of maps in the exercise of power', maps which fused 'polity and territory', at different scales, 'ranging from global empire building, to the preservation of the nation state, to the local assertion of individual property rights' – maps bequested power to those who had them (Harley, 1988a: 281–282).

In all then, Harley showed how the map and mapping was open to suspicion, that neither geographers or anyone else could, or should, take them at face value. Instead, he argued that maps are ideological, and always have been, that they are embedded with cultural and social values and beliefs that say as much about the mapper as the mapped. This was the work that Harley was undertaking in the latter years of his academic life, opening up new ways of looking at maps and map-making. He saw no difference between, for example, the overtly political 'propaganda' maps of, say, Nazi Europe and the apparently 'neutral' topographical maps of the US Geological Survey, or between maps of the British Empire and maps produced by using modern technologies of Geographic Information Systems – to Harley all were open to the same critical interpretation, to which he sought the advice of particularly those cultural theorists that in the 1980s were having such a hold on geographical enquiry, such as Foucault and Derrida.

KEY ADVANCES AND CONTROVERSIES

Harley's work on maps and mapping has long influenced geographical thought, from his earlier historical studies of cartographers and early maps, to the more recent work he did from a more theoretically informed position on the ideology of maps and map-making. So it is not unusual to see geographers and others still making reference to Harley's empirical studies of the Ordnance Survey of Great Britain, as well as to his later probing into the 'science' of cartography. In terms of his contribution as a key thinker on space and place, however, it is his more theorized work that singles Harley out as a 'thinker', for he engaged with the conceptual changes that were characterizing geographical thinking more broadly in the 1980s, developing positions from critical theory and drawing upon postmodern and postcolonial literatures to raise geographers' and cartographers' awareness of maps and mapping in the themes that were becoming more significant, such as geography's history, imperialism, modernity, nationalism, identity, landscape, power and so forth (see Cosgrove and Daniels, 1988; Daniels, 1993; Graham and Nash, 2000). So when it comes to discussing these issues, it becomes necessary to study maps too. Maps are implicated in the making of the geographical world. To ignore them was perilous at best and naïve at worst. In this sense, in the 1980s and 1990s Harley helped to put maps back on the 'map' of geographical thinking, and in so doing he helped 'to make cartography exciting again' (Harley, 1989a: 88).

As a key thinker, then, Harley made a significant contribution to how geographers understand the world. But there are those who have suggested his arguments were overstated and that his theorizing hindered more than it helped in supporting his views (Belyea, 1992; Andrews, 2001). Indeed, Harley did not delve too deeply into the theoretical insights of Foucault and Derrida. Instead he called upon them to help him open up arguments and see things differently. In this context, it may said that Harley did not himself make much of a conceptual contribution to geography through his work – he did not seek to advance the theoretical arguments in themselves as part of his work. What he was doing was engaging with theoretical debates, and in so doing looking differently at his subject of cartography where others had not, while at the same time talking about maps and map-making in a language that was concurrently being used by other geographers to discuss the world around them. This was Harley's contribution, then. He was able to challenge orthodoxy in cartographic history, and able to show geographers that they could not afford to leave maps and map-making behind in their theorizing and writing.

Harley himself recognized the difficulties – and perhaps risks – of his use of theoretical work. With Foucault and Derrida in particular, he acknowledged that their 'theoretical positions . . . are sometimes incompatible', and yet he sees these philosophies as ways in which he could begin to 'search for the social forces that have structured cartography and to locate the presence of power – and its effects – in all map knowledge' (Harley, 1992: 232). So he proceeds by calling upon theoretical arguments even though he rarely cited their authors first-hand, and even though he did not see his work as theoretical in and of itself. Harley's contributions, though theorized, therefore remained largely empirical. Indeed, it has been suggested that in his last works, those written shortly before his death, Harley was deliberately downplaying the theoretical arguments and literatures he had been citing, and reflecting more fully on what he meant by an 'ideology' of maps and mapping (see Andrews, 2001). Nevertheless, Harley's more theoretically informed work certainly changed the way geographers think about maps and map-making, and also helped to steer human geographers through the 'cultural turn' of the 1980s and 1990s.

As Laxton (2001: x) has said, there is a risk that Harley's work – particularly his later 'philosophy of maps' – might simply be reduced to 'an unquestioned orthodoxy or, worse, a catechism'. This risk is perhaps heightened because of Harley's immediate writing style, 'fluent, zestful, and compulsively readable', and his ability to capture in just a few words, in a memorable passage, something inherently complex and uncertain, as he did with 'deconstructing the map' (Andrews, 2001: 2; Harley, 1992). Whatever the criticisms levelled at him and his approach since his death, there is no doubt that 'Harley's work continues to excite great interest among geographers, historians, and the substantial number of cultural and literary historians currently exploring the relationship between maps, literature, and the visuals arts' (Laxton, 2001: xiv). That this influence is still palpable ten years after Harley's death is testament to his contribution as a key thinker on space.

HARLEY'S MAJOR WORKS

Harley, J. B. (1988a) 'Maps, knowledge, and power', in D. E. Cosgrove and S. Daniels (eds) *Iconography of Landscape: Essays on the Symbolic Representation, Design and Use of Past Environments*. Cambridge: Cambridge University Press, pp. 277–312.

Harley, J. B. (1988b) 'Silences and secrecy: the hidden agenda of cartography in early modern Europe', *Imago Mundi* 40: 57–76.

Harley, J. B. (1989a) 'Historical geography and the cartographic illusion', *Journal of Historical Geography* 15: 80–91.

Harley, J. B. (1989b) 'Deconstructing the map', *Cartographica* 26: 1–20.

Harley, J. B. (2001) *The New Nature of Maps: Essays in the History of Cartography*. Baltimore: Johns Hopkins University Press.

Secondary Sources and References

Andrews, J. H. (2001) 'Meaning, knowledge, and power in the map philosophy of J. B. Harley', in J. B. Harley (ed.) *The New Nature of Maps: Essays in the History of Cartography*. Baltimore: Johns Hopkins University Press, pp. 1–32.

Belyea, B. (1992) 'Images of power: Derrida/Foucault/Harley', *Cartographica* 29: 1–9.

Cosgrove, D. E. and Daniels, S. (eds) (1988) *Iconography of Landscape: Essays on the Symbolic Representation, Design and Use of Past Environments*. Cambridge: Cambridge University Press.

Daniels, S. (1993) *Fields of Vision: Landscape Imagery and National Identity in England and the United States*. Cambridge: Polity.

Daston, S. (2001) 'Language of power', *London Review of Books* 1 November: 3–6.

Driver, F. (1993) *Power and Pauperism: The Workhouse System, 1834–1884*. Cambridge: Cambridge University Press.

Driver, F. (2001) *Geography Militant: Cultures of Exploration and Empire*. Oxford: Blackwell.

Edney, M. E. (2001) 'Works by J. B. Harley', in J. B. Harley (ed.) *The New Nature of Maps: Essays in the History of Cartography*. Baltimore: Johns Hopkins University Press, pp. 281–96.

Graham, B. and Nash, C. (eds) (2000) *Modern Historical Geographies*. Harlow: Pearson.

Gregory, D. (1994) *Geographical Imaginations*. Oxford: Blackwell.

Harley, J. B. (1960) 'Population and land-utilisation in the Warwickshire Hundreds of Stoneleigh and Kineton, 1086–1300', unpublished PhD thesis, University of Birmingham.

Harley, J. B. (1961) 'The Hundred Rolls of 1279', *Amateur Historian* 5 (Autumn): 9–16.

Harley, J. B. (1964a) 'The settlement geography of early medieval Warwickshire', *Transactions of Institute of British Geographers* 34: 115–130.

Harley, J. B. (1964b) *The Historian's Guide to Ordnance Survey Maps*. London: National Council of Social Service for the Standing Conference for Local History.

Harley, J. B. (1972) *Maps for the Local Historian: A Guide to the British Sources*. London: National Council of Social Service for the Standing Conference for Local History.

Harley, J. B. (1975) *Ordnance Survey Maps: A Descriptive Manual*. Southampton: Ordnance Survey.

Harley, J. B. (1982) 'The Ordnance Survey 1:528 Board of Health town plans in Warwickshire' in T. R. Slater and P. J. Jarvis (eds) *Field and Forest: An Historical Geography of Warwickshire and Worcestershire*. Norwich: Geobooks, pp. 347–384.

Harley, J. B. (1986) 'Introductory essay', in *Central England, volume four – The Old Series Ordnance Survey Maps of England and Wales*. Lympne Castle (Kent): Harry Margary, pp. vii–xxxv.

Harley, J. B. (1992) 'Deconstructing the map', in T. J. Barnes and J. S. Duncan (eds) *Writing Worlds: Discourse, Text and Metaphor*. London: Routledge, pp. 231–247.

Laxton, P. (2001) 'Preface', in J. B. Harley *The New Nature of Maps: Essays in the History of Cartography*. Baltimore: Johns Hopkins University Press, pp. ix–xv.

Livingstone, D. (1992) *The Geographical Tradition*. Oxford: Blackwell.

Philo, C. (1992) 'Foucault's geography', *Environment and Planning D: Society and Space* 10: 137–161.

Philo, C. (2000) 'More words, more worlds', in I. Cook, D. Crouch, S. Naylor and J. Ryan (eds) *Cultural Turns/Geographical Turns*. Harlow: Pearson, pp. 26–53.
Taylor, P. (1992) 'Politics in maps, maps in politics: a tribute to Brian Harley', *Political Geography* 11: 127–129.
Wood, D. (1992) *The Power of Maps*. London: Routledge.

Keith Lilley

27 David Harvey

BIOGRAPHICAL DETAILS AND THEORETICAL CONTEXT

David Harvey was born in England in 1935, and raised in Gillingham, Kent. He completed his BA and PhD at Cambridge University in the Department of Geography, leaving in 1960. During the 1950s, as was common in British geography departments, Cambridge stressed the importance of regional study – that is, the analysis of geographical particularity and difference – in its degree programmes. The Cambridge Geography Department was also known for its research into historical geography. Accordingly, Harvey's doctoral thesis focused on the changing location of hop production in nineteenth-century Kent. However, no sooner had the results of this research been published – in the flagship journal of British geography, the *Transactions of the Institute of British Geographers* (Harvey, 1963) – than Harvey started to move in an altogether different intellectual direction. Following a post-doctoral year at the University of Uppsala (where he encountered Gunnar Olsson), Harvey spent the 1960s working as Lecturer in Geography at Bristol University.

It was during his time at Bristol – which was punctuated by study leave at Pennsylvania State University, a major centre for scientific and quantitative geography – that Harvey first made his name as a philosopher and theorist of positivism. His landmark book *Explanation in Geography* (1969) gave the profusion of statistical and numerical methods and techniques in 1960s' human geography a robust and consistent intellectual foundation. In the book Harvey explored systematically the nature of scientific explanation in general; the role of theories, hypotheses, laws and models; the use of mathematics, geometry and probability as languages of explanation; the nature of observation, data classification and data representation; and different forms of scientific explanation. *Explanation in Geography* served two purposes. First, it gave Harvey's generation of geographers a heavyweight justification and manifesto for their project of identifying spatial patterns. Unlike an earlier generation concerned with what Fredric Schaefer (1953) had called 'exceptionalism' – the generation that had educated Harvey and his cohort at Cambridge and elsewhere – the so-called 'spatial scientists' who rose to prominence during the 1960s were concerned to identify geographical regularities within and between places. Second, *Explanation in Geography* served to align the discipline with the so-called 'real' sciences like physics and, for some geographers, boosted the discipline's self-image. Harvey achieved this by drawing upon the ideas of philosophers of science like Karl Popper.

Explanation in Geography, ironically, marked both the high-point and the end-point of Harvey's desire to fashion the foundations of a properly scientific geography. Almost immediately after the book's publication, Harvey left England for the USA, where he took up an Associate Professorship in the Department of Geography and Environmental Engineering at the Johns Hopkins University, Baltimore. This was triply significant. First, though a port city like Bristol, Baltimore was far poorer and had suffered major economic decline throughout the post-war years. Second, Harvey's new department was less narrowly disciplinary than Bristol had been and this, arguably, gave him licence to roam intellectually. Third, at Johns Hopkins, Harvey encountered a group of graduate students and academics who wanted to understand the radical ideas of the nineteenth-century political economist Karl Marx. It was in this context that Harvey wrote *Social Justice and the City* (1973), a book of equal – if very different – significance to its predecessor. As its title suggests, Harvey's second book focused on how to alleviate urban ills like poverty. Where *Explanation in Geography* was strangely detached from its wider socio-political context – namely, the attempts of Prime Minister Wilson's Labour government to make Britain a fairer society, and the more distant civil rights and anti-Vietnam protests in the USA – *Social Justice* was thoroughly immersed in it. The book charts Harvey's rapidly changing intellectual and political trajectory. Part One, titled 'Liberal formulations', contains essays that focus on how existing mechanisms of urban planning could bring about a more just urban society in Western countries. But Part Two, called 'Socialist formulations', moves beyond the reformism of Part One and proposes a revolution in urban affairs predicated on the ideas of political economists like Marx. One of the book's key essays, 'Revolutionary and counter-revolutionary theory in geography and the problem of ghetto formation', proposed nothing less than a wholesale rejection of the way human geographers both studied cities and thought about solving urban social problems at that time.

Social Justice established Harvey as a major dissenting voice in urban studies and caused quite a stir upon its publication. It did so for three reasons. First, because it chronicled Harvey's reinvention as a 'radical geographer', it took many spatial scientists by surprise. Like Gunnar Olsson's (1975) *Eggs in Bird/Birds in Egg*, *Social Justice* was deliberately schizophrenic. Second, the book challenged the prevailing view in scientific geography that 'facts' and 'values' can be kept separate. Harvey called for a geography that could study the world in order to *change it* along more socially just lines. Finally, *Social Justice* was the first major example of a left-wing style of geography. Specifically, it opened the door for the subsequent development of Marxist geography, of which Harvey was to be a pioneer.

Harvey's turn to Marxism meant that he became preoccupied with *capitalism* and its workings. Yet he did not entirely abandon earlier themes. In *Explanation in Geography*, Harvey was interested in how social processes produce spatial forms, and he ended the book with the declaration that 'Without theory we cannot hope for . . . consistent, and rational, explanation of events' (Harvey, 1969: 486). These two themes animate his essay publications during the 1970s, which are an attempt to fashion a Marxian theory of how geography is both produced by and alters the workings of the capitalist economy. By 'geography' Harvey did not mean the discipline of that name, but the material landscape of towns, cities and transport networks that act as the 'arteries' of capitalism. To fashion this 'historical-geographical materialism' Harvey spent the mid-to-late 1970s reading Marx's later works, like *Capital* and *Grundrisse*. This effort culminated in Harvey's most sophisticated book *The Limits to Capital* (1999; first published in 1982). This book is nothing less than a reconstruction and extension of Marx's theory of capitalism; the extensions are primarily geographical because Marx paid little attention to how geography influences capitalism's operation.

In the mid-1980s, with his reputation as the leading Marxist geographer (indeed, human geographer) established, Harvey returned to the specifically urban issues that were the focus of *Social Justice*. Inquiring into the specific nature and problems of capitalist cities, his two 'Studies in the history and consciousness of urbanization' take the perspectives of 'structure' and 'agency' respectively (to simplify rather). *The Urbanization of Capital* (1985a) unfolds a Marxian theory of how capitalism produces cities with characteristic material properties and contradictions; *Consciousness and the Urban Experience* (1985b) theorizes how real people respond to having to live in capitalist cities, especially those who lack economic and political power (see *The Urban Experience*, 1989a, for an accessible collection of Harvey's writings on cities).

Shortly after publishing these two books, Harvey returned to England as Halford Mackinder Professor of Geography at Oxford University. Leaving Reagan's America for Thatcher's Britain, Harvey felt himself confronted with an increasingly conservative, anti-Marxist British Left and said so in an outspoken critique of UK urban studies (Harvey, 1987a). In retrospect, the late 1980s represented the start of a sharp decline of interest among Anglophone academic leftists in Marxism. While at Oxford, Harvey also wrote his best-selling book *The Condition of Postmodernity* (1989b), a critical analysis of the rise of postmodernism in several walks of contemporary cultural life (including urban consumption, architecture and even academia). Returning to Johns Hopkins in 1993 for academic and personal reasons, he then revisited the theme of social justice and linked it to environmental issues in his fifth single-authored book as a Marxist, *Justice, Nature and the Geography of Difference* (1996). More recently, Harvey moved to City University New York to become a Distinguished Professor, publishing two collections of essays: one a 'greatest hits' volume (*Spaces of Capital*, 2001); the other about the continued relevance of Marxism to the wider project of creating a better socio-environmental future (*Spaces of Hope*, 2000). As the latter book recounts, Harvey's designation as a 'Marxist' represents something very different today in the Anglophone world compared with the early 1970s. Today Marxism is seen as rather 'old-hat' among radicals, which makes Harvey's continued commitment to it all the more

remarkable given the number of post- and ex-Marxists that litter the intellectual landscape (such as **Manuel Castells** and **Doreen Massey**).

SPATIAL CONTRIBUTIONS

Harvey's contribution to contemporary human geography is inestimable, while his broader influence on Marxists and critical thinkers more widely has grown steadily since the 1980s. His work has been as ambitious in scope as it is fundamental in its thematic concerns. It is not possible to offer a comprehensive review of Harvey's contributions to the (re)theorization of space and place (never mind his contributions to other topic areas). The four focused on here should be seen in the context of Harvey's membership of the community of academic geographers, the interdisciplinary community of Marxists, and the community of critical theorists of space and place more generally. It should also be noted that while Harvey's work over the last 30 years is very consistent – intellectually and politically – it is wrong to assume that there have been no cognitive or normative shifts in his thinking. This is not, however, the place to assess the balance of continuity and change in Harvey's *oeuvre*.

Harvey was among the first geographers to insist that space is not given and absolute or a 'container' into which intrinsically 'non-spatial' things are stuffed. As he put it in *Social Justice* (1973: 14), 'The question "What is space?" [must] . . . be replaced by the question "how is it that distinctive human practices create and make use of distinctive . . . space[s]?"' For Harvey, social practices and process create spaces and these spaces, in turn, constrain, enable and alter those practices and process – what **Ed Soja** (1989: 78) called a 'socio-spatial dialectic'. This means that Harvey has long rejected the polar beliefs that space has no social effects (a belief common among twentieth-century critical theorists) or that it has effects 'in itself' – what Sack (1974) called 'spatial separatism'. This latter belief sometimes crept into the work of Anglophone geography's spatial scientists, who now and then fixated on spatial pattern over social process. Harvey's mid-way position entails arguing that space – the material form that processes assume 'on the ground' as buildings, infrastructure, consumption sites and so on – is *both* cause and effect in/of social life – what he once called an 'active moment' (Harvey, 1985a: 3) in human affairs (see also Harvey, 1997).

Harvey's general belief that space is relative and constructed, leads us to his reading of capitalist space. For Marx, capitalism is a contradictory economic system based on three 'logics': accumulation for accumulation's sake (i.e. the quest for profit as an end in itself); competition between rival producers fighting for market share; and technological innovation in production processes and products. This trinity leads to internal contradictions in the capitalist system, as Marx's *Capital* explains. Periodically, these erupt in major 'crises of overaccumulation' in which pools of unutilized workers and commodities can find no productive use. One reason for these crises is that it is 'rational' in a capitalist system for businesses to replace workers with machinery in order to make production cheaper or more efficient. The paradox is that if many or most businesses do this then a collective problem for all businesses arises: since workers are also consumers, then laying off workers removes a major market for commodities, thus precipitating an economic crisis.

What has any of this to do with space? In *The Limits to Capital* Harvey shows how splurges of investment in buildings, roads and other infrastructure can provide capitalism with a mechanism for crisis displacement. Two of the unique features of built environments is that they are exceedingly expensive to build and the return on investment from this expenditure is strung out over a number of years. Thus a new convention centre might cost $80 million, the returns on which will trickle back to investors in the form of repeat conference fees over many years. What this means, according to Harvey, is that at times of incipient economic crisis, the capitalist system tends to 'switch' investment from current production into long-term fixed capital investments in the geographical landscape. These investments stave off economic crisis by tying up excess capital, and amount to what Harvey has famously called a 'spatial fix' for capitalism's contradictions. However, at a later date, these physical investments can become barriers to further accumulation because they are so hard to replace once made. Thus, when cars challenged railways in the early twentieth century as a principal way of moving people and commodities, the enormous investments in rail networks acted as temporary hindrance to new rounds of fixed capital investment. In sum, for Harvey space is *produced* within capitalism and expresses the system's inner contradictions. Harvey is among the few Marxists to insert space into Marx's theory of capitalism by way of these and other ideas. Others include **Neil Smith**.

Another important theme in Harvey's writings relates to the particular–general or space–place relationship. This theme has an objective and a subjective side, especially relevant to academic geographers and critical theorists respectively. By 'place' Harvey means the unique conjunction of built environments, cultures, peoples, etc., that distinguish one locality from another. However, unlike the regional geographers of the pre-1960s era, Harvey does not think that places are *singular* – that is, absolutely different – because in a capitalist world economy *different* places are linked within a *common* economic framework. For geographers, one of the major intellectual challenges is thus how to understand the persistence of place distinctiveness amidst heightened place-interdependence (that is, heightened ties across a wider space). Harvey's answer to this challenge is disarmingly simple. In *The Urbanization of Capital* he argues that we can view capitalism as a circulation process where money is advanced by businesses to purchase labour, inputs and means of production to produce commodities that are sold for the amount of their production cost plus an increment (profit). This process involves commodity production in different places becoming subject to a logic that is indifferent to the qualitative specifities of these places. Thus a rug made in Calcutta becomes a means of making money just as much as a microchip made in Silicon Valley does and, indeed, the two different places are united when a chip designer in the latter place buys a rug from the former place. Hence, because capitalism practically brings different places within the same economic universe it becomes possible 'to make universal generalizations about the evident unique particularities of [pl]ace' (1985a: 45) – see Merrifield (1993).

So much for the 'objective' or material side of the particular–general/place–space relationship. What about the 'subjective' side; that is, the way it is perceived and acted on by people living in real places? To be aware of how lives in one place are affected by the unseen actions of distant strangers elsewhere is to possess what Harvey (1990) has called a 'geographical imagination'. To lack such an awareness is to fall into the trap of inter-place competition and rivalry that is part of the geographical logic of capitalism (Harvey, 1985c, 1988). In *Justice, Nature and the Geography of Difference*, Harvey highlights the geographical dilemmas that beset all attempts to make progressive anti-capitalist links among oppressed peoples in different places. One of these relates to 'militant particularism', which is the elemental fact that any wider movement against global capitalism will necessarily be composed of lots of local actions by locally based people. The problem, Harvey argues, is that what makes sense at the local scale may not make sense for

people at a broader geographical scale. Harvey has illustrated this key problem in a book co-edited during his time at Oxford, called *The Factory and the City* (Harvey and Hayter, 1993). This book is about the fight by Rover car workers near Oxford to keep their jobs in the late 1980s. In the conclusion Harvey argues that while keeping car jobs creates muchneeded local employment, at a broader scale it is hardly a worthy socialist objective to make polluting cars for middleclass consumers. The question then becomes at which scale – the local or the global – should political objectives be set?

Ultimately the key contribution that Harvey has made to contemporary conceptions of space and place is to insist that the two can be theorized. This is not as trite a statement as it at first sight seems. **Andrew Sayer** (1985) has famously argued that space and place cannot be meaningfully theorized because the difference they make to social practices and processes is contingent. In other words, for Sayer the material importance of place and space cannot be specified at the level of theory but only at the level of empirical research. This risks relegating geography as a discipline (and geographical matters more generally) to largely factual concerns. Harvey disagrees with this view of space, theory and the empirical. Since at least *Social Justice* he has shown that one can make substantive theoretical statements about the role of space in modern life (see Castree, 2001). More particularly, he has been an exponent of what is called 'grand' or 'meta-theory'. This is a form of theory that presumes there to be broad logics at work in the world and it is the task of the theoretician to identify conceptually and produce a mental map of them. For Harvey, capitalism possesses fundamental logics of this sort, even when it appears otherwise, and space is *internal* to these logics. In *The Condition of Postmodernity* Harvey argues that the idea that we are now in a 'postmodern era' where difference, complexity and plurality have replaced similarity is a fiction. Underlying various forms of postmodern expression –

in architecture and consumption, for example – Harvey sees the same old logics of capitalism using the production of difference (real and symbolic) in the form of commodities as a means of making money.

KEY ADVANCES AND CONTROVERSIES

Harvey's writings as a Marxist, for which he is justly celebrated, have arguably achieved three things. First, they gave Marxist geography the intellectual foundation it needed through the 1970s and 1980s, along with the work of other emerging Marxists like **Neil Smith,** Erik Swyngedouw and Richard Walker (three of Harvey's graduate students at Johns Hopkins). Second, they set a precedent for radical geography more generally, emboldening younger human geographers to take up the cause of women, gays and lesbians, people of colour and so on. Finally, Harvey's writings have persuaded many Marxists at large that space and place matter to capitalism and are thus integral to any project to overthrow it.

Examples of how Harvey's ideas have been developed abound. Two key trajectories in contemporary human geography serve to illustrate of this. First, in economic geography, several writers have applied ideas of the spatial (and temporal) fix to specific research issues – like the way farmers use credit to finance long-term investments in capitalist agriculture (Henderson, 1999). Second, in urban geography, Harvey's idea of how inter-place competition is a necessary feature of the capitalist mode of production has inspired detailed studies of so-called 'growth coalitions' that vie for jobs and inward investment (e.g. Cox and Mair, 1988). Indeed, the literature on urban entrepreneurialism and place promotion draws substantially on Harvey's theoretical imagination

in its attempt to make sense of the lengths that cities will go to entice capital to 'come to town'.

However, Harvey's work has been subject to increasing criticism since *The Condition of Postmodernity* was published, especially from human geographers. In part this is because the intellectual *zeitgeist* has changed, with Marxism now being seen by a younger generation of radical academics as a rather outdated critical theory. Three criticisms of Harvey's work loom large. Firstly, he has been upbraided for the 'muscular' nature of his theoretical ideas. Feminists such as Deutsche (1991) and Morris (1992) have criticized Harvey's apparent belief that his Marxism offers a privileged perspective on the 'truth' of the world. They maintain that his confidence in his theoretical claims bespeaks a distinctively masculine desire for cognitive mastery and control, one that thereby excludes other ways of knowing the world. The result, they maintain, is that Harvey is blind to all manner of important real world processes and events that do not fit with his Marxist world-view. This links, secondly, with the criticism that Harvey's Marxism cannot accommodate 'difference'. Over the last decade those on the academic left have accentuated social 'difference' as a way of escaping the Marxist obsession with class as the primary axis of oppression and resistance in modern societies. The differences celebrated included those within and between women, gays and lesbians and people of colour. Harvey (1996) has responded by clarifying his position, arguing that non-class differences between people must be linked to, but are not reducible to, any project to overthrow capitalism. Thirdly, some (e.g. Dennis, 1987) have worried that Harvey's grand theoretical abstractions are insufficiently grounded in the messy complexities of the real world. Harvey (1987b, 1989a) has responded with the argument that theory is not intended to explain exhaustively any given situation but rather to open a window onto the key processes at work. Because of these (and some other) criticism, many geographers have developed new arguments in direct reaction or opposition to Harvey's – effectively using them as a 'negative launchpad'. Examples include Herod's (2001) work on how wage labourers fight back against business in place and across space, whereas Harvey tends to be preoccupied with the power of capital (*Spaces of Hope* being an exception); and **Peter Jackson**'s (1999) work on the consumption of commodities, which gives consumers a more active role than Harvey is seemingly prepared to.

HARVEY'S MAJOR WORKS

Harvey, D. (1969) *Explanation in Geography*. London: Edward Arnold.

Harvey, D. (1973) *Social Justice and the City*. London: Arnold.

Harvey, D. (1999) *The Limits to Capital*. Oxford: Blackwell, London: Verso (first edition 1982).

Harvey, D. (1985a) *The Urbanization of Capital*. Oxford: Blackwell.

Harvey, D. (1985b) *Consciousness and the Urban Experience*. Oxford: Blackwell.

Harvey, D. (1989a) *The Urban Experience*. Oxford: Blackwell.

Harvey, D. (1989b) *The Condition of Postmodernity*. Oxford: Blackwell.

Harvey, D. (1996) *Justice, Nature and the Geography of Difference*. Oxford: Blackwell.

Harvey, D. (2000) *Spaces of Hope*. Edinburgh: Edinburgh University Press.

Secondary Sources and References

Castree, N. (2001) 'From spaces of antagonism to spaces of engagement', in A. Brown *et al.* (eds) *Marxism and Critical Realism*. London and New York: Routledge, pp. 187–214.

Castree, N. and Gregory, D. (eds) (2004) *David Harvey: A Critical Reader*. Oxford: Blackwell.

Cox, K. and Mair, A. (1988) 'Locality and community in the politics of local economic development', *Annals of the Association of American Geographers* 78: 307–325.

Dennis, R. (1987) 'Faith in the city', *Journal of Historical Geography* 13: 310–316.

Deutsche, R. (1991) 'Boy's town', *Environment and Planning D: Society and Space* 9: 5–30.

Harvey, D. (1963) 'Locational change in the Kentish hop industry and the analysis of land use patterns', *Transactions of the Institute of British Geographers* 33: 123–140.

Harvey, D. (1981) 'The spatial fix', *Antipode* 13 (3): 1–12.

Harvey, D. (1985c) 'The geopolitics of capitalism', in D. Gregory and J. Urry (eds) *Social Relations and Spatial Structures*. London: Macmillan, pp. 128–163.

Harvey, D. (1987a) 'Three myths is search of a reality in urban studies', *Environment and Planning D: Society and Space* 5: 367–376.

Harvey, D. (1987b) 'The representation of urban life', *Journal of Historical Geography* 13: 317–321.

Harvey, D. (1988) 'The geographical and geopolitical consequences of the transition from Fordism to flexible accumulation', in G. Sternlieb and J. Hughes (eds) *America's New Market Geography*. New Brunswick: Rutgers University Press, pp. 101–135.

Harvey, D. (1990) 'Between space and time: reflections on the geographical imagination', *Annals of the Association of American Geographers* 80: 418–434.

Harvey, D. (1997) 'Contested cities: social process and spatial form', in N. Jewson and S. MacGregor (eds) *Transforming Cities*. London: Routledge, pp. 19–27.

Harvey, D. (1999) 'Foreword' to *The Limits to Capital*, 2nd edition. London: Verso.

Harvey, D. (2001) *Spaces of Capital*. Edinburgh: Edinburgh University Press.

Harvey, D. and Scott, A. (1987) 'The practice of human geography: theory and empirical specificity in the transition from Fordism to flexible accumulation', in B. Macmillan (ed.) *Remodelling Geography*. Oxford: Blackwell, pp. 217–229.

Harvey, D. and Hayter, T. (1993) *The Factory and the City*. London: Mansell.

Henderson, G. (1999) *California and the fictions of capital*. Oxford: Oxford University Press.

Herod, A. (2001) *Labor Geographies*. New York: Guilford.

Jackson, P. (1999) 'Commodity cultures: the traffic in things', *Transactions of the Institute of British Geographers* 24: 95–108.

Jones, J-P. (2003) *David Harvey: Living Theory*. New York: Continuum.

Merrifield, A. (1993) 'Space and place: a Lefebvrian reconciliation', *Transactions of the Institute of British Geographers* 18: 516–531.

Morris, M. (1992) 'The man in the mirror', *Theory, Culture and Society* 9: 253–279.

Olsson, G. (1975) *Eggs in Bird/Birds in Egg*. London: Pion.

Sack, R. (1974) 'The spatial separatist theme in geography', *Economic Geography* 50: 1–19.

Sayer, A. (1985) 'The difference that space makes', in D. Gregory and J. Urry (eds) *Social Relations and Spatial Structures*. London: Macmillan, pp. 49–66.

Schaefer, F. (1953) 'Exceptionalism in geography', *Annals of the Association of American Geographers* 43: 226–229.

Soja, E. (1989) *Postmodern Geographies*. London: Verso.

Noel Castree

28 bell hooks

BIOGRAPHICAL DETAILS AND THEORETICAL CONTEXT

With over 20 published books of theory, poetry, fiction and autobiography, bell hooks is one of the most prominent and influential feminist philosophers. Her devotion to black activism, intellectual life and cultural politics in the United States has shaped her commitment to dismantling what she terms the 'white supremacist capitalist patriarchy', and explaining the liberatory politics inherent in blackness, community and healing. It is from the margins – specifically the experiences of subaltern and disenfranchised communities – that she offers progressive ways of working through the difficulties of race, class, gender and sexual domination(s).

bell hooks/Gloria Watkins was born in Hopkinsville, Kentucky, in 1952. She has argued that her familial and household politics forced her, as a young child, to theorize the problems of black patriarchy, sexism and gender subordination (hooks, 1994b: 59–76). hooks received her BA from Stanford University, and her MA from the University of Wisconsin. She received her PhD from the University of California at Santa Cruz, writing her dissertation on Toni Morrison. hooks has taught at Yale, Oberlin, the City College of New York and Southwestern University. Her experiences as an undergraduate and graduate student advanced her earlier feminist thinking and her contributions to black feminist thought. First, hooks recognized the lack of historical and contemporary writings by and about black women; second, she noticed that second-wave feminism – although not formally named as 'white' or 'white supremacist' – unjustly marginalized, and in some cases explicitly dismissed, non-white women. This discursive and experiential racism and sexism led her to produce two texts that 'spoke back' to these interconnected forms of black marginalization. Prior to receiving her PhD, hooks (1981) completed *Ain't I A Woman: Black Women and Feminism*, a text she had been preparing for several years. Contributing to the long tradition of North American black feminist history and black feminist thought – such as the work produced by Michele Wallace, Audre Lorde, Angela Davis and Zora Neale Hurston, among others – *Ain't I A Woman* maps out US black women's contemporary and historical experiences. In addition to this, as the title suggests, *Ain't I A Woman* calls into question the white-racialized meaning of womanhood, asking the reader to consider alternative ways of conceptualizing gender, feminism and political unity *vis-à-vis* black women's struggles and histories.

hooks' (1984) *Feminist Theory: From Margin to Center* focuses specifically on feminist politics and progressive struggles within and outside black communities. Drawing on her activist and feminist experiences, as well as black and white feminism, *From Margin to Center* outlines a series of feminist issues – such as parenting, sexual oppression, paid and unpaid labour and political allegiances. The text suggests that black women's 'special vantage point' (hooks, 1984: 15) necessarily enriches feminist theory and praxis. In addition to this, *From Margin to Center* also suggests that racism, sexism and exclusion have not prevented black women from feminist theorizing in the academy and the everyday.

Following *Ain't I A Woman: Black Women and Feminism* and *Feminist Theory: From Margin to Center*, hooks produced a series of theoretical texts that sought to strengthen her commentary on exclusion, feminism, black feminist thought and cultural politics. Slowly moving away from explicit critiques of white feminism, but not abandoning her fundamental political commitments, hooks' later texts address pedagogy, representational politics, popular culture, desire, internalized racism, black femininity, stereotypes and visual art. Her many texts, the most popular (and/or cited) being *Talking Back: Thinking Feminist, Thinking Black* (1989), *Yearning: Race, Gender, and Cultural Politics* (1990), *Black Looks: Race and Representation* (1992), *Outlaw Culture: Resisting Representations* (1994a) and *Teaching to Transgress: Education as the Practice of Freedom* (1994b), couple black experiences and knowledges with the disabling processes of racial, sexual and economic domination. Drawing on thinkers such as Frantz Fanon, Toni Morrison, Paulo Freire, Cornel West, Audre Lorde, Zora Neale Hurston and Kobena Mercer, among others, hooks outlines how pervasive processes of domination are evident in national and local politics, education, employment, mass media and the self.

For hooks the existence of pervasive forms of racial, economic and sexual domination suggests that black politics, and activism in general, must be radically open and oppositional. In addition to this, the Self must be repaired, and this reparation is often constituted by being radically open and oppositional (hence, 'talking back' to, and looking back at, the intricacies and practitioners of oppression). The underlying messages in many of her texts, which have become enhanced in later arguments developed texts such as *Salvation: Black People and Love* (2001), are processes of self-renewal and resistance, and practices of love and desire which must be produced in a landscape that has systematically discouraged loving blackness and social differences. This incites 'the deepest revolution, the

turning away from the world as we know it, toward the world we must make if we are to be one with the planet' (hooks, 2001: 225; see also hooks, 1992: 9–20). These arguments, concerning healing and ethical-love-politics, while not taken up extensively by other academics, interestingly weave together most of her texts by infusing her political agenda with psychic decolonization, critical thinking, black feminist subjectivity and the combination of care, knowledge, responsibility, respect, trust and commitment (hooks, 2001).

SPATIAL CONTRIBUTIONS

Like many feminist and critical race theorists, hooks has contributed to the ways in which 'identity' is understood in social theory. Identity, for hooks, is formed according to white supremacist capitalist patriarchy. This means that the black sense of self is constructed according to broader ideological and material patterns that inscribe 'difference' or 'otherness' on the black psyche, the black body and the black community. Identity, then, is shaped by processes of marginalization and *it is within* this marginal space that diverse black identities are produced. What hooks gives a sense of is the doubleness of black identities – identities that are produced according to crude forms of race/racism *and* identities that are diversely challenging the meaning of these forms of race/racism. This understanding of identity, also explored by several other non-white theorists, is nuanced by hooks' commitment to 'talking back' and oppositional strategies. Thus, the black subject (re)produces identity and black politics by implicitly or explicitly 'coming to voice . . . where one moves from being object to being subject' (hooks, 1989: 12). hooks' conceptualization of black identities contributes to the geographical shift away from stable, white Euro-American, and androcentric identity formations (*cf.* Kobayashi and Peake, 1994). By positing the possibility of, for example, diverse black feminine subjectivities that are actively producing their identities *vis-à-vis* domination, resistance and the margin, hooks suggests that the production of space is simultaneously contextual and hegemonic.

In 'Choosing the margin as a space of radical openness', in *Yearning* (hooks, 1990: 145–153), hooks builds on her work in *From Margin to Center* (1984). hooks argues that the margin and the centre are neither antithetical nor an indication of a white–non-white disconnection. Instead, she suggests that racial, sexual, economic and social differences shape and determine a response to, and therefore a connection with, existing cultural norms. This connection, as it is understood from the margins, is significantly oppositional:

> . . . marginality [is] much more than a site of deprivation . . . it is also the site of radical possibility, a space of resistance. It was this marginality that I was naming [in *From Margin to Center*] as a central location for the production of counter-hegemonic discourse that is not just found in works but in habits of being and the way one lives. As such, I was not speaking of a marginality which one wants to lose – to give up or surrender as part of moving to the center – but rather a site one stays in . . . It offers to one the possibility of radical perspective from which to see and create, to imagine alternatives, new worlds.
> (hooks, 1990: 149–150)

hooks thus notes the interplay between margin–centre relations, suggesting that the black subject's positionality informs political alternatives. Furthermore, hooks suggests the margin is a legitimate location from which to produce knowledge and confront pain. Spatializing black identities and knowledges, hooks' work on space, place and the margin is advanced by 'the politics of location' (hooks, 1990: 145; see also Rich, 1986) and **Stuart Hall**'s 'politics of articulation' (hooks,

1990: 146; Hall, 1986: 45–60). While the politics of location attaches one's distinct social identity to a distinct spatial location (race, class, gender, sexuality, for example, inform one's environment and choices), the politics of articulation, while differing slightly from Stuart Hall's (1986) position, suggests that one's social and spatial identity is also shaped by the ways that ideological forces do, or do not, offer openings for alternative voices. Together, location and articulation – how one verbally and non-verbally negotiates one's political self – produce a space of radical openness: a space 'which affirms and sustains our subjectivity, which gives us a new location from which to articulate our sense of the world' (hooks, 1990: 149, 153).

In *Yearning* (1990), the chapter 'Homeplace: A site of resistance' reconsiders white feminist approaches to home, the family and community. Second-wave feminists critiqued the home, suggesting that it can be a feminized space where patriarchal norms are reproduced, violence is enacted and women's unpaid work is de-valued. Thus, the home was not a 'safe' or liberating space for those (white) women who performed paid and unpaid labour in that it institutionalized female subordination through gender socialization, and entrenched the material base of patriarchy (*cf.* Hartmann, 1981). Implicitly opposing the ways in which black women have been constructed in relation to the home – in which their roles as wives, mothers, sisters, domestic workers and paid and unpaid labourers are undermined by stereotypes of overpowering matriarchs, 'natural' house-servants, and nurturers of whiteness – homeplace, for bell hooks, is a subversive and feminist space.

hooks' reconceptualization of home begins with her childhood journeys through unfamiliar white neighbourhoods. These travels disclosed different forms of white supremacy, danger and fear that were diminished upon arriving 'home': 'the feeling of safety, of arrival, of homecoming . . . the warmth and comfort of shelter, the feeding of our bodies, the nurturing of our souls' (hooks, 1990: 41). By positing that the homeplace was (and is), for black women and black communities, a potential (but often unrecognized) site of resistance, hooks overturns the oppressive nature of domestic space. Rather than suggesting that black communities neatly replicate white feminist theories, hooks revisits the history of black homeplaces: spaces that were most often created by black women in order to provide an environment in which the black community can learn to love and respect blackness and heal from the wounds inflicted by white supremacy. Although sexist norms are certainly evident in black homes, hooks calls for the reclamation of the kind of place where resistance is possible and these norms are discouraged.

The concepts that lie at the heart of hooks' work – identity, the margin, and homeplace – are necessarily linked to space and place, with the spatialization of difference being a key motif in hooks' writing. Feminist and human geographers have thus used hooks' writing not only to clarify the ways that space and place adversely shape black choices and subjectivities, but also to explore the complexities inherent in spatial production. There are three ways this is done. First, geographers turn to hooks' understandings of the ways 'whiteness' and racism shape identity formation. hooks' work, primarily in *Black Looks* (1992) and *Yearning* (1990), describes the historical and everyday terrors that black people encounter in a white, capitalist, patriarchal world. Spatial power and domination therefore produce the outer world, while black subjects are forced to negotiate a world that offers few opportunities for an 'open-minded public space' (Ruddick, 1996: 136). Geographers argue that this lack of public open-mindedness and the legacy of race, racism and (in)formal segregation, spatialize difference, thus repeating and renewing the ways in which non-white racial bodies can, or cannot, occupy space. In addition to this, the non-white racialized body exposes how landscape,

space, place, infrastructure, mapping and knowledge (in the academy and the everyday) can, or cannot, be open to blackness or difference (Jackson, 1992; Gregory, 1994).

Second, feminist and human geographers have turned to hooks' politicization of location. hooks' discussion of black homeplace(s) and the margin as a site of resistance have been taken up by feminists and other geographers to underscore 'a different sense of place [one which is] no longer passive, no longer fixed, no longer undialectical' (Keith and Pile, 1993: 5). Homeplace(s) and the margins have contributed to how some geographers theorize the mutual flexibility of identity and place. Feminist geographers such as Nancy Duncan (1996), **Doreen Massey** (1994) and Linda McDowell (1999) have briefly touched on hooks' analysis of home in order to clarify the contradictory nature of femininity, race and domestic/private spaces. Duncan, furthermore, suggests that hooks' analysis can speak to other female/feminist spaces – specifically the spaces produced by, and in relation to, abused women and sexual minorities. These spaces, although sites of deep sexualization and gendering, also hold in them the possibility of politicized resistance (Duncan, 1996: 142–143). Both **Derek Gregory** (1994) and **Ed Soja** (1993, 1996) have also used hooks to identify how social differences challenge geographical theory and the production of space. Soja, in particular, unpacks hooks' textual and experiential geographies and suggests that they are indicative of a 'postmodern' way of understanding space and difference. Her positionality is therefore evidence of the complex political possibilities that coincide with contemporary, or 'postmodern', time–space patterns. Soja and Hooper thus write hooks into a 'spatial turn', arguing that her geographical contributions 'move beyond modernist binary oppositions of race, gender and class into the multiplicity of *other* spaces that difference makes; into a revisioned spatiality that creates from difference new sites for struggle and for the construction of interconnected communities of resistance' (Soja and Hooper, 1993: 189; emphasis in the original).

Finally, some feminist and human geographers bring together hooks' *subjective experiences* in, and descriptions of *her* spaces and places, with her more conceptual/metaphorical advances – such as the margin. This underscores the ways that geography and hooks herself are, together, 'multiple and intersecting, provisional and shifting' (Rose, 1993: 155). hooks' experiential and oppositional spaces are therefore simultaneously real and metaphorical. By bringing together the margins, spaces of resistance, blackness and experience, **Gillian Rose** identifies the complex potential of hooks' black feminist spaces:

> ... her argument develops from her experience of growing up in a Southern black community, segregated from the white centre of the town by the railway tracks. Its segregation on the margin of the town belies its importance to the town's economy, however, for many of its service workers – without which the town would cease to function – live there ... its poverty is structured by those social relations. But hooks refuses to comprehend its geography by seeing it mapped only in these terms, a margin defined only by its relation to the centre ... [instead] hooks describes this as a place of transgression ... (Rose, 1993: 156)

KEY ADVANCES AND CONTROVERSIES

In their introduction to *Place and the Politics of Identity*, Michael Keith and Steve Pile open up the possibility for mapping the politics of identity 'by turning to bell hooks, who is fast becoming a shibboleth for white academic men – including us – who want to prove beyond a shadow of a doubt their radical credentials' (Keith and Pile, 1993: 5). Despite their self-depreciation, and their position-

ing of bell hooks as an object-subject-catchword who necessarily provokes radical (and arguably racially 'safe') theorizing, Keith and Pile accurately spatialize hooks herself. hooks' extensive and accessible production of theory and texts about blackness and black femininity, her additional writings in *Vibe*, *The Village Voice*, and *Z Magazine*, her canonization into introductory women's studies classes, and the breadth of her subject matter – slavery to Spike Lee – have made her one of the most cited, and often the only cited, black female academics. Unfortunately, this disregards the work of other non-US black feminists, the work of hooks' US black colleagues, her forefathers and foremothers, other black geographical investigations, and the work of black geographers in general (*cf.* Gilmore, 2002; Woods, 1998; Sanders, 1990; Wilson, 2002). What this does, then, is reduce hooks/black womanhood to an authentic embodiment of blackness who is cited to lend texture to theory. As Valerie Smith (1989) argues, this use of black womanhood risks re-colonizing blackness and homogenizing all black women's experiences and identities.

Although hooks is occasionally cited alongside Frantz Fanon, **Edward Said**, Gloria Anzuldúa, **Gayatri Spivak** and Chandra Talpade Mohanty, it is also important to think about how 'the margin' and arguably hooks herself (as a site, theory and subject for geographic investigation) has effectively undermined other black feminist geographies because she is seemingly the only authentic black female voice available. Although there are exceptions, this is evidenced in the way critical engagements with hooks often only *describe* her experiences and work; many theorists resist confronting her deeper concerns, such as the struggles and meanings inherent in loving blackness, and often fail to rigorously critique her positionality or framework (an exception is Pratt and Hanson's (1994: 9) critique of 'the margin'). Geographies of difference, and in particular those geographies incited by hooks herself, not only call for a working through of how the production of space is simultaneously material, subjective, empowering and unjust, but also how multiple subaltern geographical knowledges open up new and diverse spaces for consideration.

HOOKS' MAJOR WORKS

hooks, b. (1981) *Ain't I A Woman: Black Women and Feminism*. Boston: South End Press.
hooks, b. (1984) *Feminist Theory: From Margin to Center*. Boston: South End Press.
hooks, b. (1989) *Talking Back: Thinking Feminist, Thinking Black*. Toronto: Between the Lines Press.
hooks, b. (1990) *Yearning: Race, Gender and Cultural Politics*. Toronto: Between the Lines Press.
hooks, b. (1992) *Black Looks: Race and Representation*. Toronto: Between the Lines Press.
hooks, b. (1994a) *Outlaw Culture: Resisting Representations*. New York: Routledge.
hooks, b. (1994b) *Teaching to Transgress: Education as the Practice of Freedom*. New York: Routledge.
hooks, b. (2001) *Salvation: Black People and Love*. New York: HarperCollins.
hooks, b. and Gilroy, P. (1993) 'A dialogue with bell hooks', in P. Gilroy *Small Acts*. London: Serpent's Tail.

Secondary Sources and References

Duncan, N. (1996) 'Renegotiating gender and sexuality in public and private spaces', in N. Duncan (ed.) *BodySpace: Destabilizing Geographies of Gender and Sexuality*. New York and London: Routledge, pp. 127–145.
Gilmore, R. W. (2002) 'Fatal couplings of power and difference: Notes on racism and geography', *The Professional Geographer* 54: 15–24.
Gregory, D. (1994) *Geographical Imaginations*. Cambridge and Harvard: Blackwell.
Hall, S. (1986) 'On postmodernism and articulation: An interview with Stuart Hall', *Journal of Communication Inquiry* 10: 45–60.
Hartmann, H. (1981) 'The unhappy marriage of marxism and feminism: Towards a more progressive union', in L. Sargent (ed.) *The Unhappy Marriage of Marxism and Feminism: A Debate on Class and Patriarchy*. London: Pluto, pp. 91–107.
Jackson, P. (1992) 'The politics of the streets: A geography of Caribana', *Political Geography* 11: 130–151.
Keith, M. and Pile, S. (1993) 'Introduction: The politics of place', in M. Keith and S. Pile (eds) *Place and the Politics of Identity*. London and New York: Routledge, pp. 1–21.
Kobayashi, A. and Peake, L. (1994) 'Unnatural discourse: "Race" and gender in geography', *Gender, Place and Culture* 1: 225–243.
Massey, D. (1994) *Space, Place and Gender*. Minneapolis: University of Minnesota Press.
McDowell, L. (1999) *Gender, Identity and Place: Understanding Feminist Geographies*. Minneapolis: University of Minnesota Press.
Pratt, G. and Hanson, S. (1994) 'Geography and the construction of difference', *Gender, Place and Culture* 1: 5–29.
Rich, A. (1986) 'Notes toward a politics of location', in A. Rich *Blood, Bread and Poetry: Selected Prose, 1979–1985*. New York: W. W. Norton, pp. 239–256.
Rose, G. (1993) *Feminism and Geography: The Limits of Geographical Knowledge*. Cambridge and Oxford: Polity Press.
Ruddick, S. (1996) 'Constructing difference in public spaces: Race, class and gender as interlocking systems', *Urban Geography* 17: 132–151.
Sanders, R. (1990) 'Integrating race and ethnicity into geographic gender studies', *The Professional Geographer* 42: 228–231.
Smith, V. (1989) 'Black feminist theory and the representation of the "Other" ', in C. A. Wall (ed.) *Changing Our Words*. New Brunswick and London: Rutgers University Press, pp. 38–57.
Soja, E. and Hooper, B. (1993) 'The space that difference makes: Some notes on the geographical margins of the new cultural politics', in M. Keith and S. Pile (eds) *Place and the Politics of Identity*. London and New York: Routledge, pp. 183–205.
Soja, E. (1996) *Thirdspace: Journeys to Los Angeles and Other Real-and-Imagined Places*. Cambridge: Blackwell.
Wilson, B. M. (2002) 'Critically understanding race-connected practices: A reading of W. E. B. Du Bois and Richard Wright', *The Professional Geographer* 54: 31–41.
Woods, C. (1998) *Development Arrested: The Blues and Plantation Power in the Mississippi Delta*. New York and London: Verso.

Katherine McKittrick

29 Peter Jackson

BIOGRAPHICAL DETAILS AND THEORETICAL CONTEXT

Peter Jackson is a leading voice in British social and cultural geography. He has been central in the formulation of a 'new cultural geography' that focuses on the social construction of race and other markers of identity, and that explores the role of consumption in social life. Jackson was born in 1955 in Stoneleigh, Surrey, England. After grammar school in nearby Epsom, he went to Keble College, Oxford, where and earned a First Class BA honours degree in 1976. A year later he received a Diploma in Social Anthropology and left for a Fulbright Scholarship year in the Department of Geography at Columbia University in New York. He returned to Oxford in 1979 and earned a DPhil in Geography a year later. His thesis was entitled, 'A social geography of Puerto Ricans in New York', a good indication of his early orientation and interests in geography. Finishing his degree, Jackson took a job as Lecturer in Geography at University College, London. He was promoted to Senior Lecturer in 1992. In 1993 he became Professor of Geography at the University of Sheffield. He has served as visiting lecturer at universities in the US, Canada, Finland, New Zealand, Greece and the UK.

Jackson has written, co-written or edited eight books and published nearly 70 articles and book chapters. In addition to his prolific writing, Jackson has been an influential supervisor of postgraduate students. Among these have been scholars who have gone on to play central roles in the development of queer geographies, postcolonial geographies and studies of the geography of consumption.

SPATIAL CONTRIBUTIONS

Jackson's work has been influential in four main, overlapping areas of human geography: social geographies of race and racism; new cultural geographies; geographies of masculinity; and geographies of consumption. In all of these there is an overriding interest in issues of identity, especially its social and geographical construction.

The evolution of Jackson's interest in social geography and issues of race and racism can be traced through three prominent edited collections. *Social Integration and Ethnic Segregation* (1981b) and *Exploring Social Geography* (1984), both co-edited with Susan Smith, examine patterns of social interaction and segregation, the social processes at work in ghettoization, and in the promotion of inequality. The first book is framed in part as a reckoning with the legacy of Chicago School sociology in geography. Jackson and Smith (1981a: 6) argued that social geography concerned with segregation and integration required 'a clear conception of social *structure*', and thus required a move away from the 'natural orders' sociology of the Chicago School and an encounter with Marxian theories of racism as 'an institutionalized ideology' (Jackson and Smith, 1981a: 11). Therefore, Jackson and Smith (1981a: 12) argued, ethnicity had to be understood as a 'negotiated variable' and not a biological given. The second collections of essays explored the dynamics behind such social structures and negotiated variables, suggesting that ethnicity had to be understood as contingent rather than essential (Jackson and Smith, 1984). Ethnicity was a product of social relations and contextual variables. Jackson (1985) suggested this was clearly apparent in the 1981 riots in the London suburb of Brixton, and the official report that examined their causes. Such events, together with the Greater London Council's 1984 'London Against Racism' initiative, induced a 'quite remarkable' volume of research on the sociology and geography of race and racism (Jackson, 1985: 99). Indeed, around this time the discourse in social geography shifted clearly from one about ethnicity and segregation to one about race, racism and ideology. Jackson was a leader of this move (though he acknowledges that schoolteachers were also pivotal in challenging racism; see Jackson, 1995), codifying arguments about the geography of racist ideology in his introduction to the third edited collection, *Race and Racism: Essays in Social Geography* (1987).

Simultaneously with completing his PhD, Jackson began to question why cultural theory had been so underemphasized in British human geography (Jackson, 1980). In part, Jackson found that this was due to the way that cultural geography had been reduced to the study of the visible landscape (especially in American cultural geography). He argued that 'cultural geography can finally only be of interest to the British geographical profession if it can successfully accomplish a rapprochement with social geography, in a joint commitment to study the spatial aspects of social organization and human culture – not just those aspects which are directly observable in the landscape' (Jackson, 1980: 113). Such a rapprochement required a focus on social interaction and social relations (see also Duncan, 1980). Jackson went beyond the dominant interactionism of the time to

suggest that cultural geography required a much more direct focus on, and analysis of, the social construction of *cultural identity*.

Jackson thus turned to the work of **Stuart Hall** and the Birmingham Centre for Contemporary Cultural Studies (CCCS). He argued that social and cultural geography needed to focus on 'popular subcultural forms, interpreting their specific contemporary meanings in relation to their specific materials context' (Cosgrove and Jackson, 1987: 98). The CCCS had argued that subcultures should 'not be seen as an autonomous cultural domain, but as involving the *appropriation* of certain artifacts and significations from the dominant . . . culture, and their *transformation* into symbolic forms which take on new meanings and significance for those who adopt these styles' (Jackson, 1987: 9; original emphasis). Subcultural groups fashion identity through appropriation and transformation. Jackson thus advocated a Gramscian-style analysis of hegemony that showed how subcultures are structured – given form and direction – through power. Jackson (1989: 2; original emphasis) argued that cultural geography had to shift its emphasis from 'culture itself to the domain of *cultural politics*' and to understanding how '*the cultural is political*'.

Jackson first made these arguments (with **Denis Cosgrove**) in an article outlining the development of a new cultural geography, and then in a major book, *Maps of Meaning* (1989) that established a new 'agenda for cultural geography'. *Maps of Meaning* drew on the work of **Raymond Williams** as well as that of CCCS to develop a *cultural materialism* appropriate for geography. In this book, Jackson brilliantly exemplified this turn to cultural studies and cultural materialism, while also developing the need for a geography of identity *per se*. 'Maps of meaning' is a term borrowed from Hall, who uses it to describe how people make sense of the work they are thrown into: 'cultures are *maps of meaning* through which the world is made intelligible'

(Jackson, 1989: 2). Jackson traced out these maps of meaning first by establishing the need for a geographical cultural materialism, and then by exploring the use of such a cultural materialism in the understanding of ideology, popular culture and class (a classic CCCS topic), gender and sexuality, race and racism, and the politics of language. In each of these there is a close focus on the *discourses* or *languages* of social and cultural categorization. Discourses, Jackson (1989) argues, are central to cultural materialism, even if, as he makes clear, they are not all of it. Jackson argues (by example) that social constructionism, which during the 1980s and 1990s became closely wed to discourse theory, had to be at the heart of cultural geography. Even so, Jackson was perfectly clear in his warning that:

> There are serious problems with the social constructionist approach. In challenging the *naturalness* of categories like 'race' and gender, we run the political danger of evacuating the very concepts around which people's struggles against oppression are being organized.
> (Jackson, 1991a: 193)

That is why cultural geography's focus had to be on cultural politics, and not just culture as such.

Maps of Meaning is a truly important book. It was published during a remarkable year, rocked politically in ways that have transformed the sorts of meanings through which we structure our identities. It appeared on the scene in the midst of political turmoil and it spoke directly to many of the cultural issues that arose out of that turmoil. Within the discipline it was part of a rash of books published that year (Harvey, 1989; Peet and Thrift, 1989; Soja, 1989; Storper and Walker, 1989) that contributed to multidisciplinary debates about how to best understand the rapid social, political and cultural restructuring brought on by the massive transformation of the global capitalist political economy over the preceding two decades and the consequent rise of postmodernism in art,

philosophy and culture more widely. They were all also prescient interventions into the scalar politics of geopolitical and cultural realignment attendant upon the Tiananmen uprising and the collapse of state socialism. Jackson (1989: 5) locates the development of his own focus on cultural politics in a reaction to the 'current politics of fiscal retrenchment, privatization, and economic recession in Thatcher's Britain [and] Bush's America'. In tandem with other, now classic works of the 'new cultural geography' (e.g. Cosgrove, 1984; Daniels, 1989; Duncan, 1990; Philo, 1991; *Society and Space*, 1988), *Maps of Meaning* served to transform cultural geography into a politically and socially relevant field. Further, it helped to name and shape the much broader 'cultural turn' that has reoriented so much human geography and the social sciences.

A chapter of *Maps of Meaning* was devoted to gender and sexuality. In it Jackson (1989: 128–129) presented a reminder that today merely sounds like commonsense, but at the time, in geography at least, seemed revolutionary: 'gender relations are embedded in a matrix of social relations involving *both men and women*, while the study of sexuality cannot be confined to the study of gay men'. Feminism's entry into geography had come as part of a socialist feminist movement to understand the structural relations of inequality under patriarchal capitalism, and from there explored the domination of women in public and private spheres. Only during the late 1980s did it begin to focus on questions of identity, drawing heavily from feminist psychoanalytic approaches. Much of this work was focused on *women* as the bearers of gender, while men remained something of an unmarked category. Missing was a focus on the construction of masculinity as a contested, gendered identity. Similarly, studies of sexuality moved from examinations of gay ghettos to, by the late 1980s, studies of gay male identity. Only in the 1990s did focus in geography shift to sexual identities, and broaden to include the social construction of heterosexuality (rather than seeing it only as the norm against which homosexuality was fashioned).

Making good his own critique in *Maps of Meaning* Jackson (1991b), launched a programme designed to analyse the geographical construction of masculinity that eventually culminated in a book that, firmly entrenched in standard practices of cultural studies, showed how men's magazines were a central location for the negotiation of gendered identities (Jackson *et al.*, 2001). As with his work on 'race', Jackson is concerned in his studies on masculinity to show how gender is socially and ideologically constructed, and how, at the same time, it becomes incorporated into affective identity and thus is a means by which people fashion themselves. Gender is not just an imposition: it is a negotiation.

Jackson has focused on the role of advertizing in establishing the contours of the cultural politics of masculinity. Jackson's interests in advertizing and identity have also come together and been broadened into a research programe on the geographies and cultural politics of consumption. With a number of colleagues, Jackson has explored how shopping has become a central practice in fashioning identity and the production of cultural meaning. This research effort seeks to 'transcend the cultural and economic' (Jackson, 2002b), and thus is both of a piece with his theories of geographical cultural materialism and a negation of them. As part of the cultural turn in human geography, which understands the economy to be a cultural production and culture to be a central explanatory force in economics, Jackson's studies of consumption and identity use ethnographies of the sites of consumption to show how both local cultures and economies are formed and reproduced through acts of consumption (Holbrook and Jackson, 1996; Jackson, 1999a). However, as with the forms of racialization he studied earlier, local cultural identities formed through consumption are never only lo-

cal. How can they be when identity is fashioned through commodities produced around the world, and often by transnational corporations? Identity, as constructed through consumption, is, in this sense, a local constellation of multi-scalar forces and processes. Or, to connect this argument to his earlier development of subcultural theories, Jackson remains concerned with the appropriation and transformation, by different groups, of the materials of everyday life into something meaningful (Jackson, 1995; Jackson and Holbrook, 1995; Jackson, 1999b).

Jackson's work in the cultural politics of consumption, though always attendant to geographies of power, is nonetheless highly affirmative. It is concerned with the positive making, rather than the negative imposition, of meaning and identity. He is, nonetheless, also concerned with the ways in which identities are exclusionary, and often the product of powerful impositions. This comes out especially in his studies of race and the racialization process, and in an edited collection of essays on *Constructions of Race, Place, and Nation* (Jackson and Penrose, 1993). But it also serves as a foil to some of his most recent work, on the production of transnational spaces and identities. Along with Phil Crang and Clare Dwyer, Jackson has completed an edited book on *Transnational Spaces* (Jackson *et al.*, 2002) which is the result of a project examining 'commodity culture and South Asian transnationality'. In common with his other studies on consumption, the focus here is on the construction of identity through consumption, but it also explores the processes of exclusion and inclusion that constitute the 'transnational'.

KEY ADVANCES AND CONTROVERSIES

Over his career Jackson has been instrumental in introducing into critical human geography a range of arguments that are now central in cultural and social geography. He helped to introduce the field both to sociological and anthropological work on the social construction of race and the materialist politics of racist ideology, and to cultural studies as practised by the CCCS. He established an agenda for cultural geography that was materialist in theoretical orientation, topically focused on cultural politics, and integral to the development of the study of geographies of identity. He pushed the frontiers of identity study to include the analysis of masculinity. And he helped to shift human geography's orientation from the study of geographies of production to the study of geographies of consumption.

None of these moves is without certain costs, as Jackson (e.g. 1996, 2002a) himself freely admits. Focus on the racialization process can too easily sidestep the social agency of those subject to it, even as it can fail to recognize the (positive) importance of racial categorization to those who are racially marginalized. Examinations of subcultural study often degenerate into celebrations of ingenuity rather than critical analyses of power-laden processes. And the focus on consumption might bend the stick back too far in its obsession with the agency of shoppers, failing to register the perhaps even more powerful agency of the social and geographical relations of production.

Direct criticism of Jackson's work has been aimed at his development of the 'maps of meaning' metaphor for culture. Jackson argues that culture can best be understood:

> as the medium or idiom through which meanings are expressed. If one accepts . . . arguments for a plurality of cultures, then 'culture' is the domain in which these meanings are contested.
> (Jackson, 1989: 180)

This way of framing culture has been condemned as being neither political nor materialist enough (Mitchell, 1995). The notion that culture is plural, it can be argued, elides the power relations that are always ongoing in the formation of culture as ideology (which is Jackson's starting point). Moreover, metaphors of 'medium', 'sphere', 'domain' and 'map' all reify what is always an ongoing struggle. Or more accurately, these metaphors turn away from the question of *who* reifies, by not asking directly why the map is of this and not that, why the domain exists here and not there, or why the limits of the sphere are drawn at this circumference and not that one. These are pre-eminently social questions. They are also economic, and economic in a certain way. They require less an analysis of consumption than of the geographies of production. Jackson's (1996, 2002b) response to this criticism has been to agree with the problematic nature of the metaphors he used, to restate his commitment that cultural *politics* needs to be at the forefront of analysis, and most importantly, to argue for a deeper and ongoing commitment to the ethnographic study of these cultural politics. His commitment to ethnography derives from exactly the argument that he made in *Maps of Meaning*: namely, that it is the appropriation and transformation of the materials of social and cultural life that matter.

And yet such a defence, correct as it may be, also points to a weakness in the conceptual categories of cultural studies-inspired geography. This weakness is a relative disinterest in the language of expropriation, fetishization, and especially alienation that is also part of Raymond Williams' and other Marxist approaches to the politics of culture. Studies of geographies of consumption – ethnographic or otherwise – that do not reckon with the possibility that processes of alienation and fetishization are as integral to the commodity form are as the politics of appropriation and transformation, threaten to miss exactly the political power and importance of the cultural materialist theoretical framework that Williams helped to establish (and Jackson promotes). As importantly, it turns attention too far from the role of culture-as-ideology. In particular, even as focus turns to issues like transnationalism as a context for geographies of consumption and identity, it reduces the *scale* of appropriation and transformation to the individual and local, which undermines the role that critical, materialist cultural geography could play in explaining and contesting the global-scale cultural politics of capitalist globalization, the 'war on terrorism' and the geopolitical/geocultural/geoeconomic imperialism that links them.

JACKSON'S MAJOR WORKS

Jackson, P. (ed.) (1987) *Race and Racism: Essays in Social Geography*. London: Allen and Unwin.
Jackson, P. (1989) *Maps of Meaning*. London: Unwin Hyman.
Jackson, P. and Penrose, J. (eds) (1993) *Constructions of Race, Place and Nation*. London: UCL Press.
Jackson, P. and Smith, S. (eds) (1981b) *Social Interaction and Ethnic Segregation*. London: Academic Press.
Jackson, P. and Smith, S. (eds) (1984) *Exploring Social Geography*. London: Allen and Unwin.
Jackson, P., Lowe, M., Miller, D. and Mort, F. (eds) (2000) *Commercial Cultures*. Oxford: Berg.
Jackson, P., Stevenson, N. and Brooks, K. (2001) *Making Sense of Men's Magazines*. Cambridge: Polity.
Jackson, P., Crang, P. and Dwyer, C. (eds) (2002) *Transnational Spaces*. London: Routledge.
Miller, D., Jackson, P., Thrift, N., Holbrook, B. and Rowlands, M. (1998) *Shopping, Place, and Identity*. London: Routledge.

Secondary Sources and References

Cosgrove, D. (1984) *Social Formation and Symbolic Landscape*. London: Croom Helm.
Cosgrove, D. and Jackson, P. (1987) 'New directions in cultural geography', *Area* 19: 95–101.
Daniels, S. (1989) 'Marxism, culture and the duplicity of landscape', in R. Peet and N. Thrift (eds) *New Models in Geography*, Volume 2. London: Unwin Hyman, pp. 196–220.
Duncan, J. (1980) 'The superorganic in American cultural geography', *Annals of the Association of American Geographers* 70: 181–198.
Duncan, J. (1990) *The City as Text: The Politics of Landscape Interpretation in the Kandyan Kingdom*. Cambridge: Cambridge University Press.
Harvey, D. (1989) *The Condition of Postmodernity*. Oxford: Blackwell.
Holbrook, B. and Jackson, P. (1996) 'The social milieux of two North London shopping centres', *Geoforum* 27: 193–204.
Jackson, P. (1980) 'A plea for cultural geography', *Area* 12: 110–113.
Jackson, P. (1985) 'Social geography: race and racism', *Progress in Human Geography* 10: 118–124.
Jackson, P. (1991a) 'Repositioning social and cultural geography', in C. Philo (ed.) *New Words, New Worlds*. Aberystwyth: Cambrian Printers.
Jackson, P. (1991b) 'The cultural politics of masculinity: towards a social geography', *Transactions of the Institute of British Geographers* 16: 199–213.
Jackson, P. (1995) 'Manufacturing meaning: culture, capital and urban change', in A. Rogers and S. Vertovex (eds) *The Urban Context: Ethnicity, Situational Analysis and Social Networks*. Oxford: Berg.
Jackson, P. (1996) 'The idea of culture: a reply to Don Mitchell', *Transactions of the Institute of British Geographers* 21: 572–573.
Jackson, P. (1999a) 'Commodity cultures: the traffic in things', *Transactions of the Institute of British Geographers* 24: 95–108.
Jackson, P. (1999b) 'Consumption and identity: a cultural politics of shopping', *European Planning Studies* 7: 25–39.
Jackson, P. (2002a) 'Ambiguous spaces and cultures of resistance', *Antipode* 34: 326–329.
Jackson, P. (2002b) 'Commercial cultures: transcending the cultural and the economic', *Progress in Human Geography* 26: 3–18.

Jackson, P. and Holbrook, B. (1995) 'Multiple meanings: shopping and the cultural politics of identity', *Environment and Planning A* 27: 1913–1930.

Jackson, P. and Smith, S. (1981a) 'Introduction', in P. Jackson and S. Smith (eds) *Social Interaction and Ethnic Segregation*. London: Academic Press, pp. 1–17.

Mitchell, D. (1995) 'There's no such thing as culture: towards a reconceptualization of the idea of culture in geography', *Transactions of the Institute of British Geographers* 20: 102–116.

Peet, R. and Thrift, N. (eds) (1989) *New Models in Geography*, 2 Volumes. London: Unwin Hyman.

Philo, C. (ed.) (1991) *New Words, New Worlds*, Aberystwyth: Cambrian Printers.

Society and Space (1988) Special issue on New Cultural Geographies, 6 (2).

Soja, E. (1989) *Postmodern Geographies*. Oxford: Blackwell.

Storper, M. and Walker, R. (1989) *The Capitalist Imperative: Territory, Technology, and Industrial Growth*. Oxford: Blackwell.

Don Mitchell

30 Bruno Latour

BIOGRAPHICAL DETAILS AND THEORETICAL CONTEXT

Bruno Latour was born in 1947. He comes from a well-established wine-growing family in Burgundy (*not* Bordeaux, home of 'Chateau Latour'). One of his more unusual ambitions, for an academic, is 'that people would say "I read a Latour 1992" with the same pleasure as they would say "I drank a Latour 1992"' (cited in an interview with Crawford, 1993: 248). From the outset he has taken the less-travelled path in French intellectual life. He was educated in the provinces of Dijon rather than Paris, and after starting training in the philosophy of religion he developed a belief in social science and switched to anthropology. His initial anthropological fieldwork study was in the Ivory Coast, followed up by what has become recognized as an iconoclastic study of a laboratory in California. For the larger part of his career since then he has, instead of living the lone life of French intellectual, been based in a collective 'laboratory' at the *École Nationale Supérieure des Mines* and collaborated widely with other researchers, policymakers, managers and philosophers. He is most widely recognized for the part he has played alongside Michel Callon and John Law in the initiation and remarkable spread of actor-network theory (ANT). Though firmly based in science studies, he and his work have travelled very widely; passing through sociology, art history, law, ecology, public transportation, fiction, geography and primatology, among other disciplines.

Frequently mistaken for a social constructivist, Latour *is* a constructivist, just not a *social* constructivist. This is best understood with reference to the initial 'strong programme' in the sociology of scientific knowledge that sought to *symmetrically* explain successes *and* failure in scientific progress in terms of social explanations (Barnes, 1974; Bloor, 1976). This programme accordingly suggested that when certain scientific fields (phrenology versus neuroscience) or certain phenomena (such as X-rays versus N-rays) came to be taken as fact or, conversely, were discredited, this was associated with social and cultural factors. Histories of scientific discoveries and technological breakthroughs had, until Bloor's initiative, tended to act as if scientific disciplines, facts, proofs, the number 'zero', statistics and various technologies existed independently of cultural norms, departmental struggles over funding, state armaments programmes, the cost of equipment, project cancellations, educational institutions, professional regulations and the influence of charismatic figures. Conventionally, each of these was acknowledged as an external force, with truth being internal to science, and thus guaranteed to come out and vanquish falsity. In contrast, the so-called symmetrical programme conceived of society as of one of the internal forces that gives science the shape that it has, rather than a force that bent true science or technology out of shape from the outside. Latour took this already remarkable programme a step further. Rather than allowing the social sciences a privileged vantage point, he used scientific activity to *symmetrically* explain failures and successes in the *social* sciences. In fact, Latour often shows that science is a far better analyser of society than social science; he shows scientists making facts, objects and networks to be, in effect, practical sociologists.

SPATIAL CONTRIBUTIONS

There are good reasons why Latour has been enthusiastically received in geography, not least that his work straddles the divide between science and society, a division echoed in the split between physical and human geography. At first taken up for his studies of science in action (Hinchliffe, 1996; Latour, 1987; Whatmore, 1997), Latour gained renown and his widest audience in geography by way of his most polemical book, *We Have Never Been Modern* (Latour, 1993), which argued not just against the existence of postmodernity but against modernity itself as any kind of separate age from the pre-modern (or non-modern).

In Latour's work geographers have been pleased to find an abiding attention to the connecting up, assembling, centring and distributing of all manner of things in space (Bingham, 1996; Murdoch, 1998). In describing how actor-networks are gradually extended, stabilized and sometime collapsed, Latour radically shifts away from a Euclidean concept of space and time as universal abstract axes that contain and constrain events (Latour, 1997b). For him, as for other researchers in ANT, space and time come about as consequences of the ways in which particular heterogenous elements are related to one another. The term 'topological' is therefore used to capture this sense of space as being made out of relations between its parts.

There are many controversial arguments in Latour's delightful books, not least his attribution of social agency to 'actants' that can as easily be nonhuman as human. This democratization of who can act, away from the anthropocentrism of the social sciences, has raised an awareness of the 'agency of things' previously restricted to debates over animal rights and the nature of artificial intelligence. It is characteristic of the flattening out of subjectivity found elsewhere in poststructural thought which has critiqued the Enlightenment's position of 'man' at a privileged level above all other life forms (or in Latour's case, 'things'). Geographers have been equally inspired and perplexed by Latour's extension of social agency, rights and obligations to automatic door-closers, sleeping policemen (Latour, 1992), bacteria (Latour, 1988), public transport systems (Latour, 1997a), and sheep dogs and fences (Latour, 1996). Referring to diverse objects and life forms that make up the world as 'the missing masses', Latour argues that they have been ignored socially, politically and philosophically, even as we clearly attend to, care for and depend on them in our everyday lives. Moreover it is, once again, scientists and engineers who pay special attention to the things of the world, providing extraordinary devices whereby we can listen, look or feel their wants, their characteristics and their actions.

A proverb often recited by historians is that 'there is nothing new under the sun'; for Latour 'there are many things new under the sun' since every once in a while something special happens: new things come to exist in the world. Their existence is in no way *inevitable*; they may perish as quickly as they came to gain a foothold on the earth (1997a). Without this historically assembled support of a multitude of perishable things, Latour suggests, we would live in a socially *unstable* world akin to that of baboons, where trials of strength have to be re-settled daily. In the endless busy proliferation of things as mediators, delegates, boundaries, 'immutable mobiles', we achieve the *complicated* places we live in. Geographers inspired by this attention to lowly devices and the emergence of new socio-technical-scientific agents have investigated the role of prions (i.e. protein chains) in the BSE crisis (Hinchliffe, 2001), measuring instruments in taxation (Ogborn, 1998) and financial systems in the city of London (Thrift, 1996).

Latour is an unusual figure in that cultural geographers of a highly theoretical bent have embraced his work, as have those whose inquiries are based primarily in field studies. It would be hard to imagine this kind of dual popularity for, say, either Jacques Derrida or Bronislaw Malinowski. Although critical of the reflexive strategies of postmodern anthropologists and textual experimenters like Steve Woolgar (e.g. Woolgar and Cooper, 1999) or Malcolm Ashmore (1989), he is nevertheless similarly creative, humorous and stylized in the construction of his texts. In *Aramis, or the Love of Technology* (Latour, 1997a) he writes polyphonically – mixing together a murder mystery, an ethnographic case study, philosophical reflections and the imagined voices of machines. Hence, just as he crosses the theorist/fieldworker divide, Latour also crosses the conventional/avant-garde writing divide in the social sciences and humanities.

KEY ADVANCES AND CONTROVERSIES

The longstanding problem of *structure–agency* is one to which Latour offers a novel solution. Where many social theorists and political philosophers, from Hobbes onwards, set up a binary opposition between social structure and individual agency, Latour pursues impure entities that have characteristics of structure *and* agency. They are, in other words, actors *and* networks or actor-networks. Latour suggests that those who employ an empty gulf between agency and structure do so by ignoring the dark matter of material objects that articulate, embody, coordinate and, even, author actions.

Just as Latour uses the 'actor-network' to fill the gap between agency and structure, so he uses 'hybrids' to refer to the proliferating entities that are made and remade as mixes of culture *and* nature. In doing so he responds to the endless controversies based in culture *versus* nature that have been at their most symptomatic in the 'science wars'. Rather than accepting culture or nature as explanations at face value, Latour, like many others in science and technology studies, turns them over from being explanations to being topics for his inquiries. Where the argument at its starkest uses, say, 'bacteria' as a source of explanation, Latour makes 'bacteria becoming an explanation for X happening' the topic of his inquiry. From his studies what we then find are the connections that associate specific explanations and ensuing courses of action (e.g. building the networks of pasteurization, practice of sterilization in hospitals, changes in food production, etc.) with specific kinds of bacteria. His studies convincingly describe a world where there is no pure nature nor pure culture. There are only fibrous webs gradually extending and contracting, erasing one another, copying one another and producing the shape of space and time in doing so. It is in this concern with how different assemblies of actants can connect up that Latourian spaces are often called 'topological'.

As was noted at the outset, Latour's extension of the symmetry principle deprived society from being the explanation of successes and failures in science. *A priori* favourites of the social sciences like class, race, gender and politics cannot be assumed as relevant in scientific and technological events, nor can everyday unexplicated terms like 'hard facts', 'geniuses' nor 'bias' be brought in to explain how the world moves or what moves the world. So what does Latour offer us, having denied the traditional explanatory terms for how the pure will of the subject or the blind force of the object gets bent out of shape by other effects? In typically elegant prose Latour delivers his credo of *irreductions*:

> Nothing can be reduced to anything else, nothing can be deduced from anything else, everything may be allied to everything else.
> (Latour, 1988: 163)

To make anything similar to or different from anything else requires translation, deformation, reformation or other forms of alteration. To make one thing identical to another, to make one place the same as another place requires building relations between them 'out of bits and pieces with much toil and sweat' (Latour, 1988: 162). With this leap away from the various reductions of various theories, Latour sets the 'things' free to do what they do, to ally with what they ally. As analysts we can follow their movements as they grow and shrink, associate, locate one another, become aligned, produce insides and outsides, subjectify and objectify. All of this sounds rather abstract, but Latour is never far from perspicuous examples:

> We neither think nor reason. Rather, we work on fragile materials – texts, inscriptions, traces, or paints – with other people ... The butcher's trade extends as far as the practice of butchers, their stalls, their cold storage, their pastures, and their slaughterhouses. Next door to the butcher – at the grocer's, for example – there is not butchery. It is the same with psychoanalysis, theoretical physics, philosophy, social security, in short all trades. However, certain trades claim that they are able to extend themselves potentially or 'in theory' beyond the networks in which they practice. The butcher would never entertain the idea of reducing theoretical physics to the art of butchery, but the psychoanalyst claims to be able to reduce butchery to the murder of the father and epistemologists happily talk of the 'foundations of physics'. Though all networks are the same size, arrogance is not equally distributed.
> (Latour, 1998: 187)

He is showing us here an example of the actor-network of butchering to remind us that all actors only gain agency by being part of particular networks made of more or less durable materials. If we take the butcher out of the assembly of farmers, delivery companies, freezers, trucks, sharp knives and saws, cash registers and banking, then we have a weak actor able to do very little for his trade. Latour in his studies of scientists brings them down from their privileged position to place them on a level with butchers and grocers. While retaining the greatest respect for the toil of science, he dispels its fairytale and miraculous existence in favour of taking seriously its rootedness and routedness in practices and things.

After several years of debunking 'Science' and 'Great Scientists' while almost heroicizing lowly lab workers and bacteria, Latour was bemused to find himself taken to task by a pair of scientists interested in debunking theorists from the humanities and social sciences. In a now legendary book, Sokal and Bricmont (1998) picked out several ripe targets for their loose deployment of terms derived from mathematics or physics. They selected Latour, along with other French social theorists and philosophers (Jacques Lacan, Julia Kristeva, Luce Irigaray, **Jean Baudrillard, Gilles Deleuze**, Félix Guattari and **Paul Virilio**) for being scientifically illiterate, relativistic, ambiguous, irrational and politically confusing. Given his own constant efforts to bridge the gap between science and other academic disciplines, Latour (1999b) has returned recently to attempting to make clear to scientists in what ways he might not be quite what some of them think he is as a result of his being placed alongside a theorist like Lacan. From the quote above it is clear that he is in fact on Sokal and Bricmont's (1998) side in respecting science for what it is and not attempting to theorize it. As he sharply puts it, 'In theory, theories exist. In practice, they do not' (Latour, 1998: 178).

Latour is, it should always be born in mind, anti-theory. This is a good reason, as he notes (Latour, 1999a) for ditching the term 'actor-network *theory*' since it has led many to believe it is yet another theory to add to the social sciences' extensive and perhaps excessive collection:

There is no metalanguage, only infralan-
guages. In other words there are only
languages. We can no more reduce one
language to another than build the tower
of Babel.
(Latour, 1988: 179)

An *infralanguage* for Latour holds the
promise of being able to write and reveal
things about science, engineering and society without claiming that he is laying
foundations, nor knows better than those
he is studying what it is that they do, nor
is socially critiquing their community. Yet
he does not wish to simply describe
scientific practice in detail, and this is

where he differentiates himself from eth-
nomethodological studies of science
(Lynch, 1985, 1993). Akin to Latour, ethnomethodology describes the practical activities of scientists (e.g. utilizing
equipment in laboratories) while also being critical of blanket social constructivist
explanation. However, Latour parts with
ethnomethodology since he wishes to
map out his infralanguage of paths, connections, displacements, associations,
topologies and networks, strands of ordering that are otherwise invisible since they
are hidden behind terms like 'science',
'genius' and 'society'.

LATOUR'S MAJOR WORKS

Latour, B. (1987) *Science in Action: How to Follow Scientists and Engineers Through Society*. Milton Keynes: Open University
    Press.
Latour, B. (1988) *The Pasteurization of France*. London: Harvard University Press (first published 1984).
Latour, B. (1992) 'Where are the missing masses? The sociology of a few mundane artefacts', in W. L. J. Bijker (ed.) *Shaping
    Technology/Building Society*. London: MIT Press, pp. 225–258.
Latour, B. (1993) *We Have Never Been Modern*. London: Harvester Wheatsheaf.
Latour, B. (1996) 'On interobjectivity', *Mind, Culture and Activity* 3: 228–245.
Latour, B. (1997a) *Aramis, or the Love of Technology*. London: Routledge.
Latour, B. (1997b) 'Trains of thought: Piaget, formalism and the fifth dimension', *Common Knowledge* 6: 170–191.
Latour, B. (1999a) 'On recalling ANT', in J. Law and J. Hassard (eds) *Actor Network and After*. Oxford: Blackwell with the
    Sociological Review, pp. 15–25.
Latour, B. (1999b) *Pandora's Hope, Essays on the Reality of Science Studies*. London: Harvard University Press.

Secondary Sources and References

Ashmore, M. (1989) *The Reflexive Thesis: Writing the Sociology of Scientific Knowledge*. Chicago: University of Chicago Press.
Barnes, B. (1974) *Scientific Knowledge and Sociological Theory*. London: Routledge and Kegan Paul.
Bingham, N. (1996) 'Object-ions: from technological determinism towards geographies of relations', *Environment and Planning
    D: Society and Space* 14: 635–657.
Bloor, D. (1976) *Knowledge and Social Imagery*. London: Routledge and Kegan Paul.
Crawford, T. Hugh (1993) 'An interview with Bruno Latour', *Configurations* 1 (2): 247–268.
Hinchliffe, S. (1996) 'Technology, power and space – the means and ends of geographies of technology', *Environment and
    Planning D: Society and Space* 14: 59–682.
Hinchliffe, S. (2001) 'Indeterminacy in decisions: science, policy and politics in the BSE crisis', *Transaction of the Institute of
    British Geographers* NS 26 (2): 182–204.
Lynch, M. (1985) *Art and Artifact in Laboratory Science, A Study of Shop Work andShop Talk in a Research Laboratory*. London:
    Routledge.
Lynch, M. (1993) *Scientific Practice and Ordinary Action: Ethnomethodology and Social Studies of Science*. Cambridge:
    Cambridge University Press.
Murdoch, J. (1998) 'The spaces of actor-network theory', *Geoforum* 29: 357–374.

Ogborn, M. (1998) *Spaces of Modernity: London's Geographies 1680–1780.* New York: Guilford Press.

Sokal, A. and Bricmont, J. (1998) *Intellectual Impostures, Post-modern Philosophers' Abuse of Science.* London: Profile Books.

Thrift, N. (1996) *Spatial Formations.* London: Sage.

Whatmore, S. (1997). 'Dissecting the autonomous self', *Environment and Planning D: Society and Space* 15: 37–45.

Woolgar, S. and Cooper, G. (1999) 'Do artefacts have ambivalence? Moses' bridges, Winner's bridges and other urban legends in S&TS', *Social Studies of Science* 29: 433–449.

Eric Laurier

BIOGRAPHICAL DETAILS AND THEORETICAL CONTEXT

Henri Lefebvre (1901–1991) was a neo-Marxist and existentialist philosopher, a sociologist of urban and rural life and a theorist of the state, of international flows of capital and of social space. He was a witness to the modernization of everyday life, the industrialization of the economy and suburbanization of cities in France. In the process, the rural way of life of the traditional peasant was destroyed. One indicator of Lefebvre's influence is the appearance of some of his signature concepts in left-intellectual discourse. Although not exclusively 'his' of course, Lefebvre contributed so much to certain lines of inquiry that it is difficult to discuss notions such as 'everyday life', 'modernity', 'mystification', 'the social production of space', 'humanistic Marxism', or even 'alienation' without retracing some of his arguments. Lefebvre's relevance and impact on late twentieth-century Anglo-American human geography cannot be overstated, but he cannot be fitted into a geographical straightjacket. Indeed, he was a critic of disciplinary overspecialization such as that between economics, geography and sociology, which 'parcelled-up' the study of space.

After his initial schooling on the west coast of France at Brieuc, in Aix-en-Provence and in Paris, he was profoundly affected by post-World War I *malaise* of the French populace, who felt alienated from the new industrialized forms of work and the bureaucratic institutions of civil society. This spurred him to focus on alienation and led him to the social criticism of Marx and Hegel. Although he published a number of groundbreaking translations of Marx, Hegel and texts on Nietzsche and on dialectical materialism in the 1930s, his career was disrupted by World War II. Because of this, his doctoral thesis focusing on rural sociology was not defended until the early 1950s.

He obtained a permanent University Professorship in Strasbourg in the mid-1950s, identifying with the political avant-garde and applying the critiques of an earlier generation of surrealists and communists to the counter-culture of the 1960s. He moved to the new university of Nanterre in suburban Paris, where he was an influential figure in the May 1968 student occupation of the Sorbonne and Left Bank. Nanterre provided an environment in which he developed his critique of alienation obscured by the mystifications of consumerism, suburban sprawl and the mythification of Paris by the heritage and tourism industries. These critiques of the city were the basis for Lefebvre's investigation of the cultural construction of stereotypical notions of cities, of nature and of regions. During his international travels from the early 1970s he developed one of the first theories of what came to be referred to as 'globalization'. Although retired, he continued to write and to lecture internationally. Until his death in mid-1991, he participated in the lively debates of the 'Groupe de Hagetmau' published in the magazine *M-Marxismes, Mensuels, Mensonges*.

Before discovering the work of Hegel and Marx, Lefebvre was influenced by Schopenhauer to develop a romantic humanism that glorified 'adventure', spontaneity and self-expression. Lefebvre was part of the group 'Philosophes' (including also Nisan, Friedman and Mandelbrot) who were loosely connected with André Gide, surrealists such as Breton and Dadaists such as Tristan Tzara. In turn, the Philosophes' rejection of metaphysical solutions in favour of action influenced Sartre and his circle (Short, 1966, 1979). Lefebvre's Marxist primer on the theory of *Dialectical Materialism* (1968b) made him internationally famous. He collaborated with Norbert Guterman to publish the first European translations of the work of the young Marx (1934) and one of the first introductions to Hegel (1938).

SPATIAL CONTRIBUTIONS

The core of Lefebvre's humanism is his critique of the alienating conditions of everyday life which he developed together with Guterman as a critique of the alienation and false consciousness of 1930s' popular and consumer culture (1936). In 1947 he published the first of what were to be three volumes of *Critique of Everyday Life* (1991a). In this, Lefebvre presented a Marxist materialist critique of 'Everydayness' (*quotidienneté, Alltäglichkeit* or 'banality') as a soul-destroying feature of modernity, social interaction and the material environment. Against 'mystification', against the banality of the *'metro-boulot-dodo'* ('subway-work-sleep') life of the suburban commuter, Lefebvre proposes that we seize and act on all 'Moments' of revelation, emotional clarity and self-presence as the basis for becoming more self-fulfilled (*l'homme totale*). This concept of 'Moments' reappears throughout his work as a theory of presence and the foundation of a practice of emancipation. Experiences of revelation, *déjà-vu* sensations, but especially love and committed struggle are examples of 'Moments'. By definition, 'Moments' have no

duration, but can be relived. Lefebvre argues that these cannot easily be reappropriated by consumer capitalism and commodified; they cannot be codified. **David Harvey** has taken up Lefebvre's thinking about urban social life in both its economic and symbolic dimensions. This work is imbued with a keen awareness of the temporality of urban life both in the sense of long-term accumulation in the cycles of finance, industry and infrastructure and in the shorter term of memory and meaning at the scale of individuals and communities.

Lefebvre's collaboration with the Situationniste International (SI) group lead by Guy Debord was crucial in directing his attention to urban environments as the contexts of everyday life and the expression of social relations of production. Lefebvre extended his critique of domestic life of the household to neighbourhoods and urban life. What is 'the urban', Lefebvre asked. The urban is not a certain population, a geographic size, or a collection of buildings. Nor is it a node, a transhipment point or a centre of production. It is all of these together, and thus any definition must search for the essential quality of all of these aspects. Lefebvre understands the urban from this phenomenological basis as a Hegelian *form*. The urban *is* social centrality, where the many elements and aspects of capitalism intersect in space despite often merely being part of the place for a short time, as is the case with goods or people in transit. This position can be contrasted with **Manuel Castells'** dichotomy of place and the space of flows. 'City-ness' is the simultaneous gathering and dispersing of goods, information and people. Some cities achieve this more fully than others – and hence our own perceptions of some as 'great cities' *per se*.

Thus, every person has a 'right to the city' – that is, to the city understood as the pre-eminent site of social interaction and exchange, which Lefebvre refers to as 'social centrality'. Lefebvre analysed the impact of changing social relations and economic factors under capitalism upon the quality of access and participation in the urban milieu. Echoing Debord, he argued that this interaction should not degenerate into commodified spectacles or into simply 'shopping' but should be the social form of self-presence in which individuals enjoy the right of association into collectives and self-determination.

The Production of Space (1991b) forms the keystone of the all-important 'second phase' of Lefebvre's analysis of the urban that began around 1972 (see Lefebvre, 1996; Kofman and Lebas, 1996). This later phase deals with social space itself as a national and 'planetary' expression of modes of production. As restated later in *De l'Etat* (Vol. 4, 1978), Lefebvre moved his analysis of 'space' from the old synchronic order of discourses 'on' space (archetypically, that of 'social space' as found in sociological texts on 'territoriality' and social ecology) to the manner in which understandings of geographical space, landscape and property are cultural and thereby have a history of change. Rather than discussing a particular theory of social space, he examined struggles over the meaning of space and considered how relations across territories were given cultural meaning. In the process, Lefebvre attempted to establish the importance of 'lived' grassroots experiences and understandings of geographical space as fundamental social. This is proposed as a critique of theories of space promulgated by disciplines such as planning or geography or the everyday attitude that took spatiality for granted. **Neil Smith** and other critics of homelessness and private property have drawn on the many sections of *Production of Space*, which was devoted to developing a radical phenomenology of space as the humanistic basis from which to launch a critique of the denial of individuals' and communities' 'rights to space'. In capitalist societies, for example, geographical space is 'spatialized' as lots – always owned by someone. Hence a privatized notion of space anchors the understanding of property which is a central feature of all capitalist societies.

Historical notions of space are analysed on three axes. These three aspects are explained in different ways by Lefebvre. Simplified for the purpose of introducing them, we might say that the 'perceived space' (*'le perçu'*) of everyday social life and commonsensical perception blends popular action and outlook but is often ignored in the professional, and theoretical 'conceived space' (*'le conçu'*) of cartographers, urban planners or property speculators. Nonetheless, the person who is fully human (*l'homme totale*) also dwells in a 'lived space' (*'le vecu'*) of the imagination which has been kept alive and accessible by the arts and literature. This 'third' space not only transcends but has the power to refigure the balance of popular 'perceived space' and official 'conceived space'. Gottdiener (1985) takes up Lefebvre's argument that this sphere offers lived space at its richest and most symbolic. Although suppressed in the abstract space of capitalist societies, it remains in art, literature and fantasy. Lefebvre cites Dada, the work of the surrealists, and particularly the works of René Magritte as examples challenging taken-for-granted understandings and practices of space. Also included in this aspect are clandestine and underground spatial practices that suggest and prompt revolutionary restructurings of institutionalized discourses of space and new modes of spatial praxis, such as that of squatters, illegal aliens and Third World slum dwellers, who fashion a spatial presence and practice outside of the norms of the prevailing (enforced) social spatialization.

The three axes or aspects of space are the elements of a 'triple dialectic' (*dialectique de triplicité* – the details of which Lefebvre does not fully sketch). The shifting balance between these forces defines what might be understood as a historical 'spatialization' (Shields, 1990, 1999). Lefebvre's multidimensional thesis is in direct contrast to the more customary reduction of space to part of one of production, exchange or accumulation (as in Castells, 1977). In addition to these,

[handwritten margin note: Herb's quote values – daily life rituals]

Lefebvre argues that space is a fourth and determining realm of social relations – one in which the production, exchange and accumulation of wealth and surplus value take place.

Ed Soja (1996) tentatively envisions this three-part dialectic as not 'an additive combination of its binary antecedents but rather ... a disordering, deconstruction, and tentative reconstitution of their presumed totalisation producing an open alternative that is both similar and strikingly different'. What he derives from Lefebvre's position as 'Thirding', 'decomposes the dialectic through an intrusive disruption that explicitly spatializes dialectical reasoning ... [it] produces what might best be called a cumulative *trialectics* that is radically open to additional othernesses, to a continued expansion of spatial knowledge' (Soja, 1996: 61).

In this analysis, Lefebvre broadened the concept of production to 'social production' (unaware of social constructivist theories that had been developed by non-Francophone writers such as Berger and Luckman or by Garfinkel). Contemporaneously with Poulanzas in the mid-1970s, he later refined his analysis with an assessment of the role of the state. This included his interest in the changing historical relations of capitalism and the globalization of socio-economic relations. This was also a turn to rhythm and to space–time (Lefebvre and Régulier, 1985).

A history of spatializations or 'modes of production of space' emerges that completes Marx's vision in urban, environmental and attitudinal terms. A true Communist revolution, Lefebvre suggested, must not only change the relationship of labourers to the means of production, but also create a new spatialization – shifting the balance away from the 'conceived space' of which private property, city lots and the surveyor's grid are artefacts. Embracing the 'lived space' of avant-gardes is a device for harnessing its potential and redirecting the 'perceived space' of everyday practice. This theory provides an early bridge from Marxist thought to the formative positions of the German Green Party, not to mention the punk counter-cultures of the 1970s and anti-globalization protests of the new millenium.

KEY ADVANCES AND CONTROVERSIES

After the failure of the student occupations of May 1968, Lefebvre's *oeuvre* was eclipsed by Louis Althusser's 'scientific Marxism' in which the base–superstructure division was a privileged element of a structural analysis of the repressive forces and institutions of capitalist states. Ironically, Lefebvre first became well known to English-speaking theorists through the critiques of his work by Althusserians, such as Manuel Castells, whom, in *The Urban Question*, criticized Lefebvre's urban works for their vagueness and anti-structuralist bias.

Lefebvre's patriarchal approach to the household, his gender-blindness and celebration of heterosexuality limit the usefulness of his theories for feminists and theorists of the body. He remained within the patriarchal tradition, dividing bodies and spaces heterosexually into male and female. These are conceived on the basis of a simple negation (A/not-A; that is, male/not-male) and Lefebvre, like most French theorists, was untouched by Commonwealth and American writers' theories of gay and lesbian identities as a 'third' alternative (A/not-A/neither) outside of a heterosexual dualism (Blum and Nast, 1996). Late twentieth-century postcolonial writers developed alternative theories of ethnic and race identity without reading Lefebvre (with some exceptions – see Gregory, 1994; Soja suggests links between **bell hooks** and Lefebvre).

Reliance on the dialectic has been surpassed by theories of alterity as complexity rather than contradiction or negation. Although he championed global underclasses of the landless poor, Lefebvre did not foresee the emerging politics of multiculturalism and ethnicity. Lefebvre has little to say on the question of discrimination, or on 'insiders and outsiders' and the ethics of their relationships. He tends to conceive of the state as a once-authentic instrument of a single people which has been seized by the capitalist class for itself.

Nonetheless, Lefebvre goes beyond twentieth-century philosophical debates on the nature of space which considered people and things merely 'in' space, to present a coherent theory of the development of systems of spatiality in different historical periods. These 'spatializations' are not just physical arrangements of things, but spatial patterns of social action and embodied routine, as well as historical conceptions of space and the world (such as a fear of falling off the edge of a flat world). These regimes operate at all scales. At the most personal, we think of ourselves in spatialized terms, imagining ourselves as an ego contained within an objectified body. People extend themselves – mentally and physically – out into space much as a spider extends its limbs in the form of a web. We become as much a part of these extensions, as they are of us. Arrangements of objects, work teams, landscapes and architecture are the concrete instances of this spatialization. Equally, ideas about regions, media images of cities and perceptions of 'good neighbourhoods' are other aspects of this space that are necessarily produced by each society as it makes its mark on the Earth.

What is the use of such an 'unpacking' of the production of the spatial? Lefebvre uses the changing types of historical space to explain why capitalistic accumulation did not occur earlier, even in those ancient economies that were commodity- and money-based, that were committed to reason and science, and that were based in cities (see Merrifield, 1993). One well-known explanation is that slavery stunted the development of age-labour. Lefebvre finds this unconvincing. No: it was a secular space, itself commodified as lots and private property, quantified by surveyors and stripped of the old local gods and spirits of place, that was a necessary precondition for the separation of people from the means to their own subsistence other than by work in return for wages.

As well as being a *product*, Lefebvre reminds us that space is a *medium*. Changes in the way we understand and live spatially provide clues to how our capitalist world of nation-states is giving way to a unanticipated geopolitics at all scales – a new sense of our relation to our own bodies, own world and the planets as a changing space of distance and difference.

LEFEBVRE'S MAJOR WORKS

Elden, S., Lebas, E. and Kofman, E. (trans. and eds) (2003) *Henri Lefebvre: Key Writings*. London: Continuum.

Lefebvre, H. (1968a) *Sociology of Marx*. Trans. N. Guterman. New York: Pantheon.

Lefebvre, H. (1968b) *Dialectical Materialism*. Trans. J. Sturrock. London: Cape (original edition 1939).

Lefebvre, H. (1969) *The Explosion: From Nanterre to the Summit*. Paris: Monthly Review Press (original edition 1968).

Lefebvre, H. (1991a) *The Critique of Everyday Life, Volume 1*. Trans. John Moore. London: Verso (original edition 1947).

Lefebvre, H. (1991b) *The Production of Space*. Trans. N. Donaldson-Smith. Oxford: Basil Blackwell (original edition 1974).

Lefebvre, H. (1996) *Writings on Cities*. Trans. and eds E. Kofman and E. Lebas. Oxford: Basil Blackwell.

Lefebvre, H. (2002) *Critique of Everyday Life II*. London: Verso (original edition 1961).

(NB: A complete index of Lefebvre's major works is available in Shields' *Lefebvre Love and Struggle* (1999) with annotations regarding reprints and editions collecting separate parts of previous publications).

Secondary Sources and References

Blum, V. and Nast, H. (1996) 'Where's the difference? The heterosexualization of alterity in Henri Lefebvre and Jacques Lacan', *Environment and Planning D: Society and Space* 14: 559–580.

Castells, M. (1977) *The Urban Question: a Marxist approach*. London: Edward Arnold.

Gottdiener, M. (1985) *Social Production of Urban Space*. Austin: University of Texas.

Gregory, D. (1994) *Geographical Imaginations*. Oxford: Blackwell.

Hess, R. (1988) *Henri Lefebvre et l'aventure du siècle*. Paris: Editions A. M. Métailié.

Home, S. (1988) *The Assault on Culture: Utopian Currents from Lettrisme to Class War*. London: Aporia Press and Unpopular Books.

Jameson, F. (1991) *Postmodernism or, the Cultural Logic of Late Capitalism*. London: Verso.

Kofman, E. and Lebas, E. (1996) 'Lost in transposition – time, space and the city', in E. Kofman and E. Lebas (eds) *Writings on Cities*. Oxford: Basil Blackwell, pp. 1–60.

Lefebvre, H. (1978) *Les contradictions de l'État Moderne*. Paris: UGE.

Lefebvre, H. and Régulier, C. (1985) 'The rhythmonalytical project', *Communications* 41: 141–149.

Marcus, G. (1989) *Lipstick Traces*. Cambridge, MA: Harvard.

Martins, M. (1983) 'The theory of social space in the work of Henri Lefebvre', in R. Forrest, J. Henderson and P. Williams (eds) *Urban Political Economy and Social Theory: Critical Essays in Urban Studies*. Gower, pp. 160–185.

Merrifield, A. (1993) 'Space and place: a Lefebvrian reconciliation', *Transactions of the Institute of British Geographers* 18: 516–531.

Poster, M. (1975) *Existential Marxism in Postwar France: From Sartre to Althusser*. Princeton, NJ: Princeton University Press.

Ross, K. (1988) *The Emergence of Social Space: Rimbaud and the Paris Commune*. New York: Macmillan.

Sartre, J.-P. (1958) *Being and Nothingness*. Trans. H. E. Barnes. New York: Methuen/Philosophical Library (originally published as *L'Etre et le Néant*, Gallimard, 1943).

Shields, R. (1990) *Places on the Margin: Alternate Geographies of Modernity*. London: Routledge.

Shields, R. (1999) *Lefebvre: Love and Struggle: Spatial Dialectics*. London: Routledge.

Shields, R. (2000) *Henri Lefebvre Webpage* Carleton University. http://www.carleton.ca/rshields/lefebvre.htm

Short, R. S. (1966) 'The politics of surrealism 1920–1936', *Journal of Contemporary History* 1: 3–26.

Short, R. S. (1979) 'Paris, Dada and surrealism', *Journal of European Studies* 9: 1–2 (March/June).

Smith, N. (1984) *Uneven Development: Nature, Capital and the Production of Space*. Oxford: Blackwell.

Soja, E. (1989) *Post-modern Geographies, the Reassertion of Space in Critical Social Theory*. London: Verso.

Soja, E. (1996) *Thirdspace: Journeys to Los Angeles and Other Real-And-Imagined Places*. Oxford: Blackwell.

Rob Shields

32 David Ley

An urban social geographer noted for his research on Canadian cities, David Ley completed his BA in Geography at Oxford University in 1968 and his PhD at Pennsylvania State University in 1972. These formative years coincided with significant controversy and change in Anglo-American geography, as the 'paradigm' of spatial science was rejected by a 'new generation' drawn to the alternative perspectives of social theory and humanistic philosophies. At the time when the positivist spatial theorist William Bunge (1969) was returning to research 'on the streets' in Detroit, and **David Harvey** (1973) was opening up the question of 'Social justice in the city' with a move towards Marxist social theory, David Ley was exploring a Philadelphia neighbourhood in empirical detail for his postgraduate research. Later published as *The Black Inner City as Frontier Outpost: Images and Behaviour of a Philadelphia Neighbourhood* (Ley, 1974), this innovative study was quickly to establish his reputation. In many ways this 'classic' study (see Jackson *et al.*, 1998) is not only indicative of the focus of his future career – urban social geography – but also illustrates his distinctive research strategy, one which draws upon diverse quantitative and qualitative sources, and critically combines and synthesizes different theoretical perspectives.

David Ley has spent much of his career in research and teaching at the University of British Columbia, Van-couver, and a significant amount of his work has focused on the study of urban transformation in Canada. His interest has included inner-city subcultures, housing and other community land-uses (e.g. Ley, 1974), the emergence of the 'new middle class' and related issues of gentrification (eg. Ley, 1987, 1996), immigration, ethnicity and globalization (e.g. Ley, 1995, 1999; Ley and Smith, 2000).

There are four particular characteristics of his career and contribution to human geography worthy of note. The first is his continued commitment to detailed empirical research combining critical development of theory conditioned by a sensitivity to local circumstance and its particular historical, geographical and political context (e.g. Ley 1996; Ley and Tutchener 2001). Second, Ley has continued to demonstrate an interest in issues-based research and a commitment to policy relevance. Third, he has employed an eclectic, critical and synthetic approach to theoretical perspectives, sources and methods, drawing on both qualitative and quantitative evidence: as he puts it, 'I am not persuaded by the view that "interpretation" and "measurement" are in some manner incompatible' (Ley, 1998: 79). Fourth, throughout his career he has demonstrated a strong commitment to collaboration, as manifest in a large number of co-authored articles and several co-edited collections of essays.

In broad theoretical terms, Ley has critically and cumulatively engaged with three key perspectives in human geography: humanistic perspectives, and in particular a phenomenological informed interest in the social (inter-subjective life world) (e.g. Ley 1977); a critique of Marxism and structuralism (e.g. Ley, 1982); and most recently poststructural and postmodern ideas (e.g. Ley, 1993). In *A Social Geography of the City* (1983), Ley provided a popular undergraduate text for the field, and illustrates the scope and critical synthesis of his conception of social (and cultural) urban geography – an interest in everyday life, the social basis of urban life, the city as home and as a human experience.

SPATIAL CONTRIBUTIONS

Ley's career has coincided with a period of considerable critical debate and pluralism in theoretical frameworks for explanation of urban social phenomena. By the 1970s, positivist models of urban structure with their basis in 'general laws' and reliance on quantitative measures, devoid of any real interest in the people who lived there or the specifics of locale, were largely rejected by a new generation of geographers who sought understanding of the 'real' problems of contemporary cities in all their variety and vibrancy. For some, critical social theory – and in particular that inspired by Marxism – was to provide the framework for a deeper level of analysis of the social geography of cities situated within an understanding of social structures (class), the flows of capital (and in particular globalization), and mechanisms of urban politics (or power). Other urban geographers took a more empirical turn, and drew upon earlier more humanistic traditions in the study of the social geography of cities. Here the unique historical and cultural experience of individual cities was seen as of critical importance, together with attention to how local communities perceived and behaved in these spaces – though local particularity was set within broader contexts and explanatory frameworks.

Ley adopted this more humanistic approach in his initial work on Philadelphia (Ley, 1974), drawing upon the diverse but not unrelated French tradition of *geographie humaine*, the Chicago School of urban ecology, behavioural science and environmental psychology and the emerging geographical interest in phenomenology. Rather than using a theoretical framework as a kind of lens through which to view the city, Ley started with direct experience of Philadelphia neighbourhoods and the experiences of its residents, 'triangulating' a range of sources and methods of analysis, drawing upon a complex of theoretical frameworks. He juxtaposed the personal and the reflective, with the more abstract and theoretical language of urban analysis. **Peter Jackson** (Jackson *et al.*, 1998: 75) notes that 'the reason for our enthusiasm when the book was first published is clear. It has a directness born of experience and spoke to readers with the voice of authority. The author had been there and was telling it like it is. Only later did we learn to be more sceptical about such claims to "ethnographical authority". Then as now, the book signalled a turning point in the development of social geography . . .'

Continuing throughout his career to engage critically with current theoretical frameworks, Ley has nevertheless maintained a consistent interest the humanistic perspective and the urban experience as a socio-cultural one. This is facilitated in large part by his continued commitment to combining both qualitative and quantitative methodologies, including the use of participant observation and in-depth interviews in combination with statistical records, official documents and even urban and regional novels. For Ley, space has history as well as a location and, above all, space has a range of meanings for the communities who live there. His own explanations therefore draw upon multiple factors – economic, political, social, personal, historical and cultural – to explain urban transformation, both generically and within specific cities and neighbourhoods.

Interestingly, when completing a review for *Progress in Human Geography* in

1985, he chose to title it 'Cultural/humanistic geography', perhaps in recognition of both relationships and also tensions between the emerging 'new' cultural geography and the earlier humanistic geography. This also reflects Ley's own theoretical development, as he has absorbed and developed in response to the progressive theoretical developments in the discipline, yet the continuing thread is perhaps a humanistic and liberal one. He has shown a scepticism of the efficacy of theory *per se*, but contributed to its critical development through a keen eye for the empirical, the actual geography as experienced by communities. It is in this sense that Hamnett (1998) describes Ley as essentially an exponent of 'grounded theory'.

The frequency of citations of his publications, the extensive range of his collaborations and his senior position in learned societies, suggest that Ley continues to have a significant impact upon the discipline, both within urban social geography and beyond. His contribution is first as an urban geographer who has conducted extensive empirical studies and drawn on a plurality of contemporary theories to develop a synthetic and distinctive explanation of the restructuring of cities in the last half of the twentieth century; and second, as a social and cultural geographer with an applied interest in the relevance of research to public policy, and respect for the complex interplay between socio-economic space, communities and cultures, local governance, and wider (national and international) economic and political processes. Third, though perhaps more fleetingly, Ley has made important contributions to the philosophical and theoretical development of geography. Among these contributions are his critique of positivism and the related promotion of a more phenomenological perspective in social (and cultural) geography (Ley, 1977), his critique of structuralist/Marxist approaches (Ley and Duncan, 1982), and debates about the 'postmodern city' (Ley, 1989; Duncan and Ley, 1993).

KEY ADVANCES AND CONTROVERSIES

There is a continuity and critical development in Ley's contribution from the early 1970s to the present in his interest in urban transformation as a social and cultural experience. He has made a distinctive and sometimes contested contribution to current debates. This is reflected in his work on the restructuring in contemporary Canadian cities, and in particular the process of gentrification and more recently the impact of immigration, through his role in the Canadian Metropolis Project (Ley, 1995; Ley and Tutchener, 2001). Ley has developed an explanation that brings together both economic and cultural factors, the global and the local. This synthesis is a significant achievement born of detailed empirical research and a critical stance on scope, contribution and potential integration of different theoretical perspectives.

For instance, in *The New Middle Class and the Remaking of the Central City* (Ley, 1996), he draws upon the theories of the transition to post-industrialization to develop a convincing explanation of the differential development of gentrification in contemporary cities. Although the study is based on detailed analysis of six Canadian cities, his explanation can be applied to develop understanding of urban change in other national contexts. Ley argues that gentrification is the result of changes in the local labour market, in particular the development of the financial and service sector. This sector is concentrated in the central areas of cities and has a disproportionate share of professional and managerial employees. His argument combines both local and cultural factors with wider economic and global forces. He notes in particular that cities that develop districts of gentrification are also those places that are most culturally vibrant and creative. He argues that hippies and artists are the 'storm troopers of the new middle class' (Ley, 1996: 194), moving into run-down inner-city districts, renovating and revitalizing them. This is followed by a transition in which the new middle classes – the professional and managerial groups – start to move in, perhaps initially frequenting the new cafes and cultural venues, and then in due course impacting on the local housing market.

Ley, in humanistic vein, thus describes the gentrifiers as responding to the 'structure of feeling', that is the emerging cultural identity of these changing inner-city districts. Gentrification therefore results from the coming together of both wider economic and political forces (the development of the financial and business services, which is itself dependent on innovation) and the development of local cultural identity and creativity grounded in the specifics of a given locality (now commodified and marketed to the incoming new middle-class residents). While some commentators – notably **Neil Smith** (1996) – remain sceptical of the importance that Ley places on the agency of the new middle class in promoting gentrification, his focus on the cultures of consumption implicated in the making of the postmodern city centre has proved widely influential in urban geography (see Hamnett, 1998).

In approaching the concept of the 'postmodern city', Ley avoids the excessive theorizing and linguistic gymnastics of 'postmodern geographers' (e.g. **Ed Soja**), and explores the postmodern through an empirical study of the cultural politics of the everyday urban environment. For instance, in exploring cooperative housing in Vancouver, Ley argues that 'post-modern landscapes, like others, also need to be seen as duplicitous, or better, ambivalent, not simply showcases for a new bourgeousie but also capable of supporting humane, indeed moral, public values' (Ley, 1993: 130). He subsequently goes on to refer to a 'postmodernism of resistance' and to the opportunity for 'sensitive urban place-making'. Here, he compares his own empirical analysis with the 'highly partisan discourse over architectural postmodernism, which frequently avoids careful definition of its concepts and engages in a façadism offering the thinnest of landscape interpretations' (Ley, 1993: 145). In conclusion, he thus argues for an 'oppositional post-modernism in the built environment in terms of an ontology of difference, a multi-vocal epistemology and a politics of participation'. Justifying this stance on postmodernism, Ley argues that to apply the term merely as a generic descriptor of contemporary culture and urbanism is to fall into the very trap of formulating it as a totalizing and uncontested entity. Rather, he argues that if we see landscape as a process, we recognize the contingent geographical and historical contexts that enable particular urban landscape forms to emerge (cf. Harvey, 1989).

LEY'S MAJOR WORKS

Duncan, J. and Ley, D. (eds) (1993) *Place/Culture/Representation*. London: Routledge.

Ley, D. (1974) *The Black Inner City as Frontier Outpost: Images and Behaviour of a Philadelphia Neighbourhood*. Washington DC: Association of American Geographers.

Ley, D. (1983) *A Social Geography of the City*. New York: Harper and Row.

Ley, D. (1996) *The New Middle Class and the Remaking of the Central City*. Oxford: Oxford University Press.

Ley, D. (1999) 'Myths and meanings of immigration and the metropolis', *Canadian Geographer* 43: 2–19.

Ley, D. and Duncan, J. (1982) 'Structural Marxism and human geography: A critical assessment', *Annals of the Association of American Geographers* 72: 30–59

Ley, D. and Hasson, S. (1994) *Neighbourhood Organisation and the Welfare State*. Toronto: Toronto University Press.

Ley, D. and Samuels, M. (eds) (1978) *Humanistic Geography: Problems and Prospects*. London: Croom Helm.

Secondary Sources and References

Bunge, W. (1969) 'The first years of the Detroit Geographical Expedition: A personal report'. Field Notes 1: 1–9. Reprinted in R. Peet (ed.) *Radical Geography* (1977). London: Methuen, pp. 31–39.

Hamnett, C. (1998) 'The new urban frontier: Gentrification and the revanchist city (Neil Smith), The new middle class and the remaking of the central city (David Ley), Gentrification and the middle class (Tim Butler): Book Review Essay', *Transactions of the Institute of British Geography* 23: 412–416.

Harvey, D. (1973) *Social Justice and the City*. Oxford: Blackwell.

Harvey, D. (1989) *The Condition of Postmodernity*. Oxford: Blackwells.

Jackson, P., Palm, R. and Ley, D. (1998) 'Classics in human geography re-visited: David Ley (1974) The Black Inner City as Frontier Outpost – Commentaries 1 and 2, and Response from David Ley', *Progress in Human Geography* 22: 75–80.

Ley, D. (1977) 'Social geography and the taken for granted world', *Transactions of the Institute of British Geographers* 2: 498–512.

Ley, D. (1982) 'Discovering man's place', *Transactions of the Institute of British Geographers* 7: 248–253

Ley, D. (1985) 'Cultural/humanistic geography', *Progress in Human Geography* 9: 415–423.

Ley, D. (1987) 'Styles of the times: Liberal and neo-conservative landscapes of inner Vancouver 1968–1996', *Journal of Historical Geography* 113: 40–56.

Ley, D. (1989) 'Modernism, postmodernism and the struggles for place', in J. Agnew and J. Duncan (ed.) *The Power of Place*. London: Unwin Hyman pp. 44–55.

Ley, D. (1993) 'Co-operative housing as a moral landscape: Re-examining "the post-modern city" ', in D. Ley and L. Bourne (eds) *The Changing Social Geography of Canadian Cities*. Montreal: McGill University Press, pp. 128–148.

Ley, D. (1995) 'Between Europe and Asia: The case of the missing sequoias', *Ecumene* 2: 187–210.

Ley, D. and Olds, K. (1988) 'Landscape as spectacle: world fairs and the culture of heroic consumption', *Environment and Planning D: Society and Space* 6: 191–212.

Ley, D. and Smith, H. (2000) 'Relations between deprivation and immigrant groups in large Canadian cities', *Urban Studies* 37: 37–62.

Ley, D. and Tutchener, J. (2001) 'Immigration, globalisation and house prices in Canada's gateway cities', *Housing Studies* 16: 199–223.

Smith, N. (1996) *New Urban Frontier: Gentrification and the Revanchist City*. New York: Routledge.

Paul Rodaway

33 Doreen Massey

BIOGRAPHICAL DETAILS AND THEORETICAL CONTEXT

Born in Manchester, England, in 1944, Doreen Massey is one of the most influential geographers writing today. She gained her MA in Regional Science from the University of Pennsylvania in 1972, and from 1968 to 1980 worked at the Centre for Environmental Studies in London (a research institute, now closed, that was established by former prime minister Harold Wilson to focus on urban and regional questions). She followed this with a two-year stint as a Social Science Research Council (SSRC) Industrial Location Research Fellow. Massey has been a Professor of Geography in the Faculty of Social Sciences at the Open University (United Kingdom) since 1982, and in 2002 was made a Fellow of the British Academy.

Massey's numerous publications stretching back to the early 1970s have greatly affected the directions that human geography and those disciplines close to it (particularly sociology and cultural studies) have taken. To put it simply, Massey's work has been central in transforming human geography into a disciplinary domain dedicated to the project of social theory, while encouraging the social sciences to take on board the complexity of space within their formulations. The significance of Massey's work arises, in other words, not simply from its content and methodology, but, more fundamentally, from its insistence on the importance of *conceptualizing* space and place. Massey is adamant that *how* one formulates an object of study is crucial for the theoretical and empirical claims one makes about that object of study. *How*, in other words, one formulates the concept of space or place radically shapes one's understanding of the social world and how to effect transformation in and of it. In the course of three decades of writing, Massey's reconceptualizations of a suite of key terms – space, place, region, locality – have helped revolutionize geographical thinking within the social sciences as a whole. Massey, in working through the implications of the rallying call 'geography matters', has produced a powerful and nuanced set of theories, frameworks and empirically based studies with which to understand spatial differentiation, uneven development, and historical and geographical change.

Massey shot to academic prominence by virtue of several path-breaking articles that displaced the dominant orthodoxies framing discussions of industrial location and of British regional 'problems' in the 1970s. She showed how aspatial, neoclassical accounts of industrial location were fundamentally disrupted when the spatial dimension was addressed (Massey, 1973); she also launched a powerful attack on the reliance within urban and regional studies on statistical techniques and hence on the field's implicit endorsement of a naïve empiricism. Asking '[i]n what sense . . . are "regional" problems *regional* problems' (Massey, 1979: 241), Massey disputed the usual logic whereby inner cities and more peripheral regions were somehow themselves responsible for their declining fates. Instead she suggested that '[d]ifferent modes of response by industry, implying different spatial divisions of labour within its overall process of production, may ... generate different forms of "regional problem" ' (1979: 234). Massey demonstrated that since the development of these new divisions of labour 'will be overlaid on, and combined with, the pattern produced in previous periods by different forms of spatial division', this will result in a series of geographically differentiated economic 'layers' that will reshape patterns of inequality and affect ensuing rounds of investment (Massey, 1978: 115–116).

In 1984, *Spatial Divisions of Labour* was published. The book transformed economic geography, and would become Massey's most widely cited monograph. It aimed to reconceptualize how the very sphere of 'the economic' was understood, and was an extension and refinement of Massey's earlier arguments. Massey argued that economic geography had to address the spatial organization of the relations of production rather than simply describe and render visible geographical distributions. And addressing spatial organization entailed paying adequate attention to the constitution of particular, regionally differentiated places rather than focusing exclusively on the general tendencies of capitalist accumulation:

> ... behind major shifts between dominant spatial divisions of labour within a country lie changes in the spatial organization of capitalist relations of production, the development and reorganization of what we shall call spatial structures of production. Such shifts in spatial structures are a response to changes in class relations, economic and political, national and international.
>
> (Massey, 1984: 7)

Massey developed her general claims about 'layers of investment' and 'spatial structures of production' by setting out three examples of spatial structures that she put to work in an analysis of how the British economic landscape was transformed in the 1960s and 1970s: (1) *single region* (where the whole process of production is confined within a single geographical area); (2) *cloning* (one branch is

the headquarters, but the production process itself is geographically undifferentiated such that the whole process takes place at each branch); (3) *part-process* (multi-locational, with a managerial hierarchy and a complex production process that is spatially stretched out across different plants). Of critical importance here was Massey's insistence that spatial structures of production could not be determined *a priori*, but rather that they were solidified, combated and transformed by various political and economic strategies and actors working at differing scales. Important too was Massey's acknowledgement of how ideologies of gender and race complicated her analysis of the labour process. For although her analysis worked around the axis of class, she made clear how changes in spatial divisions of labour were both affected by, and in part brought about by, social definitions of skill categorizing certain jobs as characteristic of certain types of people. But of perhaps greatest importance, and hence subject to greatest debate, was Massey's conviction that the existence of spatial variety must not be seen as a deviation from inexorable laws of capitalist accumulation. '[W]hat lies behind the whole notion of uneven development is the fact of highly differentiated and unique outcomes', Massey insisted (1984: 49): the book argued for and offered a rejuvenated and radically transformed regional geography.

It is invidious to represent a smooth lineage linking an author's early and later work. Nonetheless, it is perhaps helpful to see Massey's work after *Spatial Divisions of Labour* as extending some of the central claims of that book concerning specificity, transformation and spatial connectedness. Massey's research since the mid-1980s might, then, be collected under three broad headings. The first relates to gender in economic and social processes. Massey has been keen to analyse how the construction of gender relations is central to the spatial organization of social relations (Massey, 1994). Her writings in this area include further elaborations of how industries make strategic use of regional differences in systems of gender relations; relational accounts of identity formation; and analyses of how conceptualizations of time and space have often been problematically mapped onto the dualism masculinity–femininity. Her accounts of the high-technology industry around Cambridge (United Kingdom) have argued, in this regard, that the masculinization of the work of scientists and engineers is buttressed by the temporal and spatial flexibility of that work – a flexibility in which the masculinized workplace trumps the feminized home, and 'transcendent' mental labour trumps domestic labour and the social sphere (Massey, 1995a). The second important trajectory relates to theorizations of 'place'. Massey has produced a rich body of writings that refuses the easy association of place with nostalgia, inertia and, by implication, regressive politics. These writings include further elaborations of the concept of the region (Allen et al., 1998), and interventions into the debate over 'time–space compression'. All these writings show how places might be understood as 'porous networks of social relations' (Massey, 1994: 121), and Massey has developed the term 'power geometry' to emphasize how groups and individuals are differently positioned within these porous networks. A third key trope of Massey's work is the concept 'space–time'. Massey has demonstrated how the cemented divide between time and space is problematic in its flawed association of change with the temporal, and stasis with the spatial. She has therefore developed an alternative view of space – as 'space–time' – in which space and time are seen as inseparable (Massey, 1992). The term aims to reinforce her conviction that 'the spatial is integral to the production of history, and thus to the possibility of politics, just as the temporal is to geography' (1994: 269).

Massey is passionate about communicating across disciplinary divides and beyond the academy. She was, for example, a regular contributor to Marxism Today (a key locus for British New Left thought in the 1980s), and has long taken an active role in policy discussions concerning British urban and regional questions (for example, in relation to the regeneration of the London Docklands). In 1995, she, together with **Stuart Hall** and Michael Rustin, founded Soundings: a Journal of Politics and Culture, a journal dedicated to thinking the 'radical democratic project'. That a significant number of Massey's publications have either been co-authored (Richard Meegan and John Allen are particularly important in this regard), or have emerged from joint research projects, further demonstrates Massey's commitment to intellectual collaboration and exchange.

Many of Massey's writings have become standard readings for new generations of geographers and other social scientists. One should note here the institutional role of Massey's academic home, the Open University: the distance teaching that that university conducts is dependent on its faculty publishing co-authored course books, and these have provided an important conduit for the dissemination of Massey's ideas.

SPATIAL CONTRIBUTIONS

Massey (1995b: 317) has stated that Spatial Divisions of Labour was committed to 'reinterpreting "objects in space" as products of the spatial organization of relations'. This phrase elegantly captures what is characteristic of Massey's oeuvre as a whole: its dedication to understanding 'things' relationally – whether those 'things' be places, identities or socio-spatial formations such as the nation-state. Massey's most fundamental contribution to thinking space and place is arguably her conviction that the social and the spatial need to be conceptualized together.

But this does not imply the mechanical insertion of 'space' as a motivating or explanatory factor: '[i]t is not spatial form in itself (nor distance, nor movement) that has effects, but the spatial form of particular and specified social processes and social relationships' (Massey and Allen, 1984: 5).

Massey's intricate analyses both of the 'spatial' and, later, of 'space–time' have been inspirational for countless geographers. Massey can therefore be characterized as part of a cohort of geographers – that might include **David Harvey**, **Derek Gregory**, **Nigel Thrift** and **Gillian Rose** – that have refused to imagine space and time as neutral, *a priori* categories and instead have developed rich accounts of space and time as sutured within and productive of the formations and deformations making up the 'natural' and the 'social' world.

Many of the concepts and frameworks elaborated in *Spatial Divisions of Labour* are now central to the fields of economic, industrial and labour geography. The book broke with the models of economic change provided both by empiricist neoclassical economics and by overly rigid structuralist Marxist accounts, and presented an elastic and powerful analytic framework that was taken up by numerous academics and regional specialists. The book was undoubtedly a catalyst for an intensified cross-disciplinary dialogue between urban/social/economic geographers and sociologists (see **Andrew Sayer**) *vis-à-vis* how spatial structure might be understood as a structuring medium through which social relations unfold (see, for example, Gregory and Urry, 1985). The theoretical terrain of *Spatial Divisions of Labour* prefigured what would in the late 1980s and early 1990s become key discussions within geography and social theory: the role of local uniqueness, the usefulness of the terms 'flexibility' and post-Fordism, and critiques of universal and universalizing explanations and theories.

Massey's neologisms 'power geometry' and 'space–time' have also been

enormously influential within human geography – both in their theoretical impact, and for their ability to provide structuring frameworks for those keen to understand how fights over space and place might be understood as fights about spatialized power. Massey herself has recently directed her interrogations of spatiality towards the globalization debates to mount a powerful critique of the dominant, commonsense, *'aspatial* view of globalization' (1999: 34). In arguing that such an account turns 'real spatial difference into the homogeny of temporal sequence (we'll all be globalized in this way eventually)', she shows how such modes of thinking render impossible the thinking of 'difference' (1999: 40).

KEY ADVANCES AND CONTROVERSIES

Massey's ideas have, on occasion, been both subject to, and developed out of, vigorous contestation. The frequency and willingness with which Massey has made specific responses to her interlocutors' concerns and criticisms indicates her belief in the productivity of exchange. Alongside the huge excitement generated by the publication of *Spatial Divisions of Labour* came several refutations and criticisms of Massey's framework. Cochrane (1987) argued that her desire to create a 'new regional geography' was hampered by the book's tendency towards a 'microstructuralism' that ended up reneging on the project of tackling capitalism at national and international levels. **David Harvey** also disagreed with Massey's reworking of Marxism – excoriating the book's 'rhetoric of contingency, place, and the specificity of history' which, he claimed, reduced 'the whole guiding thread of Marxian argument . . . to a set of echoes and reverberations of inert Marxian categories' (Harvey, 1987: 373).

Others accused Massey of overemphasizing the national framework of industrial transformation at the expense of the global. Massey's deliberate decision to focus her analysis on the workplace (and the book's concomitant underemphasis on the reproductive realm) came in for criticism by several feminist geographers. (**Gillian Rose**, 1993, helpfully shows how Massey's framework acted as a catalyst for further feminist-geographical research on how home and community, as well as the workplace, are key sites for the consolidation and contestation of capitalist and patriarchal social relations.)

Spatial Divisions of Labour also became enmeshed in the heated 'locality debate' of the 1980s. The British Economic and Social Research Council (ESRC) set up three large 'locality' research programmes in the mid-1980s to explore the wide-ranging social and economic changes affecting Britain in the 1970s and 1980s by focusing attention on various small-scale localities. Of these, the Changing Urban and Regional System (CURS) programme received by far the most academic attention. Massey herself had initiated the original proposal and drawn up the research outline, though she was not directly involved in the undertaking of research herself. The CURS programme was seen by many as an abdication of 'theory' (usually meaning Marxist theories of capital accumulation) in favour of empiricism. Massey's book was often seen as a prime mover in this abdication, and participants in the debate frequently failed to distinguish 'locality' as a concept used by Massey from its use as a descriptor of the research programmes. The locality debate circled around the problematic of 'contingency' and how it should be understood in relation to the concepts of specificity, the local, the general and the abstract. The debate was also very much a political and institutional one: what kinds of research projects should geographers be developing and what theoretical frameworks should they use to guide them? Massey, reflecting on the locality studies debate, has argued

that the couplet 'specific–general' was frequently and mistakenly equated to that of 'concrete–abstract', such that it was generally assumed that only localities are 'concrete'. She has, in contrast, argued that 'the current world economy . . . is no less concrete than a local one' (Massey, 1991c: 270). Arguing that the study of local areas need not entail a return to individual, descriptive portraits of geographical regions, she has averred that studying localities does not necessitate fetishizing the local. Massey has reflected at length on the locality debate, and other criticisms relating to *Spatial Divisions of Labour*, in the second edition of that book (1995b).

Massey's long-held interest in understanding regional particularity received new impetus by virtue of the debates over 'time–space compression' and postmodernism following the publication of **David Harvey**'s *The Condition of Postmodernity* (1989). Massey profoundly disagreed with several of the overarching formulations in that book, and her famous essay 'Flexible sexism' (1991b) was an extended critique of both Harvey's and **Ed Soja**'s conceptualizations of postmodernity and the postmodern. Massey argued that while neither author 'would want to be thought of as anti-feminist', both books, she claimed, 'are in fact quite fundamentally so' (1991b: 32). Massey was particularly exercised by what she saw as Harvey's relegation of feminism to the position of a 'local' struggle in comparison with the 'general' struggle of class. She cautioned that Harvey, in desiring to construct unity by subsuming all struggles within the umbrella of class politics, was urging 'a unity enforced through the tutelage of one group over others' (1991b: 55).

Massey's difficulties with Harvey's account of 'time–space compression' provided the driving force for another of her famous essays 'A global sense of place' (Massey, 1991a). Massey countered the claim that thinking in terms of place was necessarily reactionary – a claim implied by Harvey – and argued that a sense of place 'adequate to this era of time–space compression' demanded developing an account that was not wedded to the lure of introversion, but that understood a place's specificity as 'constructed out of a particular constellation of social relations, meeting and weaving together at a particular locus' (1991a: 28). Massey's use of the term 'power-geometry' to understand this 'meeting and weaving' in terms of differential mobilities, and her famous call for 'a global sense of the local, a global sense of place' (1991a: 29) were brought to life in a vignette of her own London neighbourhood, Kilburn. Massey's several interventions concerning 'place' and the 'local' register the continuing and unresolved debates within geography over the relations between the 'general' and the 'particular' and the terms space and place – debates that are, it should be noted, far more wide-ranging than the specific differences separating Massey and Harvey.

Massey's emphasis on mobility, openness, flow and differential power relations has contributed to those terms becoming guiding principles in human geography. It is worth considering whether the ready assimilation of a 'global sense of place' within geography is in the process of establishing an orthodoxy of the contingent and the open. What might an (over-) emphasis on the contingent and open occlude, and are particular spaces and places valorized at the expense of others? McGuinness, for example, has interrogated Massey's reliance on a particular representation of 'difference' in 'A global sense of place', suggesting that her vignette of the Kilburn High Road 'could easily be seen as a very particular white Western construction of a world of difference' (McGuinness, 2000: 228). Precisely because 'difference' – in the sense of visibly marked difference – is so easy to notice in a multi-ethnic area such as Kilburn, McGuinness wonders whether 'for the normalizing white eye' hybridity becomes fundamentally associated with blackness. Massey herself would doubtless agree with McGuinness that the concept of hybridity must not function to sideline consideration of postcolonial 'white' identity formations, since much of her own oeuvre has been dedicated to unsettling, rather than reinforcing, 'normalizing' frames of geographical thinking.

MASSEY'S MAJOR WORKS

Allen J., Massey, D. and Cochrane, A. (1998) *Rethinking the Region*. London and New York: Routledge.

Massey, D. (1973) 'Towards a critique of industrial location theory', *Antipode* 5: 33–39.

Massey, D. (1984) *Spatial Divisions of Labour: Social Structures and the Geography of Production*. London and Basingstoke: Macmillan.

Massey, D. (1991a) 'A global sense of place', *Marxism Today* June: 24–29.

Massey, D. (1991b) 'Flexible sexism', *Environment and Planning D: Society and Space* 9: 31–57.

Massey, D. (1994) *Space, Place, and Gender*. Cambridge: Polity Press.

Massey, D. (1995a) 'Masculinity, dualisms and high technology', *Transactions of the Institute of British Geographers* 20: 487–499.

Massey, D. (1995b) *Spatial Divisions of Labour: Social Structures and the Geography of Production*, 2nd edition. Basingstoke: Macmillan.

Massey D. (1999) 'Imagining globalization: power-geometries of time–space', in A. Brah, M. J. Hickman and M. Mac an Ghaill (eds) *Global Futures: Migration, Environment and Globalization*. Basingstoke: Macmillan, pp. 27–44.

Massey D., Quintas, P. and Wield, D. (1992) *High-tech Fantasies: Science Parks in Society, Science and Space*. London: Routledge.

Secondary Sources and References

Cochrane, A. (1987) 'What a difference the place makes: the new structuralism of locality', *Antipode* 19: 354–363.

Environment and Planning A (1989) Special issue – '*Spatial Divisions of Labour* in practice', 21: 655–700.

Environment and Planning A (1991) Special issue – 'New perspectives on the locality debate', 23 (2): 196–266.

Gregory, D. and Urry, J. (eds) (1985) *Social Relations and Spatial Structures*. Basingstoke: Macmillan.

Harvey, D. (1987) 'Three myths in search of a reality in urban studies', *Environment and Planning D: Society and Space* 5: 367–376.

Harvey, D. (1989) *The Condition of Postmodernity*. Oxford: Blackwell.

Martin, R., Markusen, A. and Massey, D. (1993) 'Classics in human geography revisited: Massey D (1984) *Spatial Divisions of Labour*', *Progress in Human Geography* 17: 69–72.

Massey, D. (1978) 'Regionalism: some current issues', *Capital and Class* 6: 106–125.

Massey, D. (1979) 'In what sense a regional problem?', *Regional Studies* 13: 233–243.

Massey D (1991c) 'The political place of locality studies', *Environment and Planning A* 23: 267–281.

Massey, D. (1992) 'Politics and space/time', *New Left Review* 196: 65–84.

Massey, D. and Allen, J. (eds) (1984) *Geography Matters! A Reader*. Cambridge: Cambridge University Press in association with the Open University.

Massey, D. and Allen, J. (eds) (1988) *The Economy in Question*. London: Sage.

McGuinness, M. (2000) 'Geography matters? Whiteness and contemporary geography', *Area* 32: 225–230.

Rose, G. (1993) *Feminism and Geography*. Minneapolis: University of Minnesota Press.

Felicity Callard

34 Gearóid Ó Tuathail (Gerard Toal)

An innovative and prolific political geographer, Gearóid Ó Tuathail (Gerard Toal) was born in the Republic of Ireland in 1962. Growing up in County Monaghan on the border with Northern Ireland was among the influences that pushed Ó Tuathail to study political geography. He graduated with a joint BA in History and Geography in 1982 from St Patrick's College in Maynooth (now the National University of Ireland). Ó Tuathail completed his Master's degree at the University of Illinois in Champaign-Urbana under the direction of John O'Loughlin, and moved to Syracuse University where, supervised by John Agnew and political scientist David Sylvan, he completed his PhD entitled *Critical Geopolitics: The Social Construction of Place and Space in the Practice of Statecraft* (Ó Tuathail, 1989). Ó Tuathail has taught at the University of Southern California, University of Minnesota, Virginia Polytechnic Institute and State University and England's University of Liverpool.

His first publication argued for a 'new geopolitics' that was 'much more critical' than traditional evaluations of national interest and policy recommendation (Ó Tuathail, 1986: 73). Defining geopolitics as political discourse structured 'by either explicit reference to geographical location and concepts or by use of certain implicitly geographical policy rationalizations (e.g. *Lebensraum*, domino theory, containment, expansionism)', Ó Tuathail (1986: 73–74) examined US–El Salvador relations

from the 1823 Monroe Doctrine to the 1980s' Reagan administration. 'American foreign policy', Ó Tuathail (1986: 83) concluded, 'aims to perpetuate, secure and reaffirm the American way of life. Part of insuring the survival and prosperity of large industrial states such as America involves dominating, controlling and influencing.' His subsequent article assessing US foreign policy, co-authored with John Agnew (1992), 'precipitated a research agenda, which conceptualized geopolitics as a form of political discourse rather than simply a descriptive term intended to cover the study of foreign policy and grand statecraft' (Dodds, 2001: 469).

In their paper, Ó Tuathail and Agnew (1992: 192) argued that geopolitics must be studied as 'a discursive practice by which intellectuals of statecraft "spatialize" international politics in such a way as to represent it as a "world" characterized by particular types of places, peoples and dramas'. Focusing on international relations and foreign policymaking, Ó Tuathail and Agnew (1992: 194) maintained that the speeches and writings of politicians, diplomats, policy advisors and the media comprise 'geopolitical reasoning'. These statements can be analysed, not to see whether they are truthful, but rather to critically examine the effects that using certain terms and language have on the practice and impact of international relations. Painter (1995: 146) thus argues that the research agenda precipitated by Ó Tuathail's work is 'concerned particularly with the "texts" of international politics', what they mean and how they are used, rather than political events in themselves.

Evaluating geopolitics stimulated a reshaping of political geography in the 1990s, and contrasted with examinations of the geographical facts of politics and state relations. Ó Tuathail and Agnew (1992) argued there was a need to assess how 'geopolitical reasoning' constructs representations of states, territories and political regimes through discourse and how people utilize these discursive understandings to explain events, envision international relations and justify foreign policy actions. This research agenda, therefore, was a departure from existing studies within political geography concerned with state formation, contested national borders and territories, nationalism and secession, voting patterns, geographical impacts of wars and concepts such as world-systems theory, state theory and sovereignty (Painter, 1995; Atkinson and Dodds, 2000; Dodds, 2001).

SPATIAL CONTRIBUTIONS

Ó Tuathail's key contribution to debates on space and place has been his espousal of a critical theory of geopolitics. Traditionally, geopolitics is how state analysts, military or other, interpret the territorial operation of state power and visualize spatial control. In contrast, Ó Tuathail argues for a *critical* geopolitics that recognizes and exposes geopolitical assertions and makes 'informed critiques of the spatializing practices of power' (Ó Tuathail and Dalby, 1994: 513). Influenced by the end of the Cold War and postmodern, poststructuralist, feminist and psychoanalytic theories, critical geopolitics problematizes political discourses, examines their spatial assumptions, questions power relations and challenges the role of the state and how its institutional analysts envision the world. Ó Tuathail draws on **Foucault**'s understanding of governmentality to argue that the articulation of 'geo-power' over both people and territory is a critical function of modern

statehood. '[M]y concern', states Ó Tuathail (1996a: 11), 'is the power struggle between different societies over the right to speak sovereignly about geography, space and territory.' Utilizing Derrida to assess and deconstruct political discourse, Ó Tuathail (1996a: 66–67) proposes that geo-graphy and geo-politics can be hyphenated to emphasize the process of discourse in writing or 'scripting ... global space by state-society intellectuals and institutions'.

Drawing on such diverse theoretical traditions and conversant in contemporary international relations theories, Ó Tuathail productively integrated these approaches to generate analyses that interrogated contemporary international political discourse and stressed the importance to statecraft of geographical representations. Indeed, this is one of Ó Tuathail's most significant geographical contributions. Critical geopolitics made issues of space and political geography pertinent to the discipline of international relations and its practitioners, introducing geographical analyses to intellectual debates and scholars that had largely ignored these perspectives. Critical geopolitics, therefore, is interdisciplinary and Ó Tuathail has been at the forefront of developing this field of study, editing books and special issues of major journals on the topic (e.g. Ó Tuathail and Dalby, 1994, 1998a; Dalby and Ó Tuathail, 1996; Herod et al., 1998; Ó Tuathail et al., 1998).

In his book *Critical Geopolitics*, Ó Tuathail (1996a) deconstructs the canon of late nineteenth- and early twentieth-century geopolitical texts by Rudolf Kjellen, Friedrich Ratzel, Karl Haushofer and Halford Mackinder, exposing their assumptions of political power and paying attention to their constructions of space, race and gender. Traditional geopolitics was a science for men who sought to know and control territory and Ó Tuathail (1996a: 82) contends that Mackinder envisioned British East Africa (Kenya) 'as a feminized space to be penetrated, a territory breached by others but not yet conquered'. Further, Ó Tuathail (1996a: 111–

129) examines both the fact and fictionalization of Haushofer and Nazi geopolitics in US magazines *Life* and *Reader's Digest* and the movie *Plan for Destruction* (1943). These examples are contrasted with other geopolitical texts produced at the time, such as those by US foreign policy analyst Robert Strausz-Hupé, an Austrian émigré to the USA. *Critical Geopolitics* also includes studies of the writings by conservative (post-) Cold War US intellectuals Samuel Huntington and Edward Luttwak (Ó Tuathail, 1996a).

The range of topics across Ó Tuathail's numerous publications suggests that many can find material to resonate with their own interests. Alongside re-evaluations of major figures within the geopolitical canon (e.g. 1992, 1994, 2000a), Ó Tuathail typically examines contemporary and pertinent issues. These include case studies of the intersection of politics and control over territory, such as the 1991 Gulf War (e.g. 1997) and the Balkan Wars of the early 1990s (e.g. 1996a, 1996b, 1999, 2002a). One focus, centred on reports by Maggie O'Kane in the British daily newspaper *The Guardian*, argues that O'Kane disturbed standard Western geopolitical narratives about the Balkan Wars. The result, Ó Tuathail (1996b: 182) suggests, was 'anti-geopolitics' which scripted Bosnia as a place of horrors where the West must intervene, but implicitly maintained that Bosnia remains a place that is beyond the Western political sphere. Elsewhere, Ó Tuathail (e.g. 1996a: 202; 1999, 2002a) demonstrates the USA and Europe's multiple geopolitical perceptions regarding Bosnia wavering between disregard and proclamation that combatants were exhibiting the 'primitivism of "blood feud" cultures'. Ó Tuathail's examinations of Bosnia in US policy discourse are widely applauded, Smith (2000: 365) claiming that this 'represents the most fertile and adventurous critique of a geo-political tradition'.

Ó Tuathail has also developed a number of concepts for critically analysing geopolitical reasoning, dividing geopolitical discourse into 'popular geopolitics' –

evident in the mass media, movies and popular culture; 'practical geopolitics' – apparent in foreign policy and state bureaucracy; and 'formal geopolitics' – produced in think-tanks and academic venues (Ó Tuathail and Dalby, 1998b). Diagrammatically outlining how this tripartite division intertwines to 'comprise the geopolitical culture of a particular region, state or inter-state alliance', to produce a 'spatializing of boundaries and dangers' and to construct 'geopolitical representations of self and other', Ó Tuathail and Dalby (1998b: 5) maintain that geopolitics are socio-cultural phenomena evident in everyday life.

Geopolitical representations are produced and consumed in myriad ways, from tabloid newspaper headlines to presidential speeches advocating military action. Examining the 'geopolitical condition' of contemporary international politics, Ó Tuathail (2000b, 2002b) argues that processes like globalization, telecommunications and the 'world risk society' are challenging extant ways of thinking about state borders, territory, power, defence and security. With world leaders lauding the possibilities of the internet, biotechnology and telecommunications for capital, industry and science, many are simultaneously worried that these advances could get into the 'wrong hands' and generate new threats to state security. Ó Tuathail (e.g. 2002b) expands critical geopolitics beyond discourse analysis, to address the 'geopolitical world order' of state alliances, global relations of production, consumption and the spatial processes of trade or 'geopolitical economy' and world 'techno-territorial complexes' that, through scientific advances, the acceleration of transportation and communications and their utilizations, reshape power relations within, between and beyond states.

KEY ADVANCES AND CONTROVERSIES

In the name of heterogeneity and flexibility, Ó Tuathail frequently avoids defining 'critical geopolitics', 'geopolitics', 'territory', 'space' and 'sovereignty'. Some critics question such definitional malleability, claiming that this, coupled with the diverse philosophical sources drawn upon by Ó Tuathail, produce 'theoretically inconstant' assessments (Stephanson, 2000: 381). Others have commended Ó Tuathail, *Critical Geopolitics*, for example, being very well received. Heffernan (2000: 347) comments that the book is '[i]maginative, intellectually ambitious ... engaging [and] outstanding' and Sharp (2000: 361) claims *Critical Geopolitics* to be 'vital'.

Critical Geopolitics stimulated a symposium at the annual meeting of the Association of American Geographers in 1997, subsequently published in *Political Geography*. In the ensuing debate, three contentions emerged. Firstly, Ó Tuathail is challenged for over-relying on textual data to the detriment of other empirical materials, such as maps, something that is curious given the importance of visual representation to both the geographical imagination and foreign policy strategy (see Heffernan, 2000; Smith, 2000; Sparke, 2000; Stephanson, 2000). Secondly, issues of embodiment and positionality of both author and subjects were raised; as were, thirdly, contentions that Ó Tuathail's text is inadvertently elitist, focusing on a few 'great men' in the field of geopolitics, some suggesting that Ó Tuathail does not do enough to locate himself outside this canon, and Dodds (1998) adding that Ó Tuathail's focus is overwhelmingly Anglo-American.

In sum, critics suggest that Ó Tuathail is guilty of what he accuses in others – an assertion of a transcendental viewpoint from where the world and its political order can be viewed – the difference being that Ó Tuathail takes a counter-hegemonic rather than hegemonic perspective. Resultantly Ó Tuathail's, 'attack on totalization' in geopolitical discourse and foreign policy itself 'turns out ... to be a totalization' (Stephanson, 2000: 382). This reduces the power of Ó Tuathail's interrogation to intellectual games of deconstruction rather than empirical assessments of the material impacts of geopolitics on people's lives. Sharp (2000) returns to this contention, maintaining that Ó Tuathail's concentration on the statecraft of great men elides geopolitical discourses in other fields, such as popular culture. Further, Smith (2000: 367) maintains that although Ó Tuathail is sensitive to '[r]eading race and gender into the texts of geopolitics', this 'is simultaneously here a way of reading class out'.

Responding to these challenges, Toal/Ó Tuathail (2000) explains that his subject choices such as Northern Ireland (in the opening chapter) and the corridors of power within Washington DC (the closing) represent a strategy of self-positioning and embodiment without recourse to autobiography. Noting that he regards with some suspicion processes of self-situation in academic texts, Ó Tuathail maintains that his work is neither disembodied nor masculinist but aims to assess these aspects in the hegemonic tradition of 'statesmanship' – the topic of his book. Accepting that *Critical Geopolitics* does not go beyond textual sources, Ó Tuathail argues that this is something others could pursue, and are pursuing. Stimulated by Ó Tuathail's influential contribution to 'critical geopolitics' – he magnanimously credits **Peter Taylor** with coining this term during discussion at the University of Illinois (Ó Tuathail, 2000) – analysis of political discourses and their constructions of spatial power relations are advancing research within political geography into new arenas.

Ó TUATHAIL'S MAJOR WORKS

Herod, A., Ó Tuathail, G. and Roberts, S. (eds) (1998) *An Unruly World? Geography, Globalization and Governance*. London: Routledge.
Ó Tuathail, G. (1996a) *Critical Geopolitics: The Politics of Writing Global Space*. Minneapolis: University of Minnesota Press.
Ó Tuathail, G. and Agnew, J. (1992) 'Geopolitics and discourse: Practical geopolitical reasoning and American foreign policy', *Political Geography* 11: 190–204.
Ó Tuathail, G. and Dalby, S. (eds) (1998a) *Rethinking Geopolitics*. London: Routledge.
Ó Tuathail, G., Dalby, S. and Routledge, P. (eds) (1998) *A Geopolitics Reader*. London: Routledge.

Secondary Sources and References

Atkinson, D. and Dodds, K. (2000) 'Introduction to geopolitical traditions: a century of geopolitical thought', in K. Dodds and D. Atkinson (eds) *Geopolitical Traditions: A Century of Geopolitical Thought*. London: Routledge, pp. 1–24.
Dalby, S. and Ó Tuathail, G. (1996) 'Editorial introduction: The critical geopolitics constellation: problematizing fusions of geographical knowledge and power', *Political Geography* 15: 451–456.
Dodds, K. (1998) 'Review of *Critical Geopolitics*', *Economic Geography* 74: 77–79.
Dodds, K. (2001) 'Political geography III: critical geopolitics after ten years', *Progress in Human Geography* 25: 469–484.
Heffernan, M. (2000) 'Balancing visions: comments on Gearóid Ó Tuathail's critical geopolitics', *Political Geography* 19: 347–352.
Ó Tuathail, G. (1986) 'The language and nature of the 'new' geopolitics: The case of US–El Salvador relations', *Political Geography Quarterly* 5: 73–85.
Ó Tuathail, G. (1989) *Critical Geopolitics: The Social Construction of Place and Space in the Practice of Statecraft*, Unpublished PhD dissertation, Syracuse University, Syracuse.
Ó Tuathail, G. (1992) 'Putting Mackinder in his place: Material transformations and myth', *Political Geography* 11: 100–118.
Ó Tuathail, G. (1994) 'The critical reading/writing of geopolitics: Re-reading/writing Wittfogel, Bowman and Lacoste', *Progress in Human Geography* 18: 313–332.
Ó Tuathail, G. (1996b) 'An anti-geopolitical eye? Maggie O'Kane in Bosnia, 1992–94', *Gender, Place and Culture* 3: 171–185.
Ó Tuathail, G. (1997 [1993]) 'The effacement of place? US foreign policy and the spatiality of the Gulf Crisis', in J. Agnew (ed.) *Political Geography: A Reader*. London: Edward Arnold, pp. 140–164.
Ó Tuathail, G. (1999) 'A strategic sign: The geopolitical significance of "Bosnia" in U.S. foreign policy', *Environment and Planning D: Society and Space* 17: 515–533.
Ó Tuathail, G. (2000a) 'Spiritual geopolitics: Father Edmund Walsh and Jesuit anticommunism', in K. Dodds and D. Atkinson (eds) *Geopolitical Traditions: A Century of Geopolitical Thought*. London: Routledge, pp. 187–210.
Ó Tuathail, G. (2000b) 'The postmodern geopolitical condition: States, statecraft, and security into the twenty first century,' *Annals of the Association of American Geographers* 90: 166–178.
Ó Tuathail, G. (2002a) 'Theorizing practical geopolitical reasoning: The case of U.S. Bosnia policy in 1992', *Political Geography* 21: 601–628.
Ó Tuathail, G. (2002b) 'Post-Cold War geopolitics: Contrasting superpowers in a world of global dangers', in R. J. Johnston, P. Taylor and M. Watts (eds) *Geographies of Global Change*. Oxford, Blackwell.
Ó Tuathail, G. and Dalby, S. (1994) 'Editorial: Critical geopolitics – unfolding spaces for thought in geography and global politics', *Environment and Planning D: Society and Space* 12: 513–514.
Ó Tuathail, G. and Dalby, S. (1998b) 'Introduction: rethinking geopolitics – towards a critical geopolitics', in G. Ó Tuathail and S. Dalby (eds) *Rethinking Geopolitics*. London, Routledge, pp. 1–15.
Painter, J. (1995) *Politics, Geography and 'Political Geography': A Critical Perspective*. London: Arnold.
Sharp, J. P. (2000) 'Remasculinising geo-politics? Comments on Gearóid Ó Tuathail's *Critical Geopolitics*', *Political Geography* 19 (3): 361–364.
Smith, N. (2000) 'Is a critical geopolitics possible? Foucault, class and the vision thing', *Political Geography* 19: 365–371.
Sparke, M. (2000) 'Graphing the geo in geo-political: *Critical Geopolitics* and the re-visioning of responsibility', *Political Geography* 19: 373–380.
Stephanson, A. (2000) 'Commentary on Gearóid Ó Tuathail's *Critical Geopolitics*', *Political Geography* 19: 381–383.
Toal, G./Ó Tuathail, G. (2000) 'Dis/placing the geo-politics which one cannot not want', *Political Geography* 19: 385–396.

Euan Hague

35 Gillian Rose

BIOGRAPHICAL DETAILS AND THEORETICAL CONTEXT

Gillian Rose was born in 1963 in England. She completed her BA in Geography from the University of Cambridge in 1985, and went on to complete a PhD in 1989, entitled 'Locality, politics and culture: Poplar in the 1920s', also at Cambridge. Her first teaching post was at the University of London at Queen Mary College. She then became a Lecturer in Geography at the University of Edinburgh for six years, before moving to the Open University in 1999. Throughout, Rose's contributions to geography have encouraged geographers to consider the gendered constructions of geographical knowledge. Her disciplinary history in geography has very much shaped her perspective. In writing about her academic trajectory and her relationship with feminism within and outside geography, she states:

> The only thing I was successful at apparently was academic work. I was bright at school, I won a place to Cambridge University, my initial stumblings around feminist historians' accounts of public and private space were encouraged by my Director of Studies ... I've had it relatively easy in the academy, helped by my Cambridge connections, [and] the recent fashionableness of a certain kind of feminism ...
> (WGSG, 1997: 35)

Drawing from 'a feminism that [felt] intensely personal' (WGSG, 1997: 35), Rose's research spans many subject areas, from the cultural politics of landscape to notions of the performative, including empirical accounts of community arts projects to critiques of visual methodological approaches. She has collaborated with a variety of geographers, including Steve Pile (Rose and Pile, 1992), Nicky Gregson (Gregson and Rose, 2000) and **Doreen Massey**.

Rose is perhaps best known for her book *Feminism and Geography*, published in 1993. The scope of the book has been seen as both courageous and ambitious. Through an explicit critique of geography's intrinsic masculinist approach to the discipline, Rose revealed the ways in which geographers have constructed a geography that legitimates masculine forms of geographical knowledge, effectively isolating women's ways of knowing. Connected to this privileging of masculine over feminine knowledge is the construction of a culture–nature dualism. Rose demonstrates the pervasive nature of the nature–culture binary through her discussion of two strands of thought: social scientific, which lays claims to rational and objective truths; and the aesthetic, which feminizes places and landscapes. By addressing the ways in which cultural geographers' work on landscape embodies masculinist perspectives, Rose (1993a) effectively demonstrates how understandings and experiences in places have been marginalized in the discipline.

As well as challenging the masculinist nature of the discipline, Rose has also intervened in feminist methodological debates. For example, drawing on insights from **Judith Butler**'s (1990) work on performativity, Rose (1997b) has critiqued some of the assumptions in feminists' use of reflexivity as a strategy for situating geographical knowledges. Additionally, her work on visual methods has provided the basis for *Visual Methodologies* (2001), a remarkably lucid text written for a cross-disciplinary audience.

SPATIAL CONTRIBUTIONS

Rose's treatise on geography's overarching masculinist approach, *Feminism and Geography*, provided the first cohesive and comprehensive analysis of human geography's resistance to work on and by women. Citing geography's deep reluctance to listen to feminism and its focus on women, Rose (1993a: 4) explored the idea that 'it is the specific notion of knowledge through which geographers think that marginalizes women in the discipline'. Rose's book was met with an uneasy mixture of criticism and praise when it was initially published, especially from feminist geographers, despite its decidedly pro-feminist stance. While many agreed that the book finally 'br[ought] academic geography up to date with . . . feminist theory, something geography badly needed' (Morin, 1995: 415), others insisted that '[Rose] does not interrogate the categories . . . of gender, class, ethnicity, nationality or sexual preference, nor is there much idea of change or struggle' (Hyndman, 1995: 197). Yet others still challenged its seemingly dry, academic tone in what would become a recurring critique of the book (Burgess, 1996). Even Rose herself admits that she has concerns with the narrative style of the book in an early chapter: 'I have tried to make my prose sound differently from the unmarked tone of so much geographical writing, but, as you have probably noticed, I have found it extraordinarily difficult to break away . . .' (see also Penrose *et al.*, 1992).

Rose's last chapter of the book, evocatively titled 'A politics of paradoxical space', has sparked ongoing debate among feminist geographers about the meaning and value of the metaphor. Rose never provides a precise definition of paradoxical space (Desbiens, 1999), nor does she 'specify how the [concept can] displace the master subject' (Hyndman, 1995: 201). As such, some commentators have criticized the concept of paradoxical space for being 'elusive and multiple in meaning', claiming that it represents an overzealous attempt by Rose to end her book on a high note (Desbiens, 1999). Yet, at the same time, it has also been suggested that it is a 'tantalizing' concept which has the potential to provide a radical framework for feminist geography (Katz, 1997; Desbiens, 1999). Mahtani (2001), for example, has used the concept of paradoxical space to provide a useful theoretical underpinning for her own research on 'mixed race women' in Canada, highlighting the subversive potential of such a position. Mahtani (2001) further insists that it is precisely the fluid nature of paradoxical space that serves to provide feminist geographers with a nuanced theoretical model for understanding complex forms of oppression (*vis-à-vis* race and class) in the academy. Despite its mixed reviews, *Feminism and Geography* remains an essential reference for the amplification of a feminist geographical perspective (Desbiens, 1999) and is considered required reading for human geographers. It opened up new forms of engagement about the varied forms of patriarchy within the discipline (see WGSG, 1997) and also outside geography in the area of cultural studies and literary theory (see Friedman, 1998; George, 1996).

Perhaps unwittingly responding to concerns about her lack of engagement with race and class in *Feminism and Geography*, Rose's second book, *Writing Women and Space: Colonial and Postcolonial Geographies*, co-edited with Alison Blunt, explicitly articulated the complicit and complicated relations between gender, race, class and sexuality through an edited collection about women's multiple, contested, and shifting senses of subjectivities as experienced through written representations of spatial differentiation. The 1994 collection complemented Rose's first book by effectively providing grounded empirical examples of feminist geographical research, addressing the paucity of woman-centred knowledges in the discipline.

Rose continued to elaborate on the themes of patriarchy inside and outside of geography well into 1995, notably through a paper, 'Rethinking the history of geography' in the *Transactions of the Institute of British Geographers*. This article explored issues of sameness and difference (versus her discussion of the same and other in *Feminism and Geography*). Paying particular attention to the ways that gendered sameness and difference are constituted through geographical traditions, Rose explained that 'given th[e] persistent erasure of women, the construction of geographical traditions might . . . be described as the construction of geography's paternal lines of descent' (Rose, 1995e: 414). Asking 'how can feminists place women in relation to this paternal geographical tradition?' (Rose, 1995e: 415), Rose emphasized that 'feminists . . . need to critique the transparent territorialization of tradition . . . we need to focus on the boundaries at which difference is constituted . . . this project . . . entails thinking about geography through a different spatiality: a multiple space' (Rose, 1995e: 416).

In *Writing Women and Space* Rose provided a feminist geographical critique of the work of the critical 'race' theorist **Homi Bhabha**, cautioning geographers that his theory was often dressed up in an enigmatic and masculinist vocabulary, couched in language that makes it more inaccessible than it need be. She noted, 'the optimism of Bhabha's brave new hybrid world . . . is perhaps gendered in both its practice and its writing' (Rose, 1995b: 372). Rose argued that Bhabha tended to overaccentuate hyperfluency in favour of a negation of the social. She writes: '[Bhabha's] spatialities remain analytical, not lived . . . he remains resolutely disembodied, unbloodied if not

bowed . . . his violence remains epistemic and not bloody' (Rose, 1995b: 372). Reminding us of the importance of the grounded material realities, Rose's critique of Bhabha's theory effectively demonstrated that she had not forgotten about the lived day-to-day, as some had suspected with the publication of her first book (see also McDowell, 1999).

That same year, Rose contemplated the 'spatial subversions' of Jenny Holzer, Barbara Kruger and Cindy Sherman in her chapter entitled 'Making space for the female subject of feminism' (1995f) in the edited collection *Mapping the Subject*. Rose suggests that the 'process of representation is central to everyday space and to the en-gendering of subjects in that space' (Rose, 1995f: 348) and points out that:

> to think about the geography of the female subject of feminism is not to be able to name a specific kind of spatiality which she would produce; rather, it is to be vigilant about the consequences of different kinds of spatiality, and to keep dreaming of a space and a subject which we cannot yet imagine.
> (Rose, 1995f: 354)

Hence, a recurring theme in much of Rose's writing is her concern with voice and authority in the narrative of the text. In her 1996 article, 'As if the mirrors had bled: Masculine dwelling, masculinist theory and feminist masquerade', Rose experiments with style and voice, adopting several personas in the piece. Rose literally 'performed' the piece at Syracuse University's symposium on 'Place, Space and Gender.' Providing one of geography's first engagements with the work of **Judith Butler** and her notion of performance, and enthusiastically drawing from the work of Luce Irigaray, Rose self-consciously articulated different identities to explore the spaces of movement and fluidity that are possible through a displacement of the opposition between real and imagined space, emphasizing that this distinction reflects a performance of masculinist power. Geographers were particularly motivated by Rose's attempts not to prioritize the 'real' over 'metaphorical' space (Brown, 2000).

Rose (1991,1992,1993a, 1993b), makes politically strategic use of feminist discourses on the maternal body, or woman as mother, in order to subvert masculinist structures of knowledge in geography. Her work has been strongly influenced by psychoanalysis and this has largely shaped her understanding and theorization of space. The sophistication of Rose's arguments, especially her continual questioning of the epistemology and ontology of the discipline, suggests that the notion of an 'explicit sexualisation of knowledges' (Grosz, 1993: 188) must be taken seriously by geographers.

Rose's 1997 article, 'Situating knowledges: Positionality, reflexivities, and other tactics', served as a critique of feminist geography's engagement with reflexivity. In a poignant example of Rose's intimate style, the article begins with the statement 'This is an article written from a sense of failure' (Rose, 1997b: 305). Through a case study of her own challenges in completing fieldwork, Rose suggests that the notion of self-reflexivity as understood among feminist geographers relies on an assertion of transparency – a 'knowable agent whose motivations can be fully known' (Rose, 1997b: 309). She ends by suggesting that we 'inscribe into our research practices some absences and fallibilities while recognizing that the significance of this does not rest entirely in our own hands' (Rose, 1997b: 319). Lise Nelson (1999) challenged this view, suggesting that Rose cannot conceive of a self or subject that is constituted by both discursive processes and at the same time being potentially aware of them (Nelson, 1999: 350). According to Nelson, Rose is 'haunted by humanist idealizations, arrested by an ability to conceptualize self-reflexivity as anything but transparent . . . [and is] condemned to a cycle of critique without exit' (Nelson, 1999: 350).

Rose returned to examine the notion of the performative in the late 1990s, explaining how space could be seen as performative. Her chapter 'Performing space' in *Human Geography Today* considers space as a discursive practice – as not only fantasized but also corporealized, thus making the spatial contradictory, complex and multi-textured (Rose, 1999). By focusing on the spatial articulations of sexual difference, Rose again acts out a performance reminiscent to the one performed in her chapter in 'As if the mirrors had bled' (1996), this time critically engaging with the work of Butler, Irigaray and de Lauretis. She ends by emphasizing that bodily performances produce space. A year later, Rose would offer a collaborative and grounded discussion of these themes with feminist geographer Nicky Gregson. Their article, 'Taking Butler elsewhere: Performativities, spatialities, and subjectivities', unveils two case studies of car-boot sales and community arts programmes. Suggesting that we should consider spaces as performative of power relations, Gregson and Rose insist upon the importance of contemplating subjectivities produced through specific performances of knowledge production, focusing on the rich complexity and 'messiness' of performances and performed spaces (Gregson and Rose, 2000). In a case study analysis of interviews with community arts workers, Rose bravely acknowledges that she found the 'practice' of doing interviews difficult, noting that she continually questioned her own role in the acquisition of narratives and explained that she did not plan any more interview-based research anytime soon (Gregson and Rose, 2000).

Her 2001 book, *Visual Methodologies: An Introduction to the Interpretation of Visual Material*, gave Rose an opportunity to consider the analysis of images rather than interviews. Rose puts forward a comprehensive study in methods to examine visual culture, from discourse analysis techniques to the study of audiences. This book offered geographers a way to understand culture through visual representations. Examining how we interpret images and how these processes are laden with power and cultural meaning, Rose provides a critical visual methodology where she examines the visual in terms of cultural significance, social practices and complex power relations. Drawing again on her understanding of psychoanalytical theory, Rose persuasively discusses the ways that sexual difference is articulated through visual practice (Rose, 2001). Rose's argument about critical visual methodologies has already been effectively extended into the arena of GIS imaging (Kwan, 2002) and has proven an important contribution in geography and representation (Hubbard et al., 2002).

KEY ADVANCES AND CONTROVERSIES

It can be argued that the greatest appeal in Rose's work lies in her efforts to remain transparent about her own power, positionality, and knowledge within, and outside, the academy in her writing. Her research in the last decade in the arenas of performance and visual culture has demonstrated a recurring interest in understanding the theoretical and methodological possibilities for comprehending power and knowledge production, while indicating all the time how her own role as a researcher influences the structure of geographical knowledge. It has been said that her musings are often open-ended, and offer more questions than answers through her continual focus on fissures (Rose, 1997b), paradoxes (Rose, 1993a) and contradictions (Rose, 1995e). It is a criticism that Rose herself acknowledges. In her article 'Situating knowledges: positionality, reflexivities and other tactics' (1997b) she writes: 'I've

tried to produce a gap in my own interpretative project . . . I'm not sure I succeeded, and I don't think I can or should be sure' (Rose, 1997b: 305). It is precisely this kind of transparency and honesty that many geographers find both courageous and attractive (Desbiens, 1999; Mahtani, 2001). Her research always focuses on the possibilities for more thoughtful discussions in feminist geography. Her final words in *Feminism and Geography* sum this up well:

> Space itself – and landscape and place likewise – far from being firm foundations for disciplinary expertise and power, are insecure, precarious and fluctuating. They are destabilized both by the internal contradictions of the geographical desire to know and by the resistance of the marginalized victims of that desire. And other possibilities, other sorts of geographies, with different compulsions, desires and effects, complement and contest one another. This chapter has tried to describe just one of them. There are many more.
> (Rose, 1993a: 160)

Rose's words have inspired feminist geographers to consider other ways of telling stories about women and space outside of masculinist narratives, and have thus served to see those contemplations as legitimate within the discipline. Feminist geographers have responded by mapping their own complex understandings of space through a diverse array of feminist geographies that pay particular attention to the masculinist nature of geographies rife with power relations. Gillian Rose's research has self-consciously articulated the challenges, as well as the risks, of unveiling the masculinism of the discipline and has thus created productive spaces for new cartographies of diversity and difference in geography.

ROSE'S MAJOR WORKS

Rose, G. (1993a) *Feminism and Geography*. Minneapolis: University of Minnesota Press.
Rose, G. (1995a) 'Geography and gender: cartographies and corporealities', *Progress in Human Geography* 19: 544–548.

Rose, G. (1995b) 'The interstitial perspective: a review essay on Homi Bhabha's *The Location of Culture*', *Environment and Planning D: Society and Space* 13: 365–373.

Rose, G. (1995c) 'Review of "The man question: visions of subjectivity in feminist theory and Sexing the Self: gendered positions in cultural studies"', *Environment and Planning D: Society and Space* 13: 241–243.

Rose, G. (1995d) 'Distance, surface, elsewhere: a feminist critique of the space of phallocentric self/knowledge', *Environment and Planning D: Society and Space* 13: 761–781.

Rose, G. (1996) 'As if the mirrors had bled: Masculine dwelling, masculinist theory and feminist masquerade', in N. Duncan (ed.) *Bodyspace*. London: Routledge, pp. 56–74.

Rose, G. (1997b) 'Situating knowledges: positionality, reflexivities and other tactics', *Progress in Human Geography* 21: 305–320.

Rose, G. (1997c) 'Performing inoperative community: the space and the resistance of some community arts projects', in S. Pile and M. Keith (eds) *Geographies of Resistance*. London: Routledge, pp. 184–203.

Rose, G. (1999) 'Performing space', in D. Massey, J. Allen and P. Sarre (eds) *Human Geography Today*. Cambridge: Polity, pp. 247–259.

Rose, G. (2001*) Visual Methodologies*. London: Sage.

Secondary Sources and References

Brown, M. (2000) *Closet Space*. London: Routledge.

Burgess, J. (1994) 'Review of "Feminism and Geography"', *Geographical Journal* 160: 225–226.

Butler, J. (1990) *Gender Trouble*. London: Routledge.

Desbiens, C. (1999) 'Feminism "in" Geography: elsewhere beyond and the politics of paradoxical space', *Gender, Place and Culture* 6: 179–185.

Friedman, B. (1998) *Mappings*. London: Routledge.

George, R. M. (1996) *The Politics of Home*. Cambridge: Cambridge University Press.

Gregson, N. and Rose, G. (2000) 'Taking Butler elsewhere: Performativities, spatialities and subjectivities', *Environment and Planning D: Society and Space* 18: 422–452.

Grosz, E. (1993) 'Bodies and knowledges: Feminism and the crisis of reason', in L. Alcoff and E. Potter (eds) *Feminist Epistemologies*. New York: Routledge, pp. 187–216.

Hubbard, P., Kitchin, R., Bartley, B. and Fuller, D. (2002) *Thinking Geographically*. London: Continuum.

Hyndman, J. (1995) 'Solo feminist geography: A lesson in space', *Antipode* 27: 197–207.

Katz, C. (1997) Review of '*Feminism and Geography: the Limits of Geographical Knowledge*', *Ecumene* 4: 227–230.

Kwan, M. (2002) 'Is GIS for women? Reflections on the critical discourse in the 1990s', *Gender, Place and Culture* 9: 271–279.

McDowell, L. (1999) *Gender, identity and place: understanding feminist geography*. Minneapolis: University of Minnesota.

Mahtani, M. (2001) 'Racial remappings: The potential of paradoxical space', *Gender, Place and Culture: A Journal of Feminist Geography* 8: 299–305.

Morin, K. (1995) Review of '*Feminism and Geography: the Limits of Geographical Knowledge*', *Postmodern Culture* 5: 78–79.

Nelson, L. (1999) 'Bodies and spaces do matter', *Gender, Place & Culture* 6: 331–353.

Penrose, J., Bondi L., McDowell L., Kofman, E., Rose, G. and Whatmore, S. (1992) 'Feminists and feminism in the academy', *Antipode* 24: 218–237.

Pile, S. and Rose, G. (1992) 'All or nothing? Politics and critique in modernism and postmodernism', *Environment and Planning D: Society and Space* 10: 123–136.

Rose, G. (1990) 'The struggle for political democracy: emancipation, gender, and geography', *Environment and Planning D: Society and Space* 8: 395–408.

Rose, G. (1993b) 'Speculations on what the future holds in store', *Environment and Planning A* 25: 26–29.

Rose, G. (1994) 'The cultural politics of place: local representation and oppositional discourse in two films', *Transactions of the Institute of British Geographers* 19: 46–60.

Rose, G. (1995e) 'Tradition and paternity: same difference?', *Transactions of the Institute of British Geographers* 20: 414–416.

Rose, G. (1995f) 'Making space for the female subject of feminism', in S. Pile and M. Keith (eds) *Mapping the Subject*. London: Routledge.

Rose, G. (1997a) Review of '*Evictions: art and spatial politics*', *Environment and Planning D: Society and Space* 15: 627–632.

Women and Geography Study Group (1997) *Feminist Geographies: Explorations in Diversity and Difference*. London: Longman.

Minelle Mahtani

36 Edward W. Said

BIOGRAPHICAL DETAILS AND THEORETICAL CONTEXT

Until his death in September 2003, Said was one of the most prominent intellectuals in the United States, known for his groundbreaking works on the relationship between culture and imperialism – which laid the foundation for the field of post-colonial studies – and for his unrelenting political fight for Palestinian self-determination. During his prolific career Said lectured at over 200 universities in the US, Canada, Europe and the Middle East; published more than 20 books; was translated into 37 languages; and was regularly interviewed in print and on television, including in several documentaries. Said, whose last academic position was as University Professor of English and Comparative Literature at Columbia University, received numerous distinguished awards and honours, including Harvard's Bowdoin Prize in 1963 and a Guggenheim Fellowship in 1972. Said served as president of the Modern Language Association in 1999, and was a member of the American Academy of Arts and Letters, American Academy of Sciences, American Philosophical Society and the Royal Society of Literature. He received approximately 20 honorary doctoral degrees. His book *Orientalism* (1978), a runner-up for the National Book Critics Circle Award, is now in its 40th American edition and has become a Penguin Classic in the UK. Said's articles regularly appeared internationally – in the US in *The Nation*, *The New York Times*, *Wall Street Journal*; in London's *The Times*, *The Observer* and *The Guardian*; the French *Le Monde Diplomatique*; in the Arabic *al-Hayat;* and in Madrid's *El Pais*.

When Said was born in 1935 to a wealthy, Christian-Arab family in Jerusalem, Palestine had been under British administration for 15 years. Said was baptized at an Anglican mission school in Jerusalem, and following his family's exile to Cairo during the 1947 partition and war, Said attended British schools. During his youth Said read avidly, listened to classical music, learned several languages (he was fluent in English, Arabic and French, and literate in several others), and played the piano. He finished his secondary education at Mount Hermon Preparatory School in Massachusetts, attended Juilliard School of Music, and went on to receive an American university and postgraduate education. He received his BA from Princeton University in 1957, and an MA (1960) and PhD (1964) from Harvard University. His doctoral dissertation in comparative literature focused on the interplay between Joseph Conrad's fiction and his correspondence. While serving many visiting positions and fellowships at other institutions, Said remained at Columbia from 1963 onwards, thus living most of his life in the United States. He died of leukemia in 2003.

Said's life as an exiled Palestinian from the age of 13 fundamentally shaped his career as academic, public intellectual and activist. The themes of exile and displacement, and especially his advocacy for Palestinian rights, all grew out of his personal experiences. While much of Said's work prior to 1967 focused on 'high' canonical literature, especially from England, the Arab–Israeli war that broke out that year marked a point in Said's life when his own identity as a

Palestinian became inseparable from his scholarly pursuits. He lived and worked in a pro-Israeli environment hostile to Arabs, leading Said to shift attention to the West's distorted view of the Middle East and Arab world. As an academic but also a major voice in the mass media, Said contested the caricature of Arab people as 'terrorists' and 'barbarians'. He was a major dissenting voice during the Persian Gulf War, and has been perhaps the most frequently cited and interviewed critic of American foreign policy in the Middle East over the past quarter-century. Said was a member of the Palestinian National Council and supporter of the Palestinian Liberation Organization (PLO), if also its critic. He turned down an invitation to attend the White House signing of the Oslo agreements in 1993, arguing that they ignored the majority of Palestinians residing outside of Gaza and the West Bank. Not surprisingly, Said received harsh criticism and even death threats for his support of Palestine.

Said's self-construction as an exiled Palestinian was central to his scholarly arguments, and it is the resultant contradictoriness – between his personal location and his scholarly works – upon which some of his most caustic critics have focused. He was born into a Christian-Arab, not Islamic, family (though he is secular in orientation); he lived in the Middle East only during his youth and was educated in the West's most elite institutions; he was a political exile but a wealthy, privileged, 'cosmopolitan' one; and he was a scholar who advocated concern for 'narratives of the forgotten' but who himself devoted much attention to the West's high canonical literature, music and culture (Ahmad, 1993; Ashcroft and Ahluwalia, 1999). Though not to be dismissed, the politics of his own paradoxical location do little to undermine Said's extraordinary contributions to cultural studies and cultural and social geography.

SPATIAL CONTRIBUTIONS

Edward Said's intellectual contributions are difficult to categorize in disciplinary terms. He wrote on subjects as diverse as the role of the intellectual, music criticism (he was *The Nation*'s music critic), Palestinian politics, the experience of exile, as well as on culture and imperialism. His work initiated debates across the humanities and social sciences, from literature to music to anthropology to political science to geography. Much, if not all, of Said's work was explicitly geographical, although it would be highly un-Saidian to attempt to separate his 'geographical contributions' from the rest of his literary and social theory. Indeed, while much of Said's work can be classified as literary criticism, he strongly advocated against disciplinary boundaries and the excessive specialization they encourage. To him, for example, all texts, literary or otherwise, are political, and must be 'worlded' – located in the world and exposed for the geographical imaginations from which they arise. Much of Said's career was devoted to analysis of texts (literary, political, journalistic, and so on) – the ways they are located in the world and contribute to colonial and imperial relations of power. Said was at his most geographic in such attempts to 'world' texts, writers and audiences, although other of his arguments on literary criticism and theory have been highly influential in cultural studies. At the broadest scale, his analysis of the politics of cultural representation, and cultural imperialism, brought together his various interests.

Several writers astutely describe Said's impact on geography as one of an inspirational 'talisman' at this historical juncture (e.g. Clayton, 2003: 357). In fact, Said's influence in human geography, particularly his theory of Orientalism, has been so compelling that one might describe it as almost transparent to practitioners today. Said not only initiated a spatial turn in postcolonial and cultural studies, but within the field of geography itself his work has transformed the terms around which the histories of geography, critical historical geographies, geographies of empire, and analyses of territory and land dispossession are discussed today.

While geographers have long understood that imperialism and colonialism ought to be conceptualized geographically, Said's impact on the field has been in theorizing the cultural processes and discursive formations that aid colonial or imperial control over people and place – including control that is non-territorial. The complex cultural, ideological and intellectual processes involved in domination and control that accompany the political, economic and territorial have now become standard in geographical studies, thanks in part to Said's groundbreaking works (see, for a good example, Godlewska and Smith, 1994). The journal *Environment and Planning D: Society and Space* has done much to bring Saidian analysis to an audience of geographers (e.g. Barnett, 1997; Driver, 1992; Kasbarian, 1996; Rogers, 1992). Said's work has also been brought to bear on the role of travel writing in the creation of imaginative, popular geographies of empire (Blunt, 1994; McEwan, 1996; Duncan and Gregory, 1999; Morin, 1998). Yet because of geography's traditional focus on the morphology of the built environment as well as the central place that the economic has in geographical analysis, the argument that literary theories such as Said's should be better grounded in material practices has found popular backing in geography. For example, Smith's (1994: 492–493) complaint that Said problematically slid into aberrant textualism mirrors Marxist and neo-Marxist debates in other fields (see Gregory, 1995).

Said's notion of 'imaginative geographies' embedded in colonial discourse forms the basis of two of his books, *Orientalism* (1978) and *Culture and Imperi-*

alism (1993). This concept has done much to invigorate postcolonial and cultural studies with a spatial sensitivity. Imaginative geography to Said refers to the invention and construction of geographical space beyond a physical territory, which constructs boundaries around our very consciousness and attitudes, often by inattention to or the obscuring of local realities (Said, 2000b: 181). To him, the world is divided and structured according to this imperial imagination or ideology, with the ultimate objective to control people and place.

Said laid the foundation for postcolonial studies with publication of *Orientalism*, and it was this book that launched his international reputation and established the terms of debates that other postcolonial critics, such as **Gayatri Spivak** and **Homi Bhahba**, have engaged. *Orientalism* moved the concepts of colony and empire to centre stage in the American academy in the 1970s, in addition to infusing it with the critical methods of French poststructuralism. 'Orientalism' as a field of study pre-dated Said, taking in two millennia of studies of Eastern culture by the West. What distinguished Said's work was his attention to the totalizing essentialism, ethnocentrism, and racism embedded in studies of the Orient. Said examined the works of an array of eighteenth and nineteenth-century British and French novelists, poets, journalists, politicians, historians, travellers and colonial administrators (applying the same type of analysis to twentieth-century American discourses and hegemony in later works). He argued that through a series of oppositions, these works systematically represented the Orient/East as irrational, despotic, static and backward, while the Occident/West was rational, democratic, dynamic and progressive. Importantly, it did not matter to Said whether the stereotype was positive or negative (and many were in fact positive), since either is equally essentializing and originates out of a desire for domination. In a Foucauldian move, Said showed how such stereotyped representations stood for 'knowledge' itself and were deeply implicated in the exercise of power. Where the focus of Said's work becomes cultural *imperialism* is in showing how the political or economic or administrative fact of dominance relies on this legitimating discourse.

Orientalism formed the first part of a trilogy that subsequently included *The Question of Palestine* (1980) and *Covering Islam* (1981). Together these books demonstrate how European colonialism, Zionism and American geopolitics all work together to dispossess Palestinians of their homeland, and even a past. In the latter two works, Said moved away from literary scholarship to a more political and historical investigation of Palestinian dispossession. Probably the most powerful argument in *The Question of Palestine* is that Israeli Zionism is itself an Orientalist (i.e. racist) discourse, with critiques of Israeli politics being too easily dismissed as anti-Semitic. *Covering Islam* shows how the US media and foreign policy continue to work in an Orientalist fashion in the late twentieth century. In this book Said took his ideas about cultural representation into the arena of practical politics. At the time (1979) the US was engaged with the Iranian 'hostage crisis' when students seized the American embassy in Teheran. Said asserted that journalists' uninformed reports seemed devoid of any historical contextualization (such as previous US involvement in Iran, including helping to train their secret police), and simply reinforced images of Islamic barbarism and terrorism and, in contrast, American innocence and heroism. Said followed up on Palestine's unique position as the 'victim of victims' in several later books. No other contemporary academic has so passionately attempted to tell a counternarrative of Palestine. Others of his books, including *The End of the Peace Process* (2000a), document Said's political position on the Israel–Palestine conflict. He consistently argued against partition and for a bi-national state in Israel, insisting that the terms of citizenship must be made inclusive, democratic and not based on principles of racial or religious difference.

The World, the Text, and the Critic (1983) followed Said's trilogy on cultural imperialism. This book, a collection of essays written between 1968 and 1983, received much notoriety for two pieces in particular, 'Secular criticism' and 'Traveling theory.' In 'Secular criticism', Said laments what he sees as an apolitical cult of professionalism that has saturated intellectual life, arguing for a more politically charged, oppositional literary criticism and role for the intellectual. In 'Traveling theory', Said developed a geographical model of how ideas or theories 'travel' from place to place, and what happens to them when they do. He argued that because theories develop within particular socio-historical contexts, they lose their oppositional weight when moved and 'domesticated' into other spatio-temporal contexts. He later revised this position, conceding that possibilities exist for theories to be effectively reconstituted in new political situations (Said, 2000b).

Said described his next major work, *Culture and Imperialism* (1993), as the sequel to *Orientalism*. In this series of essays he addressed some of the intellectual conundrums to which *Orientalism* had given rise. Here he adopted a musical term for literary criticism, offering a strategy of 'contrapuntal' reading of texts. To read a text contrapuntally 'is to read with a simultaneous awareness of both the metropolitan history that it narrates and of those other histories against which (and together with which) the dominating discourses act' (1993: 51). These essays, while continuing to focus primarily on 'high' canonical literatures of the First World, were thus more attuned than was *Orientalism* to 'dialogues' between colony and empire – cultural hybridity, intertwined histories and spaces of resistance to European metropolitan culture. His controversial essay on *Mansfield Park*, for example, fundamentally shifted the terms of discussion about Jane Austen's novel by politicizing and grounding it in the colonial geography of the Caribbean.

In some of his later works, Said continued to reflect on his own personal experience of exile and on relationships between memory and place. He considered the pain of exile in several of his works, including in his autobiography, *Out of Place* (1999). While he struggled with his own displacement, Said recognized the empowering potential of it to one who remains distanced from partisan politics and the habitual order. In 'Invention, memory and place' (2000b), Said further reflected on the importance of narratives to collective memories and senses of place, and on the new constellations of power and identity around which 'invented' memories in place are taking shape. Using Palestine as a case study, Said showed the overlapping and competing place memories that arise there for Christians, Jews and Muslims, based as they are on historical narratives, landscape features and physical structures. Part of the 'problem' with Palestine, according to Said, is that its leaders have failed to articulate an effective, collective, national narrative as part of its independence struggle.

KEY ADVANCES AND CONTROVERSIES

The full force of Said's scholarship, the numerous studies, books, theses, and discussions which it has inspired, can only be hinted at via the selected secondary source list below. Much of the criticism of Said's work can be thought of as *extensions* of his thinking rather than attacks on it. Nonetheless, there have been a number of debates and controversies to which it has given rise, again attesting to its provocative power more than anything. Such arguments have flourished principally around Orientalism; the contradictions

brought about by Said's own social-spatial positionality; and the basic theoretical tensions in his studies of cultural imperialism across humanism, Marxism and poststructuralism.

Some unconvincing criticism of *Orientalism* has come from Islamic and Arabic specialists who see Said as unnecessarily politicizing 'innocent' scholarship on the Orient. Such critics claim that knowledge of the Orient produced by Western scholars over the last couple of millennia is well intentioned and 'disinterested'. This line of argument is unconvincing because it refuses the notion that all knowledge is political, situated and produced to serve particular purposes, whether consciously intended or not.

Conversely, an in-depth, fruitful discussion of Said's work has emerged from scholars examining the cultural aspects of empire and theories of representation. *Orientalism* in particular provoked a number of controversies and debates (e.g. Clifford, 1988; Hussein, 2002). One of its most contested aspects was Said's inclination to commit 'Orientalism in reverse' – that is, he failed to recognize vast differences within Orientalist discourses about the Orient and Middle East. Critics assert that Said himself produced a (counter-) stereotype of an homogenized, racist and ethnocentric Westerner. Texts, discourses and representations about the Orient are considerably more ambivalent, heterogeneous and dynamic than Said at first allowed, especially if looking across Western academic discourses and disciplines (literature versus social science, for example), and across national cultures (Britain, France and the US), which Said had tended to downplay in his attempt to prove resonances among them. Critics, especially those outside the West, were also quick to point out Said's failure to consider resistance and opposition to Orientalist stereotypes, which itself reinforced an (Orientalist) image of an Eastern subject as passive, inarticulate and lacking self-determination. Said attempted to correct this silence in later works, including *Culture and Imperialism* (1993).

Edward Said for the most part neglected gender as an analytical framework, and feminists in particular have made some key advances on *Orientalism*, especially those focusing on colonial women travellers, missionaries, educators, and so on (albeit going beyond the spaces of the 'Orient' to which Said had limited his own discussions). In fact it is arguable that the recent proliferation of scholarship on travel writing as a genre of colonialist discourse is attributable to Said's theories of representation and cultural imperialism. Much of the feminist work demonstrates the contradictory, heterogeneous and sometimes counter-hegemonic positions that women occupied with respect to colonial and imperial discourses and structures of power.

Ahmad (1993) submitted what came to be one of the most infamous critical assessments of Said's writing, though other, more discerning critiques came from Lazarus (1999) and Hussein (2002). Working from a materialist standpoint, Ahmad challenged Said on the relationships between texts and representations and their associated social and material practices. Ahmad criticized Said's excessive 'textualism', arguing that we come to know the world through the effects of global capitalism, not through texts and representations. This is a dichotomy that most postcolonial critics would dismiss as a false one, since representations are materially located in the world, and, as Said notes, maintain 'a web of affiliations' in the world. Among other complaints, Ahmed also noted that Said failed to 'world' himself in his particular historical, cultural and institutional frameworks that are, just as are the subjects that Said studied, governed by dominant ideologies and political imperatives. Said's own privileged class position and his affiliation with elite American schools (which, as Ahmad points out, reproduces the current international division of labour), as well as his position as part of the metropolitan elite more generally (even if an exiled one), arguably limited Said's ability to challenge the status quo.

Numerous critics maintain that Said's methodological and theoretical framework was at best eclectic and justifiably impatient with received dogmas; at worst inconsistent, arbitrary and sometimes at odds with itself. Some critics see strength in Said's ability to bring together diverse theoretical orientations, as situated provocatively between 'the West Bank and the Left Bank' (e.g. Gregory, 1995: 448). Hussein (2002: 4) in particular stresses the positive aspects of Said's methodology, his 'technique of trouble', which he argues allowed Said to subject received wisdom to 'theoretical and historical insight and to what might be called the controlled anarchism of critical consciousness'. However, Said's eclecticism raised thorny, not easily resolvable issues for others. Said has been criticized, for example, for his ambivalence about whether a 'real' Orient exists beyond its representation. On one hand, Said argued that Orientalism is a misrepresentation of the 'real' Orient, in which case Orientalism is a type of ideological knowledge in a Marxian sense. On the other, Said followed the logic of discourse theory in implying that no 'real' Orient exists – that such an idea is a Western construct, an 'imaginative geography'. (See Clifford, 1988: 255–276, for an extended discussion of what he calls Said's 'confusion' between several incompatible designations.) For Said, the issue was not so much 'a dominant representation hiding a reality, but of the struggle between different and contesting representations' (Ashcroft and Ahluwalia, 1999: 4).

Said's frequent recourse to his own lived experience and thus traditional humanist arguments about human agency also raised questions about his alliance with Foucauldian discourse theory and relationships between discourse, knowledge, and power. Said followed **Foucault** in privileging discourse and language as prime determinants of social reality, and knowledge as a type of power or force that works impersonally through a multiplicity of sites and channels. (He parted with Foucault, though, in the notion of authorless texts, instead being more concerned to locate the will of individuals and parameters of imperialism as producing discursive violence; Moore-Gilbert, 1997). Yet Said also adopted a more conventional realist approach when discussing American foreign policy in Israel and Palestinian rights. In addition, Said can be considered Marxian in orientation, particularly in his development of Gramsci's theories of cultural hegemony, the dynamics of domination, and possibilities for resistance. Yet here too, he aligned with other postcolonial critics in pointing out the limits of (an ethnocentric) Marxist theory in confronting the needs and experiences of the colonized world. As such, Said's work consistently implied that a non-dominating, non-coercive mode of knowledge is possible and desirable. Ultimately, his work suggests that such knowledge must be self-critical, must open itself disciplinarily, and must enter into dialogue with the people and places it represents (Kucich, 1988: 255).

SAID'S MAJOR WORKS

Said, E. (1978) *Orientalism: Western Conceptions of the Orient.* London: Kegan Paul.
Said, E. (1980) *The Question of Palestine.* New York: Routledge.
Said, E. (1981) *Covering Islam: How the Media and the Experts Determine How we See the Rest of the World.* New York: Pantheon Books.
Said, E. (1983) *The World, The Text, and the Critic.* Cambridge: Harvard University Press.
Said, E. (1993) *Culture and Imperialism.* New York: Alfred A. Knopf
Said, E. (1999) *Out of Place: A Memoir.* New York: Alfred A. Knopf.
Said, E. (2000a) *The End of the Peace Process: Oslo and After.* New York: Pantheon Books.
Said, E. (2000b) 'Invention, memory, and place', *Critical Inquiry* 26: 175–192.

Secondary Sources and References

Ahmad, A. (1993) *In Theory: Classes, Nations, Literatures.* London: Verso.
Ashcroft, B. and Ahluwalia, P. (1999) *Edward Said: The Paradox of Identity.* London: Routledge.
Barnett, C. (1997) ' "Sing along with the common people": Politics, postcolonialism, and other figures', *Environment and Planning D: Society and Space* 15: 137–154.
Bayoumi, M. and Rubin, A. (eds) (2000) *The Edward Said Reader.* New York: Vintage Books.
Blunt, A. (1994) *Travel, Gender and Imperialism: Mary Kingsley and West Africa.* New York: Guilford Press.
Chrisman, L. (1998) 'Imperial space, imperial place: Theories of empire and culture in Frederic Jameson, Edward Said and Gayatri Spivak', *New Formations* 34: 53–69.
Clayton, D. (2003). 'Critical imperial and colonial geographies', in K. Anderson, M. Domosh, S. Pile and N. Thrift (eds) *Handbook of Cultural Geography.* London: Sage, pp. 354–368.
Clifford, J. (1988). 'On Orientalism', in *The Predicament of Culture: Twentieth-Century Ethnography, Literature, and Art.* Cambridge, MA: Harvard University Press, pp. 255–276.
Contemporary Literary Criticism (2000) 'Edward Said', Volume 123, Detroit Gale Group.
Driver, F. (1992) 'Geography's empire: Histories of geographical knowledge', *Environment and Planning D: Society and Space* 10: 23–40.
Duncan, J. and Gregory, D. (eds) (1999) *Writes of Passage: Reading Travel Writing.* London and New York: Routledge.
Gandhi, L. (1998) *Postcolonial Theory: A Critical Introduction.* New York: Columbia University Press.
Godlewska, A. and Smith, N. (eds) (1994) *Geography and Empire.* Oxford: Blackwell.
Gregory, D. (1995) 'Imaginative geographies', *Progress in Human Geography* 19: 447–485.
Hussein, A. A. (2002) *Edward Said: Criticism and Society.* London: Verso.
Kasbarian, J. A. (1996) 'Mapping Edward Said: Geography, identity, and the politics of location', *Environment and Planning D: Society and Space* 14: 529–557.

Kennedy, V. (2000). *Edward Said: A Critical Introduction*. Cambridge: Polity Press.

Kucich, J. (1988) 'Edward Said', *Dictionary of Literary Biography* 67: 249–259.

Lazarus, N. (1999) *Nationalism and Cultural Practice in the Postcolonial World*. Cambridge: Cambridge University Press.

McEwan, C. (1996) 'Paradise or pandemonium? West African landscapes in the travel accounts of Victorian women', *Journal of Historical Geography* 22: 68–83.

Moore-Gilbert, B. (1997) *Postcolonial Theory: Contexts, Practices, Politics*. London: Verso.

Morin, K. M. (1998) 'British women travellers and constructions of racial difference across the nineteenth-century American West', *Transactions of the Institute of British Geographers* 23: 311–330.

Rogers, A. (1992) 'The boundaries of reason: the world, the homeland, and Edward Said', *Environment and Planning D: Society and Space* 10: 511–526.

Smith, N. (1994) 'Geography, empire and social theory', *Progress in Human Geography* 18: 491–500.

Sprinker, M. (ed.) (1992) *Edward Said: A Critical Reader*. Oxford: Blackwell.

Karen M. Morin

37 Andrew Sayer

BIOGRAPHICAL DETAILS AND THEORETICAL CONTEXT

Andrew Sayer completed his undergraduate degree, BA (Hons) Geography, at Cambridgeshire College of Arts and Technology (later Anglia Polytechnic University) in the late 1960s, completing an MA and DPhil in Urban and Regional Studies at Sussex University in the early 1970s. He was subsequently appointed to a lecturing post at the same university. In 1993, he moved to the Department of Sociology at Lancaster University where he was to become Professor of Social Theory and Political Economy. Sayer's scholarly work is wide-ranging but has two major themes: social theory and political economy, and philosophy and methodology in the social sciences. The first strand is illustrated by empirical work on topics in economic geography – for example, *Microcircuits of Capital* (Morgan and Sayer, 1988), and *The New Social Economy: Reworking the Division of Labour* (Sayer and Walker, 1992) – as well as in more theoretical discussions that have discussed the restructuring of the political economy (Sayer, 1995) and the ensuing relationships between political and moral economies (Ray and Sayer, 1999). The second strand is clearly indicated by *Method in Social Science* (Sayer, 1992) and *Realism and Social Science* (Sayer, 2000).

Andrew Sayer's best known contribution to the field of human geography, and subsequently to the social sciences, has been his explication and development of critical realism. His arguments about space are not his central point, but they flow directly from this project. It is important here to appreciate the context in which the debate about critical realism was played out if we are to understand the appropriations of, and reactions to, Sayer's position by those exploring the constitutive role of space (and time) in contemporary life. In the late 1980s critical realism became, in the Anglo-American context, a major touchstone for economic geographers struggling with two issues: first, the waning of interest in Marxist structuralism (and/or the perceived lack of analytical rigour in structuralist accounts of economic restructuring); second, the perceived lack of explanatory power accorded to positivist descriptions of economic restructuring. Critical realism seemed to address these lacunae and offer a new way of approaching economic geography; accordingly, critical realism briefly shifted into a hegemonic position in the discipline of geography, as well as the subdiscipline of economic geography, only to be quickly displaced by postmodern and poststructural critiques in the 1990s. Even so, it is a moot point whether the popular support for critical realism was translated into practice and understanding, or whether it simply acted as a flag of convenience for 'business as usual'. Sayer (2000) has subsequently argued that not only was/is critical realism unfairly painted as oppositional to poststructuralism, but also that the popularly assumed critical realist position that 'space matters' is a misinterpretation of its key tenets. Critical realists argue that the central point is *how* space matters. However, an answer to this question can only be reached via a re-theorization of space.

Sayer's intellectual trajectory is significant. He began as a geographer, and then was a student, lecturer and researcher in a multidisciplinary setting; he is currently located in a Sociology Department where he defines himself as 'post-disciplinary'. Sussex University, whose organizational structure is characterized by broad over-arching schools, such as Social and Community Studies, rather than traditional disciplines and faculties, presented an opportunity to develop and test Sayer's ideas in a broader social science context than that afforded most geographers. Moreover, teaching in the Graduate Division of Urban and Regional Studies, brought together a diverse range of staff in what turned out to be a productive and challenging environment (his colleagues included, among others, Peter Saunders, Kevin Morgan, Simon Duncan, Mike Savage, Susan Halford, James Barlow, Peter Dickens, Peter Ambrose and Tony Fielding). Sayer's rigorous attention to social theory and methodology while at Sussex led him to question a range of issues surrounding the formation of knowledge and how we both acquire and apply it. The attention to internal relationships and causality has led him to work against what he views as disciplinary parochialism that is prone to 'reductionism, blinkered interpretation and misattribution of causality' (Sayer, 2000: 7). As we will note below, this position has not tended to court easy support for his ideas among those more wedded to disciplinary norms; nor among those who have sought to mobilize notions of 'space' to strengthen the discipline of geography. Nonethless, Sayer's commentary on critical realism and the relations between theory and empirical work in the context of the social sciences has been particularly influential within the discipline of geography. His ideas have also found their way directly into sociology, and, to a lesser extent, economics; furthermore, methodologically, they have had an impact across the whole of the social sciences by popularizing and demonstrating the application of critical realism.

SPATIAL CONTRIBUTIONS

Sayer's position on 'space' is inextricably woven with his exposition of critical realism, and the critique of positivism. Sayer's 'turn to realism' was prompted by an attempt to resolve questions in his doctoral thesis about 'urban modelling'. Urban modelling was, at the time, a dominant mode of conceptualizing social and economic, and spatial, interaction in the form of systemic modelling that could be operationalized with quantitative measures, often through the application of 'social physics' models such as the gravity model (in passing, we might note that such approaches experienced something of a revival in the late 1990s). Sayer's (1976) reaction against closed systems and the search for causality in regularity that these models assumed, led him to explore critiques of positivism upon which such approaches were based (albeit unacknowledged). Sayer looked to the newly minted literature on critical realism (for example, Bhaskar, 1975; Harré, 1972; Keat and Urry, 1975) for insight. Simultaneously, this philosophical framework provided him with a means of distancing himself from the limitations of the 'grand narratives' and 'over-determination' that characterized the dominant structural Marxism of the time.

Sayer's point of entry was via a discussion of 'abstraction'. The 1981 paper of that title, the 1982 paper 'Explanation in economic geography', and finally the book first published in 1984, *Method in Social Science*, clearly laid out the case for critical realism for a social science audience. Critical realism places central importance on the notion of a 'depth ontology' that admits the possibility of a debate about the 'necessary', or 'internal', relations that constitute the 'causal powers' of things, which then can be seen to have the potential to be mobilized in particular settings (i.e. they are contingent relations). Critically, this ontologically rich account of social reality implies a generative, as opposed to a successionist, view of causality. In contrast, positivist approaches have an atomistic (as opposed to relational) view of interaction; they have no ontological depth (what you see is what you get); and causality is inferred from regularity: the whole is based upon the idea of closed, rather than open, systems.

This philosophical grounding allows Sayer to make some significant points about 'space'; the main one being that there is very little that one can say about space *in the abstract*. Therefore, there is no recourse to a theory of space that might, *pace* **David Harvey** (1985), create bedrock, and a justification, for geography. Thus, there was no magic spatial insight that geographers might claim knowledge of, which, *pace* **Ed Soja** (1985), if applied to other social theories (which are notably *aspatial*), would revolutionize them.

This does not mean that Sayer had nothing positive to say about 'space'. His point about abstract space, particularly the tendency of geographers to 'rake over' sociology and to accuse it of aspatiality, is that the degree of 'violence of abstraction' is variable. For example, the injury done by ignoring space in an abstract debate about social stratification may be negligible. However, the discussion of particular instances of stratification will require the consideration of 'physical space'. It is this notion of 'physical space' that Sayer stresses: the manifestation of a 'space–time–matter' combination. Sayer argues that abstraction tends to 'scramble' our understanding of causality. We can point to two types of examples: on one hand, empiricist accounts that look for regularity in relation to an abstract model; and, on the other hand, abstract theories of space that seek selectively to abstract concrete situations (substance and space) and recombine them in inappropriate ways. Both instances are akin to taking apart a machine and putting it back together incorrectly; metaphorically speaking, causality is disrupted: the machine doesn't work. By way of explication, we can point to the common tendency in geography to recombine processes in discrete spatial units (global, nation, region and locale); spatial or analytical units that may not be the relevant ones to the processes under investigation. It should be pointed out that geographers have been less culpable than those working in other disciplines who commonly uncritically recombine processes at the national scale. Thus, Sayer contends, it is crucial for empirical work to be carried out attending to the specificity of concrete processes and their temporal ordering lest one falls into the trap of spatial fetishism and reductionism.

Sayer (2001) has also turned his attention to the problems of cultural political economy, and specifically the new economic geography. Here he calls for a more critical analysis of the social and cultural embedding of economic activities that considers both embedding *and* disembedding. First, he urges caution against the tendency to either 'flip' from systemic analyses of traditional political economy and its attendant concerns with the politics of distribution to the 'lifeworld' focus of cultural political economy and its focus on questions of difference. He argues for a re-theorization of the relations between these two dimensions of society. Second, he points out that system and lifeworld should not be seen as having a necessary correspondence with particular physical spaces. For example, he argues that 'lifeworld' is not simply limited to the private sphere of the home, and that the public realm is not the only space of political discourse. On this basis, he insists that economic organizations, such as firms or businesses, should not be considered as lying only in the system, but should also be considered part of the lifeworld. This leads him to assert that critical geographical endeavour needs to reconsider, rather than ignore, 'classic' political economy at the same time that it embraces cultural issues concerned with the politics of identity and difference.

KEY ADVANCES AND CONTROVERSIES

The reactions to Sayer's ideas have been varied. At one level his interventions have been widely cited; the notion of critical realism, and the need for a particular kind of theoretically informed concrete re-search, certainly struck a chord. Notable development of realist work was carried out by John Allen and Linda McDowell; as well as **Doreen Massey** at the Open University (Massey, Allen and Sayer met as an informal group, 'the Brighton Pier Space and Social Theory Group'). Duncan and his collaborators (Savage and Good-win) based at Sussex/London also were key propagandists. Finally, the 'Lancaster Regionalism' group (among them, Urry, Bagguley, Mark-Lawson, Sharpiro, Walby and Warde) also adapted both the ideas and terminology of critical realism, al-though most of this group would not see themselves as realists. Yet one has to be more sceptical as to whether geographers, or social scientists, do social science any differently as a result of Sayer's espousal of critical realism. One of the infuriating points for many has been that critical realism requires some basic philosophical rethinking on behalf of users, and that there is no 'off-the-shelf' 'toolkit' (see Pratt, 1995). But, it may well be argued, this is the point: to rethink the way we do research rather than follow 'business as usual'.

It is here that we should note that Sayer's critical comments on space were of interest to sociologists in the wake of sociology's 'turn to space' – a 'turn' most notably flagged up by **Anthony Giddens'** (1984) structuration project, which was given significant momentum by Urry (himself a realist sociologist, although he moved away from realism in the late 1980s). The edited book, *Social Relations and Spatial Structures* (Gregory and Urry, 1985), pulled together a number of key writers to debate the role of space in the constitution of social life (namely: Cooke, Giddens, Gregory, Harvey, Massey, Saun-ders, Sayer, Soja and Urry). Interestingly, we can perhaps date the arrival of human geography as a 'new kid on the block' of social science to this time. As noted above, Soja (1985) was seeking to rad-ically overhaul the social sciences by 're-thinking them spatially'; Harvey (1985) continued to aspire to a geo-histori-cal materialism; Giddens (1985) pointed the way to mid-range theories. The tenor of the times was almost an inoculation metaphor; to inject some geography into sociology in particular, and the social sciences in general. The whole debate was given an extra spin by the growing awareness of space by postmodernist writers, especially urban theorists (includ-ing members of the so-called LA School – Davis, Dear, Scott, Storper and Soja). Due to the configuration of the debate, it was not long before cultural studies began to look to geography for some spatial con-cepts. In this context, the geography cup-board was not exactly full of ideas, so some frantic writing followed. However, Sayer's (1985) response was a let-down for the incipient geographical colonizers: as we noted above, there is no 'bolt-on' spatial theory, and abstract spatial con-cepts will not help us either.

The application of critical realism to practice, and to physical space, should have been they key moment in the de-bate. It turned out to be so, but in a rather complex manner that reminds us that ideas never simply pass through the world as their makers may have imag-ined. The key debate to focus on here is that of 'localities'. This debate has its roots in an Economic and Social Research Council research project that was to examine the differential economic and social restructuring on place (see Cooke, 1989). The study was based on five case studies: the 'localities'. Cooke, the project director, had drawn upon **Massey's** (1984) *Spatial Divisions of Labour* for both an empirical focus and a broad conceptual steer. However, at the first airing of the debate at an Institute of British Geographers conference (before concrete research actually began), Cooke drew withering attacks, notably from **Neil Smith** (1987). The main charge was one of empiricism, and that this represented a return of geography to its idiographic roots. This attack redoubled the theoretical debates about space (does it matter, how and why?), as well as increasing the tension between theoretical and empirical work. The debate ran through economic (and, to a lesser degree, social and political) geography for much of the next decade, although the localities project lasted just two years: Duncan's (1991) special edited issue of *Environment and Planning A* sought to have the last word on the topic.

The 'Localities' project thus became a lightning conductor for a number of debates. Smith's castigation of the project was allied to an attempt to defend a mode of theorizing that did not admit the possibility of contingency, and was exclusively concerned with internal relations (see Smith, 1984); yet at the same time wanted to include every diverse event. It is not surprising that Smith (1987) not only criticized Cooke, but also along with Archer (1987) and Harvey (1985, 1987), sought to demolish the case for critical realism too (with Harvey aspiring to construct a theory of the concrete and particular in the context of the universal and abstract determinations of Marx's theory of capital accumulation). Cooke's pragmatism (practical, and later philosophical) led him to propose local labour markets as the key template for the locality studies. This view was heavily criticized by Duncan (1989), from a position close to Sayer's, arguing that local labour markets were not always the relevant causal structure for the examination of phenomena. Interestingly, Sayer's early work had been spurred on not only by an attempt to

distance himself from positivist urban modelling, but also the structuralism of **Manuel Castells** (1977). In *The Urban Question*, Castells rejects the notion of the urban as an example of spatial fetishism; he argues that spatial effects should be considered social effects. Sayer (1983) agrees with part of this critique of spatial fetishism, but still argues that space matters. Space only has effects via the particular objects, with causal powers, that constitute it.

Sayer (2000) has also sought to consider further another theme raised in the localities debate: the tension between analysis and narrative, or between law-seeking and contextual approaches. Here he notes that the 'old' debate of regional geography was characterized by these problems, and that the same charge was laid at the door of critical realists analysing localities. Specifically he accuses positivists of focusing upon temporal succession but neglecting synchronic relations. This neglect leads to positivists making a fallacious link between the unique and the independent, and assuming that regularity between events equals interdependence. Sayer argues instead for a notion of causality that recognizes variety and interdependence, while at the same time cautioning that many interdependencies tend to be unique and not transferable. Sayer's argument is that just because spatial relations are constituted by social and natural objects it does not follow that spatial relations can be reduced to their constituents. We have to examine *how* space makes a difference: this requires attention not only to the abstract theory and the specification of necessary relations, but also to the particularity of the contingent conditions that may or may not combine to produce a spatial 'effect'.

SAYER'S MAJOR WORKS

Sayer, A. (1981) 'Abstraction: a realist interpretation', *Radical Philosophy* 28: 6–15.
Sayer, A. (1982) 'Explanation in economic geography', *Progress in Human Geography* 6: 68–88.
Sayer, A. (1983) 'Defining the urban', *GeoJournal* 9: 279–285.
Sayer, A. (1985) 'The difference that space makes', in D. Gregory and J. Urry (eds) *Social Relations and Spatial Structures*. London: Macmillan, pp. 49–66.
Sayer, A. (1991) 'Behind the locality debate: deconstructing geography's dualisms', *Environment and Planning A* 23: 283–308.
Sayer, A. (2000) *Realism and Social Science*. London: Sage.
Sayer, A. (2001) 'For a critical cultural political economy', *Antipode* 33: 687–708.

Secondary Sources and References

Archer, K. (1987) 'Mythology and the problem of reading in urban and regional research', *Environment and Planning D* 5: 384–393.
Bhaskar, R. (1975) *A Realist Theory of Science*. Leeds: Leeds Books.
Castells, M. (1977) *The Urban Question*. Oxford: Blackwell.
Cooke, P. (1989) (ed.) *Localities: The Changing Face of Urban Britain*. London: Unwin.
Duncan, S. (1989) 'Uneven development and the difference that space makes', *Geoforum* 20: 131–139.
Duncan, S. (1991) special issue on 'Localities', *Environment and Planning A* 23.
Giddens, A. (1984) *The Constitution of Society*. Cambridge: Polity.
Giddens, A. (1985) 'Time, space and regionalistion', in D. Gregory and J. Urry (eds) *Social Relations and Spatial Structures*. London: Macmillan, pp. 265–295.
Gregory, D. and Urry, J. (eds) *Social Relations and Spatial Structures*. London: Macmillan.
Harré, R. (1972) *The Philosophies of Science*. Oxford: Oxford University Press.
Harvey, D. (1985) 'The geopolitics of capitalism', in D. Gregory and J. Urry (eds) *Social Relations and Spatial Structures*. London: Macmillan, pp. 128–163.
Harvey, D. (1987) 'Three myths in search of reality in urban studies', *Environment and Planning D* 5: 367–376.
Keat, R. and Urry, J. (1975) *Social Theory as Science*. London: Routledge.
Massey, D. (1984) *Spatial Divisions of Labour*. London: Hutchinson.
Morgan, K. and Sayer, A. (1988) *Microcircuits of Capital*. Cambridge: Polity.
Pratt, A C. (1995) 'Putting critical realism to work: the practical implications for geographical research', *Progress in Human Geography* 19: 61–74.
Ray, L. and Sayer, A. (eds) (1999) *Culture and Economy after the Cultural Turn*. London: Sage.
Sayer, A. (1976) 'A critique of urban modelling', *Progress in Planning* 6: 187–254.
Sayer, A. (1992) *Method in Social Science*, 2nd edition. London: Hutchinson.
Sayer, A. (1995) *Radical Political Economy: A Critique*. Oxford: Blackwell.
Sayer, A. and Walker, R. (1992) *The New Social Economy*. Oxford: Blackwell.
Smith, N. (1984) *Uneven Development: Nature, Capital and the Production of Space*. Oxford: Blackwell.
Smith, N. (1987) 'The dangers of the empirical turn: some comments on the CURS initiative', *Antipode* 19: 59–68.
Soja, E. (1985) 'The spatiality of social life: towards a transformative retheorisation', in D. Gregory and J. Urry (eds) *Social Relations and Spatial Structures*. London: Macmillan, pp. 90–127.

Andy C. Pratt

38 Amartya Sen

Amartya Kumar Sen is an intellectual of global stature, highly influential in international public debate. He grew up in Dhaka (then India, now Bangladesh) in a middle-class academic family. He was born in 1933 and partly schooled at Santiniketan (in present-day West Bengal) at a school founded by the towering Bengali literary figure, Tagore. He attended the elite Presidency College in Calcutta, graduating very early with a degree in economics (during which time he survived cancer), and moved to England in 1953 to read economics at Trinity College, Cambridge University. After more undergraduate work he completed an early technical PhD in Development Economics in only a year, studying with Joan Robinson (published in 1960; see Sen, 1968), before returning to India to a Chair at Jadavpur University, Calcutta. Remarkably, therefore, he has been a Professor since his early twenties. After taking up prestigious fellowships in the USA and the UK, he became Professor of Economics at the University of Delhi from 1963 to 1971, at the London School of Economics from 1971 to 1977, at Oxford from 1977, where he later became Drummond Professor, before leaving for Harvard in 1987 as Lamont Professor of Economics and Philosophy. He then served a spell as Master of Trinity College, Cambridge, from 1998 to 2003 (the first Asian to chair an Oxbridge college) before announcing his return to Harvard in 2004. Among his many accolades, Sen was awarded the Senator Giovanni Agnelli International Prize in Ethics, the Bharat Ratna (the President of India's highest honour), and in 1998 the Nobel Prize in Economics. The Nobel was widely perceived to be long overdue. Its tardiness may be linked to the fact that Sen has 'not been captured by economics imperialism and, unlike its practitioners, he opens and is open to debate across key issues' (Fine, 2001: 13). Indeed, while never forsaking his discipline, he has repeatedly criticized some of its core neoclassical assumptions.

Sen's interests are diverse, and have led to intellectual engagements that have spread well beyond his early classical and neoclassical training. Over the years he has contributed to social choice theory (a technical field in economics), more generally to welfare economics, the understanding and measurement of poverty, explanations of famine and hunger, agrarian change and rural development issues, as well as engaging with issues relating to ethics and moral philosophy. His empirical investigations involve wide-ranging international comparisons, but with a primary focus on South Asia. Perhaps the central theme that links his writings has been his passionate concern to redress economic and social inequality and improve human well-being.

In his early years Sen was preoccupied with welfare economics, and first made important contributions to the formalistic (mathematical) expression of social choices (Sen, 1970), alongside work on Indian development (e.g. Sen, 1962). His early technical work resulted in significant challenges to Kenneth Arrow's 'impossibility theorem', specifically the notion that individuals act *purely* according to self-interest or utilitarian criteria when making welfare or income decisions (Sen, 1977). Insisting that non-utility concerns like equity, class position and family influence matter, Sen argued that the technique of 'revealed preferences' did not distinguish what drives decisions and choices. Sen thus argued that an interpersonal comparison of utility (the value of a commodity achieved in consumption) is vital. While these observations appear commonsensical to political economists, Sen's points were challenging and controverisal to neoclassical utilitarianism. Accordingly, social choice theory still figures in Sen's work given that he is interested in how aggregative judgements may be arrived at given the 'diversity of preferences, concerns, and predicaments of the different individuals within society' (Sen, 1998; no pagination).

Since the 1960s Sen has moved from the development of formalistic models, to theorizing the ethical foundations of development itself. An important bridge-builder in his intellectual journey has been the concept of human 'capability'. By the late 1980s Sen was defining capability as 'a set of functioning bundles representing the various alternative 'beings and doings' that a person can achieve with his or her economic, social, and personal characteristics' (Drèze and Sen, 1989: 18; Sen, 1985). Capability is 'tantamount to the freedom of a person to lead one kind of life rather than another' in an Aristotelian sense (Nussbaum and Sen, 1993: 2). It follows, therefore, that an individual's capability can be *repressed* (for example, by denying a person access to basic services including food, education, land, freedom of expression, or health care), or *enabled* (for example, through supportive actions by other individuals and institutions such as a 'developmentalist' state). The core of development itself is, therefore, the enjoyment of freedom – not just freedom of speech in a narrow sense, but (positive and negative) freedom for individuals to lead valuable lives.

Development as Freedom (Sen, 1999) is consistent with his earlier work on comparisons of individual preferences and values (Sen, 1992a; 1992b). Herein, Sen defines human well-being as the enhancement of capabilities, suggesting that 'development consists of the removal of various types of unfreedoms that leave people with little choice and little opportunity of exercising their reasoned agency' (Sen, 1999: xii). For Sen, freedom is not just important in and of itself; it is also *instrumental* since it enables people to attain desired ends, and *constitutive* in the making of collective moral decisions through dialogue (Gasper, 2000). Since it is consistent with these three roles, democracy is viewed as the preferred system of governance with the greatest capacity to expand basic freedoms. Achieving development, then, requires the expansion and improvement of capabilities and entitlements for the poor and underprivileged. The purpose of the expansion of capabilities ought to be the enhancement of freedom itself, because the purpose of development is ultimately, freedom. *Development as Freedom* places some weight on markets as an efficient way to allocate resources in which free choice may be exercised. In addition to expanding capability to achieve 'functionings' (the 'beings and doings' constitutive of well-being), Sen also argues for the importance of recognizing cultural values in this process (Sen, 2004).

At another stage on his intellectual journey, Sen received widespread acclaim for his attempt to improve the indicators used to measure poverty rates and human development (Sen, 1992b, 1997). The Human Development Index (HDI) and its expression in the United Nations Development Programme's (UNDP) *Human Development Report* has thus provided an alternative to the neo-liberalist 'Washington consensus' on income-related poverty measures. Sen and researchers at the UNDP have argued that development priorities should be geared towards improving human development (capabilities) – assessed through wide-ranging multivariate indices – rather than growth-centred economic policy. Sen has also argued strongly for the extension of freedom and independence to women and children in developing countries, which he says has demonstrable effects on well-being, poverty alleviation and mortality. He regards the neglect of women's nutrition and health (not least among poor African-Americans in the US) and sex-selective abortion in developing countries as criminal (Sen, 1990, 1999).

Despite these applications of his work, Sen has largely abstained from collective action or an activist role (Corbridge, 2002), although he was involved in left-wing Bengali politics as a student (Sen, 2001). He is cautious about giving policy advice and has not done so to governments, but he has used funds from his Nobel prize to set up the Pratichi Trust, concerned with literacy and schooling in India and Bangladesh (problems explored in Drèze and Sen, 2002). In Indian affairs he labels himself somewhat left of centre, opposed to the present excesses of the ruling BJP's Hindu nationalism and its role in nuclear proliferation, and remaining a strong supporter of gender equality, pluralism and pro-poor policies. He has publicly distanced himself from some more radical positions on development, particularly in relation to Gandhi's belief in local self-sufficiency and technological skepticism, and he sees an 'ugly side' to many forms of localist, communitarian and sectarian politics.

Sen is a qualified supporter of economic globalization for its potential to tackle poverty and inequality, even though it 'doesn't always work' (Sen, 2001; no pagination) and should not be reduced to the unfettered expansion of trade and markets. His point here is that globalization is nothing new, and that the poor should be able to share in its gains. His qualified support for civil society movements comes with a caveat: 'One's concern for equity and justice in the world must not carry one into the alien territory of unreasoned belief' (Sen, 2001; no pagination). His approach to these and other issues is always beautifully expressed, rational and developed through

rich and thoughtful analysis. It is important to restate that equality and justice, and the rights of the poor not to suffer 'unfreedoms', figure centrally in almost all his work (Sen, 2001; no pagination).

SPATIAL CONTRIBUTIONS

In his attempt to insert the theoretical space for a dialogue on equality into the rather introverted world of mainstream economics, and to be specific about the notions of equality and how to realize them, Sen has made some insights that have been debated and applied by geographers and in development studies. Much of his work pays attention to space, through inter-area and inter-state comparisons (e.g. Drèze and Sen, 2002) and through specific reference to places and historical events. Yet Sen holds to a cosmopolitan view of territoriality, and thus in accordance in his belief in universal rights, he argues that culture transcends place, rather than, as **Arturo Escobar** argues, that culture 'resides in places'. As Sen puts it, 'culture' is not impermeable to rational consideration and choice (Sen, 2004).

By far the greatest attention paid to Sen by geographers has been to his work on poverty and famine. Sen has not been alone in identifying insidious, quotidian hunger as a major concern, but his explanation of widespread famine has certainly seized the public interest. When Sen was a small boy he witnessed the direct effects of the 1943–45 famine in Bengal. The central argument of his most famous work, *Poverty and Famines* (Sen, 1981; see also Devereux, 2001), is that famine is not caused by a negative Malthusian relationship between population and food supply, but by the inability of famine-prone individuals to access food in times of great need, even when food supplies are adequate and the rich are still eating well. Although this is not phrased as a normative statement, the implication is that famine can be construed as a food *demand* problem, not a *supply* problem. Access to food is obtained when one has entitlement to it, and 'Famine results where access to food is reduced because of processes denying or lessening entitlement to food' (Sen, 1981). 'Endowments' are the assets and resources that people may theoretically access – 'entitlements' are those that are available, and are therefore cognate with 'acquisition power'. Survivors of serious famines have the power to acquire food – to grow it (production-based entitlement), to buy it (trade-based entitlements), by selling their labour for cash or food (own labour entitlement), by being given food by others (inheritance and transfer entitlement), or through what Sen terms 'extended entitlements' (including looting – Drèze and Sen, 1989).

Poverty and Famines was Sen's most influential work outside economics, and his 'food entitlement decline' (FED) explanation was further elaborated with the gifted economist Jean Drèze, in several volumes, using extended empirical cases (Drèze and Sen, 1989, 1990/1991). India, as Sen noted, has not suffered a major famine since 1947 (but has been close) because its leaders have been held to account through voting and a free press, which has influenced a more proactive stance on addressing demand-side constraints through food distribution and work programmes. Sen's ideas have been influential to the work of geographers and anthropologists including Piers Blaikie, Terry Cannon, Ian Davis and Ben Wisner (Blaikie *et al.*, 2003), Neil Adger, Susanna Davies, Robert Kates, Ken Hewitt, Hans Bohle (1993), and **Michael Watts** (1983). His work has been taken up in famine relief management, causing agencies to focus more on food access as well as basic food relief. Sen's theories have also aided the development of famine early-warning systems used to alert agencies and governments to impending food stress by looking at price signals in markets.

A fascinating offshoot of entitlement theory is the 'environmental entitlements' idea developed at the Institute for Development Studies, Sussex, in the 1990s (Leach *et al.*, 1999). Here entitlements and endowments thinking is reoriented to refer to natural resources, not food availability, with a specific focus on the role of institutions in mediating differentiated resource access and entitlements. This same group (along with earlier work by Robert Chambers, Gordon Conway and others) has also been instrumental in developing new ways to perceive rural livelihood systems, unpacking the dynamic and flexible ways in which rural people manage their endowments and risks in ways that echo some of Sen's core insights (Bebbington, 1999). 'Livelihoods thinking' has found a home in several international development agencies (see www.livelihoods.org), and Williams and Windebank (2003) have now applied livelihood and capability analysis to alternative economic geographies and poverty in the UK.

KEY ADVANCES AND CONTROVERSIES

Sen's work is particularly challenging to critical scholars because much of it is highly original and it cuts across disciplines and schools of thought, making it difficult to stereotype. The radical implications that may be drawn from his support for human emancipation, equality, justice and poverty alleviation have largely been taken up by others (including many PhD students), pushing his sometimes abstract ideas much further. Sen himself practices 'cautious boldness', raising issues dear to the critical left, but retaining a faith in market economics. Calls to extend his frameworks are thus more common than pleas to reject them (Fine, 2001; Alkire, 2002). For example, Ben Fine (a former student) finds an 'unresolved tension between micro-foundations of entitlement analytic and the broader recognition of famine as irreducibly macro, not least because famine is more than the sum of its individual parts . . .' (Fine, 2001: 8). Sen is aware of political-economic forces, pointing out that 'famine is dependent upon 'the exercise of power and authority . . . alienation of the rulers from those who ruled . . . the social and political distance between the governors and the governed' (Sen, 1999: 170), but Fine suggests that Sen's understanding of power and structures is 'superimposed, not built, upon the micro-foundations'.

Development geographers have likewise taken up the challenge of explaining famines. Blaikie *et al.* (2003) develop heuristic models that make multi-level causation of famine, and processes of vulnerability, more explicit. **Michael Watts**, who provided the most detailed Marxist argument for the colonial origins of famine (1983), is broadly sympathetic to Sen. He argues that if poverty (and famine) are a result of 'capability failure' and a lack of entitlements, the prime response must be to overcome vulnerability and to address demand for food. But processes of oppression can lie at famine's roots, the development of 'critical autonomy' and a strong 'sense of society' is needed to enhance freedoms (Watts, 2000: 57). Following Marx, Watts suggests that entitlements are ' . . . both constituted and reproduced through conflict, negotiation and struggle' (Watts, 2000: 62).

A powerful school of thought has developed this argument about social and political embeddedness of entitlement failures. Armed conflict and violence in famine situations serve certain elite interests, and famine can be used as a political weapon (Keen, 1994; de Waal, 1989; Middleton and O'Keefe, 1998). 'Complex emergencies' and 'war famines', where 'freedom to choose' is clearly violated beyond repair, have almost no treatment in Sen's approach. Other important criticisms include the fact that many famine

deaths occur through ill health, not starvation (de Waal, 1989), and that endowments − like communal livestock herds, or common property − lack clear property rights and can be the subject of overlapping claims, making them 'fuzzy' (Devereux, 2001: 253).

In Sen's work there is little discussion on the relationship between the production of 'opulence' resulting from growth and from wealth, and the depravation of capabilities and entitlement (Cameron and Gasper, 2000: 986). Other critics call for a richer conception of 'self' that sees human agents as much more than beings seeking to attain capabilities (Giri, 2000: 1018). Unlike **Anthony Giddens'** conception of the *autotelic self*, a 'reflective dimension of self', with values and worldviews and responsibility, is not a full part of Sen's analysis despite his call to insert value into welfare analysis (Alkire, 2002).

Lastly, there is a certain level of ambiguity in Sen's notion of 'development as freedom'. Freedom is, more often than not, accompanied by its foreclosure. Sen is less than clear about tackling the conflicts that result from people's pursuit of freedom: development seems to be about opportunity and 'choice', to which there must be limitations (Gasper, 2000: 999). But how exactly are freedoms achieved? By what political means? How may the poor marshal power to challenge vested interests other than through votes? Does Sen really support protest to support freedoms? Critics see difficulties here, and a failure to recognize the vitality of 'concerted struggles against the powers of vested interests, at all spatial scales' (Corbridge, 2002: 209). Perhaps, then, it is the quality of governance, rather than an absolute adherence to textbook democracy, that matters more for equitable development.

Several authors suspect that the reasons for certain absences in Sen's work have to do with his ethics, and his commitment to economic principles which he has extended but not overturned (Fine, 2001). Corbridge argues that 'Sen's liberalism leaves him poorly equipped to deal with questions of entrenched power and the politics of conflict and social mobilization' (2002: 203), and there is no strong political economy of capitalism in his work, which is surprising given the political and ethically charged issues that he has addressed (Fine, 2001: 12). This is not to suggest at all that he is unaware of such issues (Sen, 2001, 2003).

Amartya Sen is a beacon of common sense in the interdisciplinary terrain he occupies. Entitlements and capability stress human agency, not constraint, and they carry 'some sense of worth and of real people's lives' (Gasper, 2000: 996). His commitment to understanding and resolving inequality, and expanding basic freedoms, has placed utilitarians and many mainstream economists in an uncomfortable position. Social scientists in general should be enormously grateful that this important ambassador for 'humane economics' (Desai, 2001) can reach the ears of policymakers worldwide with his beguiling ideas.

SEN'S MAJOR WORKS

Drèze, J. and Sen, A. K. (1989) *Hunger and Public Action*. Oxford: Clarendon Press.
Nussbaum, M. and Sen, A. K. (1993) *The Quality of Life*. Oxford: Clarendon Press.
Sen, A. K. (1970) *Collective Choice and Social Welfare*. San Francisco: Holden-Day.
Sen, A. K. (1981) *Poverty and Famines: An Essay on Entitlements and Deprivation*. Oxford: Clarendon Press.
Sen, A. K. (1985) *Commodities and Capabilities*. Amsterdam: North-Holland (republished, Oxford, 1999).
Sen, A. K. (1997) *On Economic Inequality. An Expanded Edition with a Substantial Annexe by James Foster and Amartya Sen*. Oxford: Clarendon Press (1st edition 1973).
Sen, A. K. (1999) *Development as Freedom*. Oxford: Oxford University Press.

Secondary Sources and References

Alkire, S. (2002) *Valuing Freedoms. Sen's Capability Approach and Poverty Reduction*. Oxford: Oxford University Press.
Bebbington, A. J. (1999) 'Capitals and capabilities: a framework for analyzing peasant viability, rural livelihoods and poverty', *World Development* 27: 2021–2044.
Bohle, H.-G. (ed.) (1993) *Worlds of Pain and Hunger: Geographical Perspectives on Disaster Vulnerability and Food Security*. Saarbrucken/Fort Lauderdale, FL: Verlag Breitenbach Publishers.
Blaikie, P., Cannon, T., Davis, I. and Wisner, B. (eds) (2003) *At Risk: Natural Hazards, People's Vulnerability and Disasters*, 2nd edition. London: Routledge.
Cameron, J. and Gasper, D. (2000) 'Amartya Sen on inequality, human well-being, and Development as Freedom', *Journal of International Development* 12 (7): 985–1045.
Corbridge, S. E. (2002) 'Development as freedom: the spaces of Amaryta Sen', *Progress in Development Studies* 2: 183–217.
Desai, M. (2001) 'Amartya Sen's contribution to development economics', *Oxford Development Studies* 29 (3): 213–223.
Devereux, S. (2001) 'Sen's entitlement approach: critiques and counter-critiques', *Oxford Development Studies* 29 (3): 244–263.
Drèze, J. and Sen, A. K. (eds) (1990/1991) *The Political Economy of Hunger*, 3 volumes. Oxford: Clarendon Press.
Drèze, J. and Sen, A. K. (2002) *India: development and participation*, 2nd edition. Oxford: Oxford University Press.
Fine, B. (2001) 'Amartya Sen: A partial and personal appreciation', CDPR Discussion Paper 1601. Centre for Development and Policy Research. SOAS *http://www.soas.ac.uk/Centres/CDPR/DP16BF.pdf*
Gasper, D. (2000) 'Development as freedom: taking economics beyond commodities – the cautious boldness of Amartya Sen', *Journal of International Development* 12: 989–1001.
Giri, A. K. (2000) 'Rethinking human well-being: a dialogue with Amartya Sen', *Journal of International Development* 12: 1003–1018.
Keen, D. (1994) *The Benefits of Famine. A Political Economy of Famine and Relief in Southwestern Sudan, 1983–1989*. Princeton: Princeton University Press.

Leach, M., Mearns, R. and Scoones, I. (1999) 'Environmental entitlements: dynamics and institutions in community-based natural resource management', *World Development* 27: 225–247.

Middleton, N. and O'Keefe, P. (1998) *Disaster and Development: The Politics of Humanitarian Aid*. London: Pluto.

Sen, A. K. (1962) 'An aspect of Indian agriculture,' *Economic Weekly*. Annual Number, p. 16.

Sen, A. K. (1968) *Choice of Techniques*, 3rd edition. Oxford: Blackwell (1st edition 1960).

Sen, A. K. (1977) 'Rational fools: a critique of the behavioral foundations of economic theory', *Philosophy and Public Affairs* 6 (4): 317–344.

Sen, A. K. (1990) 'More than 100 million women are missing', *New York Review of Books* 20 December: 61–66.

Sen, A. K. (1992a) *Choice, Welfare and Measurement*. Oxford: Blackwell (and Harvard University Press, 1997).

Sen, A. K. (1992b) *Inequality Re-examined*. Oxford: Oxford University Press.

Sen, A. K. (1995) 'Rationality and social choice', *American Economic Review* 85 (1): 1–24.

Sen, A. K. (1998) 'Autobiography'. Swedish Academy of Sciences Nobel Prize website. *www.nobel.se/economics/laureates/1998/sen-autobio.html* (published in *Les Prix Nobel, 1999*).

Sen, A. K. (2001) Interview with Sen by David Barsamian. Alternative Radio, Colorado. *http://www.indiatogether.org/interviews/sen.htm*

Sen, A. K. (2003) *Rationality and Freedom*. Cambridge, MA: Harvard University Press.

Sen, A. K. (2004) 'How Does Culture Matter?', in V. Rao and M. Walton (eds) *Culture and Public Action*. Stanford: Stanford University Press.

de Waal, A. (1989) *Famine That Kills: Darfur, Sudan, 1984–1985*. Oxford: Clarendon Press.

Watts, M. J. (1983) *Silent Violence: Food, Famine and Peasantry in Northern Nigeria*. Berkeley: University of California Press.

Watts, M. J. (2000) *Struggles over Geography: Violence, Freedom and Development at the Millennium*. Hettner Lectures No. 3. Heidelberg: Department of Geography, University of Heidelberg.

Williams, C. C. and Windebank, J. (2003) *Poverty and the Third Way*. London: Routledge.

Simon Batterbury and Jude Fernando

39 David Sibley

David Sibley's research on socio-spatial exclusionary processes has enhanced geographers' understanding of exclusion, marginalization and difference. Born in England in 1940, Sibley completed his first degree in Geography at Liverpool University in 1962, and then travelled to America, where he was awarded an MA at Southern Illinois University. Sibley then returned to the UK to take up a planning job in Durham for two years, before returning to the USA in 1966 to teach at Temple University. While at Temple, Sibley read **Peter Haggett**'s *Locational Analysis in Human Geography* (1965) and this book influenced his decision to return to England to start a PhD in Geographical Modelling at Cambridge. His thesis consisted of the mathematical modelling of gradients, surfaces and point patterns (see Sibley, 1970, 1972, 1973). Reflecting on this, Sibley recalls, 'I was rather taken by the beauty of curves and impressed by my own ability to do maths, having been bottom of the class most of the time during high school!' (personal correspondence, 2002).

Sibley's research took on a dramatically different focus after his Cambridge days. He spent several years working outside of the academy with English Gypsies, together with his partner who started a school for Gypsy children in the Hull area. This collaborative work was to provide the basis of his first book, *Outsiders in an Urban Society* (1981). Drawing from theory in educational sociology (especial-ly that of Basil Bernstein) and social anthropology (in particular, the work of Mary Douglas), *Outsiders in an Urban Society* began to develop a distinctive take on the social construction of space. Sibley's engagement with Douglas' ideas on purity and danger was particularly noteworthy, having been introduced to her work through a radio interview with Judith Okely (author of *The Traveller Gypsies*).

Sibley would spend much of the next 20 years developing his ideas about 'outsiders' and the production of space, extending his research beyond Gypsies to include children and family dynamics, with a particular focus on domestic space. In the process he continued to cross further disciplinary boundaries. This work and thinking culminated in 1995, in the publication of *Geographies of Exclusion*, a book that has provided a key contribution to the geographic literature on identity and marginalization. Here, Sibley drew from not just social anthropology, but also feminist theory, human geography and (especially) psychoanalysis to showcase the tendency of powerful groups to 'purify' space and to see minority groups as dirty and polluting. Drawing particularly on psychoanalytical ideas about the importance of maintaining self-identity (literally, maintaining the boundaries of the Self), Sibley argued that the urge to exclude threatening Others is connected to deeply ingrained and often subconscious desires to maintain cleanliness and purity, many of which may be inculcated in early infancy:

> Experience of the world in childhood also involves the confirmation of the boundaries of the self and situating the self in the social world through the sorting of

people and things into 'good' and 'bad' categories. 'Good' and 'bad' enter the unconscious and, in the process of socialization, they are projected onto others who become the objects of fears and desires. (Sibley, 2001: 244)

Following Kristeva (1982), *Geographies of Exclusion* explored the ways in which an abject fear of the self being defiled or polluted is mapped onto those individuals and groups depicted by hegemonic society as deviant or dangerous. This engagement with psychoanalytical 'object relations' theory offered new ways for thinking about processes of marginalization, implying that the way people seek to exclude Other groups can only be understood with reference to the manner in which people identify with or against particular stereotypes on a psychic level. These ideas have been developed in Sibley's subsequent writing that considers the notion of psychogeographies and applying spatialized object relations theory in other fields outside of geography (Sibley, 2003).

SPATIAL CONTRIBUTIONS

Sibley's research has shown that varied forms of spatial exclusion reinforce and maintain social boundaries in society. He has effectively demonstrated that individuals attempt to distance themselves from 'bad' objects/subjects because of their desire to maintain purity and cleanliness. Such understandings, Sibley argues, are circulated within the larger culture through cultural representations of marginalized groups as deviant, different and dangerous. Yet Sibley's understandings of identity, marginalization and exclusion have shifted since he first considered these concepts in the early 1980s, perhaps as a result of his growing interest in psychoanalysis and the work of Melanie

Klein and Julia Kristeva. In his first book, *Outsiders in Urban Societies,* he insisted that a nuanced analysis of Gypsies and indigenous minorities in society can only be accomplished through a comparative framework that takes into account 'changes in the economy and social structure of the outsider group as they are affected by processes operating in the dominant social system' (Sibley, 1981: vii). *Geographies of Exclusion* extended this perspective on the insider/outsider binary by drawing from psychoanalytical theory to understand the relation of Self and Other.

Exploring how social and spatial boundary processes separate some groups and individuals from the mainstream, *Geographies of Exclusion* provided an elegant conceptualization of social and spatial exclusion, demonstrating that 'who is felt to belong and not to belong contributes in an important way to the shaping of social space' (Sibley, 1995a: 3). Further, it provided one of the first attempts by a geographer to explore the role of the psyche in socio-spatial process. The merits of a such an approach are summarized by Sibley thus:

Psychoanalysis provides one way of thinking about social relationships and relationships with the material world. It brings the unconscious into social theory, connecting the unconscious with the phenomenal and experiential aspects of life ... it constitutes a way of articulating anxieties and desires which are elements of the spatiality of human experience. (Sibley, 2001: 243)

Highlighting the fears and insecurities that we carry around with us as a result of the often-traumatic separation of (pre-Oedipal) Infant and Mother, Sibley's use of psychoanalytical theory thus relates the search for spatial order to the personal search for certainty and security. In discussing the merits of psychoanalysis for understanding socio-spatial process, Sibley notes that psychoanalytical theory 'has considerable value because it can help us to understand better not only the representation of others but also our own feelings about the abject, our own inse-curities about difference which affect our academic practice' (Sibley, 1995a: 185).

Sibley has also written about social exclusion within the discipline of geogra-phy itself, underscoring how geography as a whole tends to marginalize particular groups, especially women and people of colour. Extending his concerns on the exclusionary nature of academic practice, Sibley developed notions around power and the ways it is conceptualized in academe. Sibley admits that his interest in the sociology of knowledge was in part inspired by an array of rejection letters he received in regards to his article on the purification of space that was eventually published in *Society and Space* (Sibley, 1988, 1999). He suggests that local bound-ary disputes in academia are symptomatic of wider processes of gender and racial oppression. Noting the exclusions that impinge on the production of geographic knowledge, he argues for a more inclusive discipline:

> If critical ideas come from the oppressed, for example, from women or black authors in certain contexts ... they may be considered dangerous because they challenge white, heterosexual male domi-nation of the western knowledge industry. (Sibley, 1995a: 116)

Sibley maps the exclusionary nature of geography by taking urban studies to task for neglecting the research of W. E. B. Dubois, and suggesting that gender, poli-tics and epistemology worked to mar-ginalize women researchers examining the city between 1910 and 1930. Caution-ing geographers that academic post-modern writing that celebrates diversity may not necessarily signal a significant break with the masculine and racist na-ture of geography, Sibley shows how the failure or success of particular ideas in geography are profoundly affected by the contexts in which they are produced.

Within the framework of his interest on social exclusion and space, Sibley has made a significant contribution to work in social geography on family dynamics and the production of domestic space. Al-though geographical work on the home had identified it as a 'locus of power relations' (Sibley, 1995a: 92), this research predominantly focused on gender rela-tions between adult household members. Sibley (1995a, 1995b) addressed the impli-cit adultist nature of this research by highlighting the role of the home as an important site for the negotiation of child/ adult relations. In a collaborative project with psychologists Geoff Lowe and David Foxcroft, from the University of Hull (Sibley and Lowe, 1991; Lowe *et al.*, 1993), Sibley studied the psychosocial dy-namics of the home life of drinkers as a means through which to understand the patterns that contribute to adult drinking behaviour.

In doing so, Sibley and colleagues drew on the work of Basil Bernstein. In categorizing families, Bernstein distin-guishes between what he terms *positional* and *personalizing* families. In *positional* families power is vested in positions, for example in that of 'the father'; in *person-alizing* families power is more equally distributed between members. Sibley and colleagues (Sibley and Lowe, 1992; Lowe *et al.*, 1993) used this distinction to reflect on the boundaries of acceptable behav-iour in different households, including limits on the use of time and space. They argued that children's use of space and time in positional families is strongly bounded by a set of rigid rules. In con-trast, *personalizing* families are less con-strained by arbitrary rules, and children are given a greater voice in negotiating domestic spatial and temporal bound-aries.

In subsequent solo work, Sibley (1995b) further developed his interest in the bounding of domestic space. Here he drew on data about middle-class child-hoods in the UK from a Mass Observation archive to show how children, when given their own bedrooms, appropriate, transform and secure the boundaries of this space to make it their own. He compared these accounts with descriptions of other spaces within the home where children are subject to more parental control. Notably he highlighted that whether a living room is a child space or an adult space can change with the time of day, and he showed how the timing of activities and the division of space can create liminal zones within the home. Reflecting on domestic tensions between adults and children about the use of different rooms within the home, Sibley (1995b) concluded that they represent a clash between adults' desire to establish order, regularity and strong boundaries, and children's preferences for disorder and weak boundaries.

KEY ADVANCES AND CONTROVERSIES

Sibley's work has proven pivotal and inspirational for geographers who are concerned with the role of the unconscious in everyday life (see vol. 4 (3) of *Social and Cultural Geography*) as well as those seeking to construct anti-essentialist and relational theories of identity and space (Dwyer and Jones, 2000). In particular, Sibley's (1995a) *Geographies of Exclusion* has provided an important template for geographers who wished to understand how psychologically rooted fears of the Other are reproduced in various societies (e.g. Takahashi, 1997; Hubbard, 1998). However, Sibley's psychoanalytic geographies are primarily focused on understanding boundary-making and spatial transgression through the work of object relations theory (especially the work of Julia Kristeva and Melanie Klein). Others have extended this engagement by exploring the exclusionary geographies implied in the work of Freud and Lacan; for example, Wilton (1998) has utilized Freud's concept of *unheimlich* ('the uncanny') to explore processes of

exclusion, in his research on community opposition to human service facilities. Other geographers have used Freud and/or Lacan to re-theorize other aspects of the relationship between subjectivity and spatiality. For example, Steve Pile (1996) has bought the work of Freud and Lacan into dialogue with the work of **Henri Lefebvre** to think through the relationship of the body and the city, while Stuart Aitken (2001) has used both Freud and the object relations theorist Derek Winnicott to analyse the contested spaces of childhood. Moreover, given the emphasis placed on the production and repression of sexuality in many psychoanalytical theories, it is perhaps unsurprising that this geographic engagement with psychoanalysis has been particularly pronounced in feminist geography. Liz Bondi (1999), for example, has drawn on a number of psychoanalytic theories to explore the relationship between psychotherapeutic practice and an empathic human geography, while Heidi Nast (1998, 2000) has examined the spatialization of Oedipalization on a variety of different scales. Furthermore, Rose (1995, 1996) and Bondi (1997) have engaged with the ideas of Luce Irigary to critique the epistemology of geographical enquiry itself. As such, while Sibley, along with Steve Pile and **Gillian Rose**, have been hugely important for initiating a 'psychoanalytic turn' in geography, others have quickly forged new paths, adopting alternative theorists and frameworks to examine a multitude of different issues.

Sibley's work has also been influential in the development of the subdisciplinary area of children's and young people's geographies. In an article published in the journal *Area*, James asked the question 'is there a place for children in geography?' (James, 1990). Sibley wrote a response to this in which he highlighted work on children within the discipline that had been largely overlooked, and made a case for the importance of understanding children's use of space. This debate did much to spark new interest in this field. The subsequent decade saw work on children's and young people's geographies begin to reach a critical mass (Skelton and Valentine, 1998; Holloway and Valentine, 2000). Sibley, along with Chris Philo, played an important role in this development, not only through his own work about adult/child relations and the bounding of space, but also through the support and encouragement he provided to young academics who were beginning to publish and organize conference sessions in this area.

Beyond his own work on adult–child relations, Sibley's theorization of exclusion has been mobilized by others to explore young people's experiences of space. For example, Sibley's insistence that '[the] child/adult [separation] illustrates a contested boundary . . . [and that] adolescence is an ambiguous zone within which the child/adult boundary can be variously located according to who is doing the categorizing' (Sibley, 1995a: 34–35) has been invoked by cultural geographers seeking to understand the fluid and shifting ways that 'youth' has been identified and constructed as a transitional category between childhood and adulthood (Skelton and Valentine, 1998). **Doreen Massey**, in her research on the spatial identities of youth cultures, also draws from Sibley's research on the territorialization of space to demonstrate that strategies of spatial organization are always embedded within the social production of identities (Massey, 1998). Drawing from case studies in Mexico and the UK, Massey argues that youth carve out their own spaces for interaction. Tim Lucas

(1998) also finds Sibley's research useful in his own analysis of panic, anxiety and fear that is incited through local concerns with gang violence in Santa Cruz, California. Employing Sibley's insights on images of difference and the ways stereotypes collude to create landscapes of exclusion (Sibley, 1995a), Lucas points out that the social and spatial distancing of the city's Beach Flats area has been successfully invoked through oppressive representations of the community as 'an eyesore, dirty and unhealthy' (Lucas, 1998: 148). Sibley's 'language of defilement' (Sibley 1995a: 55) serves to illuminate the ways the neighbourhood is framed and defined, thus creating geographies of representation that serve to exclude residents of the Beach Flats.

Though some geographers remain indifferent or even hostile to the language and methods of psychoanalysis (see Parr and Philo, 2003), there is little doubt that Sibley's development of psychoanalytical theory has greatly contributed to geographical understandings of the Other. Furthermore, his work on exclusion within the discipline, notably the ways that minority and women's voices are marginalized within geography, have informed other geographers' critiques of the masculine and racist nature of the discipline. Overall, Sibley's insights on exclusion continue to play a formative role in creating space for new geographies to emerge that are open to different kinds of knowledges about how Self and Other are conceptualized.

SIBLEY'S MAJOR WORKS

Sibley, D. (1981) *Outsiders in Urban Societies*. New York: St. Martin's Press.

Sibley, D. (1986) 'Persistence or change? Conflicting interpretations of peripheral minorities', *Environment and Planning D: Society and Space* 4: 57–70.

Sibley, D. (1988) 'Survey 13: purification of space', *Environment and Planning D – Society and Space* 6: 409–421.

Sibley, D. (1990) 'Invisible women? The contribution of the Chicago School of Social Service Administration to urban analysis', *Environment and Planning A* 22: 733–745.

Sibley, D. (1991) 'Children's geographies: some problems of representation', *Area* 3: 269–270.

Sibley, D. (1995a) *Geographies of Exclusion*. London: Routledge.

Sibley, D. (1995b) 'Families and domestic routines: constructing the boundaries of childhood', in S. Pile and N. Thrift (eds) *Mapping the Subject*. London: Routledge, pp. 123–137.

Sibley, D. (2001) 'The binary city', *Urban Studies* 38: 239–250.

Sibley, D. (2003) 'Reflections on geography and psychoanalysis', *Social and Cultural Geography* 4: 391–400.

Sibley, D. and Lowe, G. (1991) 'Boundary enforcement in the home environment and adolescent alcohol abuse', *Family Dynamics of Addiction Quarterly* 1: 52–58.

Secondary Sources and References

Aitken, S. (2001) *Geographies of Young People*. London: Routledge.

Bondi, L. (1997) 'In whose words? On gender identities, knowledge and writing practices', *Transactions of the Institute of British Geographers* NS 22: 245–258.

Bondi, L. (1999) 'Stages on journeys: some remarks about human geography and psychotherapeutic practice', *Professional Geographer* 51: 11–24.

Callard, F. (2003) 'The taming of psychoanalysis in geography', *Social and Cultural Geography* 4 (3) 295–312.

Dwyer, O. and Jones, J. P. (2000) 'White socio-spatial epistemology', *Social and Cultural Geography* 1: 209–222.

Haggett, P. (1965) *Locational Analysis in Human Geography*. London: Arnold.

Holloway, S. L. and Valentine, G. (eds) (2000) *Children's Geographies: Playing, Living, Learning*. London: Routledge.

Hubbard, P. (1998) 'Community action and the displacement of street prostitution: evidence from British cities', *Geoforum* 29: 269–286.

James, S. (1990) 'Is there a place for children in geography?', *Area* 22: 278–283.

Kristeva, J. (1982) *Powers of Horror*. New York: Columbia University Press.

Longhurst, R. (2001) 'Geography and gender: Looking back, looking forward', *Progress in Human Geography* 25: 641–648.

Lowe, G., Foxcroft, D. and Sibley, D. (1993) *Adolescent Drinking and Family Life*. London: Harwood.

Lucas, T. (1998) 'Youth gangs and moral panics in Santa Cruz, California', in T. Skelton and G. Valentine (eds) *Cool Places: Geographies of Youth Culture*. London: Routledge, pp. 145–160.

Massey, D. (1998) 'The spatial construction of youth cultures', in T. Skelton and G. Valentine (eds) *Cool Places: Geographies of Youth Cultures*. New York: Routledge, pp. 121–129.

Nast, H. (1998) 'Unsexy geographies', *Gender, Place and Culture* 5: 191–206.

Nast, H. (2000) 'Mapping the "unconscious": racism and the Oepidal family', *Annals of the Association of American Geographers* 90: 215–255.

Parr, H. and Philo, C. (2003) 'Psychoanalytic geographies: an introduction', *Social and Cultural Geography* 4 (3): 283–293.

Philo, C. (1986) *The Same and The Other: On Geography, Madness and Outsiders*. Loughborough University of Technology, Department of Geography, Occasional Paper 11.

Philo, C. (1997) 'Of other rurals?', in P. Cloke and J. Little (eds) *Contested Countryside Cultures: Otherness, Marginalisation and Rurality*. London: Routledge, pp. 19–50.

Pile, S. (1996) *The Body and the City: Psychoanalysis, Space and Subjectivity*. London: Routledge.

Rose, G. (1995) 'Distance, surface, elsewhere: a feminist critique of the space of phallocentric self/knowledge', *Environment and Planning D: Society and Space* 13: 761–781.

Rose, G. (1996) 'As if the mirror had bled: masculine dwelling, masculinist theory and feminist masquerade', in N. Duncan (ed.) *BodySpace*. London: Routledge, pp. 56–74.

Sibley, D. (1970) 'Density gradients and urban growth', *Urban Studies* 7: 294–297.

Sibley, D. (1972) 'Strategy and tactics in the selection of shop locations', *Area* 4: 151–157.

Sibley, D. (1973) 'The density gradients of small shops in cities', *Environment and Planning* 5: 233–240.

Sibley, D. (1978) 'Classification and control in local government', *Town Planning Review* 49: 319–328.

Sibley, D. (1990) 'Urban change and the exclusion of minority groups in British cities', *Geoforum* 21: 483–488.

Sibley, D. (1994) 'The sin of transgression', *Area* 26: 300–303.

Sibley, D. (1998) 'Sensations and spatial science: gratification and anxiety in the production of ordered landscapes', *Environment and Planning A* 30: 235–246.

Sibley, D. (1999) 'Comments on "Stages on Journeys" by Liz Bondi', *The Professional Geographer* 51: 451–452.

Skelton, T. and Valentine G. (eds) (1998) *Cool Places: Geographies of Youth Cultures*. London: Routledge.

Takahashi, L. (1997) 'The socio-spatial stigmatisation of homelessness and HIV/AIDS: towards an explanation of the NIMBY syndrome', *Social Science and Medicine* 45: 903–914.

Wilton, R. (1998) 'The constitution of difference: space and psyche in landscapes of exclusion', *Geoforum* 29: 173–185.

Young, J. (1999) *The Exclusive Society*. London: Sage.

Minelle Mahtani

40 Neil Smith

BIOGRAPHICAL DETAILS AND THEORETICAL CONTEXT

Neil Smith was born in Scotland in 1954 and studied for a BA in Geography at the University of St Andrews. In 1977 he left Scotland for Baltimore, where he began PhD research into gentrification, under the supervision of **David Harvey** at Johns Hopkins University (graduating in 1982). Though younger than Harvey, Smith underwent a similar academic and political transformation to that of his supervisor. This entailed a rejection of the kind of geography taught to him as an undergraduate – a so-called 'spatial science' that Harvey's late-1960s writings embodied – and an embrace of the Marxist approach that Harvey pioneered in his 1973 book *Social Justice and the City*.

At Johns Hopkins, Smith engaged deeply with the later works of nineteenth-century political economist Karl Marx. This engagement was, initially, focused on the issue of the gentrification of inner-city neighbourhoods in Western cities. From the late 1960s a number of so-called 'ghettos' (or inner-city blight areas) underwent redevelopment in cities like New York. This redevelopment had a physical and social dimension, wherein the removal and renovation of older buildings was accompanied by an influx of urban professionals. Smith's principal concern, in his doctoral thesis, was to explain why gentrification occurs. The result was the concept of the 'rent gap' – to be explained below – which Smith popularized in some of his early publications (e.g. Smith, 1979).

However, Smith's engagement with Marxism saw his interests spiral out from the issue of gentrification alone. This issue, Smith came to realize, was just one instance of a wider phenomenon of uneven geographical development. Though a number of non-geographers had written insightfully about the causes of uneven development during the 1960s and 1970s (e.g. Amin, 1977), few had done so with an eye for how space was what Harvey (1985: 4) called 'an active moment' in the process. What is more, few of these authors had drawn out the largely implicit theory of uneven development to be found in Marx's later writings about the operations of the capitalist economic system – a system that, by the late 1970s, was increasingly global in reach. The result was Smith's landmark book *Uneven Development* (1991; first edition 1984) published while Smith was an Assistant Professor of Geography at Columbia University, New York. Like Harvey's magisterial *The Limits to Capital*, published two years earlier, *Uneven Development* was a major attempt to show how and why space matters to capitalism and capitalism to space. The book thereby served two principal purposes: it provided a theoretical basis for future work among an emergent Marxist community in the discipline of geography; and it was among the first books published within the wider, interdisciplinary community of Marxists that inserted space into the core of Marx's theory of capitalism – Harvey and **Henri Lefebvre** being the two other main proponents of a spatialized Marxism at that time.

Uneven Development contained highly fertile ideas that subsequently inspired both Smith and a generation of radical geographers. Aside from uneven development, these ideas included the 'production of nature', the 'production of space' and the 'production of geographical scale' – the first and last of which will be explored below. Shortly after *Uneven Development* was published, Smith moved to the Geography Department at Rutgers University, New Jersey, where he remained until 2000. There, he extended his earlier researches and embarked upon new ones. On the one side, he continued his theoretical and empirical inquiries into gentrification in New York and elsewhere, resulting in *The New Urban Frontier* (1996) and subsequent refinements/applications of the rent-gap thesis (e.g. Hackworth and Smith, 2001). This was accompanied by a development of Smith's interest in geographical scale in the direction of what he called a 'spatialized politics' (Smith, 1992, 1993). This is a politics that recognizes the importance of space in everyday life and shows struggles over the control of space to be flashpoints for power and resistance. On the other side, Smith developed an interest in the role of professional geographers and geography as a discipline in their wider societies (e.g. Smith, 1986a; Smith and Godlewska, 1994). Specifically, he focused on the life of American political geographer Isaiah Bowman (1878–1950) and his role in America's rise to geopolitical dominance before, during and after World War II. This resulted in *Mapping the American Century* (2002), which sets Bowman in his academic and political context and shows the links between his intellectual work and US geopolitical strategy (see also Smith, 1986a). This book is one of an emerging set of 'critical histories' of the production, circulation and use of geographical knowledge. These histories show that geographical ideas are political and material, altering the very realities they purport to describe.

Like his former supervisor, Smith's project has been (1) to insert geographical concepts like space and nature into the heart of critical thinking, and (2) to show that geography is too important to be left to geographers alone. Smith is currently a Professor in Anthropology at City University of New York and Director of the Interdisciplinary Center for Place, Culture and Politics at that institution. Within the discipline of geography he is a leading voice of the Left and is a co-founder of the International Critical Geography Group, which aims through conferences and other means to create a transnational community of scholars and activists with interests in geography, power and resistance (Desbiens and Smith, 1999). Outside geography, he is recognized as an important voice, along with Harvey and Lefebvre, in the development and defence of a spatialized Marxism. Reflecting his strong desire to make academic inquiry relevant to real-world struggles, virtually all Smith's writings are animated by political passion. Along with Harvey he is arguably the closest thing the discipline of geography has to a 'public intellectual'.

SPATIAL CONTRIBUTIONS

As noted above, Smith's main intellectual contributions are theoretical ones (which is not to say that his empirical research has lacked significance) and include concepts crucial to debates on space and place such as the 'rent gap', 'the production of space', 'uneven geographical development', and the 'production of scale'.

The first of these relates specifically to the issued of gentrification in Western cities. Smith observed, in his doctoral research, that many inner-city neighbourhoods spiral into decline until they reach a threshold point, after which a wave of capital investment leads to relatively rapid physical and social regeneration. This regeneration is not straightforwardly 'good' because it entails the control of inner-city space by wealthy professional classes at the expense of existing low-income residents – a process facilitated by property developments, local planners and credit institutions (like banks). Why, Smith asked, does this 'flip' from decline to reinvestment occur? His answer was disarmingly simple. He observed that the economic returns (or rent) to be made from land and buildings in any part of a city are only partly related to their physical properties and use value. Once a neighbourhood enters a decline – because, say, the businesses that exist in it are no longer competitive compared with rival businesses elsewhere – then the actual rent starts to diverge from the 'potential ground rent' that could be earned from a different use of the land there. Over time a 'valley' of low rent appears between city centre and outer city areas until it becomes economical for investors to reinvest in the inner area with new residences and businesses. Of all the possible types of reinvestment, expensive private property investment is typically favoured because of the high profit margins to be had. The idea of the rent gap has received empirical confirmation in several Western cities since Smith first proposed it in the late 1970s.

His work on gentrification also connects to ideas on the social production of space. Like David Harvey, Smith has long argued that space is a product of social relations and processes rather than an empty matrix to be filled. In *Uneven Development* he formalized this idea with an extended presentation of the difference between 'absolute' and 'relative' space. Conventionally, space is viewed as absolute: that is, an empty grid within which things and process are located. This absolute view of space was, until relatively recently, dominant in academic geography – especially among 'spatial scientists' who sought to identify the geographical patterning of things *in and across space* (e.g. migration patterns, disease diffusion patterns, and so on). In *Uneven Development*, Smith expounded an alternative conception of relative space. Relative space contrasts with absolute space in at least two ways. First, it is *both*

an effect and cause of socio-economic processes. That is, different socio-economic processes produce different spaces which, in turn, may reproduce or alter those processes. Second, relative space does not pre-exist – its construction and its form cannot be determined *a priori*. Smith's conception of relative space helped to open the way for a generation of critical human geographers to do research on how spaces are part and parcel of diverse forms of power and resistance in the contemporary world.

Of course, social researchers have long observed that economic and social development is geographically uneven. However, prior to Smith's *Uneven Development* it was often assumed that uneven development was a random or accidental outcome of economic growth and industrialization. What Smith argued, by contrast, is that uneven development is a *systematic* and *necessary* aspect of modern capitalism. Thus, when agglomeration of industry and people occurs in one place and region it can, over time, lead to diseconomies (traffic congestion, increased land prices, etc.). Meanwhile, non-industrialized areas might become prime candidates for investment because of cheap land, unexploited resources, etc. Equally, over time technical innovations that allowed one area to gain a competitive advantage are copied in other places, leading to an equalization in profit rates. This dialectic of geographical differentiation and equalization, according to Smith, is ongoing, leading to the constant fall and decline of cities, regions and countries within contemporary capitalism. Though it once occurred through capitalism expanding 'outwards' into new territories, today it is increasingly 'internal' to capitalism because few geographical pastures remain untouched by it.

As well as problematizing the production of space, Smith has explored the production of scale. Like space, geographical scale is often thought to be fixed and given, as with the cartographic scale on a map. However, in a number of publications, Smith has elaborated a theory of geographical scale that shows scale to be actively constructed by societies (Smith, 1993). Thus scales like 'local', 'national' and 'global' are not given in nature but contingent and historically variable outcomes of various social processes. Thus, the European Union has come to overlay the borders and regulations of individual member states since its foundation, materially altering the objective and subjective meaning of 'Europe'. For Smith, the importance of scale is that it materially 'contains' actions and events with ambivalent consequences. Thus the powerful can seek to further their agendas by limiting the scale of action of those who oppose them. For example, in the West, many national governments have assisted big business by diluting the power of national trades unions so that worker–firm negotiations now take place primarily at the local scale. On the other side, those who lack power can attempt to 'up-scale' their resistance activities to larger scales in order to be more effective. For instance, many trades unions have constructed new global networks in order to combat the power of transnational firms (see Herod, 1995).

KEY ADVANCES AND CONTROVERSIES

Smith's writings have been very influential, primarily in the discipline of geography but increasingly beyond it. First, his rent-gap thesis remains a key idea in gentrification studies and has also ensured that a Marxist voice has loomed large in these debates. Second, his notions of the production of space and uneven development helped, during the 1980s and 1990s, to bolster the growing Marxist geography movement. Finally, his notion of the production of scale has inspired Marxist geographers and non-Marxists alike. A minor cottage industry research-

ing geographical scale now exists and is in large part inspired by Smith's germinal thinking in this area (along with that of British geographer **Peter Taylor**). Key theorists in this new work on geographical scale are geographer Kevin Cox (1998) and sociologist Neil Brenner (2001).

Unlike David Harvey, Smith's work has not been the subject of too much criticism by other geographers – though he, conversely, has often been a fierce critic of others' work. Specifically, he has been involved in three debates. Firstly, in the early 1980s, he engaged with the work of the critical geographers Richard Peet (1981) and John Browett (1984) over the issues of space and uneven development respectively. The former, he argued, wrongly saw space as having effects 'in itself' and thus took Marxist geography down the wrong intellectual path (Smith, 1981). Browett accused Smith of misconstruing uneven development as a systematic and necessary dimension of capitalism, to which Smith responded uncompromisingly. *Contra* Browett, he showed that uneven geographical development is not merely a *contingent* outcome of ostensibly 'non-spatial' processes of economic growth and decline (Smith, 1986b). Secondly, during the mid-1980s, Smith chastized several British geographers for fixating on place at the expense of broader trans-local processes that, today, make places what they are (Smith, 1987a, 1987b). Like **Doreen Massey**, Smith argued that the global is now *in* the local, not separate from it. This, he argued, means that a purely localist politics is inappropriate as well as reactionary. Finally, Smith's work on geographical scale has provoked implicit critiques. Where Smith sees scales as relatively hierarchical and enduring once they are socially produced, others think the metaphor of nested scales is inappropriate. For instance, Whatmore and Thorne (1997) prefer to talk of *networks* of greater or shorter length – networks of trade, production, finance, etc. This network metaphor does away with labels like 'local' and 'global' or 'place' and 'space'. It encourages a less rigid view of how the powerful and the less powerful construct geographical landscapes to suit their purposes. Since the late 1980s, Smith's Marxism has broadened out intellectually and politi-

cally. He is recognized as a critical geographer seeking to build bridges among diverse radical causes and, arguably, has courted less disciplinary controversy than someone like Harvey for this reason.

SMITH'S MAJOR WORKS

Smith, N. (1981) 'Degeneracy in theory and practice: spatial interactionism and radical eclecticism', *Progress in Human Geography* 55, 111–118.

Smith, N. (1986b) 'On the necessity of uneven development', *International Journal of Urban and Regional Research* 9: 87–104.

Smith, N. (1991) *Uneven Development: Nature, Capital and the Production of Space*, 2nd edition. Oxford: Basil Blackwell.

Smith, N. (1993) 'Homeless/global: scaling places', in J. Bird, B. Curtis, T. Putnam, G. Robertson and L. Tickner (eds) *Mapping the Futures: Local Cultures, Global Change*. London: Routledge, pp. 87–119.

Smith, N. (1996) *New Urban Frontier: Gentrification and the Revanchist City*. New York: Routledge.

Smith, N. (2002) *Mapping the American Century: Isaiah Bowman and the Geography of Empire*. University of California Press.

Smith, N. and Godlewska, A. (1994) (eds) *Geography and Empire: Critical Studies in the History of Geography*. Oxford: Blackwell.

Secondary Sources and References

Amin, S. (1977) *Imperialism and Unequal Development*. New York: Monthly Review Press.

Brenner, N. (2001) 'The limits to scale? Methodological reflections on scalar structuration', *Progress in Human Geography* 25 (4): 591–614.

Browett, J. (1984) 'On the necessity and inevitability of uneven spatial development under capitalism', *International Journal of Urban and Regional Research* 8: 155–176.

Cox, K. R. (1998) 'Spaces of dependence, spaces of engagement and the politics of scale, or: looking for local politics', *Political Geography* 17 (1): 1–24.

Desbiens, C. and Smith, N. (1999) 'The International Critical Geography Group: forbidden optimism?', *Environment and Planning D: Society and Space* 18: 379–382.

Hackworth, J. and Smith, N. (2001) 'The changing state of gentrification', *Tijdschrift voor Economische en Sociale Geografie* 92: 464–477.

Harvey, D. (1973) *Social Justice and the City*. London: Arnold.

Harvey, D. (1985) *The urbanization of capital*. Oxford: Blackwells.

Herod, A. (1995) 'The practice of internal labor solidarity', *Economic Geography* 71 (4): 341–363.

Peet, R. (1981) 'Spatial dialectics and Marxist geography', *Progress in Human Geography* 5 (1): 105–110.

Smith, N. (1979) 'Toward a theory of gentrification', *Journal of American Planning Association* 54 (4): 538–548.

Smith, N. (1986a) 'Bowman's New World and the Council on Foreign Relations', *Geographical Review* 76: 438–460.

Smith, N. (1987a) 'Rascal concepts, minimalizing discourse and the politics of geography', *Environment and Planning D: Society and Space* 5: 377–383.

Smith, N. (1987b) 'Dangers of the empirical turn: some comments on the CURS initiative', *Antipode* 19: 59–68.

Smith, N. (1992) 'Contours of a spatialized politics: homeless vehicles and the production of geographical space', *Social Text* 33: 54–81.

Whatmore, S. and Thorne, L. (1997) 'Nourishing networks', in D. Goodman and M. Watts (eds) *Globalising Food*. London: Routledge, pp. 287–304.

Noel Castree

41 Edward Soja

BIOGRAPHICAL DETAILS AND THEORETICAL CONTEXT

American urban geographer, social theorist, and key member of the so-called Los Angeles School of Urban Studies, Ed Soja was born and raised in the Bronx. From the mid-1960s Soja undertook graduate research at Syracuse University, New York State. This provided the basis of his first book, *The Geography of Modernization in Kenya: A Spatial Analysis of Social, Economic, and Political Change* (1968). Although firmly located within the then dominant paradigm of spatial analysis, a tradition that Soja (1989: 51) was later to dismiss as little more than 'mathematized . . . description', *The Geography of Modernization* was nonetheless organized around a leitmotif that was to become central to all his subsequent work – the centrality of space in the constitution of society. As he wrote in the book's preface, *The Geography of Modernization* was underpinned 'by a strong feeling that the geographer's spatial perspective could contribute significantly to the rapidly expanding research cluster involved with the problems of social and economic change in developing areas' (Soja, 1968: v).

Teaching first at Northwestern University and then, from 1972, at the Graduate School of Architecture and Urban Planning at the University of California, Los Angeles, Soja continued to explore the geography of Third World economic and social development through much of the 1970s (see Soja and Tobin, 1974; Soja and Weaver, 1976). However, mirroring developments elsewhere in human geography, his work gradually moved away from a wholehearted embrace of modernization theory towards a radical scepticism of mainstream social science (Soja, 1979; Soja and Hadjimichalis, 1979). This scepticism was rooted in the belief that mainstream, positivistic approaches failed to describe the underlying mechanisms behind persistent geographies of uneven development. This new scepticism not only signalled the development of a self-consciously critical edge to Soja's writing; it was also to produce a major shift in the focus of Soja's intellectual project. Throughout the next two decades Soja devoted his energies to an ambitious attempt to place spatiality at the centre of radical social theoretical thinking. This search for a thoroughly spatialized, postmodern social theory has produced three major books: *Post-modern Geographies: The Reassertion of Space in Critical Social Theory* (1989), *Thirdspace: Journeys to Los Angeles and Other Real-And-Imagined Places* (1996), and *Postmetropolis: Critical Studies in Cities and Regions* (2000), as well as more than a dozen major articles.

SPATIAL CONTRIBUTIONS

The primary elements of Soja's argument for a spatiality-oriented, postmodern social theory have remained remarkably stable over the course of the past two decades, and can be summarized relatively easily. Indeed, each of his three postmodern books essentially offers refinements and repetitions of three central propositions: firstly, that the capitalist order is being reorganized in ways that profoundly privilege the spatial over the temporal; secondly, that spatiality is fundamentally constitutive of social life; and thirdly, and consequently, that critical social theory needs to take space seriously if it is to make sense of society. Here Soja's work echoes that of Marxist human geographers like **David Harvey**, Richard Peet, **Neil Smith** and **Doreen Massey** – as well as a related group of more theoretically eclectic writers such as **Anthony Giddens**, **Derek Gregory**, **Nigel Thrift**, Allan Pred, **Manuel Castells** and John Urry – who have sought to expose the profound spatiality of social life. All these writers stress that in contrast to the fundamental emphasis placed on history and temporality in traditional social theory, spatiality has been profoundly under-acknowledged and unexplored. For Soja, however, this is not enough. Spatiality does not just need to be taken into account by social theorists; it must be placed firmly back at the centre of every element of social theory.

Establishing the basis of this claim is the central narrative obsession driving Soja's (1989) *Post-modern Geographies*. In this influential and forcefully argued case for the re-spatialization of social theory, Soja asserts that critical social theory had not simply ignored space, it had actively repressed and denied spatiality. For most twentieth-century social theorists, space had been 'treated as fixed, the undialectical, the immobile. Time on the contrary was richness, fecundity, life, dialectic' (Foucault, cited in Soja, 1989: 10). What Soja seeks to uncover in *Post-modern Geographies* is an emerging counter-current within social theory that repudiates this dominant tradition of 'historicism' and instead reaches towards a profoundly spatialized mode of thinking based on 'a triple dialectic of space, time, and social being: a transformative re-theorization of the relations between history, geography and modernity' (Soja, 1989: 12). While this argument developed the work of Marxist human geographers, *Post-modern Geographies* also sought to situate a number of more disparate writers within this counter-current, including Ernst Mandel (who was praised for his recognition of the spatiality central to the recurring cyclical crises of capitalist accumulation) and Louis Althusser, as well as cultural commentators including John Berger, Marshall Berman and Fredric Jameson (who were seen as offering incisive accounts of how the experience of lived space was being profoundly reworked under conditions of contemporary capitalism). More purely theoretical sustenance was offered by **Foucault**, Sartre, Heidegger, and Giddens, each of whom had in some way sought to place spatiality at the centre of their theoretical projects.

Rising above all these thinkers is the figure of **Henri Lefebvre**. An idiosyncratic French Marxist, much of whose intellectual energies were focused on exploring everyday life and the process of urbanization, Lefebvre had since the 1960s been arguing that the production of spatiality rather than history had become the central armature of capitalist development and contradiction. As he wrote in his short book *The Survival of Capitalism*:

> The dialectic is back on the agenda. But it is no longer Marx's dialectic, just as Marx's was no longer Hegel's ... The dialectic today no longer clings to historicity and historical time, or to a temporal mechanism such as 'thesis-antithesis-synthesis' or 'affirmation-negation-negation of the negation' ... To recognize space, recognize what 'takes place' there and what it is used for, is to resume the dialectic: analysis will reveal the contradictions of space.
> (Lefebvre, 1976, cited in Soja, 1989: 43)

Lefebvre's thesis is the central logical armature of *Post-modern Geographies*, as indeed it is to Soja's postmodern project. It is through the filter of Lefebvre's epochal (and possibly seismic) claims for the centrality of spatiality in social life that Soja interprets the theoretical materials discussed in *Post-modern Geographies*. And it is through the interpretative mesh of this Lefebvrian filter that Soja can claim that the diverse collection of writers brought together in his writing can be read as providing the outline of a new kind of postmodern human geography.

For those familiar with conventional accounts of postmodernism, Soja's post-modernism is clearly highly idiosyncratic. Indeed, more than one reviewer of *Post-modern Geographies* suggested that Soja was 'a post-modernist with an identity crisis' (Resch, 1992: 146). Given that a common thread of the self-consciously postmodern philosophical project that emerged in the 1970s and 1980s was an 'incredulity towards meta-narratives' (Lyotard, 1984: xxiv), particularly those of Marxism, Soja's attempt to tie post-modernism tightly to the certainties of historical materialism appears problem-atic. For Soja, however, his postmodern geographies are postmodern because they recognize the salience of Lefebvre's claim that 'it is now space more than time that hides things from us' and, following on from this, 'that the demystification of spatiality and its veiled instrumentality of power is the key to making practical, political, and theoretical sense of the con-temporary era' (Soja, 1989: 61). He conse-quently argues that the postmodern epoch should be read as the latest manifestation of series of waves of capitalist develop-ment, and that the aim of postmodern social theory is to make sense of the capitalist restructuring that has brought this epoch into being.

Defining postmodernism in this un-usual and idiosyncratic way, Soja sought to capture the idea of the postmodern for the politically progressive traditions of the radical, socialist left. He also sought to compel leftist researchers to grasp the fundamental originality of the contempor-ary geohistorical moment. So while the bulk of *Post-modern Geographies* is de-voted to establishing the conceptual ground for a profoundly spatialized, post-modern social theory rooted within the broad traditions of historical materialism, it concludes with two chapters that illus-trate what an empirically informed post-

modern account might look like. Both chapters focus on Los Angeles, which Soja argues offers a privileged window on the ferocious restructuring processes that shape the postmodern world:

> Ignored for so long as aberrant, idiosyn-cratic, or bizarrely exceptional, Los Angeles, in another paradoxical twist, has more than any other place, become the paradigmatic window through which to see the last half of the twentieth century. (Soja, 1989: 221)

These empirical investigations present an overview of the fundamental geo-econ-omic imperatives that had been reshaping the social space of the Los Angeles metro-politan area. Yet within Soja's view of 'an emerging post-Fordist urban landscape fil-led with more flexible systems of produc-tion, consumption, exploitation, spatialization and social control' (Soja, 1989: 221) something seemed to be miss-ing. For all the rhetorical flourishes and complex narrative structure, Soja's post-modern account did not seem in essence much different to that presented by more conventional urban political economists. This raises the obvious question: if it is not that different, why expend so much trying to stake a claim for it?

KEY ADVANCES AND CONTROVERSIES

Post-modern Geographies was a hugely in-fluential book. It brought together in a lucid form a key collection of articles by an author at the forefront of the attempt to place human geography within the mainstream of social theory. It also pres-ented one of the first responses by a human geographer to the challenge of postmodernism. But while the breadth and erudition of Soja's scholarship was widely acknowledged both within human geography and the wider social science community, the vision of a postmodern geography set out in *Post-modern Geographies* met with widespread and frequently vociferous criticism. Three central criticisms were aimed at *Post-modern Geographies*. The first of these was that the basic premise of the book was simply unfounded: the world might be changing in rapid and unexpected ways but this was no reason to ditch established modes of analysis in favour of some kind of postmodernism (see Short, 1993; de Pater, 1993). A second set of criticisms more sympathetic to the idea of postmodernism was perhaps more telling. Writers including **Michael Dear** (1991), Philo (1992) and **Derek Gregory** (1994) wondered about the style of postmodernism Soja was attempting to articulate. A key attraction of postmodernism was its democratic, eclectic and inclusive ethos, yet Soja appeared to be deeply suspicious of this pluralistic spirit, or – in the case of feminism's engagement with postmodern thinking – strangely unaware of its significance (see Gregory, 1994; Massey, 1994; Rose, 1991). For all its positive attributes, *Post-modern Geographies* was therefore criticized for offering a limiting and emaciated reading of the postmodern: as Dear (1991: 653) wrote in a generally sympathetic commentary on *Post-modern Geographies*, its 'brilliant potential has been tempered by the discipline of Marxist thought'. Finally, Soja was a criticized for the way *Post-modern Geographies*, despite its claims to be profoundly concerned with everyday spatialities, constantly reduced the everyday to little more than an epiphenomenon of underlying economic processes. This is most obvious in the final chapters on Los Angeles where Soja, while stating his intention to show how the experience of space and time in the metropolis is being transformed, actually offers the reader what is very much a 'god's eye' view of the city. Rather than a living spatiality Soja offers 'a morphology of landscape that . . . is rarely disturbed by human forms' (Gregory, 1994: 301).

In *Thirdspace: Journeys to Los Angeles and Other Real-And-Imagined Places* (1996)

and *Postmetropolis: Critical Studies of Cities and Regions* (2000), Soja sought to respond to these critiques by further elaborating his vision of a radically spatialized, postmodern analysis. As such, Soja introduces a number of new concepts meant to help clarify and crystallize the key coordinates of his project. So, the 'socio-spatial dialectic' (see Soja, 1980) is refined into the notion of a *trialectics of being* – the insight that the ontology of being can only be interpreted by examining the interlocking of spatiality, historicality and sociality. In a similar way, *Thirdspace* comes to absorb the idea of a postmodern analysis, referring both to a style of analysis that places a spatialized trialectics at its centre and the particular texture of everyday lived spaces that exceeds the compartmentalized knowledges of the conventional social sciences. Furthermore, Soja is keen to demonstrate the openness of his postmodern/thirdspace project; Henri Lefebvre retains his privileged position, but a new range of postcolonial and feminist writers like **bell hooks, Homi Bhabha, Edward Said, Gayatri Spivak**, Gloria Anzaldúa and **Gillian Rose** are absorbed into the Sojaian vision. Yet, for all this, huge swathes of the social sciences are simply dismissed for their narrow focus on what is dubbed firstspace (i.e. 'the "real" material world') or secondspace (i.e. ' "imagined" representations of spatiality' – Soja, 1996: 6). For Soja, thirdspace is *the* privileged space of analysis, despite the fact for all the hundreds of pages of *Thirdspace* and *Postmetropolis* it remains a slippery term:

> *Everything* comes together in Thirdspace: subjectivity and objectivity, the abstract and the concrete, the real and the imagined, the knowable and the unimaginable, the repetitive and the differential, structure and agency, mind and body, consciousness and unconsciousness, the disciplined and the transdisciplinary, everyday life and unending history.
> (Soja, 1996: 56–57, original emphasis)

Thirdspace is claimed to encompass everything there is to say about anything (and perhaps, as a result, nothing at all?). Indeed, one reviewer (Barnett, 1997: 529) was moved to wonder, after quoting the above passage, 'Is Elvis still alive in Thirdspace?' Facetious though this comment sounds, it does point to a through-going problem within *Thirdspace* and *Postmetropolis*. In his keenness to stress just how new his 'radical post-modernism' (Soja 1996: 3) is, Soja seems to completely de-anchor himself from any established intellectual tradition. This compels him to work at a level of abstraction that undermines some of the most productive and interesting elements of his argument. As Barnett – the unsympathetic reviewer quoted above – wrote of *Thirdspace*, throughout the book terms like the trialectics of being and thirdspace 'are discussed at a very high level of "ontological" generality, which tends to obscure the fact that all Soja seems to be saying is that time, space and society are mutually constitutive' (Barnett, 1997: 528; see also Merrifield, 1999; Price, 1999; Soja, 1999).

In this light, perhaps the most compelling sections of both *Thirdspace* and *Postmetropolis* are where Soja turns to describing the contemporary metropolitan condition (the emergent age of the so-called 'postmetropolis'). As with *Postmodern Geographies*, a question remains about the connection between Soja's theoretical foundations and his empirical narratives. Nonetheless, the ambitious aim of mapping the key contours of an emergent postmetropolitan urban formation provides one of the most comprehensive summaries of what is happening to Western cityscapes. Continuing to take Los Angeles as a 'paradigmatic window', Soja presents the reader with a series of journeys through the postmetropolis: from a grand geohistorical reading of the roots of urbanization that winds back 10,000 years to the 'first urban revolution' (Soja, 2000: 4), to a collage of reactions to the 1992 LA 'justice riots' sparked by the police beatings of the Rodney King (see also Soja, 1996: 316–320). These journeys produce six discourses to help us make sense of the contemporary urban realm,

describing: the new spaces of industrialization and economic exploitation ('the postfordist industrial metropolis'); the city turned inside out ('the exopolis'); the city reshaped by globalization ('the cosmopolis'); the city of multiple and divided sociality ('the fractal city'); the privatized and sanitized city ('the carceral archipelago'); and spaces of urban simulation ('simcities'). Through a reading of these six interrelated discourses, and working within a thirdspace perspective, Soja hopes to provide both a synoptic introduction to the postmetropolitan condition, as well as a series of departure points for those interpreting the contemporary urban landscape. As he writes in the conclusion to *Thirdspace* (1996: 314), quoting Lefebvre, '*Il y a toujours l'Autre*. There is always another view.'

Soja's writing thus has an almost unique capacity to inspire and infuriate. Nonetheless, as one of a small number of human geographers whose writing has been taken up and read by cultural theorists and the wider social science community, Soja's contribution to contemporary debates around the spatiality of social life are substantial. *Post-modern Geographies* – and to a lesser extent *Thirdspace* and *Postmetropolis* – are notable achievements. For all their flaws and absences, they offer a remarkable attempt to reimagine the human geographic project. Indeed, perhaps the greatest legacy of Soja's work is not the idea of a postmodern geography *per se* (after all, relatively few appear to have been convinced by his arguments *tout court*). Rather it is how his writing has widened the theoretical and conceptual horizons of both human geography and the social sciences in general. Soja's work has not only made a compelling case for the profound importance of spatiality, it has helped generate an enormously fecund dialogue between human geography and critical social theory, while at the same time demonstrating the productiveness of experimenting with innovative forms and styles of presentation and argumentation.

SOJA'S MAJOR WORKS

Soja, E. W. (1968) *The Geography of Modernization in Kenya: A Spatial Analysis of Social, Economic and Political Change.* Syracuse, NY: Syracuse University.
Soja, E. W. (1989) *Post-modern Geographies: The Reassertion of Space in Critical Social Theory.* London: Verso.
Soja, E. W. (1996) *Thirdspace: Journeys to Los Angeles and Other Real-And-Imagined-Places.* Oxford: Blackwell.
Soja, E. W. (2000) *Postmetropolis: Critical Studies of Cities and Regions.* Oxford: Blackwell.

Secondary Sources and References

Barnett, C. (1997) 'Review of *Thirdspace: journeys to Los Angeles and other real and imagined places*', *Transactions, Institute of British Geographers* NS 22: 529–530.
De Pater, B. (1993) 'The quest for disorder: on the dogma of a contingent, chaotic world', *Tidschrift voor Economische en Sociale Geografie* 84: 175–177.
Dear, M. (1991) 'Review of Edward Soja *Post-modern Geographies: the reassertion of space in critical social theory*', *Annals of the Association of American Geographers* 80: 649–654.
Gregory, D. (1994) *Geographical Imaginations.* Oxford: Blackwell.
Lefebvre, H. (1976) *The Survival of Capitalism.* London: Allen and Busby.
Lyotard, J-F. (1984) *The Post-modern Condition: A Report on Knowledge.* Manchester: Manchester University.
Massey, D. (1994) *Space, Place, and Gender.* Cambridge: Polity.
Merrifield, A. (1999) 'The extraordinary voyages of Ed Soja: Inside the "trialectics of spatiality"' *Annals of the Association of American Geographers* 89: 345–348.
Philo, C. (1992) 'Foucault's geography', *Environment and Planning D: Society and Space* 10: 137–161.
Price, P. (1999) 'Longing for less of the same', *Annals of the Association of American Geographers* 89: 342–344.
Resch, R. (1992) 'Review of Edward Soja *Post-modern Geographies: the reassertion of space in critical social theory*', *Theory and Society* 21: 145–154.
Rose, G. (1991) 'Review of Edward Soja *Post-modern Geographies: the reassertion of space in critical social theory* and David Harvey *The Condition of Post-modernity: an enquiry into the origins of cultural change*', *Journal of Historical Geography* 17: 118–121.
Short, J. (1993) 'The "myth" of post-modernism', *Tidschrift voor Economische en Sociale Geografie* 84: 169–171.
Soja, E. W. (1979) 'The geography of modernisation: a radical reappraisal', in Obudho and F. Taylor (eds) *The Spatial Structure of Development.* Boulder: Westview, pp. 28–45.
Soja, E. W. (1980) 'The socio-spatial dialectic', *Annals of the Association of American Geographers* 70: 207–225.
Soja, E. W. (1999) 'Keeping space open', *Annals of the Association of American Geographers* 89: 348–353.
Soja, E. W. and Hadjimichalis, C. (1979) 'Between historical materialism and spatial fetishism: some observations of the development of Marxist spatial analysis', *Antipode* 11: 3–11.
Soja, E. W. and Tobin, R. (1974) 'The geography of modernization: paths, patterns, and processes of spatial change in developing countries', in G. Brewer and R. D. Brunner (eds) *Political Development and Change: A Policy Approach* New York: Free Press, pp. 197–243.
Soja, E. W. and Weaver, C. (1976) 'Urbanization and underdevelopment in east Africa', in B. Berry (ed.) *Urbanization and Counterurbanization.* Beverly Hills: Sage, pp. 233–266.

Alan Latham

42 Gayatri Chakravorty Spivak

BIOGRAPHICAL DETAILS AND THEORETICAL CONTEXT

Gayatri Chakravorty Spivak was born in Calcutta, in colonial India in 1942 to a middle-class family. The struggle for independence, the experiences of colonization, 'postcolonialism' and the meanings of 'freedom' and 'independence' have been central to her work. Spivak's middle-class status in the context of colonial educational policies provided her with the privilege of a university education. She graduated from Presidency College of the University of Calcutta in 1959 with first-class honours in English, having received gold medals for English and Bengali literature. Spivak was highly conscious that the teaching of English literature in colonial India was a central part of the 'civilizing mission' of imperialism. In 'Three Women's Texts and Critique of Imperialism' she argues that 'It should not be possible to read nineteenth-century British literature without remembering that imperialism, understood as England's social mission, was a crucial part of the cultural representation of England to the English' (Spivak, 1985: 243).

Spivak continued her education in the US, where she pursued a Master's degree (1962) in English at Cornell University. She also spent a year in Britain at Cambridge University as a fellow at Girton College before returning to the US to take up an instructor's post at the University of Iowa. At the same time she worked on a PhD (awarded in 1967). Her doctoral dissertation was on the work of Irish poet W. B. Yeats, undertaken again at Cornell

University, supervised by Paul de Man (1919–1983). Spivak was appointed Avalon Foundation Professor in the Humanities at the Department of English and Comparative Literature at Columbia University (New York) in 1992, alongside Bruce Robbins, Jonathan Arac and the influential postcolonial theorist **Edward Said**. Spivak has taught in universities across the world, including Goethe Universität in Frankfurt, Riyadh University, Jawaharal Nehru University and Hong Kong University of Science and Technology. She has held a prestigious position as Andrew W. Mellon Professor of English at the University of Pittsburgh and numerous fellowships in North America, Asia and Europe and has been invited to speak across the disciplines, in professional schools as well as outside the academy in activist circles.

For many geographers and other scholars, the name of Gayatri Chakravorty Spivak is synonymous with deconstruction and postcolonial feminism. It is no secret, however, that Spivak's work, with its dense ('opaque') and difficult style, sometimes promotes elitism or invokes dismissal. It certainly defies easy summary. Perhaps too there is sometimes an impression that those who are able to extract meaning from her works have gained access to a select circle of intellectuals who have demonstrated their intellectual capacity 'to know' by referencing Spivak (we will return to this issue of style later). Among her vast range of published work is a translation into English and critical introduction to Jacques Derrida's *De la Grammatologie* (*of Grammatology*) (1976); *In Other Worlds: Essays in Cultural Politics* (1987); *Outside the Teaching Machine* (1993) and *A Critique of Post-Colonial Reason* (1999). She has also authored dozens of articles, some of the most influential being: 'Three women's texts and a critique of imperialism' (1985); 'Poststructuralism, marginality, post-coloniality and value' (1990a); 'Can the subaltern speak?' (1988a) and 'Who claims alterity?' (1989). Throughout, Spivak's work draws on and is informed by the key conceptual frameworks of Marxism, feminism and deconstruction.

Spivak's introduction to deconstruction was via Paul de Man, who was its most prominent advocate in the US at a time when Derrida was making similar arguments in Europe. Both de Man and Derrida questioned the stability and transparency of textual meaning. Deconstruction developed as a critique of 'logocentrism', wary of the privileging of an authoritative presence held to be characteristic of much Western thought. It is – in part – an approach to reading. Yet it is not so much a method as a strategy for de-linking, destabilizing and undermining binary oppositions by revealing their mutual constitution. Barbara Johnson (1980: 5) thus notes 'the deconstruction of a text does not proceed by random doubt or arbitrary subversion, but the careful teasing out of warring forces of signification within the text'. For Spivak:

> Deconstruction does not say there is no subject, there is no truth. It simply questions the privileging of identity so that someone is believed to have the truth. It is not the exposure of error. It is constantly and persistently looking into how truths are produced. That's why deconstruction doesn't say logocentrism is a pathology, or metaphysical enclosures are something you can escape. Deconstruction, if one wants a formula, is among other things, a persistent critique of what one cannot not want.
> (Spivak, 1994: 280)

SPATIAL CONTRIBUTIONS

The sheer density of Spivak's prose means that the geographical applications of her work are not necessarily immediately obvious. Moreover, it is difficult to entirely separate her from the impact of Derrida (unsurprising given she translated *De la Grammatologie*) whom Anglophone geographers have cited rather more frequently. However, one contribution to geography relates to the so-called 'crisis of representation' and the vexed issue of 'positionality' (see Gregory, 1994; England, 1994; Radcliffe, 1994; Robinson, 1994; Barnett, 1995; Sidaway 1992). In this context, Spivak's work has been drawn on to highlight the ethnocentricity of hegemonic knowledges and to problematize Western, modernist epistemologies. Spivak's (1990a, 224) concern with ' ... the imbrication of techniques of knowledge with strategies of power' is that representations do not just describe reality but are constitutive of it. Spivak (1988a) elaborates on this when she defines representation as both 'a speaking of' that describes or depicts a perceived reality, through language and communication, and a 'speaking for' which is the same as the representation of a particular constituency. In her essay 'Can the subaltern speak?', she deals directly with such questions. Spivak takes issue with the radical claims by intellectuals to speak for the disempowered and places them alongside the benevolent colonialists who saved Hindu women from widow sacrifice in the nineteenth century. In doing, so Spivak argues that such interlocutors (even as they claim to represent and speak for the disempowered) have silenced subordinated subalterns (where the latter are distinguished in class, gender and 'caste'). The subaltern may speak, but she is misheard, ignored or her words and actions are misinterpreted by a dominant culture whose terms of meaning, truth and understanding exclude subalterns by positioning them and interpreting their actions through dominant frames of reference. Spivak points out that 'the historical and structural conditions of political representations do not guarantee that the interests of particular subaltern groups will be recognised or that their voices will be heard' (Morton, 2003: 57).

For Spivak, the experiences of 'Third World' women are neglected by general postcolonial critiques whose focus is men. At the same time, she argues that Western feminisms have sometimes propagated the 'lie of a global sisterhood', failing to recognize the differences between women (Spivak, 1986: 226). In response to such critiques, the questions of differences *between* women have occupied feminist geographers particularly in terms of the implications for the production of geographical knowledge (McDowell, 1993). This has seen feminist geographers at the forefront of debates on the implications of the politics of representation for the production of geography in terms of both fieldwork and writing (Barnes and Duncan, 1992; Duncan and Sharp, 1993; Gregory, 1994; Katz, 1992, 1996; Mohammad, 2001; Nast, 1994), leading to ongoing debates about ways out of this dilemma.

For Spivak an 'insider' is also always an 'outsider'. She argues that authorial claims based on one-dimensional experience of marginality serve to negate multiple differences which distinguish and mark the distance of those being represented from the researcher/writer. Thus 'Other' or 'Third World' intellectuals may be regarded by some as the 'authentic inhabitants of the margin' (Spivak, 1990a: 224), positioned to legitimately speak on behalf of their group, constituting and adjudicating both the 'limits of knowledge about the Other and the graphematic space the latter occupies/inscribes' (Spivak, 1990b: 175). But it is precisely the Third World intellectual's distance from the wholly Other – produced by the uneven experience of op-

pression – that positions some 'Others' closer to the 'Same', from which they derive the power to speak for the subaltern.

Another way in which questions of representation have been registered in geography is in the history of the discipline. Spivak offers inspiration to those who would seek alternative histories and lineages of geographical writing in the age of formal European imperialism. To McEwan, therefore, Spivak offers some pathways to the recovery of subaltern agency and presence in the critical re-examination of colonial geography and exploration:

> Deconstructing the colonial archive also allows for a more subtle analysis of the production of geographical knowledge, and the possibility of decentring a putatively western tradition by viewing the production of those knowledges as a complex process of cultural exchange and negotiation, in which colonized populations played an active part.
> (McEwan, 1998: 374)

Similarly, Spivak offers some departures in rethinking geographies of development by reconceptualizing development as a project whose origins and trajectory are imbued with both the potential for liberation and the reality of enduring forms of imperial, class and gender inequality (Radcliffe, 1994). Likewise, when Spivak (as in her 1998 essay on 'Revisiting the "Global Village"') interprets the former as a mode of 'cultural talk' that recodes disjuncture and difference but whose parameters derive from capitalism and imperialism, geographers have seized the deconstructive impulse in Spivak and others to problematize the discourse of 'globalization' and the assumptions about its pervasiveness and putative power (Gibson-Graham, 1996).

KEY ADVANCES AND CONTROVERSIES

For some, Spivak's density of writing and her mode of deconstruction may be off-putting – or worse. Terry Eagleton (1999), for example, called Spivak's *A Critique of Post-Colonial Reason* 'obscurantist'. **Edward Said** (1983) had earlier criticized the jargonistic style of literary theory that 'alienates' rather than communicates with the reader, something that for many is confirmed by Spivak's work. For Said, the dense jargon [of literary theory] 'obscures' social reality and 'encourage[s] a scholarship of ''modes of excellence'' very far from daily life' (Said, 1983: 4). This is part of his wider problem with 'texts', which he opposes to reality and to worldliness. Moreover, Spivak's avowed political aim is to support the oppressed, the marginalized, the excluded, the subaltern. How can this be pursued through the act of reading, of deconstruction? How can Spivak reconcile her political aims with the charges of textual idealism that have sometimes been levelled at deconstruction, apparently confirmed in the contention that 'there is nothing outside of the text' (Derrida, 1976: 158)?

Spivak (1988a) deploys deconstruction to contend with Said's charge that Derrida's criticism moves us into the text (i.e. away from the world) while **Michel Foucault**'s moves us *in* and *out*. She points out that Said constructs a false dichotomy here between the text and the wider world based on a problematic and limited understanding of texts. Instead, her deconstruction follows Derrida's understanding of the text:

> The text is not the book; it is not confined in a volume, itself confined to the library. It does not suspend reference – to history, to the world, to reality . . .
> (Derrida, 1988: 137)

Rather it is:

> . . . differential network, a fabric of traces referring endlessly to something other than itself, to other differential traces. Thus the text over runs all the limits assigned to it so far (not submerging or drowning them in an undifferentiated homogeneity, but rather making them more complex, dividing and multiplying strokes and lines) – all the limits, everything that was set up in opposition to writing (speech, life, the world, the real, history . . . every field of reference – to body or mind, to conscious or unconscious, politics, economics, and so forth).
> (Derrida, 1979: 84–85)

'There is nothing outside the text' might therefore mean that a text contains within it traces of the wider world, and that the world is rendered through a series of 'texts'; not simply literary texts and codes, but perhaps other inscriptions of difference and meaning (such as accounts and systems of value like money). Following Derrida, Spivak's work aims to make visible the constitution of the so-called 'real' world in and through the fabric and fabrication of texts; for example, those that elaborate government policies or produce legal apparatus. Thus Spivak utilizes the strategy of deconstruction, of reading texts, to pursue political goals that are also very much informed by Marxism, feminism and a postcolonial anti-imperialism.

Two key areas of Spivak's contribution have therefore been on postcolonial issues relating to class and gender. In relation to the former, Spivak reads Marx creatively. Her deconstructive approach to texts (including that of *Das Kapital* itself) is in sympathy with Marx's determination to expose the categories of value, economy and class to critical scrutiny (Castree, 1997). And in her essay on 'Supplementing Marxism', Spivak rethinks the meaning and definition of socialism, which becomes a project that:

> . . . is not in opposition to the form of the capitalist mode of production. It is rather a constant pushing away – a differing and a deferral – of the *capital*-ist harnessing of the social mobility of capital.
> (Spivak, 1995: 119)

In specific terms, Spivak (1989) illuminates the continued disempowerment of the 'subaltern' in postcolonial India. With reference to internal colonization, transnational capitalism and new forms of imperial power, she explains how the national independence movement left the existing highly polarized class structure in South Asia more or less intact. The Subaltern Studies group of radical Indian historians (see Gregory, 2000) have sought to rewrite a history from below, but Spivak points out some of the limits to this. She has drawn on the work of Subaltern Studies to argue that Western Marxism has proven unable to represent the complexity of subaltern histories of insurgency and resistance. Spivak thus questions the ability of both conventional nationalist or Western Marxism (even as developed by the Subaltern Studies group) to adequately represent disenfranchized Indians (see Morton, 2003). For Spivak, Marxism and feminism are vital political projects, but she argues these should be supplemented with deconstruction to provide those conceptual tools or vocabulary that would do justice to the everyday lives of marginalized 'Third World' (subaltern) women – both historically and in the present day – for such lives are complex, contradictory and frequently obscured from view. They are written out of history, or invoked as passive masses or victims without their subjectivity intact and are therefore represented minus their capacity to make history and geography. This can result in silencing the experiences of marginal 'Third World' subjects. Moreover, the geopolitical category of the 'Third World' is – for Spivak – to be questioned in terms that regard it as not simply as a place or space that is a mere supplement to the Western history of capitalism. Instead Spivak sees it as necessary (but neglected) constituent of that wider history; without which Western hegemony and capitalism could not have taken its exploitative course.

Rather than attempting to represent 'Third World' women, Spivak has thus sought to disrupt the codes and conventions of Western knowledges that have sought to represent the 'Third World'. In 'Literary representation of the subaltern' (1988b) Spivak provides a critical reading of the work of Mahasweti Devi, the 'Breast giver' (1995). The 'Breast giver' follows the life of the protagonist Jashoda who is employed as a wet nurse in an upper-class, high-caste Bengali household. Spivak highlights the limits of the bourgeois 'Mother India' ideal, marking the failure of Indian nationalist project to eliminate gendered–classed oppressions. She also draws on this work to point out to Western feminists that women's reproductive labour is not always unwaged and domestic. In 'Three women's texts and a critique of imperialism' (1985) Spivak reads Bronte's (1847) *Jane Eyre* against the hegemonic Western feminist readings, which she argues point to the complicity of Western feminism and imperialism. Such readings celebrate Jane Eyre as a text that privileges the narrative of its female protagonist, who is read as a liberated Western female individual. Spivak's reading finds that Jane's liberation as an individual rests on the representation of Bertha Mason (the Jamaican Creole, who is the 'mad' first wife of Jane's fiancé Rochester), as 'Other', as 'not yet human' (Spivak, 1985: 247). There, as in much of her extensive writings and interviews, Spivak's attention to the 'margins' and the occlusion of difference will continue to be instructive for those who take the time and trouble to engage carefully with her works and to tease out their mappings of enduring colonial legacies and indictments of neocolonial complicities.

SPIVAK'S MAJOR WORKS

Spivak, G. C. (1985) 'Three women's texts and a critique of imperialism', *Critical Inquiry* 12: 243–261.

Spivak, G. C. (1987) *In Other Worlds: Essays in Cultural Politics*. New York: Methuen.

Spivak, G. C. (1993) *Outside the Teaching Machine*. New York: Routledge.

Spivak, G. C. (1988a) 'Can the subaltern speak? Speculations on widow sacrifice', in C. Nelson and L. Grossberg (eds) *Marxism and the Interpretation of Culture*. London: Macmillan, pp. 271–313.

Spivak, G. C. (1988b) 'Literary representations of the subaltern', in R. Guha (ed.) *Subaltern Studies*. New Delhi: Oxford University Press.

Spivak, G. C. (1989) 'Who claims alterity?', in B. Kruger and P. Mariani (eds) *Remaking History*. Dia Art Foundation Discussions in Contemporary Culture No. 4. Seattle: Bay Press, pp. 269–292.

Spivak, G. C. (1990a) 'Poststructuralism, marginality, postcoloniality and value', in P. Collier and H. Geyer-Ryan (eds) *Literary Theory Today*. Cambridge: Polity Press, pp. 219–244.

Spivak, G. C. (1994) 'Bonding in difference', in A. Arteaga (ed.) *An Other Tongue: Nation and Ethnicity in the Linguistic Borderlands*. Durham, NC: Duke University Press, pp. 273–283.

Spivak, G. C. (1995) 'Supplementing Marxism', in C. Cullenberg and B. Magnus (eds) *Wither Marxism: Global Crises in International Perspective*. New York: Routledge, pp. 109–119.

Spivak, G. C. (1999) *A Critique of Post-Colonial Reason: Toward a History of the Vanishing Present*. Cambridge, MA: Harvard University Press.

Secondary Sources and References

Barnes, T. and Duncan, J. (eds) (1992) *Writing Worlds: Discourse, Text and Metaphor in the Representation of Landscape*. London: Routledge.

Barnett, C. (1995) 'Why theory?', *Economic Geography* 71: 427–435.

Castree, N. (1997) 'Invisible leviathan: speculations on Marx, Spivak and the question of value', *Rethinking Marxism* 9: 45–78.

Derrida, J. (1976) *Of Grammatology*. Trans. G. C. Spivak. Baltimore: Johns Hopkins University Press.

Derrida, J. (1979) *Spurs: Nietzsche's Styles/Eperons: Les Styles de Nietzsche*. Trans. B. Harlow. Chicago: University of Chicago Press.

Derrida, J. (1988) *Limited Inc*. Trans. S. Weber (Editor's foreword G. Graff). Evanston, IL: Northwestern University Press.

Devi, M. (1995) 'Breast-giver', in S. Thames and M. Gazzaniga, M. (eds) *The Breast: an Anthology*. New York: Global City Press, pp. 86–111.

Duncan, N. and Sharp, J. P. (1993) 'Confronting representation(s)', *Environment and Planning D: Society and Space* 11: 373–486.

Eagleton, T. (1999) 'In the gaudy supermarket', *London Review of Books* 21: 10.

England, K. V. L. (1994) 'Getting personal: reflexivity, positionality, and feminist research', *Professional Geographer* 46: 80–89.

Gibson-Graham, J. K. (1996) *The End of Capitalism (As We Knew It): A Feminist Critique of Political Economy*. Oxford: Blackwell.

Gregory, D. (1994) *Geographical Imaginations*. Oxford: Blackwell.

Gregory, D. (2000) 'Subaltern studies', in R. J. Johnston, D. Gregory, G. Pratt and M. Watts (eds) *The Dictionary of Human Geography*, 4th edition. Oxford: Blackwell, pp. 801–802.

Johnson, B. (1980) *The Critical Difference*. Baltimore: Johns Hopkins University Press.

Katz, C. (1992) 'All the world is staged: intellectuals and the projects of ethnography', *Environment and Planning D: Society and Space* 10: 495–510.

Katz, C. (1996) 'The expeditions of conjurors: Ethnography, power and pretense', in D. L. Wolf (ed.) *Feminist Dilemmas in Field Research*. Boulder: Westview Press, pp. 170–184.

McDowell, L. (1993) 'Space, place and gender relations, Part II: Identity, difference, feminist geometries and geographies', *Progress in Human Geography* 17: 305–318.

McEwan, C. (1998) 'Cutting power lines within the palace? Countering paternity and eurocentrism in the "geographical tradition" ', *Transactions of the Institute of British Geographers* 23: 371–384.

Mohammad, R. (2001) ' "Insiders" and/or "Outsiders": Positionality theory and praxis' in M. Limb and C. Dwyer (eds) *Qualitative Methodologies for Geographers*. London: Arnold, pp 101–117.

Morton, S. (2003) *Gayatri Chakravorty Spivak*. London and New York: Routledge.

Nast, H. J. (1994) 'Women in the field: Critical feminist methodologies and theoretical perspectives', *Professional Geographer* 46: 54–60.

Radcliffe, S. A. (1994) '(Representing) post-colonial women: Authority, difference and feminisms', *Area* 26: 25–32.

Robinson, J. (1994) 'White women representing/researching "Others": From antiapartheid to post-colonialism?', in A. Blunt and G. Rose (eds) *Writing Women and Space. Colonial and Postcolonial Geographies*. New York: Guildford Press, pp. 197–226.

Rose, G. (1997) 'Situating knowledges: Positionality, reflexivities and other tactics', *Progress in Human Geography* 21 (3): 305–320.

Said, E. (1983) *The World, the Text, the Critic*. Cambridge, MA: Harvard University Press.

Sidaway, J. D. (1992) 'In other worlds: on the politics of research by "First World" geographers in the "Third World" ', *Area* 24: 403–408.

Spivak, G. C. (1986) 'Imperialism and sexual difference', *Oxford Literary Review* 7 (1–2): 225–240.

Spivak, G. C. (1990b) 'Versions of the margin: Coetzee's foe reading Defoe's Crusoe/Roxana', in J. Arac and B. Johnson (eds) *Consequences of Theory*. Baltimore: Johns Hopkins University Press, pp. 154–180.

Spivak, G. C. (1998) 'Cultural talks in the hot peace: Revisiting the "global village" ', in P. Cheah and B. Robbins (eds) *Cosmopolitics: Thinking and Feeling beyond the Nation*. Minneapolis: University of Minnesota Press, pp. 329–350.

Robina Mohammad and James D. Sidaway

43 Michael Storper

BIOGRAPHICAL DETAILS AND THEORETICAL CONTEXT

Michael Storper was born in New York City in 1954. He was awarded a Bachelor's degree in Sociology and History by the University of California, Berkeley, in 1975 and an MA in Geography in 1979. Storper completed his PhD in Geography at the University of California, Berkeley, in 1982, supervised by Richard Walker (who was, in turn, a graduate student of **David Harvey**'s). Storper began working in the Graduate School of Architecture and Urban Planning, University of California, Los Angeles (UCLA), in 1982, being appointed as Professor of Regional and International Development in the Department of Urban Planning at UCLA in 1992. At the time of writing, Storper holds posts across three countries: Professor of Regional and International Development, UCLA; Professor of Economic Sociology, Institut d'Études Politiques de Paris and Centennial Professor of Economic Geography, London School of Economics.

Storper's writing has addressed the changing spatial dynamics of production and work; the development of territorial production complexes; and most recently the importance of regional 'worlds of production' within advanced capitalist economies. Much of his work on regional development theory has sought to critically engage with debates beyond the disciplinary boundaries of geography, most notably those occurring in institutional economics (e.g. Storper and Salais, 1997; Storper 1997). Storper also has sought cross-disciplinary audiences through publication in journals as diverse as *Futures* (Storper and Scott, 1995) and *Public Culture* (Storper, 2000). In many ways, the trajectory of Storper's work has moved roughly in tandem with broader shifts in the subdiscipline of economic geography, from an interest in the ever-changing landscapes of capital created through an incessant search for profit (Storper and Walker, 1989) and more specific concerns about the nature of flexibility and flexible specialization (Storper, 1991; Christopherson and Storper, 1989), through to current preoccupations with the nature of regional development in a globalizing world economy.

Yet Storper has been more than a follower of these trends, playing a crucial role in shaping the trajectory of economic geography. For example, the mid-1980s' volume *Production, Work, Territory: The Geographical Anatomy of Industrialised Capitalism* (Scott and Storper, 1986) quickly became *de rigueur* reading for both junior and more senior researchers in economic geography. Written with Allen Scott, the book presented a Marxist-inflected counter to existing conceptualizations of 'post-industrial' cities (Scott and Storper, 1986: vii), albeit at the same time representing a critique of some of the more 'abstract structural imperative[s]' (Scott and Storper, 1986: 4) presented in classical Marxist accounts. *Production, Work, Territory* developed ideas about the 'spatially-specific circumstances' (Scott and Storper, 1986: 13) of commodity production which shaped and were shaped by the new 'realities' of late twentieth-century capitalism. The book had a particularly strong impact given that for many English-speaking readers it represented one of the first introductions to the writings of the French Regulation School, which examines the institutional forms evolving to secure particular modes of capital accumulation (Lipietz, 1986). In the work of the so-called 'California School' of economic geography – represented by Storper, Walker and Scott – this was to involve examination of the forms emerging as rigid Fordist regimes of accumulation were giving way to more flexible post-Fordist modes. Suggesting that particular regimes of accumulation, and the mode of regulation associated with them, favoured particular production locations, Storper and colleagues began to expose the 'inconsistent' geographies of capitalism by documenting the emergence of new 'flexible production complexes' in California.

The importance of meso-level theorizing to economic geography continued to be emphasized in Storper's work (with Susan Christopherson) on flexible specialization in the film industry in Los Angeles. By the late 1990s, Storper had become more centrally interested in the regional dimensions of territorial development, largely within the context of debates about globalization. Perhaps unsurprisingly given his institutional base in Los Angeles, Storper's work occasionally has been viewed as part of a so-called 'LA School' (including **Michael Dear** and **Ed Soja**), that holds up Los Angeles as the archetypal urban formation of the twenty-first century (e.g. Curry and Kenney, 1999; see also the response in Storper, 1999). However, Storper's investigations have ranged more broadly than this, with his arguments about the formation of 'regional growth complexes' based upon consideration of a range of case studies (e.g. Northeast Central Italy and France). Furthermore, Storper has become increasingly interested in more wide-ranging debates such as the relationship between economic globalization and the rise of consumerist identities (Storper, 2000).

SPATIAL CONTRIBUTIONS

Following on from their (1986) edited collection, Storper and Walker's (1989) co-authored volume on territorial production complexes (*The Capitalist Imperative*) represented an extensive elaboration of the uneven landscapes of production created by advanced capitalism. Read alongside **Doreen Massey**'s (1984) *Spatial Divisions of Labour*, the book contributed significantly to the theorization of the spatiality of 1980s' restructuring. Primarily, *The Capitalist Imperative* was concerned to investigate the relationships between changing geographies of production and transformations in divisions of labour. In particular, Storper and Walker (1989) stressed that divisions of labour were not the simple outcomes of decisions made by individual capitalists but rather represented part of broader social processes (see Warf, 1991: 223). Like the earlier *Production, Work, Territory* (1986), *The Capitalist Imperative* sought to bring Marxian economic ideas into dialogue with 'conventional but heterodox theories of technology, external economies and industrial organization' (Storper, 2001: 159, n.18). Storper and Walker thus sought to view 'the capitalist process in broad terms' while admitting 'that Marxism had little to say about concrete issues such as technological change and industrial organisation' (Storper, 2001: 159).

Insights into the changing geographies of production that had begun to characterize the broad transition from mass production to 'post-Fordist' forms led to more specific investigations of the Los Angeles film industry. Storper and Christopherson were particularly interested in the ways in which an entire industry had shifted from mass production to flexible specialization, involving the significant vertical disintegration of large firms and the rise of small specialist production units (Storper, 1991; Christopherson and Storper, 1989). Not only did the LA industry represent a very early example of the abandonment of routinized batch production, but also it stood as a key American industry of the twentieth century, 'lacking the artisanal and regional traditions of Europe, and without the nineteenth century antecedents of other industries' (Storper, 1991: 200). As such, this consideration of the film industry enabled a close and detailed historico-geographical study of flexible specialization.

Alongside many other analysts at the *fin-de-siècle*, Storper became caught up with a wide-ranging set of debates about the broad contours of processes of 'globalization' and emergent relationships between the local, regional and global bases of economic development. Specifically, Storper's contribution has been to suggest that the global system can best be viewed as networks of local and regional economies – or regional 'worlds of production' (Storper and Salais, 1997). Such 'worlds' are defined by a set of *conventions* consisting of 'practices, institutions and material objects/tools [as well as] specific forms of cognition, theories, doctrines, institutions and rules' (Storper, 1997: 136). Individuals, markets, governments, households and firms shape these 'conventionally-based and regionalised frameworks of action', encouraging distinctive worlds of production to emerge (Storper, 1997: 137). Significant emphasis is placed here by Storper on the 'learning economy' of regions, consisting of 'an ensemble of competitive possibilities, reflexive in nature, engendered by capitalism's new metacapacities, as well as the risks or constraints manufactured by the reflexive learning of others' (Storper, 1997: 31).

By spatially locating this purposive learning in spaces where physical proximity allows for the social organization of production, Storper argues that innovation relies upon agglomeration. While he is concerned with documenting the way that new technologies are changing the importance of face-to-face contacts, Storper nonetheless argues that it is the *city-region* that potentially plays the greatest role in nurturing economic development:

> The most striking forms of agglomeration in evidence today are the super-agglomerations or city-regions that have come into being all over the world in the last few decades, with their complex internal structures comprising multiple urban cores, extended suburban appendages, and widely-ranging hinterland areas, themselves often sites of scattered urban settlements . . . These city-regions are locomotives of the national economies within which they are situated, in that they are the sites of dense masses of interrelated economic activities that also typically have high levels of productivity by reason of their jointly-generated agglomeration economies and their innovative potentials. (Scott and Storper, 2003: 6)

Acknowledging the dialectic dynamic of 'globalization and territorialization' at work in the construction of urban-regional economies, Storper (1997: 248) demonstrates that the resurgence of the regional economy in a global era can only be understood if we explore the meso-level (regional) resolution of macro- and micro-economic forces. Dispensing with traditional distinctions between the categorizations of urban and regional (and for that matter, global and local), Storper critically synthesizes ideas about learning, reflexivity and place in order to contribute to the ongoing theorization of the production of space under capitalism.

KEY ADVANCES AND CONTROVERSIES

Alongside the work of Allen Scott, Storper's work has become iconic of a new strand of inquiry in economic geography that asserts the importance of place agglomeration in economic competitiveness. Triggering major debates in economic geography, particularly concerning the 'cultural geography' of economics, it has also provided a rejoinder to those suggesting that space would become less important in a global era. It is interesting in this regard to reflect upon the movement in Storper's work from an earlier 'trinity' of production/work/territory to that of technologies/organizations/territories. His theoretical emphases now relate more distantly to transformations in the nature of work and employment and connect more closely with the broader dynamics of, for example, knowledge formation within regional economies. The landscapes of capitalism (in the form of territory/territories) are still with us, to be sure. However, as a number of critics have suggested, Storper's more recent focus on the creation of 'relational assets' or considerations of the ways in which groups of 'reflexive human actors' shape economic development in regions potentially can overshadow unequal power relations between specific actors (Pike, 1999; Sunley, 1999). Pike has argued, for example, that 'throughout [*The Regional World*] questions of ownership, power and control are skirted around with often only passing recognition that some interests shape the agenda and basis for (re)creating "relational assets" more than others' (Pike, 1999: 2). In other words, frameworks of action in particular regions may appear homogenous, but a conceptualization of the *collective* institutionalization of rules and norms may mask a diversity of ability to participate in such frameworks (Berndt, 2000: 796).

Additionally, a number of critics have expressed concerns about Storper's relatively generalized theorization of conventions. How do conventions 'work' in practice? Gertler, for example, has suggested that 'Storper is more specific about the *effects* that conventions have on economic interaction in product-based technological learning (PBTL)-based technology districts than he is about their *sources* or *origins*' (Gertler, 1997: 49; emphasis in original). Or as Sunley notes:

> At times it is hard to see where conventions actually start and stop. As all economic activity, including regional economies, involves relational conventions of some sort, further research needs to explain why certain mixes of regional convention are successful and others not. (Sunley, 1999: 493)

Storper's interest in notions of 'reflexivity' and the 'untraded interdependencies', or 'soft', non-exchange-based aspects of relationships between economic actors, has been seen by one commentator as reflecting 'the ongoing socio-cultural ("reflexive") turn in economic geography' (Oinas, 2001: 381). Certainly, an emphasis on shared conventions and cultural attributes that are created and re-created within regions does suggest sympathy with more 'culturalist' approaches to the study of economic activity. In a consideration of the transformations of consumer society, Storper has indicated that the conceptualization of conventions is important because they coordinate consumption as well as production (Storper, 2000: 392). Further, he has argued that economic actors can take on multiple positionalities: they 'are not only wage earners, but also consumers, not to mention citizens' (Storper, 2000: 404).

Somewhat ironically, however, Storper (2001) has vehemently attacked the notion of a so-called 'cultural turn' within the social sciences and humanities. His particular objection has been to what he sees as the 'celebratory relativism' that characterizes 'post-modernist and cultural-turn radicals' (Storper, 2001: 167, 170). Geographical work characteristic of such thinking is seen to include that of **Edward Soja, Trevor Barnes** (particularly Barnes, 1996) and Gibson-Graham (1996). Like many other critics of postmodernism, Storper feels that movements that have their origins in identity- or community-based politics are insufficient to achieve societal change; and that they are in danger of being co-opted by an individualistic 'right-wing' political agenda. He exhorts scholars to 'reaffirm their engagement with questions of political economy (issues of wide spatio-temporal extent)' (Storper, 2001: 173). However, the singularity with which Storper defines 'the cultural turn' is of note:

> It is seen as theory- and research-based on the overall notion that the keys to understanding contemporary society and to transforming it lie in the ways that culture orients our behaviours and shapes what we are able to know about the world ... *the* cultural turn variously blends postmodernist philosophy, cultural theories of society and poststructuralist philosophy. (Storper, 2001: 161, emphasis added)

Yet as Crang (1997: 3) has observed, 'the precise forms of [cultural] turns have been multiple and contested. There is no single cultural turn ...' Storper's (2001: 273) characterization of 'cultural turn radicals' thus at times appears to verge on caricature. Ultimately, however, Storper (2001: 173) does acknowledge that any form of radical theorizing and action should be attentive to 'sensitivities ... which have been sharpened by the cultural turn' – questions of difference – as well as a need to engage with more complex notions of human actors and actions.

STORPER'S MAJOR WORKS

Scott, A. J. and Storper, M. (1986) *Production, Work, Territory: The Geographical Anatomy of Industrial Capitalism*. London: Unwin Hyman.

Storper, M. (1997) *The Regional World: Territorial Development in a Global Economy*. New York: Guildford.

Storper, M. and Salais, R. (1997) *Worlds of Production: The Action Frameworks of the Economy*. Cambridge, MA: Harvard University Press.

Storper, M. and Scott, A. J. (1992) *Pathways to Industrialisation and Regional Development*. London: Routledge.

Storper, M. and Walker, R. (1989) *The Capitalist Imperative: Territory, Technology and Industrial Growth*. Oxford: Blackwell.

Storper, M., Tsipouri, L. and Thmodakis, S. (1998) *Latecomers in the Global Economy*. London: Routledge.

Secondary Sources and References

Barnes, T. J. (1996) *The Logics of Dislocation: Models, Metaphors and Meanings of Economic Space*. New York: Guildford.

Berndt, C. (2000) 'Review of *Worlds of production*', *Annals of the Association of American Geographers* 90: 796–797.

Christopherson, S. and Storper, M. (1989) 'The effects of flexible specialisation on industrial politics and the labour market: the motion picture industry', *Industrial and Labor Relations Review* 42: 331–347.

Crang, P. (1997) 'Cultural turns and the (re)constitution of economic geography', in R. Lee and J. Wills (eds) *Geographies of Economies*. London: Arnold, pp. 3–15.

Curry, J. and Kenney, M. (1999) 'The paradigmatic city: postindustrial illusion and the Los Angeles school', *Antipode* 31: 1–28.

Gertler, M. (1997) 'The invention of regional culture', in R. Lee and J. Wills (eds) *Geographies of Economies*. London: Arnold, pp. 47–58.

Gibson-Graham, J. K. (1996) *The End of Capitalism (as we knew it): A Feminist Critique of Political Economy*. Oxford: Blackwell.

Lipietz, A. (1986) *Mirages and miracles*. London: Verso.

Massey, D. (1984) *Spatial Divisions of Labour*. London: Macmillan.

Oinas, P. (2001) 'Review of *The regional world*', *Tijdschrift voor economische en social geografie* 92: 381–384.

Pike, A. (1999) 'Review of *The regional world*', Economic Geography Research Group Book Reviews; available at *www.econgeog.org.uk/bkreview.html* (accessed March 2003).

Scott, A. and Storper, M. (2003) 'Regions, globalization, development', *Regional Studies* (forthcoming).

Storper, M. (1991) 'The transition to flexible specialisation in the US film industry: external economies, the division of labour and the crossing of industrial divides', in A. Amin (ed.) *Post-Fordism: A Reader*. Oxford: Blackwell, pp. 195–226.

Storper, M. (1997) 'Regional economies as relational assets', in R. Lee and J. Wills (eds) *Geographies of Economies*. London: Arnold, pp. 248–258.

Storper, M. (1999) 'The poverty of paleo-leftism: a response to Curry and Kenney', *Antipode* 31: 37–44.

Storper, M. (2000) 'Lived effects of the contemporary economy: globalisation, inequality and consumer society', *Public Culture* 12: 375–409.

Storper, M. (2001) 'The poverty of radical theory today: from the false promises of Marxism to the mirage of the cultural turn', *International Journal of Urban and Regional Research* 25: 155–179.

Storper, M. and Scott, A. J. (1989) 'The geographical foundations and social regulation of flexible production systems', in M. Dear and J. Wolch (eds) *The Power of Geography*. London: Unwin Hyman, pp. 21–40.

Storper, M. and Scott, A. J. (1995) 'The wealth of regions: market forces and policy imperatives in local and global context', *Futures* 27: 505–526.

Sunley, P. (1999) 'Review of *The regional world*', *Regional Studies* 33: 493.

Warf, B. (1991) 'Review of *The capitalist imperative*', *Journal of Regional Science* 31: 222–224.

Susanne Reimer

44 Peter Taylor

BIOGRAPHICAL DETAILS AND THEORETICAL CONTEXT

Peter James Taylor was born in 1944 and grew up in Calverton, near Nottingham, England. He received his BA in Geography from Liverpool University in 1966 and his PhD in Geography from Liverpool in 1970. From 1968 to 1995, he taught at Newcastle University, and after 1995 lectured at Loughborough University. During 1990 Taylor was a research associate for a MacArthur Foundation research project on comparative hegemonies at the Fernand Braudel Center of the State University of New York, Binghamton. Significantly, Taylor was invited to participate in this project by **Immanuel Wallerstein**, head of the Braudel Center and founder of world-systems analysis.

Taylor's is best known within geography for bringing a world-systems perspective to bear on issues within political geography. In context, this must be seen as a significant contribution given that political geography was widely regarded as a moribund subdiscipline in the 1960s. Taylor's work was one of the stimuli that informed the development of a more theoretically inclined approach that revived the field in the 1980s and 1990s. At the same time, his work has contributed an analysis of states and political power to the world-systems perspective. Given that the original formulations of a world-systems perspective associated with Wallerstein's work in the 1970s and early 1980s were widely seen as providing a very limited understanding of the working of states within the global economy, Taylor's

writings since the early 1980s have helped to redress significant lacunae in political theory. In addition, and latterly, his work on world cities under conditions of contemporary globalization has sought to challenge state-centric understandings of geopolitics, creating a lively dialogue between urban and political geographers.

SPATIAL CONTRIBUTIONS

Much of Taylor's earliest scholarship focused on theoretical and methodological concerns within electoral geography (Johnston and Taylor, 1979). Yet his introduction of world-systems analysis to political geography (Taylor, 1981, 1982) demands not only a reconceptualization of electoral politics, but wider notions of territory and governance. Taylor starts with the world-systems' division of states into core, semi-periphery and periphery. This tripartite division refers to the kinds of economic processes that predominate within the boundaries of states. *Core* economic processes are relatively technologically advanced and pay high wages; *peripheral* economic processes are by contrast less technologically advanced and based on the employment of low wage labour. A small number of states, whose territories house most of the high-wage/advanced technology processes at a given time, are referred to as 'core states'. A much larger group of states whose territories predominantly house low-wage/basic technology processes are referred to as 'peripheral states'. A small but politically important group of states exist at any given time that combine a substantial amount of both core-like and periphery-like processes, and these are referred to as 'semi-peripheral states'.

Taylor's application of this tripartite division of the world-system to electoral geography (1986) suggests the need to reconsider the significance of different kinds of political-electoral processes in relation to the position of states within the hierarchical global structure. Specifically, Taylor argues that the prevalence of liberal, multi-party democracies within particular states, and the absence of this political form in others, is best explained by the position of states in the global hierarchy, rather than by attributes internal to these states such as their economic ideologies or cultural histories. Multi-party, liberal democratic systems thrive within the core countries of the global economy – forming what Taylor refers to as 'liberal democratic states' – because the high level of wealth within these states allows for the possibility of a redistributive politics, or social democracy. Broad participation in elections reflects the ability of parties within core states to satisfy certain demands of the broader populace, even while they serve to reproduce capitalist class power. In the periphery, by contrast, the absence of a large surplus to redistribute encourages elites to limit or prevent broad-based popular participation that might lead to democratic demands for redistribution. Thus, in contrast to the core, the periphery is marked by higher prevalence of military dictatorship and other political forms that limit the expression of democratic demands. Moreover, such undemocratic forms are especially prevalent in semi-peripheral states, which utilize political repression as a tool for intensifying the exploitation of labour in order to maximize capital accumulation in the attempt to 'catch up' economically with core states (Taylor, 1986; Taylor and Flint, 2000: 38, 235–285).

The significance of this argument is that it challenges the theme of 'developmentalism' prominent in orthodox development theory, where it was often assumed that peripheral countries with undemocratic governments were simply at an earlier stage of development than the already developed liberal democracies, and would eventually 'catch up'. Taylor's argument, by contrast, suggests that the socio-spatial inequalities inherent in global capitalism are likely to prevent any simultaneous process of 'catching up' by all states. From a world-systems perspective, the global capitalist economy is systematically and inevitably hierarchical, in both a social and a spatial sense. Thus, there will always be a core/semi-periphery/periphery structure to this system, even if different countries occupy the positions of this hierarchy at different points in time. Such a hierarchy will always lead to different sets of political possibilities at the top and the bottom of the system, because of the difference in possibilities for redistribution. Thus, even where there have been recent, formally democratic developments in the periphery and semi-periphery of the global economy, Taylor suggests that these are likely to be highly unstable and subject to reversal (Taylor and Flint, 2000: 281–285).

Beyond theorizing electoral politics from a world-systems perspective, one of Taylor's significant contributions to world-systems analysis has been to theorize the development of the modern territorial state and its relationship to the global economy. For world-systems analysts, the capitalist world-system is comprised of two structural features that are very different in their spatiality: (1) a capitalist market economy, based on a socio-spatial division of labour and processes of capital accumulation that take the form of global commodity chains, these having developed in Europe by the sixteenth century and expanded to become fully global in the twentieth century; and (2) a system of bounded territorial states that have more limited geographical range but nonetheless perform crucial functions in facilitating the global process of capitalist accumulation.

While Wallerstein had theorized the necessity of the territorial state as a facilitator of capitalist accumulation, he did not in his early work provide extended discussions of the nature of the international system of sovereign territorial states. Taylor's work, along with that of several other world-systems theorists (e.g. Chase-Dunn, 1989; Arrighi, 1990), has helped fill in this lacuna. In particular, Taylor has theorized the way modern states came into existence in relation to the political-territorial development of a state *system*. Taylor emphasizes that while most state theories – including Marxist theories of the capitalist state – have focused on the *internal* political economic conditions for development of territorial states (such as domestic class structures), one cannot fully understand modern states without understanding the *external* conditions necessary for their existence. In particular, the modern territorial state that developed with the rise of capitalism came to have exclusive control, or sovereignty, over the entirety of its internal territory, a condition that is only meaningful insofar as it implies the recognition of that sovereignty by other states. Thus, the development of the modern territorial state could only occur as part of the development of an international system of states in which there was mutual recognition of sovereign territories by all members of that system. This is the condition that Taylor refers to as 'inter-stateness', and his work has not only theorized this process but discussed its evolution, emphasizing that it is a historical social construct (Taylor, 1994, 1995a).

In developing a historical account of modern territorial states, Taylor has extensively addressed one particular feature of the capitalist interstate system – namely, that there have been three periods in which the system was dominated by one particular state that can be referred to as 'hegemonic'. Hegemony, following the ideas of the Italian theorist Antonio Gramsci, is taken to mean that the states in question can lead others as much through the consent they are able to

generate for their leadership as through the coercive force they are able to exercise. Taylor posits that there have been three hegemonic states in specific periods of historical capitalism, each identified with specific forms of capitalism and leading technologies of the era: (1) the United Provinces (Netherlands), in the seventeenth-century era of mercantile capitalism, based on its dominance in technologies associated with shipping and ship-building; (2) the United Kingdom, in the nineteenth-century era of industrial capitalism, based on its dominance in technologies associated with industrial machinery and production processes; and (3) the United States, in the twentieth-century era of consumer capitalism, based on its dominance in technologies associated with mass production and communications (Taylor, 1996).

For Taylor, the periods of hegemonic leadership by each of these states are also associated with particular conceptions of what it is to be modern, defined by the practices of the hegemonic power as they are presented to the rest of the world. Thus, seventeenth-century followers attempted to be like the Dutch in their commercial prosperity, while nineteenth-century followers attempted to be like the British in developing industrial prowess, and twentieth-century followers have attempted to be like the Americans through participation in consumerism (Taylor, 1999; Slater and Taylor, 1999). These 'prime modernities', as Taylor calls them, illustrate that from a world-systems perspective, culture, just like economics and politics, has to be understood through a lens that highlights interconnections at a global scale.

The fact of the international system of sovereign territorial states being a historical creation that moves through a series of hegemonies and 'prime modernities' is crucial to a final set of themes that Taylor has developed in discussing the world-system. In world-systems analysis, the state is neither an insignificant player in capitalism nor an all-powerful controller of social processes. Contemporary studies of globalization that do not take a world-systems view sometimes portray globalization as representing the outstripping of state power by transnational economic forces (see Swyngedouw, 1989). Taylor, like other world-systems analysts, rejects this 'death of the nation-state' thesis as simplistic, but also acknowledges that ongoing changes in the structure of the capitalist world-system are leading to changes in the specific forms of political power and the ways in which they operate. Utilizing a distinction made by **Yi Fu Tuan** (Tuan, 1977), Taylor argues that there is today a notable tension between *space* – the abstract areal extension across which capital and commodities flow – and *place* – the lived environments in which people create meaning and carry out their day-to-day existence. As an example, Taylor notes increasing ambiguities and tensions surrounding relationships within two major institutions of the world-system, nation-states and households. Both institutions have been portrayed historically as safe places, as havens from the oppressiveness and callous competitiveness of global capitalism. Yet increasingly, these havens have appeared to their inhabitants as cages, being transformed by processes of modernity into spaces rather than places. Thus, nation-states seem to be tools of the most egregious forms of violence and repression, especially for groups such as racial or ethnic minorities, while households are recognized as spaces of horrendous violence against women (Taylor, 1999, 2000a).

As awareness of the negative possibilities of modern institutions has grown in recent decades, movements have emerged that emphasize localism and local resistance to global capitalism, as well as to the territorial state and the patriarchal household. For Taylor, this manifests an ongoing process of transformation within the world-system. In particular, states are no longer as likely to be seen by progressive social actors as tools of liberation, in the way they had been by traditional socialist and nationalist movements. Instead, social movements increasingly work within local, place-based projects,

while also forming transnational linkages, sometimes bypassing the national state in the process.

This leads to complex, multi-scalar politics. Taylor has argued that in a general sense, the global scale has been produced as an object of human awareness by the reality of world-systems processes, while the national scale has been produced by the dominant ideology of territorial nation-states, and the local has been produced by lived day-to-day experience (Taylor, 1982; Taylor and Flint, 2000: 42–46). But with the ideological power of the national territorial state increasingly contested, the connections between local and global processes ('glocalization') become more important. Taylor highlights this transformation in two ways. First, following the 'world cities' hypothesis developed by John Friedmann in the 1980s (Friedmann, 1986), as well as **Manuel Castells'** ideas of a network society, he shows that economic interactions in the global economy are increasingly constituted by flows of capital, people and information between a group of globally dominant cities, rather than between states as territorial wholes (Taylor, 1995b, 2000b). Second, he suggests that the spread of global consumer capitalism under US hegemony is ecologically unsustainable and is destroying livable places with spaces of constantly expanding global capital flows. Thus, ecological problems constitute the 'ultimate space–place tension' and are bringing forth new responses that challenge the spatial domination exercised by global corporations and territorial states with progressive, place-based yet transnationalized forms of activism (Taylor and Flint, 2000: 366–367).

KEY ADVANCES AND CONTROVERSIES

Taylor's work has done much to highlight the importance of a holistic, global perspective in the social sciences. Where many geographers had in the past considered entities like regions to have a relatively independent existence (based, for example, on bio-physical features of the area), Taylor's work has been part of the development of 'social constructivist' approaches which suggest that many taken-for-granted geographical concepts – households, localities, regions, states, world regions, and, indeed, the idea of scale itself – are socially and historically produced (Taylor, 1991b). Most importantly, Taylor's work shows how various of these phenomena are produced *in relation to each other*. From this perspective, the only way to fully understand the phenomena in question is to understand them in relationship to the global totality of which they are a part.

Here, however, there is an important controversy surrounding the way Taylor and world-systems analysts approach the constructed character of social reality and the ways this issue has been approached in other leading social constructivist accounts. In recent decades, some of the most important of these other social constructivist accounts have been associated with postmodernist and poststructuralist philosophies. Many of these 'post-prefixed' approaches have expressed great skepticism about the existence of large-scale, unified wholes (such as global capitalist economies or unified national social formations), and thus they have eschewed attempts to theorize in terms of such wholes. Thus, world-systems analysis and 'post-prefixed' theory represent two very different approaches to social construction and historical contingency.

Taylor explicitly recognizes this difference in addressing the revival of political geography through the social theory movement and critical geopolitics. The social theory movement, exemplified by Gordon Clark and **Michael Dear**'s work on state apparatuses and their varied legal and ideological roles (Clark and Dear, 1984), is seen by Taylor as making an important contribution in areas where world-systems theory is weak – i.e., detailed analysis of the internal structures of states – but as missing the importance of states' positions in the interstate system (Taylor, 1991a: 394–397). Critical geopolitics, exemplified by the work of poststructuralist authors such as **Gearóid Ó Tuathail** (1996) on the relationship between state power and representational practices, is seen by Taylor as making a useful contribution to the understanding of specific dimensions of state power and political contestation, though it is only one part of a broader set of geopolitical analyses that need to be brought together in order to provide an adequate analysis (Taylor and Flint, 2000: 50, 102–104). Thus, from Taylor's perspective, the debate between world-systems analysts and 'post-prefixed' theorists generates a 'creative tension' that can push forward critical thinking on politicized space and place.

TAYLOR'S MAJOR WORKS

Taylor, P. J. (1981) 'Political geography and the world-economy', in A. D. Burnett and P. J. Taylor (eds), *Political Studies from Spatial Perspectives*. New York: Wiley, pp. 157–172.

Taylor, P. J. (1982) 'A materialist framework for political geography', *Transactions of the Institute of British Geographers* NS 7 (1): 15–34.

Taylor, P. J. (1986) 'An exploration into world-systems analysis of political parties', *Political Geography Quarterly* 5 (4), supplement: 5–20.

Taylor, P. J. (1991a) 'Political geography within world-systems analysis', *Review* 14 (3): 387–402.

Taylor, P. J. (1991b) 'A theory and practice of regions: the case of Europes', *Environment and Planning D: Society and Space* 9 (2): 183–195.

Taylor, P. J. (1994) 'The state as container: territoriality in the modern world-system', *Progress in Human Geography* 18 (2): 151–162.

Taylor, P. J. (1995a) 'Beyond containers: internationality, interstateness, interterritoriality', *Progress in Human Geography* 19 (1): 1–15.

Taylor, P. J. (1995b) 'World cities and territorial states: the rise and fall of their mutuality', in P. L. Knox and P. J. Taylor (eds) *World Cities in a World-system*. Cambridge: Cambridge University Press, pp. 48–62.

Taylor, P. J. (1996) *The Way the Modern World Works: World Hegemony to World Impasse*. London: Wiley.

Taylor, P. J. (1999) *Modernities: A Geohistorical Interpretation*. Cambridge, UK and Minneapolis: Polity and University of Minnesota Press.

Taylor, P. J. (2000a) 'Havens and cages: reinventing states and households in the modern world-system', *Journal of World-Systems Research* 6 (2): 544–562.

Taylor, P. J. (2000b) 'World cities and territorial states under conditions of contemporary globalization', *Political Geography* 19 (1): 5–32.

Taylor, P. J. and Flint, C. (2000) *Political Geography: World-economy, Nation-state and Locality*, 4th edition. Harlow: Prentice Hall.

Secondary Sources and References

Arrighi, G. (1990) 'The three hegemonies of historical capitalism', *Review* 13 (3): 365–408.

Chase-Dunn, C. (1989) *Global Formation*. Oxford: Blackwell.

Clark, G. L. and Dear, M. (1984) *State Apparatus*. Boston: Allen and Unwin.

Friedmann, J. (1986) 'The world city hypothesis', *Development and Change* 17 (1): 69–84.

Johnston, R. J. and Taylor, P. J. (1979) *Geography of Elections*. London and New York: Croom Helm and Holmes & Meier.

Ó Tuathail, G. (1996) *Critical Geopolitics*. Minneapolis: University of Minnesota Press.

Slater, D. and Taylor, P. J. (eds) (1999) *American Century: Consensus and Coercion in the Projection of American Power*. Oxford: Blackwell.

Swyngedouw, E. (1989) 'The heart of place: the resurrection of locality in an age of hyperspace', *Geografiska Annaler* 71B: 31–42.

Tuan, Y.-F. (1977) *Space and Place*. London: Edward Arnold.

Jim Glassman

45 Nigel Thrift

BIOGRAPHICAL DETAILS AND THEORETICAL CONTEXT

Working on the cusp between economic and cultural geography, Nigel Thrift has long been recognized as one of geography's most imaginative, articulate and productive scholars. In an exceptionally lengthy series of widely cited papers and books, Thrift has made major contributions to the history of geographical thought, introduced a variety of new perspectives into geographical thinking (ranging from structuration theory to poststructuralist concerns for identity and performativity), advocated a renewed role for regions in geographical analysis, explicated the contingent, contradictory nature of globalization and its implications for regional development, and consistently argued for the centrality of human consciousness, however bounded, in the construction and reproduction of social and spatial relations.

Born in 1949, Thrift earned his BA in Wales and a PhD from the University of Bristol in 1979. He was stationed briefly in Leeds (1976–77), spent four years (1979–83) as a Research Fellow at the Australian National University, and three years (1984–87) as a Lecturer and Reader at Saint David's University College, Lampeter, Wales. In 1987 he returned to Bristol as a Professor in the School of Geographical Sciences, and, in 2003, was appointed Dean of Life and Environmental Sciences at Oxford University. He has also taught or carried out research at a number of institutions around the world as a Fellow, Research Scholar, or Distinguished Professor, including UCLA, the University of Vienna, the National University of Singapore, the Netherlands Institute for Advanced Study and the Swedish Collegium for Advanced Study in the Social Sciences. He is an Academican of the Association of Learned Societies in the Social Sciences, and was awarded the Royal Geographical Society's Heath Award and the Medal of the University of Helsinki. He has supervised numerous doctoral students, obtained considerable research grant funding, and serves on many editorial boards, serving since 1979 as managing editor for *Environment and Planning A*, and in 1982 co-founding *Environment and Planning D: Society and Space*.

Thrift's career simultaneously reflected and spurred the enormous intellectual changes in human geography in the 1980s and 1990s, which witnessed considerable exploration and debate, particularly between various humanist and Marxist approaches. Throughout Thrift's works runs a Marxist inspiration of the non-structuralist variety, including an abiding concern for class from a non-structuralist standpoint, a position that would later be extended to a variety of non-class-based forms of social determination (Thrift and Williams, 1987). Unlike structuralist accounts, in which human consciousness is subjugated to the determining forces of social relations, Thrift's world-view takes seriously the complex conditions under which human beings take the world for granted and give meaning to it.

Thrift's career has exhibited a sustained interest in questions of time (Thrift, 1977a, 1977b), suturing time and space into an integrated totality. No longer could geographers be content with the static geometries of chorology and positivism; social theory had become self-consciously dynamic, portraying every place as a time and integrating into spatial analysis an increasingly sophisticated understanding of how time was socially constructed and thus experienced differently under varying historical and spatial circumstances. This line of work exhibited strong linkages to time-geography, which concerns itself with the movements of individual bodies through the rhythms of everyday life. Thrift also studied the restructuring of time consciousness in contexts ranging from the medieval to the contemporary, including in particular the imperatives of commodity production and capital accumulation (Thrift, 1988), a notion that closely echoed E. P. Thompson's famous portrayal of the social dynamics of factory time.

As geographers began to push the envelope of traditional social theory in the 1980s, changing the discipline from a passive recipient to active generator of new concepts, many turned to **Anthony Giddens** and structuration theory. Thrift played a key role in introducing these concepts into geography (Thrift, 1985a, 1993), simultaneously buttressing the theory's spatial dimensions. His paper 'On the determination of social action in space and time' (1983) was a landmark piece that both situated structuration theory within the broader intellectual context of 'micro–macro' divisions in the social sciences and pointed toward the emerging reconstruction of regional geography.

The publication of the two-volume *New Models in Geography* (Peet and Thrift, 1989), a play on Chorley and Haggett's (1967) positivist bible *Models in Geography*, heralded the ascendancy of political economy to a central and pivotal status within geography. Extending conventional interpretations concerned mostly with class into new domains, the essays in these volumes offered comprehensive and insightful interrogations of the production of nature, the state, gender, uneven development, the nation-state, urbanization, race, culture and postmodernism.

Over the last decade, Thrift played an influential role in moving geography into new frontiers of poststructuralism, including a variety of concerns with subjectivity, language, representation, discourse, identity and practice (Pile and Thrift, 1995). Although he has never referred to himself as a postmodernist *per se*, his theoretical orientations include similar qualities, such as a distrust of broad metatheories, an appreciation for local uniqueness, and an emphasis on social structures and change as inherently open-ended and contingent.

SPATIAL CONTRIBUTIONS

One measure of Thrift's contribution to geography, and to other disciplines, is the sheer volume of his writings: between 1975 and 2001, for example, he authored or co-authored 78 journal articles, 68 book chapters, four books, and co-edited 13 volumes. Some of the key works are recapitulated in *Spatial Formations* (1996b). Of course, mere numbers fail to convey the diversity, depth of insight and impacts of these works.

Thrift's career began as an economic geographer with analyses of industrial linkages and markets and multinational corporations (Taylor and Thrift, 1986), originally focusing on the specific instance of Australia. This thread has continued in an ongoing form through a sustained interest in organizational theory, which maintains that the specifics of individual corporations are fundamental to the understanding of how markets behave and change. Similarly, Thrift has been fascinated throughout his career by cities as dense nuclei of economic and social transactions, on the changing possibilities for economic development afforded by globalization (Thrift, 1987), and the mutations in urban form and life that such transformations entail (Pile and Thrift, 2000). Cities, particularly in the knowledge-driven economies of 'soft capitalism', both reflect and constitute the changing dynamics of social life as contingent constellations of meaning (Amin and Thrift, 2002).

Hence, if there is any theme that permeates Thrift's multitudinous works, perhaps it is his sustained interest in connectivity of one sort or another. His papers have, in different ways and contexts, addressed the relations among bodies and individuals, the diffusion of information (Thrift, 1985b), the reciprocal relations between knowledge and power, and increasingly globalized exchanges of human, symbolic and financial capital. Thus, he exhibits a consistent concern for the flows of information and knowledge that shape individuals' conceptions of themselves and one another and that play a major role in the routinized reproduction of social relations. Such linkages reflect and constitute the changing face of capitalism at a variety of spatial scales, ranging from the most intimate and local (e.g. the body) to the international and global.

Thrift's concern with the differential regional effects of globalization was materialized in a variety of case studies. He has worked extensively on the restructuring of the British space-economy in the face of globalization and mounting social and regional polarization, including the rise of the Greater London region (Thrift, 1987) and the nature and impacts of various services such as accountancy, property and real estate, finance (Leyshon and Thrift, 1997), and surveying. He also examined other European regions in various contexts (Amin and Thrift, 1992), including the rising role of dense human networks in leading innovative regions characteristic of advanced post-Fordist production systems. Thrift also worked extensively on Pacific Asia, most notably Vietnam (Forbes and Thrift, 1987; Thrift and Forbes, 1986). These empirical analyses complemented his insistence that regions mattered not only ontologically, i.e. as units of empirical analysis, but also epistemologically, i.e. that no social process was manifest in the same way in different places.

The latter line of thought would prove to hold wide appeal to the growing legions of geographers who insisted that local uniqueness mattered, many of whom were inspired by **Doreen Massey** (1984), and who sought to recover regions from the dustbin of traditional chorology. Throughout the 1980s, the 'new regional geography' developed growing linkages to top-down global political economy and to 'bottom-up' concerns for living actors and everyday life. Thrift played a key role in articulating this vision of geography, defending it against both positivist and Marxist critics for whom regional analysis could, by definition, scarcely amount to more than reconstituted empiricism. Unlike chorology, however, the reconstituted regionalism was both theoretically informed and paid great attention to the connections among regions, their linkages to the global division of labour, and the sticky issue of the production of spatial scale. The new regionalism made local uniqueness a valid object of scholarly scrutiny, demonstrating that the local *mattered* in ways that broad, generalized models could never discover and that no social process played out exactly in the same way in different time–space contexts. The recovery of regions for social theory obligated the discipline to delve into the intricate dynamics of local actors who populated them and 'made them work': thus the localities school and structuration theory's focus on everyday life were complementary. This concern with the local extended into his work on the most abstract and globalized of economic sectors, finance.

In the late 1980s, Thrift entered into an extraordinarily fecund relationship with Andrew Leyshon that generated a series of works concerning the international financial system, British financial capital and banking networks in London (Leyshon and Thrift, 1997; Thrift and Leyshon, 1992, 1994). These works both elucidated the impacts of electronic money and national deregulation and simultaneously demystified money by stressing the key roles played by interpersonal relations predicated on trust and face-to-face contact, including the critical but often overlooked roles played by local milieu, trust and 'institutional thickness', i.e. the cultural context in which financial transactions are always and everywhere embedded. Money, even in its most rarefied electronic form, is thus much more than an abstract economic relation, for its real power lies in its symbolic value and links to national and global distributions of social power (Thrift, 2001). Like all social relations, financial systems are inevitably discursively constituted through particular socio-cultural practices in different times and places. Such an approach brought a badly needed human quality to a topic dominated by ethereal understandings devoid of actors, and is vital if the new forms of 'soft' managerial capitalism are to be understood in all their complexity. Thrift's writings reveal an understanding that the contemporary forms of capitalism, dominated as they are by finance and information, simultaneously reflect the historic tendency toward increasing commodification of all activity, and the emergence of a historically unprecedented form of production and consumption, a perspective that eludes the naïve optimism and technological determinism of postindustrial social theory. The 'symbolic economy' calls for new forms of performativity for actors in order to function and succeed (Thrift, 2000).

Within this vein, Thrift has exhibited broader concerns for the implications of information technology and electronic communications on individual subjectivity, a form of what Giddens (1984) calls time–space distanciation, that are key indicators of post-Fordist 'soft capitalism' and its attendant digitized, virtual dimensions (Thrift, 1995, 1997). The massive time–space compression unleashed by digital globalization has thus altered the field of possibilities of nation-states and of individual actors in everyday life: the renewed regionalism and reconstituted rhythms of daily life are two sides of one coin. At the level of the corporation and everyday life, Thrift has also taken an interest in the possibilities extended by the widespread use of computational technology and mobile telecommunications. In sum, the contingent, simultaneously determinant interactions between culture and technology form a long-standing theme threaded throughout his papers.

KEY ADVANCES AND CONTROVERSIES

Given the length, diversity and enormous productivity of Nigel Thrift's career, any itemization of key contributions can amount to little more than a recapitulation of the major themes that have underlaid his career. There can be little doubt that he has played a key role in forcing geographers to take much more seriously than they had hitherto questions of consciousness, subjectivity and identity (Pile and Thrift, 1995). For example, he maintains that it is increasingly the relations of consumption that are critical to the negotiation and maintenance of identity, stimulating new interest in a topic long marginalized by the Marxist focus on production.

Thrift has long insisted that geographies are produced by people in pre-discursive, practical ways, and has been one of the discipline's foremost advocates for theoretical stances that emphasize the contingency of human action. This position stands as a hallmark of the contemporary 'cultural turn' in the social sciences, which has led numerous researchers to embed economies in social relations, blurring the artificial distinctions between the 'economic', the 'cultural', the 'social' and the 'political' (Thrift and Olds, 1996). More broadly, the profound role played by systems of signification, including discourses, language, texts and representations of all kinds, has led him to forge a nonrepresentational theory of action that stresses performative embodied knowledges (Thrift, 1997). Yet this relational materialist perspective maintains that the global 'space of flows' does not float in a disembodied space: rather, global flows and connections are constructed by human beings who are always embedded in networks of power and knowledge that are themselves part of an ever-changing structural context (Thrift, 2000). Invoking actor-network theory, Thrift's recent work extends the insights of **Bruno Latour** and actor-network theory, which effectively abandons the distinction between agents and structures altogether. This view focuses on the exercise of power by actors rather than just the embeddedness of power in networks. Throughout his career, Thrift has worked assiduously to portray geographies as embodied, embedded, contingent and ever changing, harnessing the fluidity of spatial relations to demonstrate how they are imbricated in changing human relations of power.

Geographers' engagements with Thrift's ideas have taken a variety of forms, of which five are suggested here. First, the popularity of structuration theory in the discipline owed no small part to his writings, sensitizing the discipline to the constructed and contingent nature of space. Thus, human geographers often came to frame their topics, questions and responses within the terms of intended actions and unintended results, an approach that elevated everyday life from the trivial to the critical.

Second, his repeated calls for an historical geography of time were important both in moving the field toward an emphatically dynamic approach to theoretical and empirical issues and underscored the social construction of time, which to all outer appearances appears 'natural'. By denaturalizing time, Thrift's work pointed towards the broader denaturalization of other phenomena, including language, power, poverty, representations and the body.

Third, economic geographers have been motivated by his books and papers to rework their subject, emphasizing the role of human actors in historical contexts, demystifying even money, the most abstract of social relations. For example, international banking networks were shown to be the very human products of business executives enmeshed in culturally specific webs of meaning. Such an approach helped geographers decisively to rupture any remaining ties to the comparative statics of neoclassical economics. Yet Thrift's corpus of writings also advanced the field beyond Marxism, exposing as they did its tendency toward teleological social predetermination and the reification of class.

Fourth, his sensitivity to regions, not simply as receivers of broader changes but as generators in their own right, has made an appreciation for local uniqueness *de rigueur* within the field. Thrift and others illustrated that no social process unfolds in the same way across different places, raising the significance of context in social explanation to a central position. In this reading, the local was far from simply a handmaiden to the global, and the critical question of spatial scale was problematized across several intellectual perspectives.

Fifth, social and cultural geographers have been inspired by his poststructuralism, and Thrift's works have fostered a critical sense of discourse and representation within the field, generating marked interest in questions of personal and social identity and how it is sutured to space. Some younger Marxists have criticized his turn towards actor-network theory, arguing that it brackets social structures to the point of meaninglessness. There can be little doubt, however, that geography's understanding of space and spatial relations has been left far more sophisticated, subtle and nuanced than it might otherwise have been. Thrift both legitimized and popularized nonstructuralist social theory among geographers, helping the discipline to shift decisively from a relatively stagnant backwater that borrowed its most original ideas from other fields into a generator of new perspectives.

THRIFT'S MAJOR WORKS

Corbridge, S., Martin, R. and Thrift, N. (eds) (1997) *Money, Power and Space*. Oxford: Blackwell.

Leyshon, A. and Thrift, N. (eds) (1997) *Money/Space: Geographies of Monetary Transformation*. London: Routledge.

Peet, R. and Thrift, N. (eds) (1989) *New Models in Geography: The Political-Economy Perspective*. Boston: Unwin-Hyman.

Pile, S. and Thrift, N. (eds) (1995) *Mapping the Subject: Geographies of Cultural Transformation*. New York: Routledge.

Thrift, N. (1981) 'Owners' time and own time: The making of a capitalist time consciousness, 1300–1880', in A. Pred (ed.) *Space and Time in Geography: Essays Dedicated to Torsten Hagerstrand*. Lund: Lund Studies in Geography Series B., No. 48.

Thrift, N. (1983) 'On the determination of social action in space and time', *Environment and Planning D: Society and Space* 1: 23–57.

Thrift, N. (1996b) *Spatial Formations*. Thousand Oaks, CA: Sage.

Thrift, N. (2000) 'Performing cultures in the new economy', *Annals of the Association of American Geographers* 4: 674–692.

Thrift, N. and K. Olds. (1996) 'Refiguring the economic in economic geography', *Progress in Human Geography* 20: 311–337.

Secondary Sources and References

Amin, A. and Thrift, N. (1992) 'Neo-Marshallian nodes in global networks', *International Journal of Urban and Regional Studies* 16: 571–587.

Amin, A. and Thrift, N. (2002) *Cities: Reimagining the Urban*. Cambridge: Polity Press.

Anderson, K., Domosh, M., Thrift, N. and Pile, S. (2002) *The Handbook of Cultural Geography*. London: Sage.

Chorley, R. and Haggett, P. (eds) (1967) *Models in Geography*. London: Methuen.

Cloke, P., Doel, M., Matless, D., Phillips, M. and Thrift, N. (1994) *Writing the Rural: Five Cultural Geographies*. London: Paul Chapman.

Crang, M. and Thrift, N. (2000) *Thinking Space*. London: Routledge.

Forbes, D. and Thrift, N. (eds) (1987) *The Socialist Third World: Urban Development and Territorial Planning*. Oxford: Blackwell.

Giddens, A. (1984) *The Constitution of Society: Outline of the Theory of Structuration*. Berkeley: University of California Press.

Massey, D. (1984) *Spatial Divisions of Labor*. New York: Methuen.

May, J. and Thrift, N. (eds) (2001) *TimeSpace: Geographies of Temporality*. London: Routledge.

Pile, S. and Thrift, N. (eds) (2000) *City A-Z*. New York: Routledge.

Taylor, M. and Thrift, N. (eds) (1986) *Multinationals and the Restructuring of the World Economy*. Kent: Croom Helm.

Thrift, N. (1977a) 'Time and theory in human geography, part one', *Progress in Human Geography* 1: 65–101.

Thrift, N. (1977b) 'Time and theory in human geography, part two', *Progress in Human Geography* 1: 415–457.

Thrift, N. (1985a) 'Bear and mouse or bear and tree? Anthony Giddens's reconstitution of social theory', *Sociology* 19: 609–623.

Thrift, N. (1985b) 'Flies and germs: A geography of knowledge', in D. Gregory and J. Urry (eds) *Social Relations and Spatial Structures*. London: Macmillan.

Thrift, N. (1987) 'The fixers: The urban geography of international commercial capital', in M. Castells and J. Henderson (eds) *Global Restructuring and Territorial Development*. London: Sage.

Thrift, N. (1988) 'Vivos voco: Ringing the changes in the historical geography of time consciousness', in T. Schuller and M. Young (eds) *The Rhythms of Society*. London: Routledge.

Thrift, N. (1993) 'Review essay: The arts of the living and the beauty of the dead: Anxieties of being in the work of Anthony Giddens', *Progress in Human Geography* 17: 111–121.

Thrift, N. (1995) 'A hyperactive world', in R. J. Johnston, P. Taylor and M. Watts (eds) *Geographies of Global Change*. Oxford: Blackwell.

Thrift, N. (1996a) 'New urban eras and old technological fears: Reconfiguring the goodwill of electronic things', *Urban Studies* 33: 1463–1493.

Thrift, N. (1997) 'The rise of soft capitalism', in A. Herod, S. Roberts and G. Toal (eds) *An Unruly World? Globalisation and Space*. London: Routledge.

Thrift, N. (2001) ' "It's the romance, not the finance, that makes the business worth pursuing": Disclosing a new market culture', *Economy and Society* 4: 412–432.

Thrift, N. and Forbes, D. (1986) *The Price of War: Urbanization in Vietnam, 1954–1985*. Boston: Allen and Unwin.

Thrift, N. and Leyshon, A. (1992) 'Liberalisation and consolidation: The single European market and the remaking of European financial capital', *Environment and Planning A* 24: 49–81.

Thrift, N. and Leyshon, A. (1994) 'A phantom state: The de-traditionalization of money, the international financial system and international financial centers', *Political Geography* 13: 299–327.
Thrift, N. and Williams, P. (1987) *Class and Space: The Making of Urban Society*. New York: Routledge & Kegan Paul.

Barney Warf

46 Waldo R. Tobler

BIOGRAPHICAL DETAILS AND THEORETICAL CONTEXT

Waldo Tobler was born in 1930, the son of Swiss parents living in the Pacific Northwest of the United States. After schooling in the US and in Switzerland and four years in the American army, he studied at the University of British Columbia, where he took his first cartography course with Ross Mackay. He transferred to the University of Washington, where he received his BA in 1955, his MA in 1957 and his PhD in 1961 for a dissertation entitled 'Map transformations of geographic space'. He was part of the group who studied under John Sherman and William Garrison at Washington in the late 1950s, a group that included **Brian Berry**, Duane Marble, Richard Morrill, John Nystuen, Michael Dacey and William Bunge. This placed him at the centre of what is often termed geography's 'quantitative revolution'. Tobler (2002) subsequently described this period as 'a most exciting time', noting, however, that his academic concern was for models and theories rather than numbers as such.

Tobler spent 16 years, from 1961 to 1977, at the University of Michigan (with John Nystuen and Olsson). Here he developed his interest in computer programming and mathematics, and learnt differential geometry, which proved invaluable for his work on map projections. From 1977, at the University of California, Santa Barbara (with **Golledge**, Smith and Goodchild). At Santa Barbara he was associated with the National Center for Geographic Information and Analysis,

taking up the position of Professor Emeritus upon retirement in 1994. Throughout, Tobler was widely acknowledged with maintaining geography's interest in cartography during the quantitative revolution and acclaimed as a pioneer in the use of computers in cartography. Well versed in mathematics, Tobler was particularly acclaimed for his development of continuous-space representations of geographical space. He developed interests in other fields, especially modelling migration flows, and completed innovative methodological work in quantitative geography. He has also worked on spectral analysis and has used vector and tensor analysis, often preferring mathematical methods to the statistical methods favoured by most geographers interested in quantitative modelling.

SPATIAL CONTRIBUTIONS

Tobler's early work was concerned with the development of new perspectives on mapping. He was a leader in the use of computers in cartography (Tobler, 1959, 1967a), especially in the design of new projections. In his PhD thesis, he introduced the concept of maps scaled by travel time or cost, rather than distance. He also developed the idea of employing map transformations in order to represent social or demographic reality rather than just the earth's surface. If a theory, such as those of Christaller or von Thünen, makes unrealistic assumptions about the study area, he suggested that this can be tackled by transforming geographic space to make the assumptions fit. This is done by the use of cartograms, which can be of many types but generally attempt to depict the earth surface in a way where the size of features reflects not their surface area but their importance. As he put it, 'In many ways these maps are more realistic than the conventional maps used by geographers ... The important point, of course, is not that the transformations distort area, but that they distribute densities uniformly' (Tobler, 1963a). In this and in other contexts, Tobler has consistently been generous in making his software available for other researchers to use.

In addition to cartograms, Tobler developed many new and unusual map projections. He has applied his approach to map projections to examine historical maps and the implicit projections that underlie them (Tobler, 1966a). Another interesting historical project was his work with Wineberg (Tobler and Wineberg, 1971) on the location of ancient Hittite settlements. The existence of records of trade links between settlements of the period was exploited to estimate the location (in many cases unknown) of the settlements involved. This was done using an inverse form of the gravity model, in which the number of times two places were mentioned in close proximity was used as an estimate of interaction, and distances were predicted on the basis of size and interaction. Subsequent archaeological investigation verified the model's predictions.

Much of Tobler's work has been concerned with adopting mathematical methods to new geographical contexts. One example of this is his paper 'Of maps and matrices' (Tobler, 1967b), in which he argues that data for regularly arranged places on the earth's surface can be regarded as forming a matrix, and that matrix multiplication can be performed with such data to yield geographically interesting results. The argument is illustrated using **Torsten Hägerstrand**'s simulation of the spread of innovations.

Tobler's development of mathematically based but empirically useful techniques is also exemplified by his development of pycnophylactic interpolation (Tobler, 1979) as a solution to the problem of areal interpolation: the problem of comparing geographical distributions (e.g. of population at two dates) when the sets of boundaries are incompatible. His solution is based on the (perhaps questionable) idea that the imposition of a set of areal units on geographical space renders the underlying continuous distribution into a discrete data set. The problem of interpolation becomes one of estimating the continuous surface and then calculating the values of the surface over the new data set. Tobler (1990) has continued to show interest in this issue, and the related 'modifiable areal unit' problem (where fine-grained spatially referenced data are effectively aggregated 'out of existence' by being published on a broader spatial scale). However, he argues that such problems result from the use of inappropriate statistics like the correlation coefficient instead of the spatial cross-coherence function. For him, geographical distributions should be considered as two-dimensional spatial series. Not everybody, however, would agree with this, nor with his conclusion that the modifiable areal unit problem goes away when geographers use the correct analytical procedures.

Tobler participated in the explosion of interest in geographical information systems associated with the establishment of the National Center for Geographic Information and Analysis at Santa Barbara, particularly with his attempts to conduct geographical analysis which is 'frame independent', i.e. independent of the areal units used. He proposes that 'all methods of spatial analysis be examined for the invariance of their conclusions under alternative spatial partitionings, and that only those methods be allowed which show such invariance' (Tobler, 1989: 115). He subsequently advocated a new type of geographical information system, like a raster system except that it is not based on regularly located pixels (picture elements) but on irregularly bounded *resels* (resolution elements) such as the administrative areas for which so much geographical information is available. Tobler has also been keen to remind geographers that we live on the surface of a sphere; he finds it 'curious that most spatial analysis done by geographers uses a flat earth' (2002: 315). Thus Tobler and Chen (1986) discussed how the 'quadtree' system of geographical information storage could be adapted to store data for the whole world.

Another innovation in geography to which Tobler has been a major contributor was the use of computer-based simulation, notably his 'computer movie' simulating the population growth of Detroit (Tobler, 1970). Tobler has also contributed to other aspects of quantitative geography, for example in work on geographical variances (Tobler and Moellering, 1972) and on geographical interpolation (Kennedy and Tobler, 1983; Tobler and Kennedy, 1985). He has introduced the concept of bidimensional regression (Tobler, 1994), a procedure for regression analysis in which both the dependent and independent variables are pairs of coordinates (such as latitude and longitude, or eastings and northings). Such a method is particularly relevant to map comparisons, and examples are given from historical cartography and cognitive mapping.

Another long-standing interest of Tobler's has been migration, viewed in aggregate terms (1981, 1987, 1995). Although interested in interaction-based models of migration, his work has been distinctive in viewing migration movements as 'a continuous vector field driven by a partial differential equation representing the attractivity potential' (2002: 315). Migration is regarded as taking place in continuous space, leading to discussion of migration 'winds', 'fields' and 'currents'. Dorigo and Tobler (1983) developed a view of migration defined in terms of pushes and pulls from the origin and destination areas, discounted by a distance deterrence effect. Their model is based on the total numbers of in-migrants and out-migrants for each place, with the total migration flow being the sum of the push effect for the origin and the pull effect for the destination, divided by the distance between. The approach is open to the criticism that such concepts do not give a satisfactory explanation of why observed migration patterns occur, or of exactly what the push and pull factors consist of. It may also seem less intuitively appealing than a gravity model approach where certain entities (cities or states) are easier to relate to economic and other theories.

KEY ADVANCES AND CONTROVERSIES

Tobler's first major achievement was the transformation of geographical cartography from a field basically concerned with visualization to one where graphical methods could be recast as mathematical operations, and where computers began to be accepted as essential tools for the map-maker. The field of analytical cartography has thus been based to a considerable extent on Tobler's ideas, and given a major boost by the development of geographical information science. Much of the agenda of contemporary geographic information science derives from issues raised or first explored by Tobler. His contribution has not been restricted to the academic world, however, as his cartographic innovations and perspectives have been adopted by government and the software industry.

Tobler's work, especially on map projections, was very relevant to the development in the 1960s of the idea that geography and geometry were closely related, found for example in works by Haggett (1965) and Bunge (1966). Like

Peter Haggett, Tobler (1963a) argued that there were interesting similarities in shape between natural and human phenomena, such as between cities and leaves. Such ideas led **David Harvey** (1969) to argue for geometry as 'the language of spatial form', and to build on Tobler's ideas to explore the relevance of non-Euclidean geometries to reflect social space. This emphasis on geometry was fairly short-lived – Cosgrove (1989: 235), for example, seeking alternatives to 'the cold geometries of spatial analysis' and Sack (1972) casting doubt on the ability of geometry to explain geographical reality. Cartograms are still widely used (e.g. Dorling, 1995) – as are other aspects of Tobler's work – but quantitative geographers nowadays use these concepts and tools without claiming any special status for geometry as a language for geography.

Tobler's concern with spatial form and with concepts and techniques from geometry lays him open to accusations that he is more concerned with methods to describe and manipulate geographical reality than he is to explain or understand phenomena. Even on topics such as migration, which have attracted a great deal of behavioural and humanistic study, Tobler's contributions often explicitly avoid such perspectives – not necessarily because he does not think them important, but more because he feels he is best able to contribute in other ways. He also tends to adopt deterministic rather than statistical approaches to modelling geographical phenomena. This may be partly because of his unusual competence (for a geographer) in the relevant branches of mathematics, but is partly because of an approach to modelling in which he is prepared to accept the explanatory limitations of a model if it illuminates an idea that is important.

Another concern of quantitative geographers from the 1960s onwards was the wish to emulate the sciences, and Tobler's use in many of his papers of methods and concepts from the physical sciences and mathematics is an illustration of this. He admits (2002: 322) that these methods yield only crude approximations when applied to human populations, but argues that 'they seem to give definite insight into the processes being studied'. Another attribute of the sciences that has been important in positivist approaches to human geography is the drive to identify 'laws' in the way that has been done in physics (Golledge and Amedeo, 1968). For example, when revisiting Ravenstein's 'laws of migration' (1885), Tobler (1995) suggested that they stand up well today, demonstrating (for him) the desirability of identifying such laws. In this context, it is fitting to conclude with what Tobler (1970) called the 'first law of geography' – 'everything is related to everything else, but near things are more related than distant things'.

TOBLER'S MAJOR WORKS

Dorigo, G. and Tobler, W. R. (1983) 'Push-pull migration laws', *Annals of the Association of American Geographers* 73: 1–17.

Tobler, W. R. (1959) 'Automation and cartography', *Geographical Review* 49: 526–534.

Tobler, W. R. (1963a) 'Geographic area and map projections', *Geographical Review* 52: 59–78.

Tobler, W. R. (1966a) 'Medieval distortions: the projections of ancient maps', *Annals of the Association of American Geographers* 56: 351–360.

Tobler, W. R. (1969) 'Geographic filters and their inverses', *Geographical Analysis* 1: 234–254.

Tobler, W. R. (1970) 'A computer movie simulating urban growth in the Detroit region', *Economic Geography* 46: 234–240.

Tobler, W. R. (1979) 'Smooth pycnophylactic interpolation for geographical regions', *Journal of the American Statistical Association* 74: 519–536.

Tobler, W. R. and Moellering, H. (1972) 'Geographical variances', *Geographical Analysis* 4: 34–50.

Secondary Sources and References

Bunge, W. (1966) *Theoretical Geography*, 2nd edition. Lund Studies in Geography, Series C, No. 1.

Cosgrove, D. E. (1989) 'Models, description and imagination in geography', in W. Macmillan (ed.) *Remodelling Geography*. Oxford: Blackwell, pp. 230–244.

Dorling, D. (1995) *A New Social Atlas of Britain*. Chichester: Wiley.

Golledge, R. G. and Amedeo, D. (1968) 'On laws in geography', *Annals of the Association of American Geographers* 58: 760–774.

Haggett, P. (1965) *Locational Analysis in Modern Geography*. London: Arnold.

Harvey, D. (1969) *Explanation in Geography*. London: Arnold.

Kennedy, S. and Tobler, W. R. (1983) 'Geographic interpolation', *Geographical Analysis* 15: 151–156.

Ravenstein, E. G. (1885) 'The laws of migration', *Journal of the Royal Statistical Society* 46: 167–235.

Sack, R. D. (1972) 'Geography, geometry and explanation', *Annals of the Association of American Geographers* 62: 61–78.

Tobler, W. R. (1963b) 'D'Arcy Thompson and the analysis of growth and form', *Papers of the Michigan Academy of Science, Arts and Letters* 48: 385–390.

Tobler, W. R. (1964) 'Automation in the preparation of thematic maps', *Cartographic Journal* 1: 1–7.

Tobler, W. R. (1965) 'Computation of the correspondence of geographical patterns', *Papers of the Regional Science Association* 15: 131–139.

Tobler, W. R. (1966b) 'Notes on two projections', *Cartographic Journal* 3, 87–89.

Tobler, W. R. (1967a) 'Computer use in geography', *Behavioral Science* 12: 57–58.

Tobler, W. R. (1967b) 'Of maps and matrices', *Journal of Regional Science* 7(2) (Supplement): 275–280.

Tobler, W. R. (1971) 'Uniform distribution of objects in a homogeneous field: cities on a plain', *Nature* 233.

Tobler, W. R. (1981) 'A model of geographical movement', *Geographical Analysis* 13: 1–20.

Tobler, W. R. (1987) 'Experiments in migration mapping by computer', *American Cartographer* 14: 155–163.

Tobler, W. R. (1990) 'Frame independent spatial analysis', in M. Goodchild and S. Gopal (eds) *Accuracy of Spatial Databases*. London: Taylor & Francis, pp. 115–122.

Tobler, W. R. (1994) 'Bidimensional regression', *Geographical Analysis* 26: 186–212.

Tobler, W. R. (1995) 'Migration: Ravenstein, Thornthwaite, and beyond', *Urban Geography* 16: 327–343.

Tobler, W. R. (2002) '*Ma vie*', in P. Gould and F. R. Pitts (eds) *Geographical Voices: Fourteen Autobiographical Essays*. Syracuse: Syracuse University Press, pp. 293–322.

Tobler, W. R. and Chen, Z. (1986) 'A quadtree for global information storage', *Geographical Analysis* 18: 360–371.

Tobler, W. R. and Kennedy, S. (1985) 'Smooth multidimensional interpolation', *Geographical Analysis* 17: 251–257.

Tobler, W. R. and Wineberg, S. (1971) 'A Cappadocian speculation', *Nature* 231, 5297: 39–42

Robin Flowerdew

47 Yi-Fu Tuan

BIOGRAPHICAL DETAILS AND THEORETICAL CONTEXT

Born in Tientsin, China, in 1930, Yi-Fu Tuan was educated in China, Australia, the Philippines and England (Oxford University), and moved to the US in 1951 (University of California, Berkeley). Although Tuan's initial research was in geomorphology – studying pediments in southwestern Arizona – his reputation has been established as a cultural geographer. During the 16 years he spent at the University of Minnesota, he produced some of his most influential work. From 1984 to his retirement in 1998, he was based at the University of Wisconsin, Madison, maintaining a continued interest in 'systematic humanistic geography' (Tuan, 1998a).

Over a career spanning some 40 years of teaching, research and writing, Yi-Fu Tuan has written 15 books, each notable for scholarship and individual voice and published numerous papers generally of a reflective and thought-provoking kind. Very much defining 'humanistic geography' for a generation, Tuan redefined our understanding of human geography as the study of 'human-environment relationships', and in particular pursued the more fundamental questions arising from a search for the meaning of existence based on an understanding of ourselves as 'being-in-the-world', that is fundamentally defined by and in relation to the world and our relationship to it, both physical and emotional (an interpretation of phenomenology). For Tuan, such research is also about self-discovery, and therefore it

is no surprise that one of his later books was *Who Am I? An Autobiography Of Emotion, Mind and Spirit* (1999), in which he reveals much of the motivation behind his work.

Tuan's career corresponded to a momentous period of change and development for the discipline of geography. He worked somewhat outside of the more immediate froth of intellectual fashions and has consistently charted his own unique course, more interested in enduring moral and spiritual truths – what he has termed 'the good life' (1986). Tuan drew from a much longer tradition of scholarship, not so much of geography itself as of the Western humanist tradition. Counter to his contemporaries in the 1960s, he was to be increasingly concerned with the emotional and intimate engagement of people, culture, environment and place. This was not a human geography of statistics and computer simulations, nor of models or critical theories, but one of personal encounter, literary reflection, and of humility and wonder.

The 1970s, and the reaction to positivist spatial science, reawakened many human geographers to the uniqueness of place, the emotional encounter with the environment, and this humanistic tradition. Tuan was to quickly become associated with this 'new geography' for which the term 'humanistic geography' gained wide currency. While Tuan has been grouped with other practitioners such as Lowenthal, Buttimer, Relph, Seamon and others, he was never really a leader of any kind of movement, but rather its emergence coincided with his already formed and developing personal perspective on the human geography project. Nevertheless, this period coincided with Tuan's own explorations into the deeper philosophical underpinnings of the 'humanistic' perspective. Seminal papers from this period, much quoted by his contemporaries, were 'Geography, phenomenology and the study of human nature' (1971) and 'Humanistic geography' (1976). Along with books such as *Topophilia* (1974), *Space & Place* (1977) and *Landscapes of Fear* (1979), Tuan established his own reputation and contributed to the rebirth of cultural geography.

Tuan absorbed the ideas of existentialism and phenomenology. In particular the ideas of Martin Heidegger – 'being-in-the-world' and 'dwelling', and the fourfold connectivity of Being with the earth, the cosmos, the body and the spirit – were to have a profound impact on his thinking. He did not apply these ideas slavishly, but rather they resonated with his own intuitive understanding of human geography. This more existential, experiential and holistic concept of the intimate connection of people and places, culture and geography, and the relationship to nature or 'geopiety' was to be a unifying theme of his work.

Although something of an outsider, Tuan's trajectory is not totally isolated from the wider debates of society and academy in the latter part of the twentieth century. His enduring interest in attitudes to nature parallels the rise of the environmental movement (Tuan, 1974, 1984, 1993). Further, his increasing interest in the nature of human consciousness, the self and society parallels developments in wider social sciences (e.g. Tuan 1982, 1996) and in many ways his last 'academic study', *Escapism* (1998b), combines his interests in the meaning of existence, nature and culture, with an implicit and somewhat indirect critical counter to the postmodern turn in human geography.

Each of his texts explore particular trajectories of the humanistic perspective. For instance, in *Topophilia* (Tuan, 1974) his interest focuses upon attachments to place; in *Segmented Worlds and Self* (Tuan, 1982) his analysis deepens to a consideration of the link between human consciousness and spatial structures; and in

Dominance and Affection (Tuan, 1984), his attention moves to the 'aesthetic exploitation' and mistreatment of nature. Throughout his work, there is a consistent exploration, reflection and refinement of his understanding of what it is to be human, that is a 'being-in-the-world', and human–environment relationships are explicated as not merely objective and material, but are affective and moral.

His achievement has been recognized through numerous prizes. As early as 1968 he was awarded a Guggenheim Fellowship; in 1973 he was given the Association of American Geographer's (AAG) award for a 'meritous contribution to geography'; and again in 1987 he gained from the AAG the Cullum Geographical Medal. He was also elected a fellow of the American Association for the Advancement of Science, and has been a respected member of professional bodies and associations. He was recently named Laureat d'Honneur 2000 of the International Geographical Union. In honour of his contribution to human geography, two volumes of essays by various writers have been published: *Textures of Place; Exploring Humanistic Geographies* (Adams et al., 2001) and *Progress: Geographical Essays* (Sack, 2002).

SPATIAL CONTRIBUTIONS

Yi-Fu Tuan has sought from his earliest work to expand concepts of geography beyond the physical towards the metaphysical, ethical and aesthetic. He was inspired by von Humboldt, whom he described as a 'hero' figure who, while predominantly known for explaining the physical world, was 'among the first to use landscape painting and poetry to extend the range of geographical experience – feeling, emotion, and concept' (Tuan, 1999). Tuan has also cited the cultural geographers Carl Sauer and J. B. Jackson as important initial influences. And in terms of Tuan's own contribution to the discipline, he has been widely quoted by exponents of the humanistic perspective, and has gained a position of respect in learned societies.

He has described himself as 'off-center even in geography ... I now enjoy a modicum of recognition in geography, but it is more a consequence of my longevity ... than of any marked intellectual influence' (1998a). Yet, at a deeper level, his impact has been far more profound, linking with a longer humanistic tradition in geography (when his contemporaries were drawn first by spatial science, and later by critical social theory), and preparing the ground, in many ways, for the resurgence of interest in cultural geography. J. Nicholas Entrikin (in a review of *Cosmos & Hearth*) writes:

> without question, Yi-Fu Tuan has been an intellectual force in contemporary cultural geography and environmental thought. His work is cited by authors across a wide range of disciplines, from the more philosophical, such as literary criticism, to the more applied, such as landscape architecture. Surely he is one of the best known geographers outside of his home discipline. As has been the case with other leading humanistic geographers this century, however, for example J K Wright, Tuan's contribution to American geography has rarely been given adequate treatment in the histories and surveys of the field.
> (Entrikin, 1997)

Yi-Fu Tuan has had a strangely paradoxically impact. On the one hand, he has produced a considerable output of books and refereed articles which have been widely read and much valued both within the discipline and beyond. Yet, apart from the surge of interest in 'humanistic geography' in the 1970s, Tuan has always been something of an outsider, following his own highly personal, and sometimes idiosyncratic, explorations of cultural values, attitudes to the environment, space and place, the self and the cosmos, emotional and spiritual geographies.

With the rise of critical and postmodern discourses in geography, especially since the 1980s, the humanistic perspective and Tuan's interpretation of it has come under much criticism for its singular authorial voice, tendency to universalize traits across cultures, sometimes essentialist assumptions, and perhaps political naivety. Yet, what is forgotten perhaps is the fundamental humility and wonder within Tuan's discourse. Through his work he revisits and reworks his themes and examples, testing and fine-tuning his interpretations. Yi-Fu Tuan has held to his underlying project, that search to understand the meaning of 'the good life'. He has not ignored the debates raging in human geography, the social sciences and the humanities at large, but he has not been enticed by superficial attractiveness of their 'new language' and concepts. In this respect, it is interesting to compare *Escapism* (Tuan, 1998b) to contemporary writing on this theme inspired by postmodern theorists, as it explores the notion of reality and the real from a 'systematic humanistic geography' perspective.

Monaghan (2001) notes that many of the contemporary interests of geography have an indirect intellectual ancestry in earlier work by Tuan but are never acknowledged as such. Adams *et al.* (2001) argue that 'a lot of people in and out of the discipline consider Yi-Fu as an inspirational figure. But it strikes me that his role has been less to inspire direct imitation than to inspire people to do things that don't look exactly or much like his work'.

KEY ADVANCES AND CONTROVERSIES

Much of the human geography of the latter half of the twentieth century was driven by the desire to explain how specific social, economic and spatial patterns arise, are sustained and ultimately change. The focus has essentially been material and functional, as geographers have sought to describe and explain the world around them in ways that are seen to be useful to society at large.

Yi-Fu Tuan has taken a different journey, fascinated by the more fundamental and difficult question of *why* – that is, the meaning of existence. In particular, he has focused upon the relationship between human beings (individuals and cultures) and their environment (nature and place), both materially and, most especially, emotionally and spiritually. Drawing on existential and phenomenological philosophies, Tuan has explored the ways in which we are, or become, beings-in-the-world, and critical to him has been the establishment of the meaning of this spatial and social existence as it is established at different times and in different cultures. He has sought to uncover more universal characteristics of being human, while at the same time not laying claim to discovering any grand theories. In the 1970s, Tuan for a brief time was central to the controversies and debates about 'humanistic geography', but for much of his career he has stood outside of such debates. His books of the 1980s and 1990s, while making a unique contribution to the discipline, have not generated discipline-wide controversy. In part this is perhaps because they do not directly engage with the contemporary debates of critical social theory, feminism or post-modernism, nor with more immediately popular concerns of environmentalism, globalization or terrorism. Furthermore, the style of analysis and the language of his prose make his work less suited to the age of soundbites and quick fixes.

His project has been far more reflective, and less concerned with immediate gain, and its influence has been often less direct. Much closer to his project has been the notion that geographic discovery is also about self-discovery. For much of his academic career this has been a controversial notion, as human geographers at large have subscribed to notions of scientific objectivity, studying the world, and peoples, as 'objects', and de-emphasizing the possibility or value of either self-reflection or the potential impact of geographic research upon the researcher. With the rise of humanistic geography in the 1970s, other geographers began to rediscover the value of the individual, the subjective and self-reflection. Although this 'humanistic' approach has largely been superseded by the ideas of critical social theory, feminism and postmodernism from the 1980s, the importance of the personal, the subjective, the affective and the moral has continued to resonate. Although there is no direct link to Tuan's work, it is interesting that a number of feminist geographers have asserted the value of personal experience, subjectivity and self-discovery, as the personal became political.

In his Charles Homer Haskins Lecture in 1998, Yi-Fu Tuan summarized his contribution enigmatically:

> I wonder about Socrates' . . . dictum 'the unexamined life is not worth living'. . . my own type of work, ostensibly about 'people and environment', draws so much on the sort of person I am that I have wondered whether I have not written an unconscionably long autobiography. By tiny unmarked steps, examination turns into self-examination. Is it worth doing? Will it lead one to the good life? Or will it, as Saul Bellow believes, make one wish one were dead? I oscillate between the two possibilities. In the end I come down on the side of Socrates, if only because the unexamined life is as prone to despair as the examined one; and if despair – occasional despair – is human, I would prefer to confront it with my eyes open, even convert it into spectacle, than submit to it blindly as though it were implacable fate.' (Tuan, 1998a; no pagination)

TUAN'S MAJOR WORKS

Tuan, Y-F. (1971) 'Geography, phenomenology and the study of human nature', *The Canadian Geographer* 15: 181–192.
Tuan, Y-F. (1974) *Topophilia: A Study of Environmental Perception, Attitudes and Values*. Englewood Cliffs, NJ: Prentice Hall.
Tuan, Y-F. (1976) 'Humanistic geography', *Annals of the Association of American Geographers* 66 (2): 266–276.
Tuan, Y-F. (1977) *Space & Place: The Perspective of Experience*. Minneapolis: University of Minnesota Press.
Tuan, Y-F. (1979) *Landscapes of Fear*. Oxford: Blackwell.
Tuan, Y-F. (1996) *Cosmos & Hearth: A Cosmopolite's Viewpoint*. Minneapolis: University of Minnesota Press.
Tuan, Y-F. (1998b) *Escapism*. Baltimore: Johns Hopkins Press.
Tuan, Y-F. (1999) *Who Am I? An Autobiography of Emotion, Mind and Spirit*. Madison: University of Wisconsin Press.

Secondary Sources and References

Adams, P. C., Till, K. E. and Hoelscher, S. D. (eds) (2001) *Textures of Place: Exploring Humanist Geographies*. Minneapolis: University of Minnesota Press.
Entrikin, J. N. (1997) 'Review: Cosmos & Hearth: Yi Fu Tuan', *Annals of the Association of American Geographers*: 176–178.
Monaghan, P. (2001) 'Lost in Place: Yi Fu Tuan may be the most influential scholar you've never heard of', http://chronicle.com/free/v47/i27/27a01401.html
Sack, R. (2002) *Progress: geographical essays*. Baltimore: Johns Hopkins Press.
Tuan, Y-F. (1982) *Segmented Worlds & Self: A Study of Group Life and Individual Consciousness*. Minneapolis: University of Minnesota Press
Tuan, Y-F. (1984) *Dominance & Affection: The Making of Pets*. New Haven: Yale University Press.
Tuan, Y-F. (1986) *The Good Life*. Madison: University of Wisconsin Press.
Tuan, Y-F. (1989) *Morality & Imagination: Paradoxes of Progress*. Madison: University of Wisconsin Press.
Tuan, Y-F. (1993) *Passing Strange & Wonderful: Aesthetics, Nature & Culture*. Washington DC: Island Press/Shearwater Books.
Tuan, Y-F. (1998a) 'A Life of Learning', *Charles Homer Haskins Lecture for 1998*; available at www.acls.org/op42/tuan.html

Paul Rodaway

BIOGRAPHICAL DETAILS AND THEORETICAL CONTEXT

Paul Virilio was born in Paris in 1932, being evacuated to Nantes in 1939, in an attempt to escape the worst of the German *Blitzkrieg* of World War II. A self-styled 'urbanist', he served as a President of the *École Spéciale d'Architecture* in Paris, where he first gained tenure in 1969, becoming Director General in 1975, and President in 1990. He had previously set up the *Architecture Principe* group with Claude Parent in 1963 (Virilio and Parent, 1996), though a rift between the two developed following Virilio's involvement in *les évenéments* (the student–worker uprisings) of 1968. Prior to training as an architect and urban planner, Virilio had served an apprenticeship in the craft of *vitrail* (stained-glass-window making), working alongside Georges Braque and Henri Matisse. In the late 1950s, he studied philosophy at the Sorbonne, where he was particularly influenced by Maurice Merleau-Ponty's teachings on phenomenology.

While Virilio's twin identity as urbanist and philosopher is sometimes (rather unconvincingly) billed as an incongruous juxtaposition, it is his unrelenting focus on war that gives his work its unique bearing. He has repeatedly claimed that 'War was my university' (Virilio and Lotringer, 1983: 24; Der Derian, 1998: 16), often recounting childhood memories in interviews. He recalls, for instance, hearing news of the German invasion on the radio only minutes before the sound of German tanks rolling into Nantes could

be heard outside his window. Later, Virilio was drafted into the French army to fight in the Algerian War (1954–62). Little wonder, therefore, that he should have developed a fascination with the relationship between war and space. His *Bunker Archaeology* project, which began in 1958, and eventually led to an exhibition at the *Musée des Arts Décoratifs* in Paris in 1975–76, focused on Adolf Hitler's Atlantic Wall – the planned 'impregnable front' running along 2,400 miles of the West European coastline (Virilio, 1994a). Beginning with *L'Insecurité du territoire* (*The Insecurity of Territory*; 1976a), Virilio's theoretical writings have proceeded to excavate the relationship between military and urban space in terms of perception, technology and speed. In addition to his many books, Virilio has been actively associated with a variety of journals, including *Architecture Principe, Cahiers du Cinéma, Cause Commune, Critique, Esprit, Le Monde Diplomatique* and *Traverses*; as well as having initiated, with Georges Perec, Galilée's 'Espace Critique' series.

As a number of commentators have recently taken pains to emphasize, Virilio's theoretical stance has frequently been misspecified as 'poststructuralist'-cum-'postmodernist' (Armitage, 2000). Although he has had some involvement with the Collège International de Philosophie, Virilio's relation to the likes of Jacques Derrida and Jean-François Lyotard is hardly straightforward. Nor does his work share many philosophical affinities with the work of **Jean Baudrillard**, despite a significant number of resonances, cross-overs and mutual borrowings. The same can be said of **Michel Foucault**, **Gilles Deleuze** and Félix Guattari. In fact, Virilio's work inherits the phenomenological mantle of

Merleau-Ponty (1962), with an added dose of Gestalt psychology (Guillaume, 1937). Despite many convergences with poststructuralism, therefore, Virilio remains very much a humanist. He is also, moreover, a committed Christian, having been inspired by the example of the Abbé Pierre, the priest who fought for the French Resistance, campaigned for the homeless, and whose dress sense was the subject of an essay by Roland Barthes. While his own work as a campaigner for the homeless and a peace activist is occasionally noted, Virilio is best known for the neologisms and phrases that litter his texts: 'oblique function', 'dromology', 'chronopolitics', the 'aesthetics of disappearance', 'speed pollution', 'infosphere', and so on.

SPATIAL CONTRIBUTIONS

Since Virilio's 'reflections on urbanism are invariably also reflections on politics' (Luke and Ó Tuathail, 2000: 362), there is little sense in attempting to disentangle Virilio's spatial imagination from his social theory. One would also do well to think of Virilio as a strictly political rather than a social thinker: 'I don't believe in sociology . . . I prefer politics and war,' he confesses (Virilio and Lotringer, 1983: 17). Given this emphasis, it is important to recognize that the 'collocation of time and conflict is of the essence of the political for Virilio' (Docherty, 1993: 19). This same collocation, in Virilio's eyes, lies at the origins of the city – an understanding which provides the launch-pad for a full-blown critique of modern society, science and technology. The logic that first gave rise to the city finally rebounds in a world 'devoid of spatial dimensions, but inscribed in the singular temporality of an instantaneous diffusion' (Virilio, 1991b: 13).

Virilio begins in the phenomenal realm, with a consideration of appearances and illusion. For Virilio (1991a: 37), the world as we know it is always 'already a kind of dissolving view, reminding us of the reflection of Paul of Tarsus . . . [A]ll is calm, and yet: *this world as we see it is passing away*'. One is also reminded of Arthur Schopenhauer, for whom 'the world was its representation' (Virilio, 1999: 6). Virilio similarly adheres to Carl von Clausewitz's (1976) precept that conflict is fundamental to the human species, which relates more closely to his concerns with appearance than is first apparent. Adopting Sun Tzu's (1963) ancient maxim that 'speed is the essence of war', Virilio develops a way of understanding conflict in aesthetic (perceptual) terms: 'a war begins with the planning of its theatre . . . the stage on which the scenario should be played out' (Virilio, 1990: 14). The crux of the matter is the sense in which '*war consists in the organization of the field of perception*' (Virilio, 2001b: 185), through a double game of playing off the visible against the invisible, and the perceptible against the imperceptible. War is the orchestrated displacement (i.e. destruction) of the real – as it *appears* (in one's sights), and, more importantly, as it *appears to appear* (as a manifestation of antagonism). Hence the centrality of strategies of deception, dissimulation and disappearance to creating the 'fog of war', such as camouflage, disinformation, virtualization and stealth technologies. This explains Virilio's ambivalent appreciation of the Italian Futurist writer, Filippo Marinetti, whose adulation of the machinery of warfare and celebration of the links between aesthetics and politics became ever more thoroughly imbricated with fascism. More importantly, however, it allows Virilio to construct an account of the urban, not in terms of 'the classic opposition of city to country but as that of stasis to circulation' (Virilio, 1986: 5). Such a conception is not without its antecedents. Lewis Mumford (1979: 13), for instance, held that 'Human life swings between two poles: movement and settlement.' For Virilio (1986: 6), however, 'there is only *habitable circulation*' – implying a dialectical and ecumenical resolution of Mumford's opposition against a generalized background of conflict.

According to Virilio (1990: 15), once 'the possibility of pastoral flight disappears, with the advent of agricultural settlements and a change in the nature of wealth (non-transportable goods)', the relationship between humans and their environment profoundly alters. Nomadic existence saw the occasional adoption of elevated vantage-points. Yet taking the high ground was less a defensive manoeuvre than a means of obtaining 'quicker information on the surroundings' (Virilio, 1990: 15), thereby displacing the immediacy of direct encounter. However, rather than simply being *informed* by the environment, a sedentary population actively began to *transform* its surroundings – adopting defensive hilltop positions, constructing observation platforms, building fences and ramparts, etc. This reorganization of space effects a manipulation of time. Such transformations institute a delay between one's own position and that of potential aggressors. If one has sight of one's enemy before one is in their sights, one can act ahead of time. 'Attack and defense then split on this terrain to form two elements of a single dialectic: the former becomes synonymous with speed, circulation, progression and change; and the latter with opposition to movement, tautological preservation, etc.' (Virilio, 1990: 15–16). Virilio thus refigures Clausewitz's dialectic of attack and defence as a dialectic of speed and slowness (the latter of which is routinely missed by those enamoured by the 'speeding up' of modern life: acceleration is necessarily accompanied by a relative deceleration). This same dialectic governs exchanges other than violent conflict, such as the economic exchanges of commerce and trade. The essence of the city itself thus lies in regulating the passage (or 'pacemaking,' as Parkes and Thrift, 1979, once put it) of others: in creating a

position from which the passage of others can be slowed down, enabling one always to act in advance. Hence Virilio's dromological conception of spatial politics (from the Greek, *dromos* – 'running', 'course'): the *polis* arises in response to the emergence of 'dromomaniacs,' whose unpredictable behaviour calls forth a 'dromocratic revolution', resulting in rule by 'dromocrats'. We thus inhabit not a democracy but a 'dromocracy' – a movement bureaucracy or speed (and slowness) technocracy. One thinks of the famous Haussmanization of Paris in the second half of the nineteenth century, which reconstructed the city with Grand Boulevards, sewer systems, and centralized abattoirs in order to expedite the free circulation of some things (goods, consumers, capital, traffic, military forces, storm-water and animals) and impede the circulation of other things (vagrants, the masses, revolutionaries and miasma).

This dromological perspective is vital to appreciating the attention Virilio lavishes on technology. He is far from being the technological fantasist he is sometimes held to be. Nor is he a technological determinist; in fact, he is a technological *indeterminist* (cf. Baudrillard). He is fascinated by the way in which technologies are 'capable of orchestrating the perpetual shift of appearances' (Virilio, 1995: 23), thus ensuring an indeterminacy that always opens up the possibility of resistance. His prime focus – and here the inevitable comparison with Marshall McLuhan (1969) is unhelpful – is the way in which the technological 'extensions of man' operate as 'prostheses of speed'. This has to be understood, Virilio maintains, in relation to the forces of destruction (war) rather more than the forces of production (industry). In contrast to **David Harvey**'s (1989) account of 'time–space compression', Virilio refuses to regard the political economy of *speed* as subservient to the political economy of *value*, the latter of which is a manifestation of the market and the state. Speed relates to the 'war machine', which is exterior to the state and is its condition of possibility (cf. Clastres, 1977; Deleuze and Guattari, 1987).

Virilio argues that the smooth course of life has always been subject to interruption – to acts of God, fate, accidents and unintended consequences – which give rise to unease, dislocation and conflict. War is the exemplary interruption of life. For Virilio, war is the *absolute* accident: its pure form. By a strange dialectical twist, however, warfare (*destruction*) has become caught up with the logic of industry (*production*). The original sense of production was to 'render visible' or to 'make apparent'. While destruction was initially opposed to production, which served to undermine it, this relation has become obscured. Not only do all manner of everyday consumer products, from tinned foods to the internet, find their origins in the military-industrial complex, Virilio suggests that the industrial mode of production *itself* is the accidental result of the progression of the means of destruction, and has come to serve as a cover for its development towards a state of 'pure war' (Virilio and Lotringer, 1983).

> The *scientific and industrial mode of production* is perhaps only an avatar or, as they say, *fallout*, of the development of the means of destruction, of the *absolute accident of war*, of the conflict pursued down through the centuries in every society, irrespective of its political or economic status – *the great time of war* that never ceases to *unexpectedly befall* us time and again despite the evolution of morals and the means of production, and whose intensity never ceases to grow apace with technological innovations, *to the point where the ultimate energy, nuclear energy, makes its appearance in a weapon that is simultaneously an arm and the absolute accident of History.*
> (Virilio, 1993: 212–213)

This is the point at which a process leading unerringly towards *total war* (the nuclear annihilation of life on Earth) mutates into the paradoxical state of *pure war* – a state of 'peace' that is merely the continuation of war by other means. If the 'logic of deterrence' by which pure war reigns was initially a product of the Cold War, Virilio's sense of a society in a perpetual state of preparation for war – and the attendant sense of *apocalypse from now on* – remains all too pertinent in the post-Cold War era.

Military intelligence continues to pervade society in all manner of pernicious ways, while 'the old illusion ... persists that a state of peace means the absence of open warfare, or that the military which no longer fights but "helps" society is peaceful' (Virilio, 1990: 36). The relations between the logistics of perception and the logistics of war point to a decisive but well-camouflaged link between 'civilian' culture and military technology, which reveals itself only incidentally – in the common ancestry of the Gatling gun and the ciné-camera, radar and video, and gunpowder and printer's ink, for example. Yet given that all technologies work as prosthetic devices, simultaneously disabling and enabling the subject, cybernetic extensions of vision inevitably entail the (calculated) loss of other ways of seeing and being in the world. By the 1930s, for instance, 'it was already clear that film was superimposing itself as a geostrategy which for a century or more had inexorably been leading to the direct substitution, and thus sooner or later the disintegration, of things and places' (Virilio, 1989: 47). Virilio (1989, 2002a) consistently conceives of technology in terms of this kind of 'substitution'. New technologies tend to nullify and displace the problems they are meant to solve and overcome, producing new configurations that redefine the direction taken by society as a whole. The world according to Virilio (2002c) is a world in which technological change all too easily follows a trajectory of its own accord, leaving us out of the picture without us even realizing the fact. On the brink of pure war, we risk failing to perceive the way in which the 'reality principle' has been usurped by an accelerating series of 'reality effects' (cf. Baudrillard and Deleuze on simulacra).

Perhaps Virilio's lasting legacy will be in terms of his explication of the geographical implications of technology. To-day, says Virilio (2001b: 84), 'we live in a world no longer based on geographical expanse but on a temporal distance constantly being decreased by our transportation, transmission and tele-action capacities'. He highlights a broadly chronological transition from *vehicular technologies*, associated with the conquest of space (which create a territorial infrastructure of road and rail networks, ports, airports, and so on – a fixity in space permitting movement over space), to *transmission technologies*, associated with the conquest of time (immaterial, electromagnetic means of communication permitting near-instantaneous contact regardless of physical distance – and retrospectively revealing that physical distances have always been a function of the speed, and cost, with which they could be overcome). Consequently, 'The new space is speed-space; it is no longer a time-space' (Virilio, 2001b: 71). The ultimate limit of these transmission technologies is the speed of light, which would imply a situation of pure telepresence.

> So after the nuclear disintegration of *the space of matter*, which led to the implementation of a global deterrence strategy, the disintegration of *the time of light* is finally upon us. This will most likely involve a new mutation of the war game, with deception finally defeating deterrence.
> (Virilio, 1994b: 72)

KEY ADVANCES AND CONTROVERSIES

War at the speed of light, employing technologies of deception, might sound like science fiction. Yet Virilio insists that this is precisely the point at which we have arrived. The development of technoscience means that, today, 'the battlefield is global' and, in consequence, there is an increasing 'feeling of confinement in the world' (Virilio, in Der Derian, 1998: 17, 21). Such insights into our globalized existence invite frequent accusations that Virilio fails to pay due attention to human corporeality. Yet nothing could be further from the truth. Virilio's concerns for the body underlie his entire critique. He fears that the internal colonization of the body by transplant technologies means that our bodies, not merely our cities, territories and planet, are increasingly the (passive) medium of technological change. 'That which favours the equipping of territories, of cities, in particular, threatens to apply to the human body, as if we had the city in the body and not the city around the body. The city ' ''at home'', *in vitro, in vivo*' (Virilio, in Der Derian, 1998: 20). In such a situation, we face the development of the 'last vehicle' – a paradoxical 'stationary vehicle' which, like a running-machine or flight simulator, induces motion but puts an end to movement by transforming the actor into a tele-actor. 'This tele-actor will no longer throw himself into any means of physical travel, but only into another body, an optical body; and he will go forward without moving, see with other eyes, touch with hands other than his own, to be over there without really being there, a stranger to himself, a deserter from his own body, an exile for evermore' (Virilio, 1999: 85). In the final analysis, then, Virilio might be regarded as having both radicalized and simplified the analysis of speed. Yet he has remained, paradoxically, an optimistic extremist – or, if you prefer, an extreme optimist.

VIRILIO'S MAJOR WORKS

Virilio, P. (1986) *Speed and Politics: An Essay on Dromology*. Trans. M. Polizotti. New York: Semiotext(e) (original edition 1977).
Virilio, P. (1989) *War and Cinema: The Logistics of Perception*. Trans. P. Camiller. London: Verso (original edition 1984).
Virilio, P. (1991a) *The Aesthetics of Disappearance*. Trans. P. Beitchman. New York: Semiotext(e) (original edition 1980).
Virilio, P. (1991b) *The Lost Dimension*. Trans. D. Moshenberg. New York: Semiotext(e) (original edition 1984).
Virilio, P. (1994a) *Bunker Archaeology*. Trans. G. Collins. New York: Princeton Architectural Press (original edition 1975).
Virilio, P. (1994b) *The Vision Machine*. Trans. J. Rose. London: British Film Institute (original edition 1988).
Virilio, P. (1995) *The Art of the Motor*. Trans. J. Rose. Minneapolis: University of Minnesota Press (original edition 1993).
Virilio, P. (2000) *The Information Bomb*. Trans. C. Turner. London: Verso (original edition 1998).
Virilio, P. and Lotringer, S. (1983) *Pure War*. Trans. M. Polizotti. New York: Semiotext(e).

Secondary Sources and References

Armitage, J. (ed.) (2000) *Paul Virilio: From Modernism to Hypermodernism and Beyond*. London: Sage. (Also published as *Theory, Culture & Society* 16 (5–6).)
Clastres, P. (1977) *Society Against the State*. Trans. R. Hurley. Oxford: Blackwell (original edition 1974).
Clausewitz, C. von (1976) *On War*. Trans. M. Howard and P. Paret. Princeton, Guildford: Princeton University Press (original edition 1832).
Deleuze, G. and Guattari, F. (1987) *A Thousand Plateaus: Capitalism and Schizophrenia*. Trans. B. Massumi. Minneapolis: University of Minnesota Press (original edition 1980).
Der Derian, J. (1998) *The Virilio Reader*. Oxford: Blackwell.
Docherty, T. (1993) *Postmodernism: A Reader*. London: Harvester Wheatsheaf.
Guillaume, P. (1937) *La Psychologie de la Forme*. Paris: Flammarion.
Harvey, D. (1989) *The Condition of Postmodernity: An Enquiry into the Origins of Cultural Change*. Oxford: Blackwell.
Luke, T. and Ó Tuathail, G. (2000) 'Thinking geopolitical space: the spatiality of war, speed and vision in the work of Paul Virilio', in N. Thrift and M. Crang (eds) *Thinking Space*. London: Routledge, pp. 360–379.
McLuhan, M. (1969) *Understanding Media: The Extensions of Man*. London: Routledge & Kegan Paul.
Merleau-Ponty, M. (1962) *The Phenomenology of Perception*. Trans. C. Smith. London: Routledge.
Mumford, L. (1979) *The City in History: Its Origins, Its Transformations, and Its Prospects*. Harmondsworth: Penguin.
Parkes, D. and Thrift, N. (1979) 'Time spacemakers and entrainment', *Transactions of the Institute of British Geographers* 4: 353–372.
Sun Tzu (1963) *The Art of War*. Trans. S. B. Griffith. Oxford: Clarendon Press.
Speed (1997) Theme issue on Paul Virilio http://proxy.arts.uci.edu/_nideffer/_SPEED_/1.4/ (accessed: 19.02.03)
Virilio, P. (1976a) *L'Insécurité du territoire*. Paris: 'Éditions Stock.
Virilio, P. (1976b) *La Dromoscopies ou la lumière de la vitesse*. Paris: Minuit.
Virilio, P. (1984) *L'Espace critique*. Paris: Christian Bourgois.
Virilio, P. (1990) *Popular Defense and Ecological Struggles*. Trans. M. Polizotti. New York: Semiotext(e) [original edition 1978].
Virilio, P. (1993) 'The primal accident', in B. Massumi (ed.) *The Politics of Everyday Fear*. Minneapolis: University of Minnesota Press, pp. 211–218 [original edition 1982].
Virilio, P. (1997) *Open Sky*. Trans. J. Rose. London: Verso [original edition 1995].
Virilio, P. (1999) *Polar Inertia*. Trans. P. Camiller. London: Sage [original edition 1990].
Virilio, P. (2000) *The Strategy of Deception*. Trans. C. Turner. London: Verso [original edition 1999].
Virilio, P. (2001a) *A Landscape of Events*. Trans. J. Rose. Cambridge, MA: MIT Press [original edition 1996].
Virilio, P. (2001b) *Virilio Live: Selected Interviews*. Ed. J. Armitage. London: Sage.
Virilio, P. (2002a) *Negative Horizon: Toward a Dromoscopy*. Trans. M. Degener. London: Continuum [original edition 1984].
Virilio, P. (2002b) *Desert Screen: War at the Speed of Light*. Trans. M. Degener. London: Sage [original edition 1991].
Virilio, P. (2002c) *Ground Zero*. Trans. C. Turner. London: Verso [original edition 2002].
Virilio, P. and Lotringer, S. (2002) *Crepuscular Dawn*. Trans. M. Taormina. New York: Semiotext(e).
Virilio, P. and Parent, C. (1996) *Architecture Principe, 1996 et 1966*. Trans. G. Collins. Besançon: L'Imprimeur.
Virilio, P. and Petit, P. (1999) *Politics of the Very Worst*. Trans. M. Cavaliere. Ed. S. Lotringer. New York: Semiotext(e).
Wark, M. (1994) *Virtual Geography: Living with Global Media Events*. Indianapolis: Indiana University Press.

David B. Clarke and Marcus A. Doel

49 Immanuel Wallerstein

BIOGRAPHICAl DETAILS AND THEORETICAL CONTEXT

Immanuel Wallerstein was born in 1930 in New York, where he attended Columbia College and then Columbia University, receiving his PhD in 1959. He went on to teach in the Sociology Department at Columbia, moving to McGill University in Montreal in 1971, and then to the State University of New York (SUNY) at Binghamton in 1976. Wallerstein has been based at SUNY-Binghamton since that time, serving as Distinguished Professor of Sociology and Director of the Fernand Braudel Center for the Study of Economies, Historical Systems, and Civilizations.

Wallerstein's work from the mid-1950s to the early 1970s focused on Africa, but he has achieved international renown for a series of works he published in the 1970s and subsequently that addressed broader issues of the global political economy. These works, beginning with a frequently cited 1974 article on the 'The rise and future demise of the world capitalist system' (2000: 71–105) and the first volume of his three-volume series *The Modern World-System* (Wallerstein, 1974, 1980, 1989), marked Wallerstein as the founder and leading figure of world-systems analysis. Wallerstein's development of world-systems analysis originated as part of a broader reaction against modernization theory, exemplified in the work of economic historian Walter W. Rostow (1960). Rostow posited a 'normal' process of development for all countries, in which they moved through a series of

'stages of development', leading from pre-industrial to industrial, much in the fashion of early industrializing countries such as Great Britain and the United States. Wallerstein, like many Latin American structuralist economists and more politically radical dependency theorists, found such a view to be inaccurate and generative of bad political prescriptions. For Wallerstein, as for later collaborators such as Terence Hopkins and the members of his 'Gang of Four' (Andre Gunder Frank, Samir Amin and Giovanni Arrighi), modernization theory was based on a perception of the world rooted in the interests of privileged social actors within the most advanced industrial capitalist countries. These actors – primarily economic and political elites and their academic supporters – asserted that it was undesirable for developing country states to either protect domestic industries (the approach championed by the structuralists) or adopt socialist development strategies (the approach championed by the radical dependency theorists). Instead, modernization theorists encouraged developing country leaders to facilitate foreign investment and trade, and to follow the putatively market-led development path of the already industrialized capitalist countries – an approach that Wallerstein and others saw as insuring continued subordination within the world-system.

World-systems analysis was thus part of the broader development of what has come to be called 'core–periphery' theorizing, this including not only world-systems analysis but also Frank's (1967) version of radical dependency theory, Amin's (1974) analysis of 'accumulation on a world scale', and the 'dependent development' approach of Fernando Enrique Cardoso and Enzo Faletto (1979). In spite of significant differences between each of these approaches, they hold in common the view that the development of capitalism cannot be seen as a process that works the same way within the earliest centres of capitalist development as it does in places that were incorporated into the global capitalist economy later, under colonialism and imperialism. Indeed, it is not accidental that a significant number of such core–periphery theorists have come from countries characterized as part of the periphery, nor that these theorists generally speak from a perspective that they claim represents the interests of the world's most marginalized populations (Porter and Sheppard, 1998: 96–118).

Wallerstein's world-systems approach has one further connection to issues of marginalization and social struggle that transcends these particular core–periphery dimensions. Wallerstein has argued that the political upheavals occurring around the world during 1968 – not only outside of the core of the global economy but within it – are representative of a crisis for both global capitalism and the movements that have arisen since the nineteenth century to challenge it (Wallerstein, 2000: 355–373). He calls these the 'anti-systemic movements' and includes among them the social democratic reform movements of Western Europe and North America, the revolutionary movements of the former socialist bloc, and the nationalist movements of the Third World. The rebellions of 1968 manifest discontent not only with capitalism but with the anti-systemic movements in power that had claimed to be forwarding viable alternatives to capitalism. World-systems analysis has developed not only to counter modernization theory but also to present a framework for analysing this crisis of the anti-systemic movements and their inability to transform the global capitalist economy through their capture of state power (Wallerstein, 2000: xxii).

SPATIAL CONTRIBUTIONS

Wallerstein (1999: 192) characterizes world-systems analysis as helping to 'clear the underbrush' for more useful approaches to the social sciences than the currently dominant disciplinary approaches. He sees world-systems analysis as doing this through raising crucial foundational issues. These include issues that he addresses under the headings of 'historicity' and 'globality' (Wallerstein, 1999: 195–196). Historicity is central to Wallerstein's analysis of global capitalism. Following the French historian Fernand Braudel, Wallerstein insists that capitalism is not a development of the last two centuries, a view common among most Marxists. Rather, as he sees it, capitalism had developed in Europe by the middle of the sixteenth century, expanding outward from that period. This makes it necessary to study capitalist development over what Braudel calls the *longue durée*, extended periods of development that encompass economic and political cycles of varying lengths, including Kondratieff cycles of 50–60 years (25–30 years of global economic growth followed by 25–30 years of global economic stagnation).

The geographical dimension of world-systems analysis, what Wallerstein calls its insistence on 'globality', is perhaps its most distinctive feature. Wallerstein's work can be called 'neo-Marxist' in that it adopts much of Marx's emphasis on class struggle and capital accumulation, along with Marx's critical perspective on capitalism as a social system, but Wallerstein both transforms certain Marxist categories and rejects certain Marxist claims. Like Marx, Wallerstein characterizes capitalism as centred on class exploitation and the perpetual accumulation of capital through this process; but unlike many Marxists, he does not abstract the capitalist mode of production from its geographical-historical context of development. This means that for Wallerstein the capitalist mode of production existed, from its birth, on a transnational (or world) scale, and integrally involved numerous forms of labour besides the theoretically free wage labour that Marx focused on in his analysis of capitalism.

Wallerstein's neo-Marxism thus differs from Marxism in the way it approaches class, since the capitalist class structure is seen within world-systems analysis to exist at a world scale right from capitalism's inception – even when capitalists and workers are only politically aware of their class position on a national scale. This difference is important because for Wallerstein capitalism on a world scale always involves a complex and geographically expansive class structure that incorporates many more non-proletarians than proletarians, with this non-proletarian labour being indirectly exploited by capitalists. Thus, for Wallerstein, forms of labour such as subsistence and semi-subsistence production, household production and simple commodity production are all integral to the actual development of historical capitalism, and one cannot meaningfully abstract from them if one is to understand the functioning of the capitalist mode of production (Wallerstein, 1979: 119–131).

In this vein, Wallerstein argues that the idealized functioning of capitalism as a free market system – which classical and neo-classical economists extolled and which Marx assumed for purposes of critique – has never occurred and could never occur without producing disaster for capitalists. For repeated and expanded capitalist accumulation to occur, capitalists must constantly generate profits, and this is not accomplished only through direct exploitation of wage labour within a context of market competition, but also by the indirect exploitation of non-proletarian labour. Furthermore, profits are generated by the development of monopolies and quasi-monopolies. A truly free and fully competitive market would reduce or eliminate profits and would thus spell the end of capitalism.

For this reason, capitalists have always favoured specific forms of state intervention in the market that allow *partial* freedom (especially for investors; less so for workers) while protecting monopolies through various measures. It is for this reason that states play a crucial and necessary role in global capitalism. Wallerstein argues that what is distinctive about capitalism as a historical system is that it involves a social division of labour (and thus class structure) that is *transnational* and built around global commodity chains, along with an interstate system comprised of *national* states. These states, moreover, have been brought into the world capitalist system (at various points in time) *in relation to one another* and with differing positions in the global capitalist hierarchy. Certain nation-states occupy *core* positions within the global capitalist economy, dominating the leading technologies and higher-value production processes of the particular era, while others occupy *peripheral* positions within this same economy, having lesser technological development and typically being dependent upon export of raw materials or lower value-added products to the global core. In between these poles, there are also *semi-peripheral* nation-states that contain some mix of core and peripheral production processes.

Core states in the world-system absorb much of the surplus value produced within peripheral states through the process of unequal exchange, which results from the fact that workers in peripheral countries get paid much less than those in core countries for the same amount of labour time. This difference means that commodities produced in peripheral areas can be sold at lower prices than comparable commodities embodying the same amount of labour but produced in core areas. As a consequence, when peripheral countries export goods (e.g. electronics components) and use the revenues to purchase imports from core countries (e.g. computers), they must give up more embodied labour time than they receive in exchange. For Wallerstein, such unequal exchange is central to global capitalism, making it inherently polarizing in a socio-spatial sense.

Within the global process of polarization, semi-peripheral states are important because they provide outlets for reinvestment of capital from the core during economic downturns in the Kondratieff cycle, and also because they provide some limited prospects for upward mobility within the system, thus potentially undercutting the appeal of anti-capitalist alternatives. What is important to recognize, for Wallerstein, is that the movement of one or another country up or down the global hierarchy – which has frequently occurred – does not change the overall structure of the system or eliminate its polarizing tendencies.

The socio-spatial structure of global capitalism therefore helps to account for both kinds of phenomena that gave rise to world-systems analysis. First, the failure of modernization approaches to provide reasonable options for peripheral countries as a whole stems from their failure to address the relational, hierarchical structure and polarizing tendencies of capitalism: even when one country successfully and rapidly 'develops' – e.g. moves from periphery to semi-periphery – this does not lift peripheral countries as a whole and in fact makes it more difficult for those that remain because it increases the competition in economic activities that other peripheral countries are attempting to develop. Second, the fact that global capitalism features a transnational economy and a system of national states accounts for the failure of the anti-systemic movements to achieve a reversal of capitalism's polarizing tendencies: the anti-systemic movements achieved power only at the level of nation-states – social democratic governments in the core areas of Western Europe and North America, formally 'socialist' governments in the semi-peripheral states of the former socialist bloc, and postcolonial nationalist governments in the states of the periphery – but not within the global capitalist economy as a whole.

Wallerstein's views have done much to explicitly spatialize thinking throughout the social sciences, and they have entered directly into geography through the work of political and economic geographers such as Colin Flint and **Peter J. Taylor** (Taylor and Flint, 2000). Taylor's work has directly utilized a world-systems analysis to approach a range of issues in political geography, including patterns of development of the inter-state system (Taylor, 1996). Urban geographer John Friedmann has used a world-systems approach to develop the hypothesis that the global economy is increasingly constituted by flows between major nodes that he designates 'world cities' (Friedmann, 1986, 1995). Other geographers have directly employed Wallerstein's tripartite division of the global economy and his approach to global commodity chains. Most importantly, perhaps, from its inception, world-systems analysis and other core–periphery approaches have directly or indirectly influenced the thinking of geographers concerned with issues of uneven development.

KEY ADVANCES AND CONTROVERSIES

The influence of world-systems analysis on interpretations of the spatiality of capitalist development has been considerable. Moreover, as the social sciences have increasingly become absorbed in debates about 'globalization', the world-systems analysts' emphasis on interconnections between processes at a global scale has been reinforced by both current events and contemporary discourse.

Nonetheless, even as many of its claims have been absorbed by the social sciences, world-systems analysis has been dismissed by some social scientists, in part because of ways in which it challenges established disciplinary boundaries and practices, but also because it raises complex and still unresolved issues regarding how to conceive the spatiality of global capitalism. Perhaps the best example of the latter is the oft-cited criticism of Wallerstein's work raised by Marxist historian Robert Brenner (1977). Brenner collectively critiques Wallerstein, Frank and Amin as representatives of what he calls 'neo-Smithian' Marxism – a deviant form of Marxism that substitutes Adam Smith's emphasis on market forces and exchange for Marx's emphasis on class relations and exploitation. For Brenner, the global connective tissue of sixteenth-century capitalism emphasized by Wallerstein is nothing more than trade between separate societies. These societies each had different kinds of local or national class structures. The class process by which labour is exploited is thus, for Brenner, fundamentally rooted *within* states, while the processes that lead to unequal exchange are *external* to such class processes. From this perspective, Wallerstein mis-identifies as capitalist exploitation the movement of surplus from periphery to core through trade arrangements that are not in actuality central to the story of capitalist exploitation.

This cuts to precisely the issue of whether or not capitalism should be seen as an abstract mode of production organized at the local or national scale and based on hypothetically free labour, or whether it should be seen as a more inclusive and concrete historical process that always involves transnationalized classes and varying types of labour. In contrast to Brenner, Wallerstein sees capitalist exploitation as taking place internationally, integrally involving production

in specific locations and trade or exchange with other locations, along with international transfer of surplus value through this trade and exchange. This difference implies a difference in political strategy. For Brenner, the central struggles against exploitation are at the points where production occurs, though these points of production may become linked in struggle through acts of international solidarity. For Wallerstein, on the other hand, struggles against capitalist exploitation exist within an already transnational space and within a complex web of economic relations including non-proletarianized labour taking place in households and elsewhere (Wallerstein, 2000). Moreover, anti-capitalist struggles may frequently take the form of national liberation movements or struggles in which race or ethnicity come to the foreground. Though this may slow the development of transnational class-consciousness and transnational anti-capitalist politics, such struggles cannot be written off as demonstrating false consciousness, since they reflect the real, racialized and ethnicized, socio-spatial polarization of the global capitalist economy (Wallerstein, 1979, 2000).

How to understand the spatial structures of global capitalism – as fragmentary or as a unity-in-diversity – is an important and ongoing debate, and it relates directly to practical questions being asked by radical geographers and social scientists about the most appropriate and effective spatial forms for struggles against capitalism. Thus, many in the social sciences are today labouring in fields prepared for them at least in part by the socio-spatial ideas of world-systems analysts and other core–periphery thinkers.

WALLERSTEIN'S MAJOR WORKS

Wallerstein, I. M. (1974) *The Modern World-system I: Capitalist Agriculture and the Origins of the European World-economy in the Sixteenth Century.* New York: Academic Press.

Wallerstein, I. M. (1979) *The Capitalist World-economy.* Cambridge: Cambridge University Press.

Wallerstein, I. M. (1980) *The Modern World-system II: Mercantilism and the Consolidation of the European World Economy, 1600–1750*. New York: Academic Press.

Wallerstein, I. M. (1983) *Historical Capitalism*. London: Verso.

Wallerstein, I. M. (1989) *The Modern World-system III: The Second Era of Great Expansion of the Capitalist World-economy, 1730–1840s*. San Diego: Academic Press.

Wallerstein, I. M. (1999) 'The rise and future demise of world-systems analysis', in I. M. Wallerstein (ed.) *The End of the World as We Know It: Social Science for the Twenty-first Century*. Minneapolis: University of Minnesota Press, pp. 192–201.

Wallerstein, I. M. (2000) *The Essential Wallerstein*. New York: The New Press.

Secondary Sources and References

Amin, S. (1974) *Accumulation on a World Scale: A Critique of the Theory of Underdevelopment*. New York, London: Monthly Review Press.

Brenner, R. (1977) 'The origins of capitalist development: a critique of neo-Smithian Marxism', *New Left Review* 104 (July/August): 25–92.

Cardoso, F. H. and Faletto, E. (1979) *Dependency and Development in Latin America*. Berkeley: University of California Press.

Frank, A. G. (1967) *Capitalism and Underdevelopment in Latin America*. New York: Monthly Review Press.

Friedmann, J. (1986) 'The world city hypothesis', *Development and Change* 17 (1): 69–84.

Friedmann, J. (1995) 'Where we stand: a decade of world city research', in P. Knox and P. Taylor (eds) *World Cities in a World-system*. Cambridge: Cambridge University Press, pp. 21–47.

Porter, P. W. and Sheppard, E. S. (1998) *A World of Difference: Society, Nature, Development*. New York: Guilford Press.

Rostow, W. W. (1960) *The Stages of Economic Growth: An Anti-communist Manifesto*. New York: Norton.

Taylor, P. J. (1996) *The Way the Modern World Works: World Hegemony to World Impasse*. Chichester: Wiley.

Taylor, P. J. and Flint, C. (2000) *Political Geography: World-economy, Nation-state and Locality*. Harlow: Prentice Hall.

Jim Glassman

50 Michael J. Watts

BIOGRAPHICAL DETAILS AND THEORETICAL CONTEXT

Michael J. Watts was born in England in 1951. He earned a Bachelor of Science degree in Geography from University College, London, where he graduated with First Class Honours in 1972, going on to pursue graduate studies in Geography in the USA at the University of Michigan, where he earned an MA in 1974, and a PhD in 1979. Watts' geographic education was influenced initially by the various dependency theories of the 1960s, and later by debates within Marxian political economy. His doctoral work at Michigan, under the direction of Bernard Nietschmann, focused initially on the limits of cultural ecology and ecological anthropology as approaches to understanding peasant production systems in semi-arid Africa. During his graduate studies, Watts was taught and advised by a remarkable cast of intellectuals, including Gunnar Olsson, Roy Rappaport, Michael Taussig and Marshall Sahlins, who together left a lasting imprint on his work.

Watts' graduate studies took him to northern Nigeria, where he conducted long-term field research on the effects of capitalist penetration on peasant agricultural systems. This work began as an attempt to understand how farmers perceived their environments in the context of drought and the vulnerability wrought by 'natural' hazards. This initial focus gave way to a study of famine and vulnerability, and led Watts to rethink the fields of human ecology and hazards, with their focus on human adaptation and cyber-netic feedback systems. Drawing heavily on the work of E. P. Thompson and Perry Anderson, this research ultimately sought to explain the manner in which the 'moral economy' of the northern Nigerian peasantry was undermined, and made more vulnerable to drought and famine, by colonial capitalism.

After receiving his PhD in 1979, Watts took a position at the University of California, Berkeley, where he became Chancellor's Professor in the Department of Geography, and Director of the Institute of International Studies. His first major work, *Silent Violence: Food, Famine and Peasantry in Northern Nigeria*, was published in 1983. Based on his doctoral dissertation research, *Silent Violence* was, as Michael Redclift (2001) has noted, 'a big book in every sense'. Its nearly 700 pages sought to explain Nigerian famine in the nineteenth century, examined the implications of British colonialism for peasant economies, and posed a major theoretical challenge to the cultural and human ecology of the 1970s. *Silent Violence* was 'an attempt to . . . move beyond the limitations of natural hazards research and cultural ecology, beyond functionalism, the pitfalls of behaviorism, and the reluctance to engage with the political economy of the market' (Watts, 2001). From a personal standpoint, Watts' book collided head-on with the cultural ecology of Roy Rappaport, with whom Watts studied anthropology at Michigan, and Barney Nietschmann, his dissertation advisor. Based on extended fieldwork in west Africa, meticulous empirical research, and a rigorous and wide-ranging theoretical engagement with Marxian political economy, *Silent Violence* has influenced two generations of geographers, and helped shape the study of society–

environment relations within geography during the 1980s and 1990s. Though to a certain extent overshadowed by **Amartya Sen**'s *Poverty and Famines*, released in the same year, *Silent Violence* remains a landmark work within geography.

Watts' subsequent work has continued to focus on the intersections of political economy, culture, nature and power. Watts' critical (some might say polemical) contribution to a key volume on hazards research (Hewitt, 1983) argued for a more sophisticated engagement with social theory within the fields of human ecology and natural hazards. Thus Watts, by this time at Berkeley, drew a sharp theoretical distinction between himself and the Berkeley School, though in many respects (such as his concern for historical relationships between society and environment, peasant agricultural systems, and resource use and degradation) his affinity with the legacy of Carl Sauer is clear. Throughout the 1980s and 1990s, he continued to publish on the political economy of agricultural systems, environmental degradation and the marginalization of the African peasantry (e.g. Watts, 1987, 1989, 1990a, 1990b; Little and Watts, 1994). Watts found himself at the centre of debates surrounding the political economy of environmental change and Marxian challenges to cultural and human ecology.

Indeed, Watts has arguably been the single most important figure in the formation and development of political ecology as a subfield, both for his own enormously important work, and for the work of his many students, who as a group have profoundly shaped the examination of the political, economic and cultural contexts of environment–society relations within geography. Watts' own interests have diversified in theme and focus, though they have retained a central commitment to Marxian political economy. In addition to his ongoing work on peasant economies in west Africa, he has extended his interests to the industrial restructuring of agricultural systems in the USA, environmental social movements and the transna-

tional networks of which they are a part, the cultural politics of modernity, and the political ecology of petroleum extraction. His work on Third World economic and cultural transformation has earned him a reputation as one of the leading thinkers on development studies within the field of geography.

SPATIAL CONTRIBUTIONS

Monumental as it was, *Silent Violence* was in some respects under-appreciated in geography. Watts later lamented that 'it was a book often cited but rarely read' (Watts, 2001: 626), perhaps because its materialist, and what some might call totalizing, narrative attempted to 'reclaim something from Althusser when the riptide of social theory was racing in the opposite direction' (*ibid.*). Indeed, it could be argued that much of the value of this work may be located in its insistence on materialist analysis, and the linking of careful empirical investigation (based largely on long-term fieldwork) with a 'willingness to range widely in the search for ideas and intellectual inspiration' (Redclift, 2001: 622), precisely at a time when much social science research was looking inward, confronted by the 'postmodern turn'. Watts' work serves as a reminder of the value of committed fieldwork, rigorous empiricism and theoretical breadth.

This is not to say that Watts' own work avoided poststructuralism or cultural theory. By the early 1990s, his writing addressed directly the themes of identity politics, globalization and the apparent 'deterritorialization of culture' about which so much was being written. Retaining his commitment to Marxian analysis, Watts sought to locate culture within a nexus of spatial, political and economic relations, attempting an interrogation of the 'complex articulations of capitalism, modernity and culture understood as a field of struggle' (Watts, 1991: 7). In this work, Watts was particularly concerned to investigate the questions of identity politics, and the contradictions that certain forms of identity construction have with political mobilization. Thus, through his work Watts sought to illuminate the spaces of globalization, not as abstract 'spaces of flows' (Castells, 1996) or 'global ethnoscapes' (Appadurai, 1991), but in terms of the territorialization, de-territorialization, and re-territorialization of cultural space. Watts showed particular concern for the interstitial spaces of frontiers, the material and symbolic meeting places of political, economic and cultural systems:

> Frontiers are, of course, particular sorts of spaces – symbolically, ideologically, and materially ... At the margins of state power, they create their own territorial form of law and (dis)order ... But frontiers are also locally encoded in symbolic terms, and often carry a powerful ideological valence, particularly when national identity itself is seen to derive from 'frontier stock', or, if economic potential ('development') is seen to be wedded to the opening of the frontier.
> (Watts, 1992: 116–117)

Watts' concern with the spatiality of cultural difference carries with it a critique of what he saw as the naïve celebration of multiculturalism that was increasingly prevalent in academic literature. Concerned by the tone of some debates over hybridity, multiplicity and contingency, and by the fact that multiple identities (e.g. within a social movement) can erode underlying unity – a fact which is used to the benefit of dominant powers – he notes that, 'Some practices of multiculturalism can, therefore, act to reinforce centralized state power,' and questions 'how identity that rests on difference and splitting can produce a common ground for politics' (Watts, 1991: 125).

These concerns were further examined in *Reworking Modernity*, co-written with Allan Pred (Pred and Watts, 1992). In the co-authored introductory chapter, as well the three chapters of which he is sole author, Watts is concerned with questions of identity and difference within the context of the shifting geographies of capitalism and modernity. Central to this work is a concern for scale: linking processes of globalization with the transformation of particular places. In this sense, this work may be seen as an ethnography of modernity, with concern for:

> ... the symbolic discontent that emerges as new forms of capital make their local appearance; as the agents and actions of capital intersect with already existing – more or less deeply sedimented – everyday practices, power relations, and forms of consciousness; as local residents simultaneously experience modernity and hegemony in new guises.
> (Pred and Watts, 1992: xiii–xiv)

Watts' longstanding concern with the discourses and practices of international development were further examined in a number of key works during the mid-1990s, including a special double-issue of the journal *Economic Geography* in 1993, co-edited with Richard Peet (see Peet and Watts, 1993). Many of the contributions to this double issue re-emerged (substantially re-worked) in *Liberation Ecologies: Environment, Development, Social Movements* (Peet and Watts, 1996a). These collections mark an explicit concern to link political ecology, broadly inspired by Marxian political economy, with a poststructural focus on identity politics, discourse analysis and 'new social movements'. The volume asserts that 'theories about environment and development – political ecology in its various guises – have been pushed and extended both by the realities of the new social movements themselves, and by intellectual developments associated with discourse theory and poststructuralism' (Peet and Watts, 1996b: 2–3). Thus, at the heart of the volume are the practices and discourses by which Third World social movements conceptualize, negotiate and contest development. In particular, Peet and Watts seek to illuminate the liberatory potential of social movements and environmental politics, arguing that contemporary environmental struggles involve not only new social movements, but extend to (and through) various transnational environmental and human rights networks, multilateral institutions and forms of global governance. Thus, a central concern of the volume is the way in which trans-scalar processes and actors influence (and are in turn influenced by) local-scale environmental struggles. While *Liberation Ecologies* marks a significant contribution to the fields of political ecology and development studies, it nevertheless reproduces some of the weaknesses of that literature by incorporating only studies of rural Third World societies (Neumann, 1998). The extension of the political ecology framework to transnational, First World and urban contexts would have to await other authors (see, for instance, Bebbington and Batterbury 2001; McCarthy 2001; Robbins *et al.*, 2001).

KEY ADVANCES AND CONTROVERSIES

Watts' work has been central to the establishment and consolidation of political ecology as a subfield within geography. In part, this is because his own work has been characterized by an emphasis on empirical and theoretical rigour, a dedication to field-based research and a sensitivity to cultural contingency and global power. The influence of his work is also in part attributable to Watts' own caustic critique of cultural and human ecology, particularly the field of hazards research. His 1983 piece, 'On the poverty of theory,' included as a chapter in an important edited collection on hazards research (Hewitt, 1983), takes head-on the func-

tionalist, cybernetic approach to the study of social adaptability (at the heart of much cultural ecological research) and of 'natural' hazards. Drawing on the work of Sahlins, Wadell and others, Watts criticizes what he saw as the mechanistic scientism of these approaches to hazards research, particularly as they were applied to peasant societies in transition, and as they are confronted by global capitalism:

> As is clear in retrospect, these [cross-cultural] field studies were ahistoric, insensitive to culturally varied indigenous adaptive strategies, largely ignorant of the huge body of relevant work on disaster theory in sociology and anthropology, flawed by the absence of any discussion of the political-economic context of hazard occurrence and genesis, and in the final analysis having little credibility in light of the frequent banality and triviality of many of the research findings.
> (Watts, 1983b)

Watts' call for the incorporation of sophisticated social theoretical analysis into the study of society–environment relations shaped subsequent research on hazards, environmental degradation, struggles over resource access, agricultural production and environmental conservation. It also helped widen an enduring gulf between various geographical approaches to the study of society–environment relations. Political ecology does not have universal appeal, to say the least, and itself came under fire by Vayda (whose own work was criticized in Watts' early work): 'Indeed, it may not be an exaggeration to say that overreaction to the "ecology without politics" of three decades ago is resulting now in a "politics without ecology" . . .' (Vayda and Walters, 1999: 168). Ironically, perhaps, Watts himself has called into question the quality of political theory found in much of political ecology, and in particular the 'regional political ecology' of Piers Blaikie and Harold Brookfield's immensely influential volume *Land Degradation and Society* (1987) (see, for example, Watts,

1990a; Peet and Watts, 1996a). Watts critiqued this work as lacking a coherent political theory and subscribing to a simplistic understanding of power. Nevertheless, *Land Degradation and Society* remains hugely influential, and with Watts' own work helped give rise to the subfield of political ecology.

In addition to his longstanding concern for the highly politicized and violent geographies of oil extraction (e.g. Watts, 1984, 1994, 1997), Watts has turned his attention to the alarmist 'environmental security' perspective of resource scarcity and violence, as popularized by journalist Robert Kaplan and scholars Thomas Homer-Dixon and Gunther Baechler. Taken up by the Clinton administration in the USA and argued by international 'strategic experts' in the context of global environmental governance, this body of work is the latest manifestation of the Malthusian nightmare: environmental concerns marshalled in defence of capital accumulation and bourgeois lifestyles, and repackaged as 'national security'. Watts and Nancy Lee Peluso take on these arguments in their edited volume *Violent Environments* (Peluso and Watts, 2001a), arguing that violence (physical, structural, symbolic) in the context of environmental conflict cannot automatically be attributed to resource 'scarcity'. Again, the arguments are material as well as symbolic, rooted in political economy as well as epistemology: 'scarcity' is socially constructed, and as such must be carefully viewed in the context of power relations, cultural meanings and the workings of capital:

> Our approach is not intended to merely identify the 'environmental triggers' of violent conflict nor does it start from a presumed 'scarcity.' Rather, *Violent Environments* accounts for ways that specific resource environments (tropical forests or oil reserves) and environmental processes (deforestation, conservation, or resource amelioration) are constituted by, and in part constitute, the political economy of access to and control over resources.
> (Peluso and Watts, 2001: 5)

This work has drawn attention to the relationship between environmental conflict and the production of space, and the fact that scarcity and conflict are far from natural phenomena, though they may become 'naturalized' in the highly charged context of resource politics. Spaces of scarcity and the particular places produced through social conflict and violence only exist in relation to other places, spaces and scales, in a context of uneven capitalist development. In this respect, Watts' work has done much to bridge the gap between the nature/society and space/society traditions in geography, which have had remarkably little interaction during the past half-century (Hanson, 1999). Watts' call for the careful contextualization of environmental degradation, peasant marginalization and resource conflict has done much to shape the type of work undertaken by contemporary political and cultural ecologists, as well as by those concerned with development studies, peasant studies and globalization. The best of this work is attentive to theories of space, place and scale; incorporates sophisticated theoretical analysis of political economy and culture; and demonstrates a thorough understanding of the complexities of environmental and social processes.

WATTS' MAJOR WORKS

Goodman, D. and Watts, M. J. (eds) (1997) *Globalising Food: Agrarian Questions and Global Restructuring*. London: Routledge.

Peluso, N. and Watts, M. J. (eds) (2001a) *Violent Environments*. Ithaca: Cornell University Press.

Peet, R. and Watts, M. J. (eds) (1996a) *Liberation Ecologies: Environment, Development, Social Movements*. London: Routledge.

Pred, A. and Watts, M. J. (1992) *Reworking Modernity: Capitalisms and Symbolic Discontent*. New Brunswick: Rutgers University Press.

Watts, M. J. (1983a) *Silent Violence: Food, Famine, and Peasantry in Northern Nigeria*. Berkeley: University of California Press.

Watts, M. J. (1999) 'Collective wish images: Geographical imaginaries and the crisis of development', in J. Allen and D. Massey (eds) *Human Geography Today*. Cambridge: Polity Press, pp. 85–107.

Watts, M. J. (2000a) '1968 and all that . . .', *Progress in Human Geography* 25: 157–188.

Watts, M. J. (2000c) 'Development ethnographies', *Ethnography* 2: 283–300.

Watts, M. J. (2000d) 'Development at the millennium', *Geographische Zeitschrift* 88: 67–93.

Secondary Sources and References

Appadurai, A. (1991) 'Global ethnoscapes: notes and queries for a transnational anthropology', in R. G. Fox (ed.) *Recapturing Anthropology: Working in the Present*. Santa Fe, NM: School of American Research Press, pp. 191–210.

Bebbington, A. J. and Batterbury, S. P. J. (2001) 'Transnational livelihoods and landscapes: political ecologies of globalization', *Ecumene* 8: 369–380.

Blaikie, P. (1985) *The Political Economy of Soil Erosion in Developing Countries*. London: Longman.

Blaikie, P. and Brookfield, H. (1987) *Land Degradation and Society*. London: Methuen.

Castells, M. (1996) *The Information Age: Economy, Society and Culture; Volume 1: The Rise of the Network Society*. Oxford: Blackwell.

Hanson, S. (1999) 'Isms and schisms: healing the rift between the nature-society and space-society traditions in geography', *Annals of the Association of American Geographers* 89: 133–143.

Hewitt, K. (1983) *Interpretations of Calamity*. Boston: Allen and Unwin.

Little, P. D. and Watts, M. J. (eds) (1994) *Living Under Contract: Contract Farming and Agrarian Transformation in Sub-Sahara Africa*. Madison: University of Wisconsin Press.

McCarthy, J. (2001) 'States of nature and environmental enclosures in the American West', in N. L. Peluso and M. L. Watts (eds) *Violent Environments*. Ithaca: Cornell University Press, pp. 117–145.

Neumann, R. (1998) 'Book review of "Liberation Ecologies"', *Economic Geography* 74: 190–192.

Peet, R. and Watts, M. J. (1993) 'Introduction: development theory and environment in an age of market triumphalism', *Economic Geography* 69 (3): 227–253.

Peet, R. and Watts, M. J. (1996b) 'Liberation ecology: Development, sustainability, and environment in an age of market triumphalism', in R. Peet and M. J. Watts (eds) *Liberation Ecologies: Environment, Development, Social Movements*. London: Routledge, pp. 1–45.

Peluso, N. L. and Watts, M. J. (2001b) 'Violent environments', in N. L. Peluso and M. J. Watts (eds) *Violent Environments*. Ithaca: Cornell University Press, pp. 3–38.

Redclift, M. (2001) 'Classics in human geography revisited: Commentary 1', *Progress in Human Geography* 25: 621–623.

Robbins, P., Polderman, A. and Birkenholtz, T. (2001) 'Lawns and toxins: an ecology of the city', *Cities: The International Journal of Urban Policy and Planning* 18 (6): 369–380.

Vayda, A. and Walters, B. (1999) 'Against political ecology', *Human Ecology*, 27: 167–179.

Watts, M. J. (1983b) 'On the poverty of theory: natural hazards research in context', in K. Hewitt (ed.) *Interpretations of Calamity*. London: Allen and Unwin, pp. 229–262.

Watts, M. J. (1984) 'State, oil, and accumulation: from boom to crisis', *Environment and Planning D: Society and Space*, 2: 403–428.

Watts, M. J. (1987) 'Drought, environment and food security', in M. Glantz (ed.) *Drought and Hunger in Africa*. Cambridge: Cambridge University Press, pp. 171–212.

Watts, M. J. (1989) 'The agrarian question in Africa', *Progress in Human Geography*, 13: 1–41.

Watts, M. J. (1990a) 'Is there politics in regional political ecology?', *Capitalism, Nature, Socialism* 4: 123–131.

Watts, M. J. (1990b) 'Visions of excess: African development in an age of market idolatry', *Transition* 151: 124–141.

Watts, M. J. (1991) 'Mapping meaning, denoting difference, imagining identity: dialectical images and postmodern geographies', *Geografiska Annaler* 73B: 7–16.

Watts, M. J. (1992) 'Space for everything (a commentary)', *Cultural Anthropology* 7: 115–129.

Watts, M. J. (1993) 'Development I: Power, knowledge, discursive practice', *Progress in Human Geography* 17: 257–272.

Watts, M. J. (1994) 'The devil's excrement', in S. Corbridge, R. Martin and N. Thrift (eds) *Money, Power and Space*. Oxford: Blackwell, pp. 406–445.

Watts, M. J. (1996) 'Mapping identities: place, space, and community in an African city', in P. Yaeger (ed.) *The Geography of Identity*. Ann Arbor: University of Michigan Press, pp. 59–97.

Watts, M. J. (1997) 'Black gold, white heat', in S. Pile and M. Keith (eds) *Geographies of Resistance*. London: Routledge, pp. 33–67.

Watts, M. J. (2000b) 'Violent geographies: speaking the unspeakable and the politics of Space', *City and Society* 13 (1): 83–115.

Watts, M. J. (2001) 'Classics in human geography revisited: Author's response', *Progress in Human Geography* 25: 625–628.

Thomas Perreault

51 Raymond Williams

BIOGRAPHICAL DETAILS AND THEORETICAL CONTEXT

As much as Raymond Williams' scholarship and politics were shaped through his engagement with the issues and thinkers of his time, they remained indelibly connected to his personal experiences of place and class. Williams' political commitments and activities, as well as his way of academic scholarship, were *aligned* – a term he used to refer to an allegiance or orientation stemming from social position, similar to class-consciousness, though also connected to place. Williams saw alignment as a critical determinant of the commitments one chooses as an individual. Alignment with a given world is neither innocent nor static. Indeed, it is political and it is worked through. The particular way that Williams simultaneously connected and developed abstract theoretical concepts, specific personal experiences and complex social history into a way of understanding the political possibilities of the world was thus distinctive, proving deeply influential in and well beyond geography.

Williams (1921–1988) was born in Pandy, a Welsh village near the English border. His father was a railway signalman and union activist. Williams' ideas and politics were deeply structured by the 'border world' in which he grew up. Pandy was rural, but his father was an industrial worker; the railroad forged modern connections but never fully displaced customary ways of life. Pandy was in Wales, but 'there was all the time a certain pressure from the East ... from

England' (Williams, 1979b: 21). When he went to Cambridge on a scholarship, these borders deepened: he was working class in the bastion of the elite, Welsh in the heart of Englishness, and radical at the centre of conservativism. At Cambridge Williams sought out other radical students and became an influential voice in the Communist Party's 'Writer's Group' (Inglis, 1995). At the same time, Williams was deeply influenced by F. R. Leavis, who had done so much to realign the study of literature towards the study, in Williams' (1989d: 9) terms, of 'art and experience'. Despite Leavis' conservativism, Williams saw a radical potential in Leavis' method of studying literature as almost a symptom of society to become an important tool in the construction of a progressive, socialist world. These commitments were deepened during World War II, when Williams served as a tank commander in the battles that followed the D-Day invasion. While in the army, Williams was an active campaigner for the Labour Left, using his role as the editor of a regimental paper to support the party (Inglis, 1995: 101–102). By this time he had left the Communist Party, though he never later sought to distance himself from his time in it (Williams, 1989f).

Following his discharge from the army in 1945, Williams returned to Cambridge to complete his degree (he later incorporated his thesis into his first book; Williams, 1952). His goal was to become a writer of both novels and political journalism and to become a teacher of working-class people. Briefly in Cambridge and then for 15 years around Oxford, Williams worked as an itinerant teacher of adult education, surveying international relations as well as English literature. In all his teaching he sought to show his working-class students how their own experiences allowed them to make sense of the world and to see that they had *made* a culture, not just received one from the political and cultural elite. During this period, Williams published five books of criticism and cultural analysis; his first novel, *Border Country* (1960); and innumerable essays and pieces of journalism. With **Stuart Hall** and others, he helped to found the *New Left Review* (*NLR*). His most influential works of this period were two books that became a foundation of the academic field of cultural studies, *Culture and Society, 1780–1950* (1958) and *The Long Revolution* (1961), and an essay, 'Culture is ordinary' (1989d). These three works bring a socialist and Marxist sensibility to the study of the pressures and determinations that make 'a culture'; and they launch two projects: one that seeks to redefine just what *counts* as 'culture'; the other to explore the role and importance of new institutions of communication in the *formation* of culture.

In 1961, Williams was appointed to a lectureship in English at Cambridge University. In 1969, he submitted his work for a DLitt, receiving it in July of that year. In 1974, he was promoted to a personal Chair and named Professor of Drama. During the 22 years he was at Cambridge, Williams was deeply involved in politics, including work with the Campaign for Nuclear Disarmament, continuing involvement with the *NLR*, and, in 1967, co-authoring the influential *New Left May Day Manifesto*, which sought to organize the growing dissent of the period to push Labour leftwards. In the 1970s and 1980s he was involved in left Labour politics and was active in the promotion of regionally based arts. He responded to the election of Margaret Thatcher, and the virulent class politics it represented, by helping to found the (short-lived) Socialist Society and by aligning himself with the Socialist Environment and Resources Association.

Between 1962 and 1983, Williams was the author, editor or co-editor of 20 scholarly and political books as well as numerous scholarly and popular articles. He served four years in the 1970s as the television critic for the *Listener*. His most influential books during this period include: *Communications* (1962), which more or less established the field of media studies in Britain; *The Country and the City* (1973), perhaps his most cited work in geography, and one that establishes the relationship between contested representations of the national landscape and its material form; *Keywords* (1976), a social etymology of theoretically and politically crucial terms; and *Marxism and Literature* (1977; see also 1980), where he develops his influential form of cultural materialism, while also most fully stating his important concept of 'structure of feeling'. Williams describes 'structure of feeling' as a 'cultural hypothesis' about the way people's practical consciousnesses ('meanings and values . . . actively lived and felt') of living in society comprise a set, or structure, with internal relations of contradiction and coordination (Williams, 1977: 132). Part of this hypothesis is that structures of feeling, in formation, are 'taken to be private, idiosyncratic, and even isolating, but which in analysis (though rarely otherwise) [have their] emergent, connecting, and dominant characteristics, indeed, [their] specific hierarchies' (Williams, 1977: 132). As a conceptual tool, structure of feeling provides a way of linking experience and feeling into organized politics.

Williams examined the implications of his structure of feeling hypothesis in his novels. While at Cambridge, Williams published three novels: *Second Generation* (1964) and *The Fight for Manod* (1979a), both sequels to *Border Country*; and *The Volunteers* (1978), an at times bitter prospective look at the emergent Britain and Europe (written in the midst of the Labour disaster of the late 1970s and on the eve of the authoritarianism of Thatcherism). In his novels, Williams played out the structure of feeling as it is experienced and lived by individuals who are structured by class, borders and constant social and individual change.

Between retirement (1983) and his death from a heart attack, Williams completed two novels (one published posthumously), and wrote enough of a third for his wife, Joy Williams, to complete the editing and see it to publication. In addition he published two books of essays and criticism, co-edited a third, and launched an important political intervention, *Towards 2000* (1983), which explored the politics of nuclear destruction, and examined the politics of spectacle at the heart of the Falklands war. In this book Williams sought 'to catch local politics in one grand theory' (Inglis, 1995: 275).

Indeed, Williams' life and work can be summarized in this phrase. His political and theoretical work (which were really one and the same) were a constant struggle against the institutions, class structures and massive condemnation of history that seek to exclude and oppress, to expropriate and deny, to thwart the development of what Marx called humans' 'species being' – their true creative capacity. Williams stood for an open, creative, worked culture against those hegemonic notions of culture that are reinforced by the dominant class, through education, structures of work and the multiple forms of practical consciousness that permeate social existence.

SPATIAL CONTRIBUTIONS

Williams' work and ideas have been directly and indirectly influential in human geography. Indirectly, Williams is one of the founding figures of the field of cultural studies. *Culture and Society* served as one of the two 'founding texts' of the Birmingham Centre for Contemporary Cultural Studies (CCCS) (the other is Hoggart, 1952). Especially under its second

director, **Stuart Hall**, CCCS sought to understand culture as: firstly, both ordinary and deeply political; secondly, a significant site of ongoing social struggle and the working out of power relations between 'sub-cultures'; and thirdly, important in the shaping of identity.

The work of CCCS began to be cited in geography during the early 1980s, but its impact has been greatest as geography has undergone a 'cultural turn' (Crang, 2000). The cultural turn is itself in part a result of the influence of what is called the 'new cultural geography'. With its radical emphasis on culture as a medium of social power' (Daniels, 1989: 196), new cultural geography clearly echoed the work of the CCCS. But it also was rooted in Williams' work more generally (e.g. **Cosgrove**, 1984; Daniels, 1989; Jackson, 1989). Williams' influence is perhaps strongest in **Peter Jackson**'s landmark text *Maps of Meaning* (1989). Jackson outlines the value of cultural materialism for geography, and in the process suggests the importance of geography to cultural materialism. A citation to Williams opens the book, and a sustained explication of his ideas follows. Jackson (1989: 35) writes that 'by focusing on "specific and indissoluble real processes", Williams offers a view of determination that is thoroughly appropriate to a reconstituted cultural geography'. In part this is because 'Williams' work comprises a thorough attack on the . . . culturalism' that was so central to existing cultural geography and related fields (Jackson, 1989: 36). 'This is the value of Williams' definition of culture' (Jackson, 1989: 38). By seeing it 'as "a realized signifying system" ', or, in other words, 'a set of signs and symbols that are embedded in a whole range of activities, relations, and institutions, only some of which are "cultural", others being overtly economic, political, or generational', Williams' theories provide a model for how *not* to artificially separate the social, political and economic from the cultural. Coupling this model with theories of *hegemony*, Jackson promoted an 'agenda for cultural geography' that was materialist *and* deeply concerned with signification, took language and discourse seriously while never ignoring those social practices that could be called 'extra-linguistic', and that saw the social analysis of *cultural politics* as its central focus.

Simultaneous with *Maps of Meaning*, Daniels (1989) established the relevance of Williams' work to landscape geography. Daniels follows Williams to argue that landscape is not only, or perhaps even primarily, a reflection of some sort of undifferentiated culture, but rather the materialization of a class-based and class-ridden social order. For Williams, the class-based nature of work and the work of representation that go into a landscape produced a duplicity: landscape seems to speak for all but really only represents the few. The point was thus to dig beneath this duplicity and to bring to light the contradictions as well as the relations of power that undergird the landscape.

This way of understanding the landscape is possibly now predominant in geography. However, in a critique of the new cultural geography, Mitchell (2000) draws on Williams to argue that much new cultural geography is still 'not materialist enough'. Mitchell (2000) takes issue with how 'new' cultural geographers had deployed Williams' arguments. In other words, by the end of the century, Williams' work was beyond mere explication in geography and instead the central debate.

The debates about how best to theorize culture in geography support one of Williams' (1976) most quoted contentions: that 'culture' is one of the two or three most complicated words in English (and 'nature' is perhaps the most complicated). Geographers are fond of quoting Williams on this point as a means of framing and legitimating their own arguments about each of the terms, and as a means of quickly developing an argument about the production – or social construction – of nature and culture. Williams' method of tracing the etymological transformation of 'keywords' over time reveals shifts in meanings as a *social* and not just an etymological history. Through etymology, Williams shows not only what has been gained, but also what has been lost over time. Often the losses are of communal social relations (what he called 'the common culture' (1958, 1961, 1989e) as the individualizing power of capitalism spread and deepened. Such losses are also at times gains in certain kinds of freedom, certain opportunities for a different kind of social life. This is a method and project most fully developed by Kenneth Olwig (1996, 2002) in his attempt to recover what he calls the 'substantive meaning of landscape'. This substantive meaning resides in archaic regionally based systems of justice that have both been eroded by and yet still persist in contemporary capitalism. For Olwig (1996: 633), like Williams (1973), this substantive meaning can be reclaimed and renewed, since 'a *landskab* was not just a region, it was a nexus of law and cultural identity'. Olwig argues that recovering such a substantive meaning of landscape also implies making a substantial argument about justice: those who make a landscape have a right to participate in it and to shape the judicial form that governs it. And here Olwig's project intersects directly with that of Williams, for running through all of Williams' work is a deep and abiding sense of right and of rights (see, e.g., Williams, 1980: 1–6).

Williams' project was a prospective one: it sought to provide the grounds, and to legitimate the need, for an ongoing socialist struggle for social justice. But these grounds and this legitimation were rooted in a retrospective, historical argument about alienation, expropriation and the uprooting of lives and landscapes – the common culture – that is so central to the success of modernizing capitalism, and that must be reclaimed for socialism. This argument is most fully, if also indirectly, developed in Williams' novels. In these he makes it clear that any common culture had to have social and geographical difference – differences in ways of life and structures of feeling – at its core.

This theme has been picked up in geography most prominently by **David**

Harvey (1996). For Harvey, Williams' writing provides a way to articulate the complex problems of injustice and oppression, as well as the alliances and actions that could serve to reconfigure the world in a better way. Considering the world as a totality – as Williams (1958, 1961, 1977) did – together with examining the significance of everyday experience as evidence of what the world is like (as in the 'structure of feeling' hypothesis), and closely analysing the politics of loyalty and alignment, require a dialectical and geographical understanding of society: this is the lesson that Harvey takes from Williams. He combines his reading of Williams with an engagement with more direct theories of social justice (e.g. **Young**, 1990), to argue that any project of social justice that takes difference (including geographical difference) seriously, and that seeks to be truly transformative, must, in fact, be *common* and *emancipatory* – two terms that are central to Williams' work, but anathema to postmodernist theories of justice that Harvey argues against. Loyalty and alignment are thus key geographical concepts for Harvey, since one's particular positionality, based on place or on social identity, is necessarily the very foundation from which solidarity – a common culture – can be structured. Williams presents the characters in his novels as struggling to balance their allegiances to the idea of socialism in the abstract while at the same time maintaining commitments to specific places (Harvey, 2001: 163–185). Harvey argues, based on Williams' characterizations, that though there may often be a tension between the particularities of people's identities and the progressive, universalist goals of socialism or Marxism, the two sets of goals – difference and commonality – often can and do work together. Uncovering the way this happens, Harvey argues, is the task of theory.

A significant part of the task of making clear the links people have to broader, common goals is advanced by understanding geography as a system that is both made and used by capitalism, and by all the people who are caught up in it. Harvey and Williams both assert the importance of place, as constructed relationally and through scale, in providing a commonality between loyalties at different levels of abstractions. Referring to Williams' notion of 'militant particularism', Harvey (2000: 55) argues that 'the universalism to which socialism aspires has . . . to be built by negotiation between different place-specific demands, concerns, and aspirations'. The geography of difference is key, since it is real material places and spatial-social relations that shape both the opportunities and constraints for the production of a socially-just world. The geography of difference is what gives form to alignment, experience, commitment and, perhaps most importantly, loyalty.

KEY ADVANCES AND CONTROVERSIES

Williams' political and theoretical advances include key new conceptualizations such as 'structure of feeling', a theorization of the 'common culture', a concern with the social history of keywords, important explications of theoretical arguments within Marxism, the development of tools for the analysis of media and, probably most significantly, a body of cultural theory, developed over four decades, that retains enormous power and subtlety and remains a model for theoretical development and empirical exploration throughout the humanities and social sciences. More generally, Williams was a key figure in post-World War II British socialism and Western leftist politics and intellectual endeavours.

His work, however, has not been received uncritically. Williams (1979b) himself admitted that he was much too blind to the difference that gender makes, and he has been criticized on this matter in a number of places. Though he was one of the first in literary studies to turn his attention to television as an important cultural medium, later practitioners of cultural studies have found him to be too pessimistic – indeed almost scornful – about the progressive potential of pop culture (see Fiske, 1989). Similarly, his cultural criticism as exemplified in *Culture and Society*, *The Long Revolution* and *The Country and the City* can sometimes seem almost anti-modern and nostalgic (see Harvey, 1996). Furthermore, some critics, such as **Edward Said** (1989, 1993) and Gilroy (1979), charge Williams with constructing a theory of commitment and loyalty that is exclusionary, rather than cosmopolitan, because of its heavy reliance on the experiences of Britain's white working class. Said (1993) further suggests, though perhaps inaccurately, that Williams (1973) was largely ignorant of the critical importance of the colonies and slave labour to the construction of the typically English countryside. Finally, Inglis (1995) takes Williams to task over his socialist politics, accusing him on several occasions of (unspecified) 'bad faith' in

his theorizing, but seeming to mean that Williams failed to live up to Inglis' own faith in a more moderate Labourism. *Marxism and Literature* (1977) was not uniformly praised when it was published. Williams' disdain for the theories of ideology advanced by Althusser was criticized by many, including **Stuart Hall**, who has been central in promoting the importance of Althusser's theories for the study of culture and identity.

Nonetheless, nearly fifteen years after his death Williams' influence in geography and other social sciences and humanities seems to be growing. Almost all of his work is still in print, and it continues to provide inspiration for new work in geography (e.g. Mitchell, 2003). Williams' own alignments and commitments created inevitable blind-spots in his theorizing. But at the same time, they are exactly what gives his work such force, and what gives it its abiding relevance. In the end, then, Williams' greatest advance was probably the most basic one: his life and work are a model of what socially and politically engaged scholarship should and can be.

WILLIAMS' MAJOR WORKS

Williams, R. (1958) *Culture and Society (1780–1950)*. London: Chatto and Windus.
Williams, R. (1961) *The Long Revolution*. London: Chatto and Windus.
Williams, R. (1962) *Communications*. Harmondsworth: Penguin.
Williams, R. (1973) *The Country and the City*. Oxford: Oxford University Press.
Williams, R. (1976) *Keywords*. London: Fontana.
Williams, R. (1977) *Marxism and Literature*. Oxford: Oxford University Press.
Williams, R. (1979b) *Politics and Letters: Interviews with New Left Review*. London: Verso.
Williams, R. (1980) *Problems in Materialism and Culture*. London: Chatto and Windus.
Williams, R. (1981) *Culture*. London: Chatto and Windus.
Williams, R. (1983) *Towards 2000*. London: Chatto and Windus.

Secondary Sources and References

Cosgrove, D. (1984) *Social Formation and Symbolic Landscape.* London: Croom Helm.

Crang, P. (2000) 'Cultural turn', in R. Johnston *et al.* (eds) *The Dictionary of Human Geography*, 4th edition. Oxford: Blackwell, pp. 141–143.

Daniels, S. (1989) 'Marxism, culture and the duplicity of landscape', in R. Peet and N. Thrift (eds) *New Models in Geography*, Volume 2. London: Unwin Hyman, pp. 196–220.

Fiske, J. (1989) *Understanding Popular Culture.* Boston: Unwin Hyman.

Gilroy, P. (1987) *There Ain't No Black in the Union Jack.* London: Verso.

Harvey, D. (1996) *Justice, Nature and the Geography of Difference.* Oxford: Blackwell.

Harvey, D. (2000) *Spaces of Hope.* Berkeley: University of California Press.

Harvey, D. (2001) *Spaces of Capital: Towards a Critical Geography.* New York: Routledge.

Hoggart, R. (1952) *The Uses of Literacy.* London: Chatto and Windus.

Inglis, F. (1995) *Raymond Williams.* London: Routledge.

Jackson, P. (1989) *Maps of Meaning.* London: Unwin Hyman.

Mitchell, D. (2000) *Geography: A Critical Introduction.* Oxford: Blackwell.

Mitchell, D. (2003) *The Right to the City: Social Justice and the Fight for Public Space in America.* New York: Guilford.

Olwig, K. (1996) 'Recovering the substantive nature of landscape', *Annals of the Association of American Geographers* 86: 630–653.

Olwig, K. (2002) *Landscape, Nature, and the Body Politic: From Britain's Renaissance to America's New World.* Madison: University of Wisconsin Press.

Said, E. (1989) 'Appendix: media, margins, and modernity', in R. Williams (ed.) *The Politics of Modernism.* London: Verso.

Said, E. (1993) *Culture and Imperialism.* London: Chatto and Windus.

Williams, R. (1952) *Drama from Ibsen to Eliot.* London: Chatto and Windus.

Williams, R. (1960) *Border Country.* London: Chatto and Windus.

Williams, R. (1964) *Second Generation.* London: Chatto and Windus.

Williams, R. (1966) *Modern Tragedy.* London: Chatto and Windus.

Williams, R. (1968a) *Drama from Ibsen to Brecht.* London: Chatto and Windus.

Williams, R. (ed.) (1968b) *May Day Manifesto.* Harmondsworth: Penguin.

Williams, R. (1970) *The English Novel from Dickens to Lawrence.* London: Chatto and Windus.

Williams, R. (1971) *George Orwell.* New York: The Viking Press.

Williams, R. (1974) *Television, Technology and Cultural Form.* London: Chatto and Windus.

Williams, R. (1978) *The Volunteers.* London: Hogarth.

Williams, R. (1979a) *The Fight for Manod.* London: Chatto and Windus.

Williams, R. (1985) *Loyalties.* London: Chatto and Windus.

Williams, R. (1989a) *People of the Black Mountains: The Beginning.* London: Chatto and Windus.

Williams, R. (1989b) *The Politics of Modernism.* London: Verso.

Williams, R. (1989c) *Resources of Hope.* London: Verso.

Williams, R. (1989d) 'Culture is ordinary', in *Resources of Hope.* London: Verso, pp. 3–18 (first published 1958).

Williams, R (1989e) 'The idea of a common culture', in *Resources of Hope.* London: Verso, pp. 32–38 (first published 1968).

Williams, R. (1989f) 'You're a Marxist, aren't you?', in *Resources of Hope.* London: Verso, pp. 65–76 (first published 1975).

Williams, R. (1990) *People of the Black Mountains: The Eggs of the Eagle.* London: Chatto and Windus.

Young, I. M. (1990) *Justice and the Politics of Difference.* Princeton: Princeton University Press.

Don Mitchell and Carrie Breitbach

52 Iris Marion Young

BIOGRAPHICAL DETAILS AND THEORETICAL CONTEXT

Born in New York City in 1949, Iris Marion Young gained her PhD in Philosophy in 1974 from Pennsylvania State University, focusing her thesis on Wittgenstein's later writings. Before becoming a Professor in the Political Science Department of the University of Chicago, she held various academic positions in the United States in philosophy and in political theory. Young is a prolific writer and has made theoretical advances within numerous cross-disciplinary debates from the 1980s to the present. She is best known for her writings on justice and the politics of difference, though her contributions extend into many other domains within the humanities and social sciences. Young's work has been widely translated, and she has held several visiting fellowships at universities and institutes internationally. These facts, coupled with the reprinting of seminal essays, help explain her international reputation and the dissemination of her ideas well beyond the fields of political theory and philosophy. Young's influence within the academy is not restricted to her monographs and sole-authored essays. In her role as editor, she has cemented the importance of feminist philosophy (Jeffner and Young, 1989; Jaggar and Young, 1998) and feminist ethics and social policy (DiQuinzio and Young, 1997). She is also committed to addressing policy debates and has intervened in discussions surrounding the location of hazardous waste, the treatment of pregnant addicts, residential segregation and institutions of global law.

Young's contributions to philosophy, political theory and feminist theory have had added resonance in the context of US academic responses to the new social movements of the 1970s – movements mobilizing for justice around such categories as gender, ethnicity, 'race', sexuality and nationality. Such movements challenged both traditional Marxist commitments to the primacy of class in analysing social formations, and liberalism's investment in particular models of equality, equity and democracy. For Young, taking those movements' demands seriously required opening up the category of justice beyond its usual restriction to questions of redistribution. Furthermore, Young argued, reformulating justice and democracy must be undertaken using *situated* analysis: she has described her work as engaging in 'critical theory' – in her words, 'normative reflection that is *historically and socially contextualized*' (1990b: 5; emphasis added).

Young's writings are underpinned by her participation in several social and environmental movements, most notably the women's movement. As Young puts it:

My personal political passion begins with feminism, and it is from my participation in the contemporary women's movement that I first learned to identify oppression and develop social and normative theoretical reflection on it. My feminism, however, has always been supplemented by commitment to and participation in movements against military intervention abroad and for systematic restructuring of the social circumstances that keep so many people poor and disadvantaged at home.
(Young, 1990b: 13)

Young also draws from a wide range of philosophical frameworks, and her writing shows its indebtedness to the liberal tradition as well as to feminism, deconstruction, phenomenology, psychoanalysis and pragmatism. Her essays have engaged with Arendt, de Beauvoir, Derrida, **Foucault**, Habermas, Irigaray, Kristeva, Lacan, Marx, Merleau-Ponty, Rawls and Sartre. Young's use of such different and often conflictual philosophical traditions makes her own work difficult to characterize with a single rubric. This has not, however, staved off attempts to compartmentalize her scholarship: she has been described variously as a radical postmodernist, a post-socialist, a devoted Habermasian, and too great a defender of liberal humanism.

Young's earlier writings wrestled with the debates surrounding socialist feminism and with the challenge of 'gynocentrism' (the revaluation of 'female gendered experience'). In *Throwing Like A Girl and Other Essays in Feminist Philosophy and Social Theory* (1990c), Young took up various feminist political projects associated with the Reagan–Bush Sr. years. The book moved from challenging the homogeneous public sphere elaborated by much political theory, to understanding women's embodied oppression (in essays on gendered movement, pregnancy, clothes and breasts). Young's yearning for 'heterogeneous public life' and her critique of the ideal of universal citizenship manifested her desire for a more inclusive polity: this preoccupation continues throughout her writings. The essay 'Throwing like a girl' was a phenomenological account of 'some of the basic modalities of feminine body comportment, manner of moving, and relation in space' (1990c: 143). Here Young argued that there are specific modalities of 'feminine spatiality' – modalities in which woman 'lives space as *enclosed* or confining ... and ... experiences herself as *positioned* in [not positioning herself in] space' (151; emphasis in original).

Although *Throwing Like a Girl* contains many geographical insights, it was

through the publication of her groundbreaking and prize-winning book *Justice and the Politics of Difference* (1990b) that she became widely known in the discipline of geography. In that book and in an allied essay (1990a), she laid out her fundamental claims about justice, community and the politics of difference. Arguing that a conception of justice 'should begin with the concepts of domination and oppression' (1990b: 3), rather than focusing on redistribution, she elaborated 'five faces of oppression' through which to understand how structural phenomena 'immobilize or diminish' particular groups in society. These she enumerated as exploitation, marginalization, powerlessness, cultural imperialism and violence. She jettisoned a logic in which 'liberation' is dependent upon a transcending of group difference – the ideal of assimilation – and instead proposed that 'justice in a group-differentiated society demands social equality of groups, and mutual recognition and affirmation of group differences' (1990b: 191). Central to her formulations concerning justice and difference were conceptualizations of spatio-temporal relations. For Young, the usual ideal of community privileges face-to-face relations, and in doing so, 'seeks to suppress difference in the sense of the time and space distancing of social processes, which material media facilitate and enlarge' (1990a: 314). Thus her account displaced the ideal of community with the ideal of '[c]ity life as an openness to unassimilated otherness' (1990b: 227), thereby presenting a vision of desirable social relations as 'a being together of strangers in openness to group difference' (1990b: 256).

Justice and the Politics of Difference gained an enormous readership and had an inflammatory effect in the social sciences. It served as a flashpoint for debates over the putative 'death', or at least displacement, of the hypostatized figure of the universal, male citizen of the Enlightenment. Of particular importance was the question of what it meant socially and politically to validate, as Young had done, the category of the differentiated 'group' in place of the individual. Thus Young, in her third monograph, *Intersecting Voices: Dilemmas of Gender, Political Philosophy, and Policy* (1997), was at pains to demonstrate that her account of the social group should not be equated with an essentializing 'identity politics'. In *Intersecting Voices*, she drew upon Habermas (in particular his model of communicative ethics) and Irigaray in developing the concept of 'asymmetrical reciprocity'. We cannot put ourselves in others' places, Young stressed, but we *can* communicate across difference: asymmetrical reciprocity points, therefore, to the acknowledgement of 'the difference, interval, that others drag behind them shadows and histories, scars and traces, that do not become present in our communication' (1997: 53).

In *Inclusion and Democracy* (2000), Young reinforced her conviction that democracy is the most productive institutional frame through which to realize 'difference'. There, her analyses extended more obviously to the scale of the global and to the debates over cosmopolitanism: she argued that her interrogation of democracy must, to have any purchase, work in the context of dense and globally extensive social, economic and legal networks. Young also devoted more attention to the role of civil society; emphasized her commitment to an agonistic rather than quiescent account of the social sphere and the democratic process; and provided another elaboration of the social group. Throughout *Inclusion and Democracy* Young developed a more explicit language of spatiality and position, particularly as a means to deflect accusations that her accounts of the social group do not provide the individual *within* the group with agency or specificity of her own:

> Understanding individuals as conditioned by their positioning in relation to social groups without their constituting individual identities helps to solve the problem of 'pop-bead' identity: A person's identity is not some sum of her gender, racial, class, and national affinities. She is only her identity, which she herself has made by the way that she deals with and acts in relation to others [sic] social group positions, among other things.
> (Young, 2000: 102)

Young, in pointing out that '[f]ew theories of democracy ... have thematized the normative implications of spatialized social relations' (2000: 196), solidifies her place as one of the few political theorists to have done so.

SPATIAL CONTRIBUTIONS

The moment at which Young's name gained widespread visibility in geography was one in which heated discussions were taking place about which conceptual frameworks geographers might – and should – use to understand socio-spatial relations. One way to exemplify that moment is to point to the debates surrounding **David Harvey**'s (1989) *The Condition of Postmodernity* – debates preoccupied with how to understand the vexed relations between Marxist theory, feminist theory and that which sits, loosely, under the umbrella of 'postmodern' theory. In short, the geographical question of that moment was: What kind of socio-spatial theory might best account for 'difference'?

Justice and the Politics of Difference spoke directly to these debates. That the book was resolutely committed to a liberatory politics extended its appeal to a variety of theoretical domains. In addition, Young's pragmatic tendencies and her commitment to theory that did not set itself apart from practical application fitted well with geography's guiding principles. The book, with its attentiveness to relational accounts of identity and affiliation, its theorizations of oppression and domination, and its drawing together of 'culture' and political economy endorsed many of the projects characterizing the burgeoning sphere of 'critical geography'. One indication of the proximity of Young's ideas to new directions being taken in geography is the pervasive – if not always explicit – influence of her scholarship in Keith and Pile's (1993) important collection *Place and the Politics of Identity*. Likewise, Young's work has been excerpted in McDowell and Sharp's (1997) feminist reader, in Kasinitz's (1994) reader of metropolitan writings, and in Bridge and Watson's (2002) city reader. These reprintings disseminate Young's ideas further within the subdisciplines of urban, feminist and social-cultural geography.

This widespread citation and excerpting of Young's writing bears testament to the fact that she has contributed to conceptualizations of space and place at a variety of scales – the body, the city, the region and the cosmopolitan global. In relation to the first, Young's work on a 'heterogeneous public' and her privileging of the city as the locus for imagining a 'politics of difference' helped open up models of social relations beyond those indebted to the Chicago School of sociology (with its frequent endorsement of social assimilation), and beyond Marxist accounts that seemed not adequately attentive to axes of difference aside from that of class. Young's reformulations of community and her validation of mediated urban relations resonated with geographers' desires to account for the complexity of city interactions, and the ways in which communities are fractured rather than organically unitary. Valentine has engaged Young's ideas in her work on lesbian communities and the 'fluid, multiple and constantly ... renegotiated and contested' structures of meaning that underpin them (Valentine, 1995: 109). Dunn has found Young's 'anti-assimilationist' approach towards group difference inspiringly progressive in his study of the concentration of Indo-Chinese Australians in Sydney (Dunn, 1998). Similarly, Fincher and Jacobs's edited collection

Cities of Difference (1998) shows the imprint of several of Young's formulations.

Young's work has also contributed to debates on the geography of the body. Young's 'five faces of oppression' (exploitation, marginalization, powerlessness, cultural imperialism and violence) moved away from a typology that alluded to named groups of 'sufferers' (as with the terms racism, ableism, heterosexism), and towards a more flexible matrix for understanding how people may be differently inhibited or constrained. In doing so, it garnered interest from geographers wanting to think the axes of gender, 'race', disability, sexuality and class together rather than simply alongside one another. **Gillian Rose** has used Young to elaborate upon female experiences of confinement in space (Rose, 1993: 146); Longhurst has incorporated Young's analyses of pregnancy and the production of 'ugly' bodies into her own examination of 'corporeographies' (Longhurst, 2001); McDowell has taken up the category of 'cultural imperialism' to help understand how women working in the financial industries in the City of London are positioned as 'inappropriate or vile bodies' (McDowell, 1997: 8); and Imrie has found the five faces of oppression a good 'heuristic device ... for situating the lived experiences of disabled people in some form of non essentialist structural context' (Imrie, 1996: 6). Nonetheless, Young's feminist reading of the ambivalences structuring the category of 'the home' (an essay included in *Throwing Like a Girl*) are perhaps less well known within human geography. Further, Young's theoretical formulations are far from static and exhibit great elasticity. Indeed, in her writing she frequently alerts the reader to how her thinking has changed. One elegant example is Young's auto-critique (1998b) nearly 20 years later of her seminal essay 'Throwing like a girl'.

In summary, Young's refocusing of questions of justice around the undermining of oppression and domination have been thought-provoking for geographers engaged in formulating just arrangements in material situations. Smith (1994), for example, endorses Young's approach in his interrogation of models of social justice, but argues that although her conceptions are appropriately spatial, they are essentially utopian. Varley (2002) has found Young's interrogation of Western thought's obsession with producing stable dualisms useful in challenging the problematic divide between 'legal' and 'illegal' that underpins debates over the occupation of land by the poor in urban Mexico. Most notably perhaps, David Harvey has directly engaged some of the theoretical implications of Young's work on justice.

KEY ADVANCES AND CONTROVERSIES

Justice and the Politics of Difference undoubtedly contributed to the setting of new agendas within social and cultural geography, feminist geography, political economy and urban studies. It did so by demonstrating the centrality of the term 'difference' for analyses of socio-spatial relations, and by providing new theoretical 'implements' (the 'heterogeneous public' and the 'five faces of oppression') through which the term 'difference' could be explicated. That Young's work takes seriously the challenges of a capitalist, urban landscape and of complex spatial-temporal positioning has endeared her to many geographers. At the same time, her work has been subject to the following critiques: (1) her account of difference rests on an inadequate account of the social group; (2) she does not adequately outline the mechanisms necessary for achieving the ideal of city life as an openness to unassimilated otherness (see Smith, 1994: 107); and (3) her formulations concerning difference end up exoticizing 'the other' by overemphasizing an attitude of 'wonder' *in the face of* difference at the expense of accounting

for shared experiences and desires *across* difference (Stavro, 2001). Below are detailed two controversies that expand upon these criticisms.

David Harvey has found Young's 'five faces of oppression' productive *vis-à-vis* his own work on justice. Though endorsing Young's critique of constrictive versions of communitarianism, he has criticized her ideal of openness to unassimilated otherness as being 'naively specified in relation to the actual dynamics of urbanization' (Harvey, 1996: 312). Furthermore, Harvey has countered her 'politics of difference' in the context of his own analysis of the kinds of *class-based* politics that he deems necessary to deal with such events as a horrific fire in 1991 in a US chicken-processing plant that killed 25 employees, many of whom were women and/or African-American (Harvey, 1993, 1996). To summarize, for Harvey a politics of difference threatens to deflect political attention away from combating the exploitation wrought by capitalist processes. Young in turn has criticized Harvey (1993, 1996) for reducing group-based social movements to identity politics, and has challenged his model according to which 'class' is able to offer a vision of commonality, whereas gender or 'race' are deemed to be mired in partiality (Young, 2000: 108; Young, 1998a).

Scholars in disciplines beyond geography have also criticized Young's formulation of 'the social group'. Some have suggested that her politics of difference cannot be clearly distinguished either from traditional interest group politics or from a simple endorsement of 'identity politics' (Squires, 2001). Mouffe (1992: 380) has argued that Young's model of the group is 'ultimately essentialist' since in Young's account, 'there are groups with their interests and identities already given, and politics is not about the construction of new identities'. Fraser (1995) has argued that Young implicitly validates a 'cultural' definition of a social group that elevates a model of an 'ethnic group' into a paradigm for *all* collectivities. This, Fraser (1995) argues, papers over the fact that there are *different* kinds of difference – and that while some should be enjoyed, others (for example 'groups' differently positioned within the capitalist division of labour) should be undermined or eliminated. Understanding collectivities on the model of the 'ethnic group' produces a strangely *American* conceptualization of the social group: '[w]here else but in the United States', Fraser (1995: 174) asks, 'does ethnicity so regularly eclipse class, nation and party'? Young (2000: 82) is adamant that her politics of difference is *not* equivalent to identity politics, arguing that 'groups do not have identities as such, but rather that individuals construct their own identities on the basis of social group positioning'.

Young, in her chapter on city life in *Justice and the Politics of Difference* (1990b), cited **Castells** and Harvey. However, the relays between Young's work and geographical scholarship went for some time largely in one direction: that is, geographers read, took up or critiqued Young's formulations. By the late 1990s that pattern had somewhat shifted: Young has not only addressed the specifics of David Harvey's arguments (Young, 1998a, 2000), but in *Inclusion and Democracy* (2000) cites the work of other geographers (for example, **Michael Storper** and Liz Bondi). However, geographers have, arguably, engaged less with this part of Young's *oeuvre*. The turn in human geography in the late 1990s towards performance and performativity perhaps made aspects of Young's work – her engagement with the legacies of liberalism and with deliberate democracy – less 'translatable' into the discipline of geography. Why Young's more phenomenological essays on embodiment, female motility and the home are not well assimilated into geography is a fascinating, and unanswered, question.

YOUNG'S MAJOR WORKS

DiQuinzio, P. and Young, I. M. (eds) (1997) *Feminist Ethics and Social Policy*. Bloomington: Indiana University Press.
Jaggar, A. M. and Young, I. M. (eds) (1998) *A Companion to Feminist Philosophy*. Malden, MA and Oxford: Blackwell.
Jeffner, A. and Young, I. M. (eds) (1989) *The Thinking Muse: Feminism and Modern French Philosophy*. Bloomington and Indianapolis: Indiana University Press.
Young, I. M. (1990a) 'The ideal of community and the politics of difference', in L. J. Nicholson (ed.) *Feminism/Postmodernism*. New York and London: Routledge, pp. 300–323.
Young, I. M. (1990b) *Justice and the Politics of Difference*. Princeton, NJ: Princeton University Press.
Young, I. M. (1990c) *Throwing Like a Girl and Other Essays in Feminist Philosophy and Social Theory*. Bloomington and Indianapolis: Indiana University Press.
Young, I. M. (1997) *Intersecting Voices: Dilemmas of Gender, Political Philosophy, and Policy*. Princeton, NJ: Princeton University Press.
Young, I. M. (1998a) 'Harvey's complaint with race and gender struggles: a critical response', *Justice, Nature and the Geography of Difference* Antipode 30: 36–42.
Young, I. M. (1998b) ' "Throwing like a girl": twenty years later', in D. Welton (ed.) *Body and Flesh: A Philosophical Reader*. Malden, MA and Oxford: Blackwell, pp. 286–290.
Young, I. M. (2000) *Inclusion and Democracy*. Oxford: Oxford University Press.

Secondary Sources and References

Bridge, G. and Watson, S. (eds) (2002) *The Blackwell City Reader*. Oxford: Blackwell.
Dunn, K. M. (1998) 'Rethinking ethnic concentration: the case off Cabramatta, Sydney', *Urban Studies* 35: 503–527.
Fincher, R. and Jacobs, J. M. (eds) (1998) *Cities of Difference*. New York and London: Guilford.

Fraser, N. (1995) 'Recognition or redistribution? A critical reading of Iris Young's *Justice and the Politics of Difference*', *Journal of Political Philosophy* 3: 166–180.

Harvey, D. (1989) *The Condition of Postmodernity*. Oxford: Blackwell.

Harvey, D. (1993) 'Class relations, social justice and the politics of difference', in M. Keith and S. Pile (eds) *Place and the Politics of Identity*. London and New York: Routledge, pp. 41–66.

Harvey, D. (1996) *Justice, Nature and the Geography of Difference*. Oxford: Blackwell.

Imrie, R. (1996) *Disability and the City: International Perspectives*. London: Paul Chapman.

Kasinitz, P. (ed.) (1994) *Metropolis: Center and Symbol of Our Times*. New York: New York University Press.

Keith, M. and Pile, S. (eds) (1993) *Place and the Politics of Identity*. London and New York: Routledge.

Longhurst, R. (2001) *Bodies: Exploring Fluid Boundaries*. London and New York: Routledge.

McDowell, L. (1997) *Capital Culture: Gender at Work in the City*. Oxford: Blackwell.

McDowell, L. and Sharp, J. P. (eds) (1997) *Space, Gender, Knowledge: Feminist Readings*. London: Arnold.

Mouffe, C. (1992) 'Feminism, citizenship, and radical democratic politics', in J. Butler and J. W. Scott (eds) *Feminists Theorize the Political*. New York and London: Routledge, pp. 369–384.

Rose, G. (1993) *Feminism and Geography*. Minneapolis: University of Minnesota Press.

Smith, D. M. (1994) *Geography and Social Justice*. Oxford: Blackwell.

Squires, J. (2001) 'Representing groups, deconstructing identities', *Feminist Theory* 2: 7–27.

Stavro, E. (2001) 'Working towards reciprocity: critical reflections on Seyla Benhabib and Iris Young', *Angelaki* 6: 137–148.

Valentine, G. (1995) 'Out and about: geographies of lesbian landscapes', *International Journal of Urban and Regional Research* 19: 96–111.

Varley, A. (2002) 'Private or public: debating the meaning of tenure legalization', *International Journal of Urban and Regional Research* 26: 449–461.

Felicity Callard

Glossary

This glossary provides brief definitions of key terms that are not fully explained in the context of individual entries. In each case, we refer to the principal entries where these concepts are referred to.

Actor-network theory: A theoretical approach that holds to the indivisibility of human and nonhuman agents, exploring the ways that different materials are enrolled in networks. Originally developed in debates about the production of scientific knowledge, actor-network theory (often abbreviated to ANT) opens up the 'black boxes' of action to explore the way that heterogeneous materials are continually assembled to allow actions to occur. See **Latour; Thrift**.

Behaviouralism: An outlook or system of thought that believes that human activity can best be explained by studying the human decision-making processes that shape that activity. Originally developed in psychology, largely as a reaction to the mechanistic excesses of experimental psychology, behaviouralism – and more particularly cognitive behaviouralism – came to prominence in the human geography of the 1960s and 1970s. Primarily based on methods of quantification, behavioural geography has been criticized for its adherence to positivist principles, as well as its unwillingness to explore the role of the unconscious mind, although it still underpins many research projects, particularly those based on survey research. See **Ley; Golledge; Hägerstrand**.

Capitalism: The political-economic system in which the organization of society is structured in relation to a mode of production that prioritizes the generation of profit for those who own the means of production. Such a structuring sees a clear division in status, wealth and living conditions between those few who own or control the means of production (bourgeois) and those that work for them (proletariat). See **Castells; Davis; Harvey; Lefebvre; Massey; Sayer; Smith; Storper; Wallerstein; Taylor; Watts**.

Consumption: In everyday use, the utilization of a good or a service until it has no value remaining. In the social sciences, consumption is deemed to be a more complex stage in a cycle of production and consumption, whereby a commodity may be transformed or even have value added. In contemporary capitalist societies, consumption is depicted as the driving force of the economy, and, given that commodity relations have penetrated into all spheres of social life, geographers have accordingly devoted increased attention to the spaces and rituals of consumption. See **Baudrillard; Bauman; Ley; Jackson**.

Corporeality: A recognition of the importance of the body/bodily, as opposed to simply the mind, in lived experience. See **Bourdieu; Butler; Young**.

Cultural politics: The study of the ways in which the cultural is political, and how social difference and conflict are bound up in complex power relations centred on categories of identity such as race, gender, disability, sexuality. See **Bhabha; Hall; hooks; Jackson; Young**.

Cultural turn: A trend in the late twentieth and early twenty-first centuries that has seen the social sciences and humanities increasingly focus on culture, and specifically the production, negotiation and contestation of meanings. In geography this has meant the integration of social and cultural theory into spatial analyses to create 'new' cultural geography, wherein culture is conceived as a process and the principal means through which society and space are constructed, providing people with their sense of identity at the same time that it maps out power-laden social and spatial hierarchies. See **Cosgrove; Jackson; Said; Sayer; Williams**.

Culture: Defined by Raymond Williams as the most disputed term in the English language, culture has been taken by geographers to refer to both the ways of life accruing to a particular group and the placed meanings of the material objects they produce and consume. See **Bourdieu; Hall; Jackson; Williams**.

Critical geography: Though diverse in its epistemology, ontology and methodology, and hence lacking a distinctive theoretical identity, critical geography nonetheless brings together those working with different approaches (e.g. Marxist, feminist, postcolonial, poststructural) through a shared commitment to expose the socio-spatial processes that (re)produce inequalities between people and places. In other words, critical geographers are united in general terms by their *ideological* stance and their desire to study and engender a more just world. This interest in studying and changing the social, cultural, economic or political relations that create unequal, uneven, unjust and exploitative geographies is manifest in engagements with questions of moral philosophy, social and environmental justice as well as attempts to bridge the divide between research and praxis. See **Harvey; Massey; Ó Tuathail; Sibley; Smith; Young**.

Deconstruction: A method of analysis that seeks to critique and destabilize apparently stable systems of meaning in discourses by illustrating their contradictions, paradoxes, and contingent nature. See **Hall; Harley; Haraway; Jackson; Spivak**.

Development studies: A broad field of analysis that examines processes of economic, political and social change and the mechanisms that maintain inequalities between regions and countries, with particular reference to divisions between the rich North and poorer South. See **Corbridge; Escobar; Sen; Taylor; Watts**.

Diaspora/disaporic: The scattering or dispersal of a population. It can also be used as a noun to refer to dispersed or scattered populations, such as the black diaspora or the Jewish diaspora. Because transnational linkages develop across diasporic communities, diaspora is also used as a theoretical concept to challenge fixed understandings of identity and place. See **hooks; Said; Spivak**.

Dialectics/al: A form of explanation and representation that emphasizes the resolution of binary oppositions. Rather than understanding the relationship between two elements as a one-way cause and effect, dialectical thinking understands them to be part of, and inherent in, each other. A dialectic approach has been an important part of structuralist accounts that seek to understand the interplay between individuals and society. See **Giddens; Harvey; Smith**.

Difference: Poststructuralist theory has emphasized the need to recognize the complexity of human social differences associated with culturally constructed notions of gender, race, sexuality, age, disability, etc. This means providing an analysis that is sensitive to the differences between individuals and avoids over-generalizations. See **Butler; hooks; Massey; Jackson; Young**.

Discourse: A set of ways of thinking about, speaking of and acting towards particular people or places. Emphasizing that language and thought are in, not outside, the world, discourse analysis has been important in developing poststructural theories. In human geography, discourse analysis has been widely used to expose the importance of representation in constructing stereotypes of place and nation. On occasion, this has required geographers interrogating their own processes of knowledge production. See **Barnes; Foucault; Gregory; Harley; Said; Spivak**.

Empiricism: A philosophy of science that emphasizes empirical observation over theory. In other words, it assumes that 'facts speak for themselves'. See **Massey; Watts**.

Ethnography: A qualitative mode of research and writing that emphasizes the importance of in-depth, contextual and intensive study in excavating the relationships between people and place. See **Davis; Ley**.

Feminism: A set of perspectives that seek to explore the way that gender relations are played out in favour of men rather than women. In human geography, such perspectives have suggested that space is crucial in the maintenance of patriarchy – the structure by which women are exploited in the private and public sphere. See **Haraway; hooks; Massey; Rose; Young**.

Fordism: A prevalent form of capitalism that was characterized by industrial and social practices pioneered within the factories of Henry Ford and which led to the creation in the West of mass production and mass consumption. See **Harvey; Massey**.

Genealogy: A method of creating a 'history of the present' by historically reconstructing the ways in which we produce knowledge that shapes social formations. Here, there is an emphasis on deconstructing at individual and local scales how society is constituted and how ideas became commonsense, taken-for-granted and universal. See **Foucault**.

Gentrification: The process by which working-class populations are displaced from an area or neighbourhood and replaced by wealthier populations. Often associated with material and aesthetic improvements in housing stock, debates concerning gentrification in rural and urban areas in the 1980s and 1990s provided a basis for exploring the way that economic, social and political processes entwine to create particular spatial arrangements. See **Bourdieu; Ley; Smith**.

Geopolitics: The analysis of how the geographical is bound up in the theory and practice of statecraft and international relations. More recently, the development of critical geopolitics has focused on the discourses and practices that underpin statecraft and shape the relations between territories and nations. See **Ó Tuathail; Taylor**.

Globalization: The suite of economic, social and political processes that are enabling connections to be made between people and places on a worldwide scale. Implicated in a process of time–space compression, there are nonetheless many arguments about whether globalization is leading to increasing homogenization or differentiation of space. Human geographers are accordingly exploring both the causes and consequences of these processes, developing new concepts of hybridity and transnationalism as they seek to develop a language and conceptual toolkit that is adequate for these 'global times'. See **Anderson; Bauman; Castells; Dicken; Giddens; Harvey; Taylor; Virilio; Wallerstein; Watts**.

Governance: Governance does not simply refer to the practice of governments. Rather it relates to the acts and processes of governing more generally. In its broadest sense, the study of governance is therefore interested in how individuals, organizations and states govern and regulate both others and themselves. See **Foucault; Ó Tuathail; Taylor**.

Hegemony: Hegemony generally refers to a situation of uncontested supremacy. The term, under the influence of Gramsci, has been reworked in the social sciences, however, to refer to the power of a dominant group to persuade subordinate groups to accept its moral, political and cultural values as the 'natural' order; as desirable, inevitable, and taken-for-granted. See **Bhabha; Corbridge; Jackson; Taylor; Wallerstein; Williams**.

Humanistic geography: An approach to understanding human geography that focuses on the creativity of human beings to shape their world and create meaningful places. Focusing on human consciousness as the basis of being in the world, humanistic geography challenged the ideas of spatial science, behavioural geography and structural approaches when it came to prominence in the 1970s. Methodologically, humanistic geographers pioneered qualitative techniques that are now widespread across the discipline, and continue to highlight issues of subjectivity in research. See **Ley; Tuan**.

Hybridity: In the social sciences, this is a term that is used to describe the results of a meeting of two apparently distinctive cultures. In human geography, there has been significant interest in the hybrid cultures and ways of life that occur on the boundaries between different nations, as well as the hybrid cultures that emerge as international migration creates cities typified by multiculturalism. There is also widespread geographical interest in the hybridity associated with the blurring of culture/nature. See **Bhabha; Haraway; Spivak**.

Identity: Early academic understandings of identity conceptualized them in terms of coherent, social categories such as gender and class, in which identity was assumed to reflect a core or fixed sense of self. More contemporary theorizations understand identities as a reflexive projects, emphasizing their multiple, fluid and contested nature. See **Butler, hooks; Jackson; Williams**.

Ideology: In general terms, ideology refers to a coherent set of meanings, ideas or values that underpin a particular course of action. In relation to research praxis, ideology concerns the underlying social/political reasons or purpose for seeking knowledge. Some philosophies posit that the production of knowledge should be an ideologically neutral activity. This view implies that it is the job of a theorist to develop ideas about how the world works and it is for others to decide how to use the ideas they uncover. Others posit that it is impossible to be ideologically neutral, and that, whether we like it or not, it is impossible to isolate personal social and political beliefs from wider theorization. On this basis, they argue that academics should use their theories to try and change the world for the better rather than leave it to others to interpret their ideas. See **Hall; Haraway; Latour; Williams**.

Landscape: As with space and place, landscape is a key concept in human geography yet has different connotations for different commentators. For some, landscape is an area that can be mapped and explored for the traces of its making; for others, it is a distinctive way of representing and making space that involves the imbrication of knowledge and power. See **Cosgrove; Harley; Rose**.

Localism: The study of local places and communities as sites (or laboratories) of social and economic relations and how such sites relate to, are affected and affect other locations. It is often criticized as ignoring the wider picture of how a specific location is situated in relation to a broad economic and political landscape. See **Massey; Sayer; Smith**.

Locational analysis: A mode of analysis that seeks to develop spatial laws and models that explain the distribution and flow of phenomena and activities across the Earth's surface. See **Berry; Hägerstrand; Haggett; Tobler**.

Marxism: A broad perspective that holds to the primacy of capitalist relations in structuring social and economic life. Based on the ideas of Karl Marx, Marxism has been responsible for notable examples of political radicalism within geography, as well as the development of a politically motivated Marxist human geography that seeks to expose the injustices wrought by capitalism. See **Davis; Harvey; Lefebvre; Massey; Smith; Soja; Taylor; Wallerstein**.

Mode of production: A Marxist-derived term that denotes the way that relations of production are organized in specific periods. Currently, it is accepted that the world is organized so as to reproduce and maintain a capitalist mode of production, though it is emphasized that feudal, socialist and communist modes of production have been (and in some cases, still remain) dominant in some nations. See **Lefebvre; Taylor; Wallerstein**.

Modelling: A method of analysis that seeks to create a realistic model of how a system works or to predict outcomes given certain parameters. It is often used extensively in locational analysis and now most often performed using a Geographic Information System. See **Berry; Haggett; Tobler**.

Modernity: A period associated with the West from the eighteenth century onwards characterized by the reorganization of society through a combination of the development of a capitalist economy, the political reorganization associated with nation-states, and the pre-eminence of cultural values such as rationality and progress arising from philosophy of Enlightenment. This gave rise to a particular social order that remained dominant in the West until the late twentieth century. See **Bauman; Beck; Giddens; Gregory; Taylor**.

Multiculturalism: A view that recognizes that different cultural or ethnic groups have a right to assert their own distinct identities rather than be assimilated into mainstream norms. See **Bhabha; Hall; hooks; Said; Watts**.

Nationalism: A political ideology that promotes among a nation of people shared, communal feelings of belonging and ownership to a particular, bounded territory. See **Anderson; Ó Tuathail**.

Neo-liberalism: Ideological project of the New Right that has grown in strength since the 1980s in the West. Neo-liberalism positions the free market, rather than the state, as the central organizing feature of society. Such a positioning has had a profound effect on social formations and economic and political policy, particularly with regards to welfare and oversees development and debt. See **Corbridge; Davis; Harvey; Smith**.

Nonrepresentational theory: A theory that seeks to move the emphasis of analysis from representation and interpretation to practice and mobility. Emphasis is placed on studying processes of becoming, recognizing that the world is always in the making, and that such becoming is not always discursively formed (framed within, or arising out of, discourse). Here, society consists of set of heterogeneous actants who produce space and time through embodied action that often lacks reason and purpose. To understand how the world is becoming involves observant participation; a self-directed analysis of how people interact and produce space through their movement and practice. See **Thrift**.

Objectivity: Used within the social sciences, objectivity refers to conducting analysis in a supposedly non-personal, disinterested fashion that makes the findings of research neutral and value-free. The pursuit of objective analysis is a central feature of much quantitative geography, particularly that which is positivist in formulation. See **Berry; Bourdieu; Haggett; Haraway**.

Other/otherness: The 'Other' refers to the person that is different or opposite to the self. 'Othering' is the process through which the Other is often defined in relation to the self in negative ways; for example, woman is often constructed as other to man, black as other to white. See **Bhabha; Hall; hooks; Jackson; Said; Sibley; Spivak**.

Paradigm: The assumptions and ideas that define a particular way of thinking about and undertaking research that become the dominant way of theorizing a discipline over a period of time until challenged and replaced by a new paradigm. See **Barnes; Ley; Sayer**.

Patriarchy/patriarchal: Literally means 'rule of the father'. This term is used to describe a social system and practices whereby men dominate women. These unequal relationships can occur at a range of scales from the household to society as a whole. See **Butler; hooks; Rose; Massey**.

Performativity: Developed within literary theory, performativity is a rhetorical device for thinking about the way in which identity is produced in non-foundational and non-essentialist ways. Most commonly associated with the writings of Judith Butler, it is argued that identities are continuously brought into being through their performance. In relation to gender, what it means to be a woman or man is produced and sustained through acts, gestures, mannerisms, clothing and so on. In other words, there are no 'natural' gendered identities; gender is what people *do* and by performing gender people reproduce the notion of what gender constitutes. See **Butler; Rose; Thrift**.

Political ecology: An approach that entwines Marxism with cultural ecology to account for the diverse ways in which nature was produced and exploited for capital gain in an era of flexible accumulation. Here, there is a recognition that resource management and environmental regulation and stability are wedded to how communities are integrated into a global economy. In other words, the pressure brought to bear on ecosystems is often due to economic and political pressures rather than simply mismanagement or overpopulation. See **Escobar; Watts**.

Political economy: Theoretical approaches that stress the importance of the political organization of economic reproduction in structuring social, economic and political life. Associated in human geography with the influence of Marxist thinking, political economy perspectives in fact encompass a variety of approaches that explore the workings of market economies. See **Harvey; Massey; Sayer; Smith; Taylor; Wallerstein; Watts**.

Positivism: A philosophy that suggests that universal laws and explanations can be constructed through repeated scientific observations and measurements. Though different packages of positivist thought argue that the burden of scientific proof rests with the researcher in different ways, all hold to the ideal of the researcher as objective and independent. In human geography's attempt to recast itself as a spatial science, the influence of positivism was paramount, and the tenets of positivism still arguably inform the majority of geographical investigation. See **Berry; Haggett; Harvey**.

Postcolonialism: A set of approaches that seek to expose the ongoing legacy of the colonial era for those nations that were subject to occupation by white, European colonizers. Emphasizing both the material and symbolic effects of colonialism, postcolonial perspectives are particularly concerned with the ways that notions of inferiority and Otherness are mapped onto to the global South by the North, though postcolonial perspectives have also been utilized to explore the race and ethnic relations played out on different scales. See **Bhabha; Gregory; hooks; Ó Tuathail; Said; Spivak**.

Postmodernism: Often used to denote playful and self-referential styles of art, architecture and literature, postmodernism is a term that captures both the logic of a particular epoch (i.e. late capitalism) and the methods required to make sense of this era. Associated with a breakdown of order, rationality and assured progress, postmodern times are often taken to be typified by fluidity and flexibility; accordingly postmodern theorists argue that grand theories such as feminism, Marxism or positivism are no longer appropriate for exploring social life. Instead, postmodern thinkers seek to embrace difference and fluidity, adopting methods and writing styles that are in tune with postmodern times. See **Baudrillard; Davis; Dear; Harvey; Ley; Soja**.

Poststructuralism: A broad set of theoretical positions that problematizes the role of language in the construction of knowledge. Contrary to structural approaches, which see the world as constructed through fixed forms of language, such approaches emphasize the slipperiness of language and the instability of text. A wide-ranging set of assertions follow from this key argument, including the assertion that subjects are made through language; the idea that life is essentially unstable, and only given stability through language; the irrelevance of distinctions between real and simulcra, ultimately, that there is nothing 'beyond the text'. In human geography, poststructural thought has provoked attempts to deconstruct a wide variety of texts (including maps) and has encouraged geographers to reject totalizing and foundationalist discourses (especially those associated with structural Marxism). See **Baudrillard; Butler; Deleuze; Foucault**.

Praxis: Praxis is how theoretical ideas are translated into the world. In relation to academia, this is primarily through research, teaching, discussion and debate. See **Haraway; Harvey; Rose**.

Producer services: Those 'knowledge-rich' businesses that provide services and expertise to corporate and governmental clients, particular those transnational corporations whose activities require careful coordination across time and space. Examples include accountancies, lawyers, advertising and promotions agencies, and management consultancies, all of which tend to cluster in world cities so as to be able to service their global clients. See **Castells; Dicken; Thrift; Taylor**.

Production: The processes by which commodities are bought into being. In human geography, production has traditionally been the remit of economic geographers, who explore how space is a formative influence on, and not just an outcome of, economic activity. Latterly, an increasing focus on consumption has questioned this analytical specialization, and has encouraged study of the way consumption and production are linked through commodity chains that extend through space in complex ways. See **Barnes; Dicken; Sayer; Stoper; Thrift**.

Psychoanalysis: Narrowly defined as a mode of therapy that seeks to resolve patients' neuroses by exploring the contents of the unconscious mind. This form of therapy involves attempts to encourage people to reconcile the unconscious and conscious mind, leading to a fuller knowledge of Self. More widely, the ideas of leading psychoanalysts such as Freud, Jung and Lacan have been interpreted as providing theories of socialization that can be used to understand how people are constructed as (repressed) subjects. In human geography, the theories – and occasionally, the practices – of psychoanalysis have been taken up in various ways, primarily by geographers exploring the relations of sexuality and space. However, the relationship between feminism and psychoanalysis remains strained in many respects. See **Rose; Sibley**.

Qualitative method: In human geography, this denotes those methods that accept words and text as a legitimate form of data, including discourse analysis, ethnography, interviewing and numerous methods of visual analysis. Mainly tracing their roots to the arts and humanities, such methods have often been depicted as 'soft' methods, and hence described as feminist in orientation. Latterly, however, such simplistic assertions have been dismissed, and such methods proliferate across the discipline in areas including economic and political geography. See **Barnes; Ó Tuathail; Rose**.

Quantitative method: In human geography, this denotes those methods that prioritize numerical data, including survey techniques, use of secondary statistics, numerous forms of experimentation and many forms of content analysis. Mainly derived from the natural sciences, such methods are often depicted as 'hard' methods, deriving their analytical rigour and validity by association with masculine modes of science and exploration. However, numerous critiques have exposed the subjectivity of quantitative methods, and suggested that its techniques cannot be understood as objective ways of looking at and understanding the world. This has led to a reappraisal of quantitative method in areas of the discipline such as social and cultural geography where quantitative method has long been anathema. See **Barnes; Berry; Hägerstrand; Haggett**.

Queer theory: A set of approaches and practices that expose and challenge the implicit and explicit heterosexism of academic knowledge production. Sometimes allied to feminist approaches, but sometimes in sharp opposition to them, such approaches demonstrate how homophobia is embedded in both the spaces of academia and the spaces of everyday life. This approach is often associated with confrontational writing styles and modes of research that emphasize the subjectivity of the researcher. See **Butler**.

Realism: A theoretical perspective that seeks to transcend many of the problems associated with positivism and structuralism by seeking to isolate the causal properties of things that cause other things to happen in given situations. Based on a methodological distinction between extensive and intensive research, this approach was widely embraced in human geography in the 1980s as a way of distinguishing between spurious associations and meaningful relations. See **Giddens; Massey; Sayer**.

Regional geography: The study of the various phenomena – social, political, economic, cultural – that make up a large area that is identifiably distinguished from surrounding regions. See **Massey; Thrift**.

Representation: A set of media or practices through which meaning is communicated. The study of representation has become a common way for researchers to examine sense of place and how people understand, construct and relate to socio-spatial relations. See **Cosgrove; Jackson; Hall; Harley; Ó Tuathail; Spivak**.

Sense of place: A central concept in humanistic geography, intended to describe the particular ways in which human beings invest their surroundings with meaning. Sense of place is seen be an elusive concept, yet human geographers seek to find its traces in a variety of texts and representations, including paintings, poetry, prose and cinema. See **Ley; Tuan**.

Situated knowledge: In a challenge to objectivity, a situated knowledge is one where theorization and empirical research are framed within the context within which they were formulated. Here it is posited that knowledge is not simply 'out there' waiting to be collected but is rather made by actors who are situated within particular contexts. Research is not a neutral or objective activity but is shaped by a host of influences ranging from personal beliefs to the culture of academia, to the conditions of funding to individual relationships between researcher and researched, and so on. This situatedness of knowledge production needs to be reflexively documented to allow other researchers to understand the positionality of the researcher and the findings of a study. See **Haraway; Rose**.

Social justice: A set of normative approaches concerned with the fair and equitable distribution of things that people care about such as work, wealth, food and housing, plus less tangible phenomena such as systems of power and pathways of opportunity. See **Harvey; Sen; Smith; Young**.

Spatiality: A term that refers to how space and social relations are made through each other. That is, how space is made through social relations, and how social relations are shaped by the space in which they occur. See **Lefebvre; Soja**.

Spatial science: An approach to understanding human geography that holds to the idea that there can be a search for general laws that will explain the distribution of human activity across the world's surface. Associated with the precepts of positivism, and mainly reliant on quantitative method, spatial science signalled geography's transition from an atheoretical discipline to one concerned with explanation rather than mere description. Emerging in the 1950s, and bolstered by the quantitative revolution of the 1960s, spatial science continues to be dominant in many areas of the discipline, though in others its philosophical underpinnings and theoretical conceits have long discredited it. See **Barnes; Berry; Hägerstrand; Haggett; Tobler**.

Structuralism: A theoretical approach that suggests that life is structured by 'deep' political, economic and social structures (such as capitalism) whose existence cannot be directly observed. This approach is particularly associated with linguistics, where structuralist thinkers argue that the meaning of language cannot be understood through the analysis of the use of individual words and phrases, but through the questioning of language as a system. Methodologically, structuralism relies on the collection of data that might be used to prove the existence of such structures: these data might give clues as to the relations between constituent elements of the system (e.g. language, practices, behaviours) and, through a process of dialectic reasoning, seek to expose their opposite and contradictory existence within that system. In human geography, Marx's ideas have been fundamental in the development of geographical accounts that highlight the role of space in perpetuating capitalist structures. See **Harvey; Massey; Smith**.

Subaltern: The figure of the subaltern refers to the collective agency of exploited, oppressed and marginalized groups. See **hooks; Spivak**.

Territoriality: The strategies adopted by a set of people to claim and govern a particular unit of land and its contents. See **Ó Tuathail**.

Time geography. An approach to the study of space that prioritizes the effects of time and temporality. Here there is an emphasis on understanding the effects of how events are sequenced in time and space, including for example, patterns of home-to-work journeys and the diffusion of diseases. See **Hägerstrand; Thrift**.

Time–space convergence: The speeding-up of the pace of life due to increasing time–space compression (shrinking of space by time) and time–space distantiation (the interconnectedness and interdependence of people and places across the globe). See **Castells; Giddens; Harvey; Massey; Virilio**.

Transnationalism: The ongoing economic, social, cultural and political links developed and maintained by migrants and trade across the borders of nation-states. See **Jackson**.

World cities: Those cities in which a disproportionate amount of the world's business is carried out. Measurements of world city importance are thus normally based on numbers of company headquarters, business transactions and size of capital markets rather than cultural vitality, population size or infrastructure (meaning that London, New York and Tokyo tend to be described as world cities, but larger cities, including Tehran, Dhaka or Khartoum, are not). See **Castells; Taylor; Thrift**.

World-systems theory: Perspectives that argue that social, economic and political analysis cannot proceed on a nation-by-nation basis, but needs to explore the way the world works as an integrated (capitalist) unit. Such perspectives have been deemed to have particular relevance in an era when globalization is seemingly undermining the sovereignty of the nation-state. See **Dicken; Taylor; Wallerstein**.

Index

action research 3
actor-network theory 9, 24, 133, 171, 202–205, 298, 344
Adorno, T. 33, 98
Agnew, J. 79, 81, 226, 227
Ahmad, A. 242
Aitken, S. 261
Allen, J. 111, 221, 248
Alonso, W. 48
Althusser, L. 29, 69, 72, 73, 161, 162, 163, 165, 211, 270, 324, 335
Amin, A. 111, 296
Anderson, B. **16–21**
Anzuldua, G. 193, 272
applied geography 3
Arendt, H. 33
Augé, M. 8
Auster, P. 105
Austin, J. 65

Badiou, A. 102
Barnes, T. 7, **22–26**, 168, 277, 285
Barthes, R. 28, 312
Bataille, G. 28, 103, 125
Baudrillard, J. 2, 8, 10, 12, **27–30**, 33, 103, 106, 205, 311, 313, 314
Bauman, Z. 8, **33–39**
Beauregard, R. 99, 100
Bebbington, A. 117, 118, 254
Beck, U. 2, 38, **40–44**
Bell, D. 67, 68
behavioural geography 4, 10, 129, 136–140, 149, 151, 215, 304, 344
Benjamin, W. 91, 92, 145
Beresford, M. 85
Berger, J. 270
Bergson, H. 104
Berkeley School 6, 12, 85, 151, 324
Berman, M. 92, 270
Bernstein, B. 258, 260
Berry, B. 2, 4, 10, **47–51**, 137, 301
Bhabha, H.K. **52–56**, 233, 239, 272
biographical approach 11–12
Blaikie, P. 118, 151, 254, 255, 326
Blaut, J. 19
Bondi, L. 261, 342
Bourdieu, P. 6, 11, 33, **59–64**
Bowman, I. 265
Brenner, N. 267
Brenner, R. 321
Bridge, G. 62, 340
Brookfield, H. 326
Browett, J. 267
Brown, M. 69
Bulkeley, H. 45
Bunge, W. 214, 301, 303
Butler, J. 6, 8, 10, 12, 53, **65–71**, 231, 233, 234
Butler, T. 61–62
Buttimer, A. 5, 152, 306

Callard, F. 67
Callon, M. 171, 202
capital 7, 8, 9, 37, 43, 59, 61, 63, 74, 76, 102, 109, 117, 118, 215, 278
capitalism 5, 7, 12, 16, 19, 20, 29, 34, 41, 42, 49, 50, 56, 73, 76, 80, 90, 91, 93, 98, 115, 116, 118, 130, 145, 147, 158, 162, 169, 177, 182, 184, 185, 186, 197, 209, 210, 212, 220, 242, 256, 266, 270, 282, 285, 289, 290, 296, 297, 318, 319, 320, 321, 325, 333, 344
cartography 3, 7, 10, 12, 18, 85, 86, 87, 174–179, 301–305
Cassirer, E. 86
Castells, M. 2, 5, 8, 12, **72–77**, 82, 92, 115, 183, 209, 210, 211, 249, 270, 291, 325, 342
Castoriadis, C. 33
Castree, N. 185, 278
Chari, S. 81–2
Chatterjee, P. 19–20, 79
Chicago School 48, 72, 74, 98, 100, 195, 215, 340
children 258, 260–262
Chorley, R. 109, 155, 156
Christaller, W. 24, 109, 149, 156, 301
Christopherson, S. 283
citationality 66
citizenship 131
Cixious, H. 106
Clark, G. 48, 111, 292
class 6, 7, 18, 20, 33, 43, 49, 56, 59, 60, 61, 62, 73, 81, 91, 92, 93, 98, 123, 130, 160, 161, 162, 186, 197, 204, 220, 223, 229, 252, 276, 278, 289, 290, 294, 318, 321, 330, 332, 333, 335
Cleveland, D. 116, 118
climate change 41, 45
Cloke, P. 17, 171
cognitive map 136, 303
cognitive spacing 36
colonialism 18, 19–20, 52–57, 59, 78, 114, 118, 146, 147, 160, 175, 176, 238, 239, 242, 255, 277, 279, 318, 323
community 33, 43, 44, 54, 62, 171, 190, 216, 223, 338, 340
consumption 28, 29, 30, 33, 35, 36, 43, 60, 74, 117, 183, 186, 187, 195, 198, 199, 200, 208, 210, 217, 271, 283, 285, 290, 297, 344
Conzen, M. 85
Cooke, P. 248
Corbridge, S. 5, **78–83**, 115, 253, 255
corporations 8, 9, 41, 109, 110, 198
Cosgrove, D. 6, **84–89**, 176, 177, 178, 196, 197, 304, 332
Cox, K. 136, 137, 139, 186

Crang, M. 2, 17, 76
Crang, P. 67, 198, 286, 332
Cream, J. 67
Cresswell, T. 61, 163
critical realism 245, 246, 248
Crozier, M. 37
cultural capital 60
cultural politics 163, 189, 190, 196, 198, 199, 217, 231, 324, 344
culture 7, 9, 17, 25, 45, 52, 53, 59, 86, 161, 164, 167, 168, 169, 171, 195, 196, 202, 204, 231, 253, 306, 307, 324, 331, 333, 345
Curry, L. 151
cyborg 167–69, 171, 172

Dacey, M. 151, 301
Daniels, S. 85, 87, 176, 177, 178, 197, 332, 333
Darby, H. 174
Davis, M. 7, 12, **90–95**, 97, 248
de Beauvoir, S. 65, 338
de la Blache, V. 145, 152
de Man 275, 276
Dear, M. 7, 10, 12, 50, 90, **96–101**, 248, 272, 283, 292
Debord, G. 92, 209, 210
deconstruction 3, 7, 12, 53, 69, 169, 172, 174–176, 229, 275–279, 338, 345
Deleuze, G. 9, 10, 12, 68, **102–107**, 205, 311, 314, 315
demography 50
Derrida, J. 52, 63, 103, 104, 106, 175, 176, 177, 178, 204, 227, 275, 276, 278, 311, 338
development 8, 43, 54, 78–82, 113–118, 251–256, 289, 345
diaspora 54, 160, 163, 164
Dicken, P. 8, **108–112**
difference 7, 9, 29, 36, 44–45, 61, 103, 105, 118, 191, 193, 222, 242, 247, 339, 340, 341, 345
diffusion 149, 150, 151, 156, 157, 266
disability 138, 140
discourse 6, 9, 27, 53, 55, 56, 66, 114, 116, 122, 123, 133, 146, 147, 161, 163, 169, 171, 197, 217, 226, 227, **228**, 240, 295, 297, 298, 333, 345
Doel, M. 29, 75, 172
Domosh, M. 84, 86
Douglas, M. 33, 258, 259
Dovey, K. 61
Dowling, R. 62
Drèze, J. 252, 253, 254
Driver, F. 12, 87, 126, 176, 177
Dubois, W.E.B. 260
Duncan, J. 87, 94, 131, 196, 197, 216, 239, 277
Duncan, N. 87, 192

Dunn, K. 340
Durkheim, E. 129
Dwyer, C. 198

Economy 7, 8, 9, 22, 25, 37, 43, 50, 75,
     76, 108, 110, 220, 223, 252, 254,
     266, 284, 288, 295, 296, 324
Elias, N. 33
Elden, S. 124, 126
embodiment 5, 6, 9, 59, 60, 62, 63,
     65–69, 132, 165, 169, 192, 211,
     212, 229, 234, 296, 298, 315, 340
Enlightenment 4, 24, 29, 45, 84, 88,
     106, 144, 147, 203
Entrikin, J.N. 308
Escobar, A. 12, 79, 82, **113–120**, 253
ethics 35, 43, 44, 126, 171, 172, 212,
     251, 255, 307, 337
ethnography 117, 198, 199, 325, 346
exchange value 28, 29
existentialism 5, 59, 121, 208, 307, 309
Eyles, J. 139

famine 251, 253, 254, 255, 323
Fanon, F. 52, 53, 190, 193
fatal theory 27–30
feminism 3, 6–7, 10, 12, 20, 24, 44,
     65–69, 86, 145, 158, 161, 167, 170,
     186, 189–194, 197, 211, 223, 227,
     231–235, 258, 261, 272, 275–277,
     279, 309, 337, 338, 339, 346
flâneur 36
Flusty, S. 50, 97, 98, 99
Foucault, M. 2, 8, 10, 12, 27, 28, 33,
     40, 53, 61, 62, 69, 79, 104, 106,
     113, **121–128**, 146, 175, 176, 177,
     178, 227, 239, 242, 270, 278, 311,
     338
Freire, P. 190
Friedmann, J. 291, 320
Freud, S. 30, 34, 69, 105, 261

Garrison, W. 301
Geertz, C. 86
gender 6, 8, 11, 20, 41, 56, 60, 65–69,
     123, 152, 168, 169, 170, 186,
     189–194, 197, 204, 220, 221, 227,
     229, 232, 242, 253, 260, 276, 278,
     291, 335, 337, 338
genealogy 10–12, 29, 123, 126, 346
gentrification 61, 62, 63, 214, 216, 217,
     264, 266, 346
geographic information systems 3,
     137, 158, 177, 302, 303
geographical imagination 6, 44, 78,
     144, 185, 235, 238
geopolitics 5, 7, 56, 78, 79, 81, 82,
     226–229, 265, 288, 292, 314, 346
Gertler, M. 111, 285
Gibson-Graham, J.K. 24, 277, 285
Giddens, A. 2, 37, 82, **129–135**, 144,
     152, 248, 255, 270, 295, 297
Gilmore, R. 193
Gleeson, B. 4, 140
globalization 3, 8, 9, 35, 37, 38, 40–44,
     59, 73, 75, 76, 79, 82, 98, 108, 110,
     131–133, 161, 200, 208, 211, 214,
     215, 222, 228, 253, 273, 277, 283,
     284, 290, 294, 295, 296, 309, 320,
     324, 325, 327, 346

glocalization 37, 291
Gold, J.137, 139, 140
Golledge, R. 2, 4, 10, 47, **136–142**,
     156, 301, 304
Gottdiener, M. 210
Gould, P. 47, 136
governance 73, 77, 123, 126, 216, 252,
     255, 288, 326, 327, 346
governmentality 61, 126, 227
Gramsci, A. 7, 33, 79, 132, 161, 196,
     243, 290
Gregory, D. 8, 77, 87, 93, 133,
     143–148, 151, 157, 168, 177, 192,
     211, 222, 239, 242, 248, 270, 272,
     276, 277, 279
Gregson, N. 68, 133, 231, 234
Guattari, F. 68, 102, 104, 105, 205, 311

Habermas, J. 33, 132, 146, 338, 339
habitus 60–63
Hägerstrand, T.4, 10, 132, **149–154**,
     156, 302
Haggett, P. 2, 4, 10, 47, 49, 109, 137,
     151, **155–159**, 258, 303, 304
Hall, P. 155
Hall, S. 2, 12, **160–166**, 191, 196, 221,
     331, 332, 335
Halliday, F. 82
Hamnett, C. 216, 217
Hanson, S. 24, 193, 327
Haraway, D. 3, 11, 44, 145, **167–173**
Harley, B. 7, 12, 88, **174–180**
Harriss, J. 80, 81, 82
Hartshorne, R. 155
Harvey, D. 5, 9, 10, 12, 25, 47, 49, 60,
     72, 81, 108, 109, 132, 144, 145,
     146, 156, 157, **181–188**, 209, 214,
     217, 222, 223, 247, 249, 264, 266,
     267, 270, 282, 304, 313, 334, 335,
     339, 341
Haushofer, K. 227
Hayter, R. 22
healthcare 96
Heffernan, M. 4, 229
hegemony 7, 37, 55, 56, 79, 81, 82,
     114, 118, 160, 162, 196, 229, 229,
     239, 243, 245, 259, 276, 279, 290,
     291, 332, 346
Heidegger, M. 12, 124, 270, 307
Held, D. 82
Herod, A. 187, 267
Hinchliffe, S. 45, 203, 204
Hoeschler, S. 20
Hoggart, R. 160, 332
homelessness 96, 210
hooks, b. 6, 11, 56, **189–194**, 211, 272
Hoskins, W. 85, 87
Houston, D. 69
Hubbard, P. 2, 5, 6, 11, 235, 261
humanistic geography 5, 8, 10, 84, 85,
     129, 131, 138, 152, 158, 214, 215,
     216, 217, 304, 306–309, 346
Husserl, E. 145
Hume, D. 104
hybridity 8, 9, 44, 52, 54, 55, 56, 115,
     116, 118, 163, 167, 168, 169, 172,
     204, 224, 325, 346
hyper-reality 30–31

iconography 85, 86

identity 7, 9, 48, 54, 55, 61, 65, 66, 67,
     73, 75, 103, 105, 117, 131, 146,
     160–62, 164, 165, 178, 190–93,
     195, 196, 199, 200, 217, 221, 241,
     247, 258, 261, 262, 276, 294, 297,
     298, 325, 332, 339, 341, 347
ideology 29, 34, 56, 85, 117, 130, 146,
     161, 162, 163, 176, 177, 178, 190,
     191, 195–200, 220, 239, 242, 289,
     291, 292, 325, 347
imagined community 16–20
Imrie, R. 140, 340
individualism 35, 41–43
Innis, H. 22
Irigaray, L. 205, 233, 234, 261, 338,
     339
Isard, W. 109, 156
interpolation 302

Jackson, J.B. 85, 87, 307
Jackson, P. 7, 9, 17, 87, 100, 163, 187,
     195–201, 214, 215, 332
Jacobs, J. 56, 57
Jameson, F. 145, 270
Jewitt, S. 81
Johns, M. 81
Johnston, L. 67
Johnston, R. 10, 11, 288

Kant, I. 104, 145
Katz, C. 232, 277
Keith, M. 192, 193, 340
King, L. 47, 136
Kirby, K. 67
Kitchin, R. 139, 140
Klein, M. 8, 259, 261
knowledge
     production of 11, 12, 23–25, 87,
          121–125, 146, 167–172, 190, 192,
          235, 241, 260, 265, 276–77, 296,
          298
     situated 11, 142, 168–170, 172, 231,
          337, 351
Kobayashi, A. 191
Kondratiev cycles 49, 318
Kristeva, J. 8, 106, 205, 259, 261, 338
Krugman, P. 48, 111
Kuhn, T. 10
Kumar, S. 80

LA School 7, 12, 90, 96, 97, 248, 269
Lacan, J. 28, 52, 69, 205, 261, 338
landscape 3, 6–7, 8, 61, 84–88, 175,
     178, 196, 212, 217, 231, 235, 307,
     308, 347
language 7, 8, 9, 20, 24, 55, 66, 67, 69,
     116, 146, 165, 197, 226, 276, 295,
     333
Lash, S. 41
late-modernity 40–45, 131
Latour, B. 3, 9, 23, 24, 44, **202–207**,
     298
Law, J. 171, 202
Law, L. 56–57
Leach, M. 61
Lefebvre, H. 5, 6, 9, 12, 27, 28, 30, 56,
     60, 62, 72, 74, 79, 92, 132,
     208–213, 261, 264, 270, 272
Leibnez, G. 104
Levinas, E. 33

Levi-Strauss, C. 33
Ley, D. 5, 10, 131, 138, 139, 158, **214–218**
Leyson, A. 133, 296
Lipietz, A. 81, 282
Livingstone, D. 2, 3, 4, 87, 177
Lloyd, P. 109, 110
locality studies 5, 9, 24, 219, 223, 248–249, 267, 347
locational analysis 23, 109, 143, 151, 155, 156, 347
Longhurst, R. 340
Losch, A. 109, 156
Lowenthal, D. 5, 138, 306
Lucas, T. 262
Lyotard, J-F. 33, 34, 103, 106, 271, 311

Mackinder, H. 227
Mahtani, M. 69, 232, 235
Malinowski, B. 104
managerialism 35, 75
Mandel, E. 270
Marinetti, F. 312
Martin, A. 19, 20
Martin, R. 111
Marx, K. 12, 28, 29, 33, 104, 109, 121, 129, 160, 182, 184, 208, 211, 255, 264, 278, 318, 321, 332, 338
Marxism 1, 4, 5, 6, 10, 16–17, 19, 22, 24, 27, 28, 30, 49, 56, 62, 72, 75, 79, 81, 92, 94, 98, 105, 117, 118, 121, 131, 132, 133, 143, 144, 146, 158, 182–83, 186, 195, 208, 211, 214, 215, 216, 222, 223, 239, 242, 245, 246, 255, 264, 268, 272, 278, 282, 290, 294, 318, 323, 331, 334, 339, 347
masculinity 195, 197, 198, 199
Massey, D. 2, 5, 10, 17, 109, 144, 164, 165, 183, 192, **219–225**, 231, 248, 262, 267, 270, 272, 283, 296
materiality 20, 56, 69, 105, 106, 162, 165, 169, 298, 309, 332, 333
Mauss, M. 28
Mawdsley, E. 81
May, J. 62, 152
McClintock, A. 20
McDowell, L. 20, 67, 192, 233, 248, 277, 340
McGuinness, M. 224
McLuhan, M. 27, 29, 313
Merleau-Ponty, M. 62, 311, 312, 338
Merrifield, A. 73, 92, 185, 212, 273
migration 18, 44, 52, 54, 55, 92, 149, 150, 214, 216, 301, 303, 304
Mitchell, D. 20, 199, 333, 335
Mitchell, K. 56
mobility 19, 37, 61, 75, 78, 150
modelling 151, 156, 158, 246, 258, 301, 304, 347
modernity 29, 30, 31, 33, 34–38, 40–45, 54, 55, 116, 131, 144, 178, 203, 208, 209, 270, 291, 324, 347
modernization theory 269, 317
modifiable areal unit problem 302
Mohanty, C.T. 193
money 78, 79, 103, 131, 171, 296, 298
Moore-Gilbert, B. 55–56, 243
morality 34–37, 78, 86, 171, 251, 252
Moss, P. 11

multiculturalism 53, 212, 325, 348
multidimensional scaling 136, 138
Mumford, L. 312, 313

Naipaul, V.S. 52, 53
Nash, C.87, 178
Nast, H.211, 261, 277
nationalism 16–20, 56, 67, 80, 81, 82, 131, 178, 227, 228, 320, 337, 348
nation-state 8, 16–20, 37, 38, 41, 43, 79, 81, 82, 102, 108, 130, 131, 132, 177, 221, 288–291, 297, 319
nature 92, 114, 115, 116, 118, 146, 167, 168, 169, 171, 183, 204, 231, 265, 295, 307, 327, 333
Nelson, L. 234
neo-classical economics 48, 109, 117, 219, 222, 251, 252, 298
network society 75, 76
Nietzsche, F. 12, 27, 104, 121, 124, 125, 208
non-representational theory 6, 9, 61, 65, 68, 87, 103, 105, 170, 297–298, 348

Ogborn, M. 87, 204
object relations theory 259, 261
objectivity 3, 4, 36, 168, 169, 309, 348
Olsson, G. 150, 181, 182, 301, 323
Olwig, K. 333
O'Connor, J. 61
Ó'Tuathail 79, 81, **226–230**, 292, 312
Orientalism 7, 53, 114, 133, 146, 237, 238, 240, 241, 242
otherness 8, 36, 54, 125, 146, 170, 259, 261, 262, 277, 341, 348

Painter, J. 60, 61, 116, 226, 227
Panofsky, E. 86
paradigm 10–11, 114, 144, 214, 348
Parr, H.262
patriarchy 66, 189–192, 197, 211, 223, 232, 233, 348
Peake, L. 191
Peck, J. 111
Peet, R. 80, 114, 116, 118, 267, 270, 325
performativity 9, 24, 25, 53, 65–69, 231, 234, 294, 297, 348
personal guidance system 137, 138
phenomenology 1, 5, 35, 121, 129, 136, 138, 144, 209, 214, 215, 216, 306, 307, 309, 311, 338
Philo, C. 174, 176, 197, 262, 272
physical geography 2, 3, 157
Pike, A. 285
Pile, S. 55, 61, 68, 192, 193, 231, 261, 295, 297, 340
place 5–10
    bounded 5
    progressive sense of 6, 9, 223–224
    non- 8, 76
    relational 6
    scaleless 5
    sense of 5, 7, 164, 192, 241, 307, 351
placelessness 8
planning 48, 113, 182, 210
Podmore, J. 61
political ecology 114, 115, 116, 118, 324–327, 348

political economy 20, 22, 25, 72, 91, 109, 247, 252, 256, 271, 285, 295, 296, 313, 317, 323, 325, 340, 349
polyvocality 8
Popper, K. 181
positionality 65, 68, 193, 229, 234, 235, 276
positivism 1, 5, 10, 12, 131, 132, 138, 139, 143, 144, 145, 157, 158, 174, 175, 181, 216, 245, 246, 247, 249, 269, 294, 295, 296, 304, 306, 349
postcolonialism 1, 3, 24, 52–57, 61, 86, 146, 147, 160, 163, 165, 170, 177, 195, 211, 224, 237–243, 272, 275, 277, 278, 320, 349
postfordism 90, 222, 271, 273, 283, 296, 297
postmodern city 7, 12, 76, 96–100, 216, 217
postmodernism 1, 3, 7, 10, 27, 48, 50, 84, 90, 96, 97, 99, 144, 145, 158, 163, 167, 177, 183, 192, 204, 215, 217, 223, 227, 245, 248, 260, 269, 271, 272, 285, 292, 308, 311, 324, 334, 339, 349
postmodernity 9, 29, 31, 33, 34–35, 93, 131, 183, 185, 203
poststructuralism 1, 3, 7, 9, 10, 12, 24, 27, 44, 66, 84, 102, 105, 117, 118, 133, 145, 165, 167, 203, 214, 227, 239, 245, 292, 294, 295, 298, 311, 349
Poulantzas, N. 72, 211
power 6, 7, 8, 9, 10, 17–18, 20, 23, 27, 34, 60, 66, 73, 74, 80, 85, 90, 92, 113, 116, 123, 126, 130, 131, 146, 168, 170, 177, 178, 190, 196, 199, 215, 222, 227, 228, 233, 234, 240, 242, 254, 255, 256, 258, 260, 265, 266, 271, 277, 288, 292, 296, 318, 324, 332
power geometries 6, 222, 224
pragmatism 59, 338
Pratt, G. 24, 193
Pred, A. 109, 144, 145, 149, 152, 270, 325
probability theory 149
Probyn, E. 69
production 22, 28, 29, 76, 108, 109, 110, 184, 199, 220, 271, 282, 283, 284, 285, 290, 295, 296, 313, 314, 319, 321, 350
psychoanalysis 3, 8, 12, 53, 56, 102, 197, 227, 234, 235, 258–262, 338, 350
Pulido, L. 69

quantification 4, 5, 136, 150, 155, 157, 158, 246, 301, 303, 350
quantitative revolution 4, 12, 23, 47, 49, 149, 151, 156, 301
queer theory 1, 65, 350

race 6, 11, 48, 49, 56, 62, 66, 69, 72, 76, 92, 93, 102, 152, 162, 163, 164, 167, 186, 189, 190–193, 195, 196, 204, 211, 214, 227, 232, 239, 260, 291, 337
Radcliffe, S. 19, 276, 277
Ratzel, F. 227

reality 27–31, 35, 66, 350
reflexivity 8, 24–25, 65, 68, 131, 143, 231, 234, 285, 309
regional geography 5, 10, 48, 109, 133, 144, 149, 155, 181, 184, 219, 220, 222, 249, 282, 284, 285, 292, 295, 296, 298, 350
Relph, E. 5, 306
representation 5, 7, 8, 9, 18, 23, 31, 69, 84–87, 96, 97, 105, 145, 146, 147, 161, 163, 164, 168, 169, 171, 190, 233, 235, 238, 240, 242, 259, 276, 277, 292, 295, 297, 312, 331, 333, 350
resistance 7, 45, 54, 56, 61, 80, 82, 91, 96, 117, 160–63, 190–93, 217, 243, 265, 266, 267, 279, 318
rhythmanalyses 6, 211
risk society 40–45, 228
Robson, G. 61–62
Rorty, R. 33
Rose, G. 6, 8, 10, 11–12, 55, 56, 67, 68, 69, 169, 193, 222, 223, **231–236**, 261, 272, 276, 340
Rose, M. 69
Rostow, W. 317
Rushdie, S. 52, 55
Rustin, M. 221

Saarinen, T. 136, 138
Sack, R. 158, 184, 304
Said, E. 2, 7, 11, 19, 53, 54, 114, 146, 193, **237–244**, 272, 275, 278, 335
Sanders, R. 193
Sassen, S. 9
Sauer, C. 6, 12, 85, 151, 307, 324
Sayer, A. 5, 132, 144, 185, 222, **245–250**
scale 6, 8, 133, 200, 212, 220, 265, 266–267, 291, 292, 325, 326
Schoenberger, E. 111
Schutz, A. 35
Scott, A. 25, 90, 97, 109, 248, 282, 283, 284–85
semiology 28
Sen, A. 2, 8, 80, **251–257**, 324
Sennett, R. 33
sexuality 8, 20, 56, 65–69, 73, 123, 125, 126, 131, 169, 186, 189, 195, 197, 198, 232, 337
Sharp, J. 67, 229, 340
Sheppard, E. 22
Shields, R. 210
Sibley, D. 8, 11, **258–263**,
Sidaway, J. 13, 276
Simmel, G. 33, 142
simulacrum 29–31, 103, 314
Slater, D. 115, 290
Smith, D. 341
Smith, N. 8, 10, 80, 184, 186, 210, 217, 229, 239, 249, **264–268**, 270
Smith, S. 87, 195
Smith, V. 193
social
  choice theory 252

construction 17, 24, 44, 85, 169, 196, 197, 199, 202, 206, 211, 292, 333
justice 12, 25, 49, 72, 182, 183, 252, 253, 333, 337–342, 351
movements 33, 43, 73, 80, 82, 113, 114, 115, 117, 118, 160, 253, 285, 291, 318, 324, 325, 326, 337
Soja, E. 5, 7, 10, 12, 56, 90, 96, 97, 144, 183, 192, 211, 217, 223, 247, 248, **269–274**, 283, 285
space 3–5
  absolute 2, 4, 5, 35, 138, 183, 266
  abstract 5, 35, 60, 247
  of flows 8, 9, 74–75, 325
  paradoxical 232
  performative 68, 234
  production of 12, 168, 183–85, 210, 265, 270
  relative 5, 184, 266
spatial
  divisions of labour 5, 220, 222, 283
  laws 4
  physics 4, 181
  science 10, 12, 47, 137, 139, 156, 157, 158, 181, 214, 219, 264, 266, 269, 306, 351
  statistics 4, 22, 156
  turn 2
spatiality 4, 5, 6, 56, 111, 165, 270, 271, 272, 339, 351
Spinoza, B. 104
Spivak, G. 8, 54, 239, 272, **275–280**
Sraffa, P. 22, 24
Stoddart, D. 157
Storper, M. 5, 7, 9, 12, 25, 97, 109, 111, 248, **282–287**,
structuration theory 129–133, 144, 147, 248, 294, 295, 296, 298
subaltern theory 1, 8, 145, 276–279, 351
subcultures 160–62, 196, 214, 332
subjectivity 8, 59, 65, 86, 122, 133, 139, 162, 164, 165, 169, 171, 172, 185, 190, 193, 203, 232, 295, 297, 309
subpolitics 43, 44
sustainability 114
Swyngedouw, E. 186, 291
symbolic exchange 28, 30

tactile maps 138
Taylor, P. 5, 8, 9, 76, 109, 157, 158, 175, 229, 267, **288–293**, 320
technology 72, 73, 115, 117, 167, 168, 171, 202, 204, 221, 297, 311, 313, 314, 315, 319
territory 5, 9, 17–20, 24, 37, 61, 104, 113, 114, 115, 162, 177, 227, 228, 239, 288, 290
territoriality 42, 227–228, 253, 351
terrorism 7, 40, 44, 80, 81, 200, 238, 309
thirdspace 54–57, 269–273
Third World 93, 113, 114, 210, 279, 318, 324, 325, 326
Thompson, E.P. 132, 143, 295, 323

Thrift, N. 2, 5, 6, 9, 60, 61, 68, 76, 87, 109, 111, 122, 133, 139, 143, 144, 147, 152, 157, 170, 204, 222, 270, **294–300**
Tickell, A. 110
time 20, 31, 35, 60, 130, 131, 133, 144, 149, 150, 151, 152, 221, 222, 270, 294–295, 298, 312, 313, 315, 351
timeless time 75
time-space
  compression 9, 37, 132, 144, 145, 221, 223, 224, 313, 352
  distanciation 132, 297
  modelling 4
transnationalism 8, 9, 41, 43, 44, 45, 79, 171, 198, 200, 267, 279, 291, 319, 321, 324, 352
Tobler, W. 2, 4, 10, 12, **301–305**
Touraine, A. 72, 73
Tuan, Y-F. 5, 10, 85, 158, 291, **306–310**

urbanism 48, 72, 74–75, 90–94, 96–100, 143, 182, 186, 209, 215, 264, 265, 271, 273, 282, 284, 311, 312
uneven development 109, 110, 219, 220, 264–268, 269, 295
Urry, J.75, 144, 222, 246, 248, 270
use value 28, 29
utility theory 22, 252

Valentine, G. 67, 69, 262, 340
Veblen, T. 28
Virilio, P. 2, 8, 205, **311–316**
von Humboldt 307
von Thünen 24, 149, 156, 301

Walker, R. 144, 186, 282, 283
Wallerstein, I. 5, 288, 289, **317–322**
Watts, M. 8, 10, 81, 82, 115, 116, 118, 254, 255, **323–329**,
Weber, M. 24, 29, 34, 73, 129, 130, 132, 133, 144, 156
Whatmore, S. 44, 171, 172, 203, 267
Williams, R. 2, 7, 196, 199, **330–336**
Wilson, A. 4
Wilson, B. 194
Wilson, E. 50
Wilton, R. 261
Winnicott, D. 8, 261
Wolch, J. 96, 97
Woods, C. 194
workplace 42, 67, 223
world cities 9, 75, 93, 28, 320, 352
world systems theory 227, 288, 290, 291, 292, 317–321, 352
Wright, J.K. 308
Wynne, B. 44
Wynne, D. 61

Yeates, M. 51
Yeung, H. 3, 108, 111
Young, I.M. 7, 334, **337–343**
Young, R. 55